Energy Storage, Compression, and Switching

Energy Storage, Compression, and Switching

Edited by

W. H. Bostick

Stevens Institute of Technology
Hoboken, New Jersey

V. Nardi

Galileo Ferraris National Electrotechnical Institute
Torino, Italy
and Stevens Institute of Technology
Hoboken, New Jersey

O. S. F. Zucker

University of California
Lawrence Livermore Laboratory
Livermore, California

PLENUM PRESS · NEW YORK AND LONDON

Library of Congress Cataloging in Publication Data

International Conference on Energy Storage, Compression, and Switching,
Asti-Torino, Italy, 1974.
Energy storage, compression, and switching.

Includes bibliographical references and index.
1. Energy storage–Congresses. 2. Energy transfer–Congresses. 3. Electric
switchgear–Congresses. I. Bostick, Winston Harper, 1916-   II. Nardi,
V., 1930-   III. Zucker, O. S. F., 1939-   IV. Turin. Instituto
elettrotecnico nazionale Galileo Ferraris. V. Stevens Institute of Technology,
Hoboken, N. J. VI. Italy. Consiglio nazionale delle ricerche. VII. Title.
TJ153.I59 1974   621.3   75-42405
ISBN 0-306-30892-4

Proceedings of the International Conference on Energy Storage, Compression,
and Switching held in Asti-Torino, Italy, November 5-7, 1974

© 1976 Plenum Press, New York
A Division of Plenum Publishing Corporation
227 West 17th Street, New York, N.Y. 10011

United Kingdom edition published by Plenum Press, London
A Division of Plenum Publishing Company, Ltd.
Davis House (4th Floor), 8 Scrubs Lane, Harlesden, London, NW10 6SE, England

Printed in the United States of America

International Conference on
ENERGY STORAGE, COMPRESSION, AND SWITCHING
November 5-7, 1974

Sponsors

Istituto Elettrotecnico Nazionale Galileo Ferraris
Torino, Italy

Department of Physics
Stevens Institute of Technology
Hoboken, N. J. 07030

Consiglio Nazionale delle Ricerche
Roma, Italy

Foreword

This international conference was organized by the sponsoring agencies with the following objectives in mind: to bring together active researchers involved in energy compression, switching, and storage who have a major interest in plasma physics, electron beams, electric and magnetic energy storage systems, and high voltage and high current switches. Areas of interest include:

Slow systems: 50-60 Hz machinery, transformers, flywheel-homopolar generators, slow capacitors, inductors, and solid state switches. Intermediate systems: fast capacitor banks, superconducting storage and switching, gas, vacuum, and dielectric switching, nonlinear (magnetic) switching, fast (10^5 - 10^6 Hz) capacitors and fuses. Fast systems: Marx, Blumlein, oil, water, and pressurized water dielectrics, switches, magnetic insulation, electron beams, and plasmas.

The Editors extend thanks to all the authors, and attendees (and their supporting institutions, and companies), every one of whom in his own measure helped to make the conference a success. The Editors further wish to thank the members of the Scientific Committee for the help they have given in organizing the conference and in editing, especially J. C. Martin and H. L. Laquer.

Special recognition is due the Lawrence Livermore Laboratory whose Electrical Engineering Department provided the Secretary of the Scientific Committee and one of the Editors, and the yeowoman services of Sharon Dodson and Cheri Johnson in all the mailings, correspondence, and receiving and organizing of the manuscripts. The LLL Technical Information Department provided the design and printing of the conference announcements and the instructional formats for the authors' manuscripts.

The Editors also express their gratitude to the sponsoring organizations and their personnel -

1. Istituto Elettrotecnico Nazionale Galileo Ferraris (IEN), Torino, Italy for providing the initial impetus and much of the financial support of the Conference. The superb work of the local committee of Torino is due singular praise: M. Canavese (Secretary), C. Cortese, A. Ferro Milone (Chairman), S. B. Toniolo and L. Franchino for the administration, E. Tessitore and E. Vecchiotti for technical assistance.

2. Department of Physics, Steven Institute of Technology, Hoboken, New Jersey for providing offices and facilities for two of the editors and two members of the scientific committee. The handling of editorial corrections and the correspondence thereof: R. Southworth, Eleanor Gehler, Diane Gioia, Janet DeGennaro, and N. Khambatta.

3. National Council of Research (CNR) Rome, Italy, for moral and financial support.

Substantial financial support was also provided by the Istituto di Fisica, Universitá di Ferrara.

Finally, on behalf of all the attendees the editors wish to express particular thanks to Il Consiglio Provinciale di Asti and its President for their cordial and elegant hospitality.

July, 1975   W. H. Bostick, V. Nardi, O. S. F. Zucker

Contents

INDUCTIVE AND CAPACITIVE STORAGE SYSTEMS

COLLECTIVE EFFECTS

SWITCHES

Electron Beams

SHORT PULSE HIGH VOLTAGE SYSTEMS

J. C. Martin
MOD (PE) Atomic Weapons Research Establishment
Aldermaston, Reading (UK)

ABSTRACT

The design features of current high speed pulse generators will be outlined. Existing single machines can provide more than 10^{13} watts at voltages above 10 million volts. Likely extensions of present design techniques to provide similar or higher wattages at a couple of million volts will be mentioned. One of the earlier limitations to the rise time of the pulses produced by the machines was the main output gap. The development of multi-channel switching will be briefly reviewed and the present position in this will be given. The inductance of the vacuum envelope and its immediate feeds presents another limit to the pulse rise time in low impedance systems and some present day limits will be indicated. Within the diode the energy can be carried by a vacuum insulated feed and some limits and solutions to the power flow that can be sustained will be discussed. Further convergence of the power flow can then be obtained by using electron beams, such as by pinches within the diode or by neutralised electron beam transportation. The present power fluxes achievable at the various stages of convergences will be indicated. The history of the development of high wattage systems of various voltages will be indicated in order to give some idea of what may be achievable in the near future. In conclusion some brief comments will be made about the possibilities and difficulties of longer pulse systems and also about those with multishot capabilities to achieve significant mean power levels.

I would like to start by briefly describing modern short pulse electron beam generators. Then I would like to mention some of the probable developments of these to higher powers and shorter pulses. I would then like to give some orders of magnitude of the flux of electrical energy at various stages in the machines, and finally mention some potential difficulties and limitations to their operation in fusion power schemes.

FIGURE 1

This shows a schematic layout of a modern electron beam generator. A Marx generator pulse charges the high speed pulse-forming lines in a time like a microsecond. Because of this, ordinary dielectric media can be used at higher electrical fields than is normal, and it also enables less common dielectrics, like water and glycerine, to be used. The pulse charging eases tracking problems and also allows the output switch to be un-triggered, where this is desired. The pulse-forming lines are designed to deliver their energy in typically 50 to 100 nanoseconds after the output switch or switches close. Between the pulse-forming lines and the load there is a feed section, and this may take the form of an impedance transformer, such as an exponentially tapered line.

In the case of a generator producing electron beams, a vacuum-supporting interface has to be included, and this is the vacuum diode. Inside this there is a further feed which carries the pulse to the cathode, where the electron beam starts. Further convergence of the flux of electrical energy may then take place, either within the anode-cathode region, or on the other side of the anode, during beam transport.

I would like to describe briefly the individual components of such a generator and to outline their present-day performance.

FIGURE 2

Modern Marx-like generators have the following properties :

High voltages - up to 10 MV in relatively compact arrangements.
High storage energy - up to 5 megajoules.
Low inductance - their inductance can be less than 1 microhenry per million volts.
Erection time can be less than 50 nanoseconds per million volts.
Jitter in erection time - this can be as low as 5 nanoseconds, which allows a number of Marx generators to be erected in parallel.
Triggering range - down to at least 40 per cent of their self-break voltage.
Overall voltage gradients up Marx - $1^{1}/2$ MV/metre even in gas insulated systems.

Not all of the above properties can be had in any one system, but most of them are compatible.

PULSE FORMING LINES

Figure 3 shows the two most usual forms of the pulse-forming lines in strip transmission line geometry. The first and simplest form is usually used in 1 million volt generators, but provides a pulse of only half the charging voltage into a matched load. The second arrangement is that due to Blumlein and provides an output equal in voltage to pulse charged line voltage, when the output switch closes. This is the circuit used in the higher voltage generators with outputs of up to 15 million volts.

Just about every possible dielectric - solid, liquid and gaseous - has been used in these and many other circuit arrangements but, in general, water is now used for million volt machines and transformer oil for higher voltage systems.

FIGURE 4

This shows the general areas where the different circuits and dielectrics are used in modern 50 to 100 nanosecond generators. Note that the impedance is plotted so that it decreases as you go up the axis. This means that the farther you go away from the origin, the more energy the generator is delivering.

MAIN OUTPUT SWITCH OR SWITCHES

In the older generators the output switch was a self-closing one which closed on the charging voltage a little before peak voltage. However, the inductance and resistive phase of a single channel switch can give a poor rise time to the output pulse of large machines and most modern machines use triggered switches, so that a number of spark channels are produced in order to give fast rise times to the output pulse. While many types of output switch have been successfully used, such as the trigatron switch or laser triggered gap, the field distortion switch has probably been most often used.

FIGURE 5

This shows, very schematically, the field-distortion switch as used with gaseous or liquid dielectrics. In the case of the axially symmetric geometry, the trigger may be placed outside the main electrodes, or within them. A linear version of the field-distortion switch is known as the rail switch. When the trigger electrode is mounted half-way between the electrodes, the switch operates in the cascade mode - first one half of the switch is closed by the triggered pulse and then the voltage across the switch

closes the second half of the switch. The electrode, however, can be displaced so that the trigger pulse essentially closes both portions simultaneously. For gases the ratio of the spacings is about 2 to 1, while for liquids it is more like 7 to 1. However, for pulse-charged, high-pressure switches there is a version - due to Ian Smith - where the spacings are in the ratio of 10 to 1, or even 15 to 1. This version is known as the V/n switch and needs only a small trigger voltage to close it.

The field distortion switch using solid dielectrics is also shown schematically and typically uses dielectric thickness ratios of 8 to 1 for high voltage operation.

While all these types of switch work very well at voltages of tens of kilovolts, their real application here is for pulsed voltage operation at voltages of 1 million volts to 12 million volts. In addition to being mechanically simple and robust, they also operate multi-channel when fed with a sufficiently fast rising trigger pulse. I personally use the term 'multi-channel switch' to mean one in which all the electrodes are continuous and where the channels are not isolated from each other by the velocity of light transit time.

FIGURE 6

This shows the voltages at which triggered multichannel switches have been operated to date and it does not appear that there are any serious limitations to raising these voltages to much higher levels if there is a need. For low voltage operation the gaps need trigger pulses of amplitude about equal to the switch breakdown voltage. For operation at several million volts they need trigger pulses of about 20 per cent of the gap operating voltage. The maximum values of $\partial i/\partial t$ and $\partial V/\partial t$ achieved to date in the output pulse are both about 5×10^{14} amps or volts per second.

The development of multichannel switches has been an important one, since for a long time the performance of the switch was the main limitation to the rate of rise of the output pulse. In principle now, given enough channels, the output of the pulse is no longer limited by the switch performance but is limited by the pulse breakdown strength of the dielectric forming the lines.

FIGURE 7

From the output of the pulse forming line there is a feed which carries the pulse

to the load. This needs to be designed so
as to transmit the pulse without increasing
its rise time. However, this feed may be
used to transform the impedance and voltage
of the out pulse. NRL pioneered the use of
a tapered impedance line to convert a 4 MV
pulse down to 1 MV. However, with the
introduction of multichannel switches, such
large transformer ratios are no longer
necessary. For high energy 1 MV water
systems, however, it is usual to include
one or two extra line lengths of lower
impedances to obtain a voltage reduction of
a factor of 2 or so. Somewhere along the
feed it is usual to include an extra multichannel self-closing switch. The operation
of this switch causes some sharpening of the
pulse front, but the main reason for its
inclusion is to reduce to the minimum the
voltage pulses which come before the main
pulse. These pre-pulses can cause the
cathodes to malfunction if their voltage is
too big. In a well-balanced Blumlein system
the pre-pulse can be as low as 3 per cent of
the main pulse; however in the simpler
circuits pre-pulses of 10 per cent, or even
15 per cent, of the main pulse are not
uncommon. The addition of a pre-pulse
isolation switch can reduce these levels to
more like 1 per cent, which is usually enough
to avoid early cathode operation troubles.

FIGURE 8

Present-day vacuum diodes mostly take
the two forms shown. For high voltage systems
a series of insulator rings are used. The
inner faces of these are sloped at about 45^0
to the field lines so that the electrons
which are emitted from them do not hit the
insulator surface again. Considerable care is
taken to ensure that the voltage gradient up
the insulator stack is as uniform as possible
and that there is a minimum impedance disturb-
ance to the pulse as it passes through the
diode. For the 15 MV machine AURORA, made by
Physics International, the vacuum diodes are
large enough to hold a meeting of some 10
people within them.

For generators working at about
1 million volts, a second form of diode has
been developed and widely used and this is
known as the diaphragm tube. In this the
metal surfaces on either side of it are carefully tailored so that the field lines leave
the insulator surface at about 50^0 and the
field is nearly uniform along the radius.
The inductance of this type of tube is about
20 nanohenries for 1 million volt pulses.
This type of tube is also easily cleaned
and maintained.

VACUUM INSULATED FEED

(Figure 9). In some generator uses
a significant length of vacuum insulated
coaxial line feed may be required. Unless
the inner surface of this is kept highly
polished, small plasma bushes may form at
its surface from exploded metal whiskers
and substantial currents can then flow
radially, shorting out the feed. However,
if the load at the end of the vacuum coax
has a resistance lower than the impedance of
the coaxial feed, the self-field generated
by the current flowing on it returns the
electrons back to the inner conductor.
This simple theory gives the right answer
for high voltage systems but for lower
voltage systems the parapotential current
flow equations of Dave Dupack seem to be
better, and cut-off is achieved at about
1 MV for a load resistance about half that
of the coaxial impedance. In addition, for
small spacing, motion of the plasma bushes
must be allowed for. The same cut-off
effect may also be achieved with externally
applied magnetic fields.

CONVERGENCE IN THE ELECTRON BEAM

At the cathode plasma bushes are
formed and the electrons from these accelerated across the anode-cathode gap; these can
then be transported in a beam beyond the
anode. A tremendous amount of work has been
done by many groups on the propagation of
such beams in low pressure gases, with and
without externally applied magnetic fields,
etc. Drift in initially neutral gas without
external fields has given current densities
of up to 50 kA/cm^2. With the addition of
external magnetic fields current densities of
greater than 100 kA/cm^2 have been achieved
with larger total currents. In addition,
pinching of the electron trajectories within
the diode has given current densities of over
100 kA/cm^2. More recently, work by a number
of people - including Willard Bennett and
Jerry Yonas and his group at Sandia Laboratories, Albuquerque, has shown current
densities of 5 MA/cm^2 and higher in very
tight pinches within the anode-cathode region.
So far there are limitations to the total
current carried in these pinches and the time
for which they can be sustained, but further
work will doubtless remove these constraints
and possibly achieve even greater current
densities.

FIGURE 10

As an example of what has been built
in the way of large machines, AURORA, working
at about 15 MV, has delivered 3×10^{13} watts.

As examples of about 1 MV machines, MAXIBEAM of Maxwell Laboratories, GAMBOL 2 of US Naval Research Laboratories, HYDRA of Sandia Laboratories, and OWL of Physics International, have all reached or exceeded 10^{12} watts.

I would now like to give a review of likely developments in the near future with regard to machines working at 1 to 3 million volts and having rather shorter pulse lengths of around 20 nanoseconds. This particular class of machine is of interest for possible e-beam fusion applications and also for pumping certain types of lasers. Another class of machines, for heating plasma in contained fusion systems, is likely to involve longer pulses and higher voltages and therefore, apart from a requirement to fire at a significant repetition rate, will not differ too greatly from existing machines. I will make some brief comments on significant repetition rate systems and their possible difficulties at the end of this talk.

Returning to the basic components of a large 3 million volt, 20 nanosecond, e-beam generator, as before, I will start with the Marx generator. By using multichannel rail switches in the Marx with very low inductance capacitors, generators with inductances of 200 nanohenries per million volts are very likely to be buildable. However it is possible to erect four or five present-day Marx generators in parallel and to achieve the same inductance. Mr Fitch of Maxwell Laboratories has been responsible for a number of very elegant Marx-like generator designs and has shown how this may be neatly done. Also the use of transfer capacity techniques - mentioned in a moment - means that present-day Marx generators may be fast enough. An alternative development may be to replace the Marx generator with inductive storage combined with opening switches. Papers later in the Conference cover this topic. However, I feel that Marx generators will continue to be built for quite a long time to come for many generator applications which do not require the very greatest energy outputs.

TRANSFER CAPACITY TECHNIQUES

The US Naval Research Laboratories group employed a transfer capacity in their first large machine, GAMBOL 1, seven years ago. In this the energy from the Marx is transferred to a pulse-charged capacitor from which the pulse-forming line is charged, in a time like 300 ns rather than a microsecond or more.

FIGURE 11

This shows the two main fast pulse line schemes, pulse charged from what is now a transfer pulse charged line. The spacing of the transfer capacity, or line, is that appropriate to a charging time of about 1 microsecond, while the spacing of the high speed lines, which can be charged in times like 100 nanoseconds, is about half as big. For the Blumlein circuit both halves of the pulse-forming lines have to be charged at the same time from a double-sided transfer capacity or line. The use of a transfer capacity with short pulse generators enables the fast pulse lines to be charged an order of magnitude more quickly and, hence, if the dielectric is a liquid, the energy storage is increased by a factor of 4.

SWITCHING

If the aim is to achieve 1 to 3 million volt generators able to deliver 10^{13} watts to a load - that is, one order of magnitude greater than present systems - and also to produce 20 nanosecond pulses, the rate of rise of current from the switches must be improved by between one and a half and two orders of magnitude. This means multichannel switching with a vengeance! While it is possible to consider triggered multichannel switches for such a job, there would be many of them and the expense and complexity would be very high. However, with rapid charging of the final pulse-forming lines, self-closing untriggered switches can again be considered. In a large machine it is calculated that of the order of 1000 current-carrying channels can be obtained with a sub-nanosecond time jitter. The feed from the pulse lines to the diode will probably have to be water-insulated and will be tapered as the diode is approached, partly to reduce the area of metal which has large fields on it. Even so, the metal will have to be highly polished or conditioned, in order to take the required stress, if the radius of the end of the feed is less than 1 metre, which seems to be desirable.

VACUUM DIODE

(Figure 12). If the cathode has to be initially limited to 25 cm or so, then at least an order of magnitude improvement must be made in the inductance of the insulators and the vacuum feed to the cathode. This Figure shows a schematic cross-section of a diode which might provide 10^{13} watts or more at voltages of 2 million and above. The height of the insulator stack is estimated to be about 6 cm per million volts and many small insulator sections are used.

Flux-excluding metal blocks are used, both inside and outside the insulator rings, to reduce the inductance. For a cathode radius of about 25 cm an inductance of the tube plus the vacuum insulated feed is estimated to be about 2 nanohenries per million volts. It should be emphasised that this number is an estimate and that the design mentioned is marginal, to say the least. However, it does suggest that even if the cathode radius has to be as small as 25 cm, it may be possible to deliver about half a megajoule to it with a pair of generators working at 3 million volts, with a 20 nanosecond pulse. Beyond this rate of energy delivery the radius of the cathode and diode must be increased, unless significant advances are made in a number of areas of voltage breakdown. Even if machine designers are forced to go to larger radii diodes to reduce the inductance further, improvements in low inductance electron beam transport and focussing may still enable yet higher rates of delivery of energy to be achieved.

I would now like to discuss approximate electrical energy flux at various stages through the generator.

FIGURE 13

The numbers given in this Figure are only approximate, as they depend slightly on the area or volume stressed by the pulse. With regard to the flux figures given for the vacuum diode and feeds, these components are best considered as inductances and as such, an indefinitely large flux can eventually flow through them without exceeding their flash-over or breakdown voltages. However, given a suitably rising input pulse, they can be regarded as small lengths of mismatched transmission line and for each transit time across them, they can add the flux quoted. Thus a diode insulator stack 6 cm wide can transmit 10^9 watts per square cm after about 0.3 nanoseconds, and 10^{10} watts per square cm after 3 nanoseconds, without exceeding the diode's flash-over voltage. This raises the point that if the pulse arriving down the line rises too quickly, the diode will flash over and short out the pulse. This can be countered by using a diode with a greater hold off voltage, but such a diode will have a larger inductance and hence not lead to a significant improvement in pulse rise time at the load. It will be a sad experimentalist who has gone to a great deal of trouble to produce an extremely quickly rising pulse only to cause breakdown to occur inside a diode which would otherwise operate satisfactorily with a slower pulse!

The Figure shows that as the electrical energy passes through the generator, the flux level rises and hence the area needed to transmit a given flux decreases. Thus the generator tapers from the Marx to the electron beam. This can be done either conically in three dimensions, like an ice-cream cone, or in two dimensions, like a washer. In the case of conically tapering generators, in principle the surface of a sphere can be covered by such generators. This, of course, ignores purely practical matters like supporting the generators, maintenance, etc. In the case of washer-like generators, two may easily be stacked to double the rate of delivery of energy to a small load, but further generators cannot easily be added without involving electron beam transport, which may present considerable difficulties.

I would now like to talk briefly about some of the problems that may be met in making generators which can fire at a few pulses a second over prolonged periods of time.

Two classes of problem can easily be foreseen. One is the sheer rate of electrical energy dissipation when, say a 1 MJ system is fired at 10 pulses per second. Normally, if 50 per cent of the electrical energy is finally delivered in the beam, the generator designer will feel he has done very well. However, in a system required to fire at a significant repetition rate, the other 50 per cent of the energy must be removed. The operation of the spark switches will lead to the generation of large volumes of rather active products, causing corrosion problems in the switches, etc.

The second class of problem can perhaps be best shown by considering this piece of coaxial cable, which is about 1 cm in diameter. For a few pulses it will transmit voltages of over 100 kV. However, its radio frequency rating is 5 kV. This is partly because solid dielectrics have a life proportional to 1 over the operating voltage to the 8th power. Thus if a given volume of plastic has a single pulse breakdown strength of, say, 1 MV, then it will break on average at $1/2$ MV after 1 thousand shots, and $1/4$ MV at 1 million shots. This sort of de-rating applies to the capacitors, and indeed to many other solid items which are repetitively stressed. With regard to liquids, the position is less clear, as there has been very little experimental data. However, it is probable that very clean liquids will have to be used and, even then, whiskers growing on the electrodes, and other effects, may force some lowering of the electrical stresses. Pressurised gases, when used as the insulating di-

electric of the lines, appear to have only a
modest degradation of properties with prolonged pulse stressing. Unfortunately, they
are an order of magnitude poorer as energy
storage materials than liquids and solids.

So far, I have been describing electron
beam generators with usefully large outputs.
However, these generators have been designed
to work rather close to breakdown and while
they have lives of several thousand shots,
with suitable maintenance, the stresses in
them may well have to be reduced by a factor
of two or more for operation at a significant
repetition rate. Such a reduction in working
electrical stresses will lead to a significant drop in performance. I have no doubt
that generator designers will win back much
of this loss of performance, but, in my
opinion, a very considerable effort and cost
will be needed to do so.

I would now like to give my personal
views about the longer range future of pulse
generators.

First, I would like to discuss possible
improvements in the electrical properties of
the dielectrics used in them. I do not consider there is much chance of improvement in
the breakdown stresses of gases and solids.
Over the last twenty years or more there have
been no really significant advances in their
basic properties. There has been some
improvement in the energy carrying capability
of solids, but this has mainly been by way of
increased cleverness in using them, such as
employing stacks of them in very thin sheets,
etc. However, with regard to liquids, I feel
there is a very real chance of obtaining considerably improved performance. This may
come about by the use of entirely new liquids
or by improving the performance of oil and,
particularly, water. With regard to water,
considerably improved performance has already
been demonstrated in small volumes by the use
of high pressures, by carefully polishing and
conditioning the electrodes, and by shielding
the electrodes by electrolytic layers. All
of these results can be accounted for by
assuming the breakdown processes originate
at whiskers on the electrodes. However, I
do not feel that any of these ways of improving the strength have any real applicability
to large machines; but I expect that simple
treatments will be found that will suppress
the formation of, or will remove, the whiskers
on the electrodes. This will enable increases
in energy storage in the more slowly charged
portions of the machines to be made, by
perhaps an order of magnitude. However, for
the case of the few million volt, 20 nano-

second, machine mentioned above, the stress
in the feed lines to the tube is already
nearing a million volts per centimetre and
a further doubling is all that can be used,
probably, as the very large dielectric
constant of water will start to decrease at
such fields and so cancel out the advantage.
Thus, while I think that improved liquid
breakdown will be forthcoming, this may only
serve to reduce the size of the intermediate
energy stores, which in fact are not very
costly items, and such increases may not
lead to short pulse generators such as the
one described earlier having a very much
improved performance.

How, then, are electron beam generators,
working at a few million volts, to achieve
power levels of 10^{14} watts and above?

One approach has already been mentioned;
that is, to use more of the solid angle
available. However, once this has been done
I do not think there will be substantial
improvements in the rate of energy delivery
unless the radius at which the cathodes are
located is increased. I think there are a
number of possibilities of coupling the
present pulse power technology to inductive
storage in a vacuum, combined with opening
switches. For instance, a DPF charged at a
voltage of 500 kV in times like 100 nanoseconds might be a very interesting device.
However, the resulting electron beam generated,
if it were to work, would have energies of
tens of millions of volts and for e beam
fusion work this looks like an undesirable
direction in which to go. Thus after systems
delivering, say, 10^{14} watts in 20 nanoseconds
have been constructed, I feel that the radius
at which the electrons are generated will
have to be considerably increased. From this
point on, progress may only be made by
obtaining more of the convergence in the
electron beams. The convergence may be in
space - to yield higher current densities or
to transport the high current density beams
over a few metres: or it may be in time -
to produce much shorter pulses and hence
higher watts. However, in view of the rather
dramatic rate of progress over the last ten
years, I feel tolerably confident that the
next decade will see similar advances being
made, although it will be the electron beam
which will provide most of the increase in
the flux of energy rather than the pulse
generator itself.

I have not given any real acknowledg-
ments to the many workers in the field
covered by this talk: I would like to
finish by doing so. While there has been

substantial progress in Russia and Europe,
the largest machines and the greatest progress
has been made in the USA. Much of this has
been stimulated by money from United States
Government organizations, such as the Defence
Nuclear Agency and the Air Force Weapons
Laboratory, and much credit is due to such
establishments for their intelligent support
of this field. While there are many laboratories and companies working on high voltage
pulsed electron beam generators, I would
particularly like to mention the groups at
Sandia Laboratories, Albuquerque, and the
group at the Naval Research Laboratories.
Regarding the commercial companies who have
been, or are, active in the field, Ion
Physics Corporation of Burlington were early
pioneers, while Physics International of San
Leandro (also pioneers) and Maxwell Laboratories of San Diego are now both highly active
in advancing the state of the art.

I would also like personally to thank
AWRE, who have generously supported me, and
the small pulse power group there who have
put up with me and have been responsible for
any progress we may have made.

REFERENCES

MARX GENERATORS

R A Fitch, IEEE Trans. Nucl. Sci. NS18
No. 4, 190 (1971)

MACHINES AND BEAMS

Papers in 10th, 11th and 12th Symp. on Electron
Ion and Laser Beam Tech., May 1969, 1971 and
1973. Pbl. San Francisco Press Inc.

Also papers in IEEE Trans. Nucl. Sci. NS18
No. 3 and NS20 No. 3.

The above references cover most of the machines
in the 50 to 100 ns pulse duration class built
in the United States and much of the technology
involved.

For recent 20 ns beam and machine work, the
paper "Electron Beam Induced Pellet Fusion"
G. Yonas, Sandia Report 74-5367, is relevant.

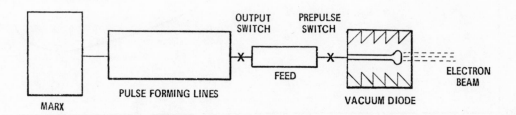

Fig. 1. Schematic of Pulsed Electron Beam Generator.

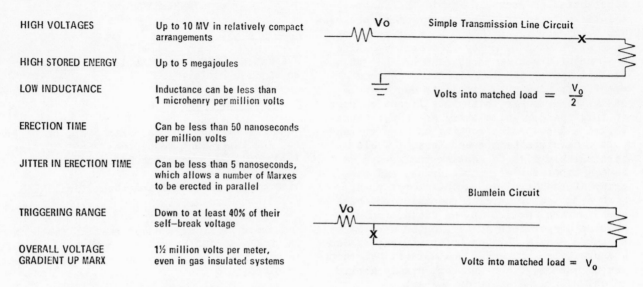

HIGH VOLTAGES	Up to 10 MV in relatively compact arrangements
HIGH STORED ENERGY	Up to 5 megajoules
LOW INDUCTANCE	Inductance can be less than 1 microhenry per million volts
ERECTION TIME	Can be less than 50 nanoseconds per million volts
JITTER IN ERECTION TIME	Can be less than 5 nanoseconds, which allows a number of Marxes to be erected in parallel
TRIGGERING RANGE	Down to at least 40% of their self—break voltage
OVERALL VOLTAGE GRADIENT UP MARX	1½ million volts per meter, even in gas insulated systems

Fig. 2.   Modern Marx Capabilities.

Fig. 3.   Principal Pulse Forming Lines.

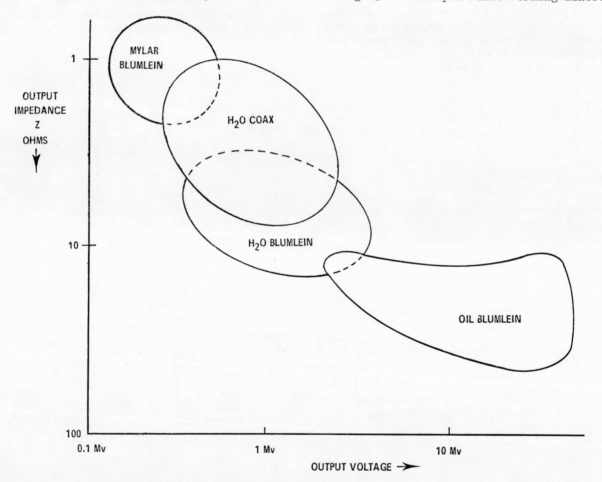

Fig. 4.   Operating Regime for Contemporary Electron Beam Genrators.

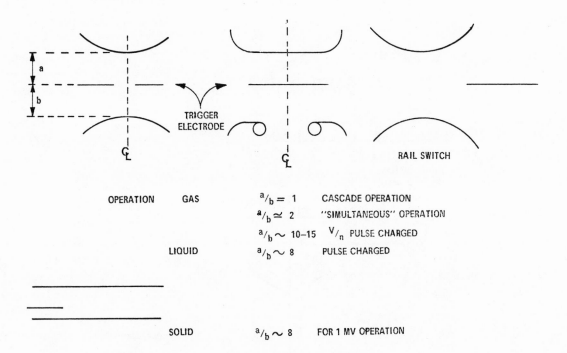

a
b
TRIGGER ELECTRODE
℄
℄
RAIL SWITCH
OPERATION GAS
a/b = 1 CASCADE OPERATION
a/b ≃ 2 "SIMULTANEOUS" OPERATION
a/b ~ 10–15 V/n PULSE CHARGED
LIQUID a/b ~ 8 PULSE CHARGED
SOLID a/b ~ 8 FOR 1 MV OPERATION

Fig. 5. Field Distortion Switch.

SOLID	LIQUID	GAS (PRESSURISED)
2 MV	4 MV H_2O 12 MV OIL	6 MV

MAXIMUM TO DATE : (NOT IN THE SAME SYSTEM)

$\frac{di}{dt} \sim 5 \times 10^{14}$ AMP/SEC

$\frac{dV}{dt} \sim 5 \times 10^{14}$ VOLT/SEC

Fig. 6. Demonstrations of Triggered Multichannel Pulse Charged Switch Operation.

OUTPUT SWITCH
Z
R = Z
PULSE FORMING LINE
FEED
Z
R < Z
TAPERED LINE TRANSFORMER FEED
Z
R < Z
STEPPED IMPEDANCE FEED

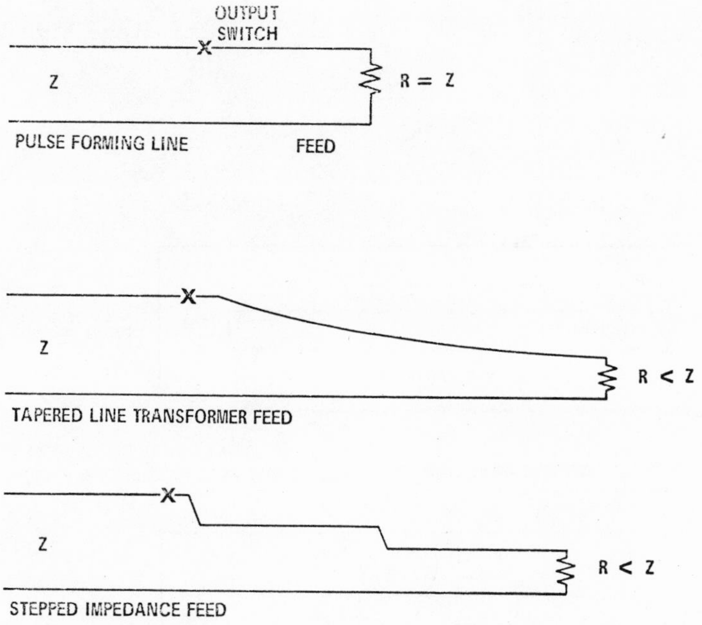

Fig. 7. Schematic of Feeds in Strip Line Geometry.

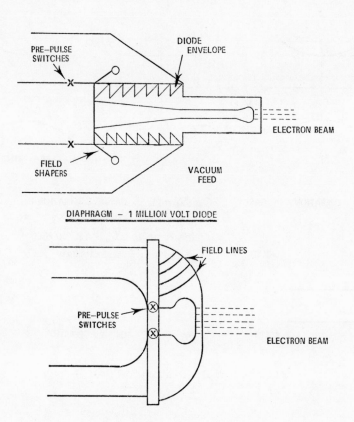

Fig. 8. Multi-Stage High Voltage Diode.

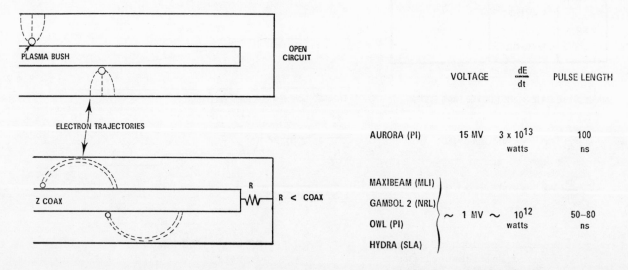

Fig. 9. Magnetically Cut Off Vacuum Insulated Feed.

	VOLTAGE	$\frac{dE}{dt}$	PULSE LENGTH
AURORA (PI)	15 MV	3×10^{13} watts	100 ns
MAXIBEAM (MLI) GAMBOL 2 (NRL) OWL (PI) HYDRA (SLA)	~ 1 MV ~	10^{12} watts	50–80 ns

Fig. 10. Examples of Large Energy Electron Beam Generators.

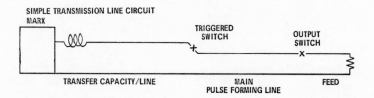

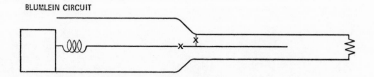

Fig. 11.   Schematic of Transfer Capacity/
Line System in Strip Line Geometry.

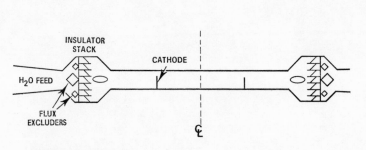

HEIGHT OF INSULATOR STACK $\simeq$ 6 cm/MV
INDUCTANCE $\simeq$ 2 nH/MV
(FOR CATHODE RADIUS = 25 cm)

Fig. 12.   Sketch of 10^{13} Watt Diode.

	ENERGY FLUX (watt/cm^2)	TIME SCALE where applicable
Marx to Pulse Line	$\sim 3 \times 10^7$	$\sim$ 1 usec
LINE DIELECTRICS		
Pressurixed SF$_6$	5×10^8	$\sim$ 1 usec
H$_2$O	2×10^9	$\sim$ 1 usec
Solid (Mylar)	10^{10}	
H$_2$O	10^{10}	$\sim$ 20 ns
VACUUM DIODE		
* Diode Interface	10^9	$\sim$20 ns
* Vacuum Feed	3×10^9	$\sim$ 20 ns
* Magnetically Cut Off Vacuum Feed	$> 2 \times 10^{10}$	
ELECTRON BEAM		
$j \sim 10^5$ amp/cm^2	3×10^{11}	
$j \sim 5 \times 10^6$ amp/cm^2	1.5×10^{12}	

* Not a limitation if the pulse length $>>$ than width of insulators or length of vacuum feed

Fig. 13.   Energy Flux at Various Stages
Through 3 MV 20 ns Generator.

LIQUID DIELECTRIC PULSE LINE TECHNOLOGY[*]

Ian Smith
Physics International Company
2700 Merced Street
San Leandro, California 94577

ABSTRACT

Liquid-dielectric pulse line technology is an important and rapidly developing aspect of short-pulse, very high power generators, such as those used to accelerate relativistic electron beams. A description is presented of recently completed systems, including the largest present examples of both oil-dielectric and water-dielectric design. Oil-dielectric lines are represented by a recently built system that delivers approximately 9 MV, 250-kA and by AURORA a 14-MV, 2 MJ-system. Water-dielectric lines are represented by reference to OWL II, a 100-kJ, 1-MV system. A brief description is given of general design principles. Mention is also made of the spark gap techniques by which the pulse lines of the systems are switched into their loads. The switches described include triggered multiple channels in the liquid dielectric itself (in both oil and water) and self-closing liquid switches. Triggered and untriggered gas switches can also be used, and these are described. Prepulse switches and the circuits used to control prepulse are also described. Brief descriptions are given of the Marxes used in each of the systems.

INTRODUCTION

The intention of this paper is to describe the fast pulse-forming sections of three recent liquid-dielectric electron-beam generators, and thereby illustrate the techniques of energy storage and switching that are at present in use in such systems. Very high power systems almost always use coaxial liquid-dielectric pulse lines to generate the output pulse, oil and water being the two most commonly used dielectrics. The pulse lines described here are those of the Pulserad 1480, a large oil Blumlein on which testing has just been completed; AURORA, a 2.5-MJ multiple oil Blumlein system; and OWL II, a 100-kJ water coaxial line. The last two constitute the largest oil and water dielectric pulse lines--in terms of electron energy delivered per pulse--that have been built to date. Descriptions of the three systems are preceded by a brief discussion of general pulse line design and a summary of the Marx generators used to charge the pulse lines.

GENERAL

Table 1 shows the important dielectric properties of oil and water. One or other of these has formed the basis for almost all of the large pulse lines built to date. This is partly because the two liquids are plentiful and cheap. It is also because they have greatly different dielectric constants, which enables a wide range of pulser impedances to be readily obtained. The formulae given in the table for the impedance of coaxial lines shows that oil is suitable for impedances in the tens of ohms range, water for less than about ten ohms. The coaxial geometry is most commonly used, since the use of round conductors simplifies engineering; the use of a closed outer conductor that contains the dielectric provides both an efficient design and an electromagnetic shield.

The electric fields that can be used in large systems with charging times of the order of 1 μsec are also shown. The fields are comparable in magnitude, but somewhat greater for

Table 1.Properties of oil and water
         as dielectrics.

	Oil	Water
Dielectric constant	2.3	80
Coaxial impedance	40 $\ln r_2/r_1$	6.7 $\ln r_2/r_1$
Useful field strength (positive electrode)	200-300 kV/cm	100-150 kV/cm
Energy density (j/ℓ)	4 to 9	35 to 80
Current density (kA/m)	80 to 120	240 to 360
Polarity effect	Variable $\approx$1.5:1	2:1

Common properties:
   Breakdown is:
   1. Self-healing
   2. Time dependent ($t^{-1/3}$)
   3. Electrode initiated
   4. Electrode-surface dependent
   5. Variable, hence area dependent

oil. The difference of dielectric constants means, however, that the energy stored per unit volume and the current that can be extracted per unit width of conductor are greater with water. Thus in applications where either liquid can be used, water will usually make a more compact design. Note that the given field strengths apply to the surfaces of positive electrodes; in both liquids higher fields can be tolerated at negative electrodes. This phenomenon is referred to as a polarity effect and is particularly pronounced in water.

The table also lists some dielectric properties common to both oil and water. Breakdown data collected by J. C. Martin and his group at AWRE, Aldermaston, yielded a $t^{-1/3}$ dependence of breakdown field on charging time for both liquids. Data collected at Physics International suggests that for very large areas charged in the 1 μsec time scale the dependence (at

least for oil) may be slightly weaker, say $t^{-1/4}$ or $t^{-1/5}$. The breakdown streamers initiate at metal electrodes, and in small scale tests the breakdown strength can be increased by a factor of the order of two by conditions that affect this interface (coating, de-gassing, pressure); it has not so far been possible to utilize such a large improvement in an operating system, however. An important aspect of scaling from small scale tests to small pulsers and thence to large generators is the fact that scatter in the breakdown process inevitably means a reduction of average breakdown strength with increasing area of electrodes.

Figure 1 shows the coaxial versions of the two circuits used in pulse line design, the simple coax with a series switch and the triax, a version of the Blumlein circuit. For the reasons indicated in the figure, the simple coax is more appropriate for generating high currents, and the Blumlein for high voltages. Thus in practice the simple coax has been associated on the whole with low impedance systems and hence with water dielectric, and the Blumlein with high impedances and oil dielectric. This general tendency happens to fit with the fact that the design of a simple coax system is aided if the liquid has (like water) a large polarity effect, which can be used to make the electric fields and hence energy densities approach the maximum throughout the generator. The polarity effect is less important in the design of the Blumlein.

Either circuit can be used in conjunction with an impedance transformer, i.e., a length of pulse line of whose impedance changes in a controlled fashion between the pulse generating circuit and the load. The transformer can be useful in reducing a particularly large current or voltage requirement on the generating circuit, or it may simply enable the generating circuit to have an impedance closer to the optimum for the dielectric employed.

The switches that generate the output pulse are most simply spark

discharges created in the liquid
dielectric itself. In many cases,
a self-closing discharge is adequate, and is formed by reducing
the electrode spacing in a controlled manner in the region where
the switching is required. If more
precise reproducibility of timing
or amplitude is required, a liquid
discharge may be triggered on
command with the aid of a field-
distortion electrode.[1] This technique can also provide multiple
parallel discharge paths and hence
a lower switch inductance.

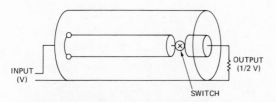

(a) SIMPLE COAX
     OUTPUT VOLTAGE = 1/2 CHARGE VOLTAGE
     OUTPUT CURRENT = SWITCH CURRENT

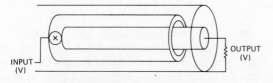

(b) BLUMLEIN TRIAX
     OUTPUT VOLTAGE = CHARGE VOLTAGE
     OUTPUT CURRENT = 1/2 SWITCH CURRENT

Fig. 1 Two types of coaxial
        pulse line.

The pulse line examples that
follow utilize both triggered and
self-closing liquid switches. Gas
switches can also be provided, and
have the advantage of less mechanical shock at the expense of some
increased complexity. Gas switches
may also be either self-closing or
command triggered. The maximum
voltage achieved with self-closing
and field-distortion switches in
oil, water, and gas, are given in
Table 2.

MARX GENERATORS

A few words will be said about
the Marx generators, since a common
Marx design is used to charge all
of the three pulse line examples
chosen here. The design of the
Marx, in particular its inductance,
affects the pulse line design. All
of the Marxes employ a 60-kV, 1.85-
μF capacitor that was developed by
Los Alamos Scientific Laboratories

Table 2. Maximum voltage for various
switch types.

| | Maximum voltage to date | |
Medium	Self Closing	Field Distortion Triggered
Oil	12.5 MV (AURORA)	12.5 MV (AURORA)
Water	7 MV (GAMBLE II)	5 MV (GAMBLE II)
Gas	6.7 MV (TEMPS)	3 MV (PROTOTYPE)

for their 10-MJ Scyllac system.
Two such capacitors in series form
a 120-kV stage; in AURORA each
stage also has two units in parallel
making four such capacitors per
stage. The Marx stages are suspen-
ded in oil by plastic straps. This
construction has enabled up to 12.5
MV to be withstood readily and could
be simply scaled to higher voltages.

The Marx switches are pressurized SF_6 spark gaps with brass electrodes, carrying up to 150 kA and
operated at about 50 percent of
self-fire voltage. A chemical reaction between the brass and the
SF_6 dissociation products stabilizes
the self-fire level, and the probability of a prefire below the intended charge voltage is estimated
as less than 10^{-5} per spark gap per
shot. The Marxes can erect over a
voltage range of more than 3:1 at
fixed pressure; to achieve this,
the Marx layout is such that some
of the Marx spark gaps are overvolted by a large factor due to capacity coupled transients, and those
which are not overvolted, are triggered by a third electrode. This
is known as a hybrid triggering
scheme, and typically results in
jitter of the order of 10 to 20 nsec
for a 100 stage Marx. It also gives
the ability to operate Marxes in
parallel with tight stray capacity
coupling, even into an open circuit.

Marx inductance is typically
350 to 400 nH per 120-kV stage.
This inductance is quite large compared to 60 to 100 nH per stage
obtained in smaller energy Marxes,[2]
but it is accepted for reasons of
economy that apply to large systems;
an AURORA Marx, for example,
stores 1.25 MJ and there are four

Marxes in the system. The reason for desiring a low inductance is that it leads to a faster charging time for the pulse line which can then be made smaller.

PULSERAD 1480

The Pulserad 1480 (Fig. 2) is a 9-MV, 200-kA, 70-nsec system recently fabricated and tested for the Commisariat a l'Energie Atomique. An oil Blumlein approach is adopted because of the high impedance and high voltage.

Fig. 2   Pulserad 1480.

Figure 3 shows a layout of the 40-foot-long, 26-ohm pulse forming section of the Pulserad 1480. The Blumlein is 14 feet in diameter. Of the three coaxial electrodes, the intermediate is charged negative by an 80-stage Marx. This polarity is favored by the polarity effect in oil, since the highest fields in the system tend to be around the output of the Marx and at the output end of the intermediate electrode. It is also convenient that the combination of negative charge and negative output locates the pulse-forming switch on axis, between the intermediate and inner electrodes.

The design charging voltage is 9.5 MV, and it is easy to calculate that the fields on the positive inner and outer electrodes are 250 kV/cm and 200 kV/cm, respectively. This small imbalance is is deliberately chosen, because the outer electrode is more subject to the presence of dirt and bubble

impurities, has a larger area over which breakdown might take place, and because a spark initiating there has a shorter distance to travel. The imbalance is controlled by choice of the intermediate cylinder diameter, which also determines the individual impedances of the two transmission lines formed by the inner and outer pair of electrodes. It transpires that the inner line is 16 ohm and the outer line is 10 ohms, conferring the additional advantage of a higher impedance at the switch. Having the two lines of unequal impedance implies a penalty in stored energy that cannot be delivered in the first output pulse time, but this is only 4 percent of the total energy.

The highest fields in the system are at the outside of the charged intermediate electrode, 225 kV/cm rising to 300 kV/cm at the output end. As already remarked, this electrode is negative.

The inner electrode is supported by nylon straps located beyond the two ends of the intermediate electrode and hence remote from the high field region. Thus, there is no chance of tracking the nylon. The intermediate electrode must be supported in the high field region; however, the straps are at right angles to the electric field where they approach the intermediate and they do not contact the outer electrode until they have entered a low-field region.

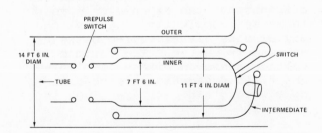

Fig. 3   Pulserad 1480 Blumlein layout.

The Blumlein pulse is generated by a self-closing oil spark in a controlled region between inner and intermediate electrodes at the

Marx end. The rms jitter of this closure is about 50 nsec when timed to occur on average at about 90 percent of peak voltage, or 1 μsec after erection of the Marx. This timing accuracy is adequate for this application, as is the few percent spread in output voltage that results from shot to shot. The breakdown voltage is controlled by adjusting the spacing in the oil at the switch point through a remotely controlled hydraulic movement of a portion of the inner Blumlein electrode.

A second switch is utilized in the 1480 system to reduce the tube prepulse, i.e., the voltage on the tube during charge of the Blumlein. Prepulse arises because the inner electrode cannot be hard grounded, i.e., a low impedance connection from the inner to ground would load the output. Instead, the inner is grounded by an inductor, across which a voltage appears during charge of the Blumlein. To minimize this voltage, the inductor is placed at the Marx end of the Blumlein, where it can be center-fed by the Marx (this location necessitates an opening in the intermediate conductor). The voltages developed on the two sections of the inductor tend to cancel out at the tube. The presence of parasitic elements, such as the Marx capacitance, prevents a true null from being obtained, however. For the 1480, during the Blumlein charge the inner electrode departs several hundred kilovolts from ground. The value of this minimum prepulse, as a fraction of total charge voltage, tends in practice to increase with system size and Blumlein charging speed.

Such a prepulse can often be tolerated at the tube, as long as the cathode design is such that it will not be preionized by the prepulse. When it is desired to reduce the prepulse at the tube, as in the case of the 1480, this can be done by introducing a series of oil switches in the inner electrode. The switches isolate the tube while the Blumlein charges, and break down on the leading edge of the output pulse. The prepulse switches can also be used to decrease the risetime of the output pulse.

A fraction of the prepulse voltage is still coupled to the tube by the capacitance of the prepulse switch. In the 1480, a 4 to 1 reduction from the Blumlein prepulse is achieved; because the goal of the 1480 is a tightly pinched electron beam, very small spacings and high fields are present between the anode and electrode. Therefore an even smaller prepulse is desirable, and a second prepulse switch was added. This is again of the overvolted type, and consists of a vacuum flash-over gap located in the cathode support itself. In this way the prepulse at the active region of the cathode is reduced to about 50 kV. This is less than 1 percent of the output voltage, a fact which contributes significantly to the impedance lifetime and reproducibility of the diode in the pinched mode. Beam currents of over 200 kA were delivered at voltages of up to 9 MV, with an effective diameter of less than 5 mm (as determined by the X-ray source). Total energy in the pinch was well over 100 kJ.

AURORA

The system described above, though fairly large, is an example of the simplest type of oil Blumlein system, and is of relatively standard design; in fact, the 1480 was designed and built in less than 8 months. AURORA, on the other hand, is a very complex system that took several years to develop and fabricate. The AURORA program was funded by the Defense Nuclear Agency.

The design output parameters of AURORA were 1.6 MA at 15 MV for 125 nsec. The high voltage dictated an oil Blumlein approach (output voltages greater than about 2 MV have still not been generated by water dielectric systems), while the impedance level required a number of pulse lines in parallel. The design chosen was four 20-ohm lines all charged to 12 MV. The use of one or two lines of lower impedance would have led to larger overall dimensions. The layout of the four lines, which are charged from a common 5-MJ Marx system, is seen in Fig. 4.

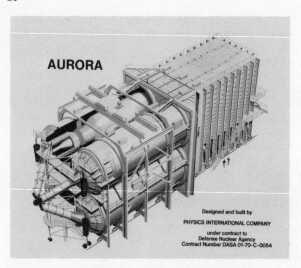

Fig. 4 AURORA overall layout.

The cross section of one Blumlein, 23 feet in diameter, is shown
in Fig. 5. The electrical design
principles are exactly as described
for the 1480, but lower fields are
used throughout. This is because
the breakdown probability is increased by the longer charging time
(2 μsec) and the very much larger
total electrode area used. Because
of the large size of AURORA, it was
felt desirable to utilize the increase of breakdown strength that
results from plastic coated electrodes, even though the increase
averages only about 10 percent for
very large areas. All Blumlein
electrodes are so coated. The
fields calculated from Fig. 5 are
190 kV/cm on the inner electrode,
160 kV/cm on the outer, and on the
intermediate 200 kV/cm rising to
270 kV/cm in places.

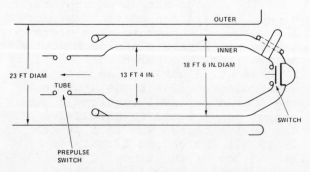

Fig. 5 AURORA Blumlein layout.

The unique feature of the AURORA
Blumleins is the use of externally-
triggered, multiple-channel oil
switches, developed especially for

AURORA. The reproducibility of
self-closing oil switches would have
been inadequate to synchronize the
four output pulses. In any case,
self-closing switches form only one
spark channel and this would have
been too inductive. The impedance
of each inner transmission line is
12 ohm and hence at 12 MV the switch
current is one million amperes.

The design of the Blumlein
switch is shown in Fig. 6. The
principle is that of the third-elec-
trode, field-enhanced trigger. The
overall switch spacing is again controlled hydraulically, this time by
the axial motion of a section of
the intermediate cylinder. The
maximum switch spacing is 24 inches.
The trigger electrode is a large
high grade steel disk, about 3 inches from the flat face of the end
of the inner cylinder. The disk
spacing is variable by a hydraulic
system in the steel shaft that can-
tilevers the trigger disk from inside the inner Blumlein electrode,

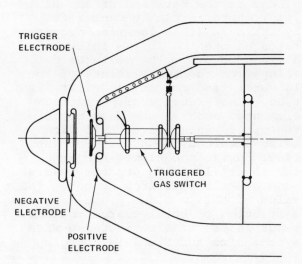

Fig. 6 AURORA oil switch layout.

where the trigger circuit components
are housed. Part of the cantilever
structure is a 2-MV gas spark gap
(a V/N switch). As the Blumlein is
charged to 12 MV, the trigger disk
rises to a capacity-divided potential of about 2 MV. The gas switch
is externally triggered and breaks
down, followed about 200 nsec later
by multiple closure of the streamers
launched in the oil by the sharp
edge of the trigger disk. The jitter is about 10 nsec; a lower value

could be achieved in a switch of similar design by the use of a larger trigger voltage.

An open shutter photograph of an AURORA switch is shown in Fig. 7. The switches can be photographed on each shot, the lower pair via TV cameras immersed in the oil. The switch trigger pulse and trigger-circuit monitor pulses travel in cables that return to ground via the inside of the inductor that charges the inner cylinder. Should the trigger circuit fail, back-up triggers in each 2 MV gas gap cause them to fire independently; the jitter is much greater than that obtained with the external trigger but it is low enough to ensure the almost uniform discharge of stored energy amongst the four switches.

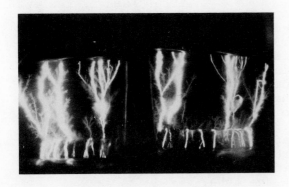

Fig. 7 Open shutter photo of AURORA oil switch.

The AURORA Blumleins generate a prepulse of about 750 kV. This is reduced to about 150 kV by an oil prepulse switch of the type already described. Because of the particular cathode design, this is well below the value that can be withstood by the tube, and no further reduction is needed.

OWL II

OWL is a simple water coax/impedance transformer system that has been in operation for about 2-1/2 years. Its present version, OWL II, was funded by DNA and has been in routine use for 1-1/2 years, e.g., as an electron beam source. It consists of a 3.8-ohm coaxial pulse line, series discharging into a nominal 1.5-ohm load, via an impedance transformer.

Originally, the 3.8-ohm pulse line generated an 80-nsec pulse, and the transformer consisted of two line sections of length equal to the pulse line, with impedances 2.9 and 1.9 ohm. Thus the voltage and current transformation is effected by three distinct mismatches including one at the switch and one at the load.

In its present form, OWL II has alternative 60-nsec and 110-nsec pulse-forming lines with the original transformer. A layout of the pulse lines with the 60-nsec driver is shown in Fig. 8. The nominal output is approximately 1.3 MV, 800 kA. OWL II was designed as a 100 kJ generator, and has produced 100-nsec electron beams of as much as 130 kJ. An additional transformer section can be added to reduce the output impedance to 1.1 ohm. The outer diameter is 5 feet 4 inches. All electrodes are of polished stainless steel. To avoid high fields on plastic supports, the high voltage electrode is cantilevered. The shorter pulse line is charged to as much as 5 MV, and the longer line to 4.4 MV, in a time of about 1 μsec. At the highest voltage the field on the positive outer is 105 kV/cm, and on the negative inner 190 kV/cm, rising to 250 kV/cm near the switch.

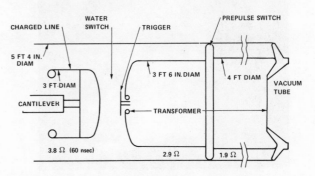

Fig. 8 OWL pulse line layout.

The 3.8-ohm pulse line stores a maximum of about 150 kJ. It is discharged into the 2.9-ohm first stage of the impedance transformer by a triggered water switch (Fig. 9). This is similar to the AURORA Blumlein switch. That the

technique of triggering by the use
of a sharp-edged electrode would be
successful in water as well as in
oil was known from work by Martin's
group, who had successfully operated two such synchronized switches
at about 2 MV. In the OWL II
switch, the trigger electrode is a
disk 9 inches in diameter, spaced
5/8 inch from the positive side of
the switch, which is near ground
potential during charge. The maximum total switch spacing is about
12 inches (controlled by hydraulic-
ally-actuated motion of the high
voltage inner cylinder of the charged line). A gas switch inside
the inner conductor of the 2.9-ohm
line provides the trigger pulse,
whicn is about 500 kV. The fraction of the total potential used
to break down the switch is thus
even smaller than in AURORA. The
breakdown time of the water switch
is about 250 nsec with a jitter of
6 to 8 nsec. These values would
be reduced by using a larger
trigger; however, OWL II has only
one pulse line and hence synchronization requirements are modest.
The main purpose of the triggered
switch is the creation of multiple
channels to reduce inductance and
energy deposition.

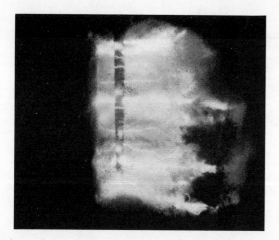

Fig. 9. Open shutter photo of
        OWL II water switch.

Prepulse is again a concern in
OWL II; in fact, in systems where
the current is high and the tube
voltage is fairly low, small anode/
cathode spacings are usually used
and then the control of the prepulse becomes very important. In
the series-switched pulse line,
prepulse is due to capacitive
coupling across the switch during
the charge of the pulse line. In
OWL II this would give rise to
several hundred kilovolts on the
tube. To avoid this, the inner
conductor of the impedance transformer is divided in two by a prepulse switch that insulates the
2.9-ohm section from the 1.9-ohm
section. The design of the prepulse switch was developed especially for OWL. A thick plastic
slab is used to decrease the capacitance, and inside this is housed
a ring of pressurized gas switches.
These operate in the same way
as oil prepulse switches at the
output of oil Blumleins; they self
close on the rise of the output
pulse.

The prepulse is further reduced by grounding the inner of the
2.9-ohm line via a   100-ohm water-
dielectric helical-transmission
line with a two-way transmit time
somewhat longer than the pulse.
The effect of the helical line is
to reduce the prepulse at the
2.9-ohm line to about 300 kV.
Capacitive division between the
prepulse switch and the 1.9-ohm
line reduces this to about 5 kV
at the tube.

The prepulse switches can be
used to sharpen the pulse risetime,
simply by increasing the pressure
well above the value needed to withstand the prepulse, so that the
switch breakdown is delayed until
the output voltage nears peak.
This is illustrated in Fig. 10.

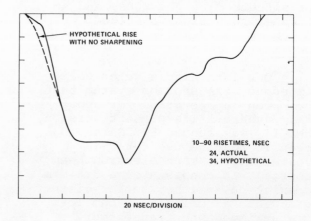

Fig. 10.   OWL II output voltage.

REFERENCES

* Work supported in part by the
Defense Nuclear Agency.

1. S. Mercer, I. Smith, and T.
Martin, "A Compact, Multiple
Channel 3 MV Gas Switch," presen-
ted at Conference on Energy Stor-
age, Compression, and Switching,
Torino, Italy, November 1974.

2. H. Aslin, "Fast Marx Generator," presented at Conference
on Energy Storage, Compression,
and Switching, Torino, Italy,
November 1974.

MEASUREMENTS OF A LOW IMPEDANCE, LARGE AREA 100 kV DIODE[*]

I.D. Smith, V.B. Carboni, G.B. Frazier, E.P. Zeehandelaar
Physics International Company
2700 Merced Street
San Leandro, California 94577
and
Daniel N. Payton, III
Air Force Weapons Laboratory
Albuquerque, New Mexico

ABSTRACT

A 25 cm by 60 cm cathode is used to produce 100 to 200 nsec electron beams of up to 350 kA at voltages less than 150 kV. A 2 to 3 nH tube is driven by a very low inductance dc-charged capacitor circuit whose characteristics are 200 kV, 220 nF and 16 nH. Cold cathodes comprising arrays of metal wires or ribbons have been used. The impedance obeys Child's Law for space charge limited planar flow at early times. If constant velocity plasma closure between the anode and cathode is assumed, the impedance remains consistent with Child's Law for times up to 250 nsec and for impedances and voltages as low as 25 milliohms and 3 kV. On this model, the apparent plasma velocity is very constant for most of the pulse duration, extremely uniform and reproducible, and depends mainly on the cathode material. Despite the agreement of impedance with a plane parallel flow model, strong pinching is observed in the diode. A uniform line pinch is formed, 60 cm long by about 1 cm wide. Pinch current has been measured as a function of time. The average electron flow is at a very small angle to the tungsten anode. Tungsten is vaporized and melted on each pulse, but the quantity of material removed is much smaller than expected. It is possible that the pinched beam deposits its energy in a depth of the anode that is much smaller than the electron range.

APPARATUS

The configuration of the diode is sketched in Fig. 1. The anode is a flat tungsten plate, 25x85 cm, in the form of a rectangle with semi-circular ends. The 25x60 cm rectangle forms the active portion of the anode, and is irradiated by electrons from a "cold" cathode. The cathode consists of many metal ribbons or wires, stretched parallel to the anode at a uniform distance that is variable up to about 1 cm. These emitters are parallel to the 25 cm width of the anode, and attach at their ends to a metal case that forms the grounded outer chamber of the X-ray tube.

The X-rays (bremsstrahlung) produced at the surface of the tungsten plate must pass through the cathode to reach the irradiation volume and hence the cathode must be highly transparent. This is best effected by using an array of wires spaced many diameters apart. If an array of ribbons is used, these are made a

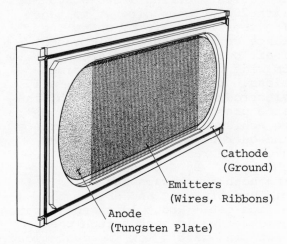

Fig. 1 Diode configuration

Cathode (Ground)
Emitters (Wires, Ribbons)
Anode (Tungsten Plate)

few millimeters wide and placed parallel to the general X-ray flow direction; one edge of each ribbon faces the anode and develops the high field that causes electrons to be emitted.

The anode is supported from the grounded chamber by an acrylic insulator that forms part of the vacuum vessel. The insulator runs around the anode perimeter in a geometry resembling a race track. It is designed to withstand 200 kV, 100 nsec pulses with the smallest possible cross section, in order to make a small inductance for the current path to the diode. A low inductance is important because it is desired to produce currents of several hundred kiloamperes rising in less than 100 nsec with a driving voltage of 100 kV. The current is fed into the tube along the full length of the two 60 cm sides of the diode, and the total inductance from the atmospheric pressure region is about 3 nH.

The pulse generator that drives the tube is novel in that it produces the required very large short pulse currents directly from dc charged capacitors without the aid of an intermediate energy store such as a pulse line. The pulser is immersed in atmospheric pressure sulfur hexafluoride gas and consists of two parallel circuits, one feeding each 60 cm side of the tube. Each circuit has two 100 kV, 0.22 µF capacitors and two 100 kV spark gaps in series, forming a two-stage Marx with 200 kV open circuit voltage. The capacitors are of a low inductance design with plastic cases. Their output terminals are about 60 cm wide to match the tube dimensions. The spark gaps have rail electrodes of a similar width, inside pressurized acrylic housings. To achieve a low pulser inductance the current must be distributed along the rails in many spark channels. By externally triggering both spark gaps and using a mixture of argon and SF_6 as the switch gas,[1] about 10 channels can be obtained in each switch when operating with a low impedance load. The inductance of each of the two parallel pulsers, up to but excluding the tube, is then about 32 nH. The total circuit comprising the two parallel pulsers and the tube constitutes 0.22 µF and about 16 nH in series with the electron beam load.

Figure 2 shows the assembly of the pulsers with the tube insulator, anode and cathode. Charging and

triggering circuits have been removed, and so has the grounded vessel that closes the tube vacuum beyond the cathode. In Fig. 2, the cathode consists of 4 mil tungsten wires, 5-mm apart. Anode damage (from about 500 shots) is clearly visible, largely confined to the central region of the diode. As will be seen, this is due to the formation of a line pinch of electrons, the most intense part of which is about 1 cm wide. Unless the cathode is very accurately made, the location of this line varies by a few centimeters from pulse to pulse.

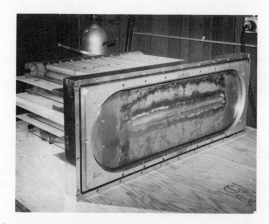

Fig. 2   Diode and pulser partly assembled.

DIODE MEASUREMENTS

The diagnostics are voltage and current monitors and time-resolved X-ray detectors of various types. The results reported here concern principally the impedance characteristics of the diode and the tendency to pinch.

Figure 3 shows a typical set of current/voltage waveforms at full charging voltage. Peak anode voltage is just over 110 kV and peak current is 350 kA. The voltage rises more rapidly than the current because the diode impedance is high at the start of the pulse and then decreases. The impedance becomes zero at late time, prior to which currents well over 100 kA can be measured at voltages as low as 2 to 3 kV.

A decrease of impedance with time is a familiar characteristic of a cold cathode system and is associated with the motion of plasma during the pulse, progressively closing the effective anode/cathode spacing. The best quantitative explanation of the observed impedance time history for the present diode is obtained starting from Child's Law for planar, space charge limited flow of non-relativistic electrons.[2] The current in a diode of area A and spacing d is given by Child's Law as

$$I = 2.34 \ Ad^{-2} \ V^{3/2} \qquad (kA)$$

where V is the applied voltage in MV. The voltage and current are known as a function of time, and using the actual diode area, the equation can be used to compute the apparent spacing (d_{CL}) of the plasmas as a function of time.

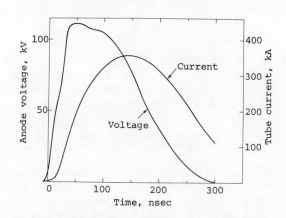

Fig. 3   Tube current and voltage waveforms at full power.

Examples of the variation of d_{CL} with time are shown in Figs. 4-6, where it can be seen that after an early phase of high and quite rapidly falling impedance, the spacing approaches zero in a straight line fashion. When the line is extrapolated back to zero time, a spacing corresponding closely with the physical anode-cathode gap is obtained.

Thus the results may be explained by an initial period when cathode emission is building up to the space charge limit, and an overall closure of plasma at a constant velocity. The smooth approach of the spacing to zero

implies a very planar plasma front; the measurements are also extremely reproducible shot to shot. The plasma velocities are largely independent of anode/cathode spacing, but are influenced by cathode material as shown by comparing Figs. 4 and 5. Under similar conditions, an aluminum ribbon cathode gives a velocity of 3.7 cm/μsec, while an identical tantalum ribbon cathode and a

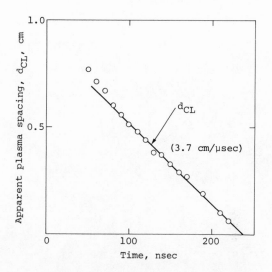

Fig. 4   Plasma spacing versus time; aluminum ribbon cathode, 8 mm spacing

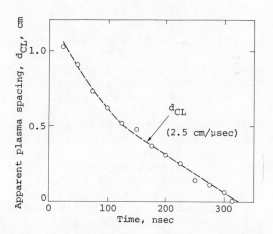

Fig. 5   Plasma spacing versus time; tungsten wire cathode, 8 mm spacing

tungsten wire cathode both give 2.4 to 2.5 cm/μsec. This is consistent with the view proposed by many workers that the motion is predominantly

that of a cathode plasma, formed
early by the explosion of micropro-
tuberances under the resistive
heating produced by field emission.
The dependence of closure velocity
on cathode material was previously
reported by Spence[3] who compared
graphite and steel. In addition,
we here note a tendency for the
velocity to increase with power
level, as shown by a comparison of
Figs. 5 and 6.

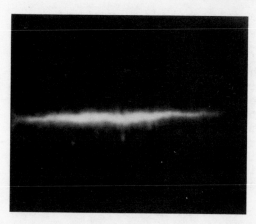

Fig. 7 Pinhole radiograph of anode
        X-rays.

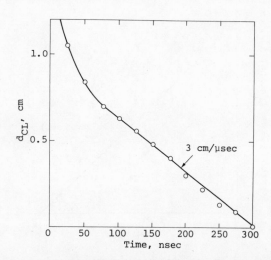

Fig. 6 Apparent spacing vs time;
       tungsten wire cathode (full
       power) 8 mm spacing.

The usefulness of Child's Law
in interpreting these experiments
is surprising in view of the observation that the electron flow is
actually pinched rather than planar.
Figure 7 is a pinhole radiograph of
the anode. The apparent fall off
in intensity of the pinch at the
ends is mostly due to pinhole camera geometry. The fact that pinching occurs only in one direction is
due to the constrained current flow
at the beam edges, which results in
a magnetic pressure that is almost
wholly perpendicular to the 60 cm
length of the diode. The results
represent what may be the first observation of strong pinching in a
low voltage diode--mean electron
energies are typically 75 keV. Note
that in the time integrated picture
the pinch is superimposed on a
fairly uniform background of radiation over the entire anode.

The onset of pinching in a rectangular diode can be predicted in
the same way that the familiar result

$$I_{crit} = 8.5 \ \beta\gamma \ (r/d) \qquad (kA)$$

is obtained for a circular diode of
radius r. If the current is assumed
to flow uniformly on the two sides
of length ℓ of the rectangle, we obtain

$$I_{crit} = \frac{8.5}{\pi} \ \beta\gamma \ (\ell/d) \qquad (kA)$$

where β, γ are again the relativistic
parameters for electrons at anode potential. Since the anode-cathode
spacing may be effectively changing
with time, we rearrange this equation
and define a critical spacing.

$$d_{crit} = 8.5 \ \beta\gamma/I \ \ell$$

If the anode-cathode spacing exceeds
d_{crit}, pinching is expected. To see
if this is so, we calculate d_{crit} as
a function of time from the instantaneous values of I, β, γ and the actual value of ℓ (60 cm) and compare
it with the apparent plasma spacing
d_{CL} obtained as previously described.
In Fig. 8 we we plot the "pinch parameter," d_{CL}/d_{crit} as a function of
time throughout a typical pulse.
The quantity exceeds unity, satisfying the conditions for pinch, from
a time early in the pulse until later on when d_{CL} nears zero. Thus
the observed pinch is predictable
on this basis.

A more detailed correlation is made by measuring as a function of time the approximate fraction of the total current carried by the pinch (the remainder of the current is uniformly spread over the anode). This measurement is made with X-ray detectors viewing different portions of the anode. The result, also plotted in Fig. 8, shows that the pinch first appears at about the time that the pinch parameter first exceeds unity; thereafter, the fraction of the beam actually pinched increases with the pinch parameter, to a maximum of about 70%. The fraction pinched also decreases late in the pulse, and it appears that the pinch may disappear entirely near to the time that the pinch parameter becomes less than unity. This correlation, obtained using the plasma spacing deduced from Child's Law, further supports the validity of the use of Child's Law to deduce the approximate plasma spacing, even though most of the electron flow is highly non-planar.

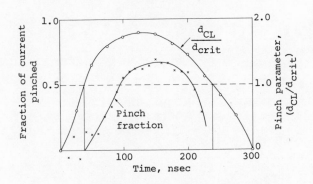

Fig. 8 Pinching as a function of time.

The on-axis pinching of electrons assumed to be emitted at the diode edge implies average electron motion at a small angle (2 to 3 degrees) with the diode plane. Electrons may well be incident on the anode at an angle very near grazing, with the resulting energy deposition profile in the anode very peaked at the front surface. With a total of 4 kJ of electrons delivered, over 2 kJ is in the 60 cm^2 or so of the pinch region, and the electron fluence is thus of the order of 10 cal/cm^2. Even if the high incidence angles are ignored, the front surface dose must be

several thousand calories per gram. The pinched beam should cause vaporization and/or melting over most of the electron range, which is about 10 mg/cm^2. Most of the melted or vaporized material should be removed, leading to an anticipated loss of hundreds of milligrams of tungsten per pulse. In practice, much less material is removed. Estimates obtained by weighing material plated out (in a form which resembles a vapor deposit) suggest only a few milligrams per pulse. It is possible that the pinched beam deposits its energy in a relatively low-mass blowoff from the anode, in which the electrons are trapped by their self-magnetic field.

REFERENCES

* This work is partially supported by the Air Force Weapons Laboratory, Albuquerque, N.M. under contract No. 74-C-0099.

1. P. Champney, "Some Recent Advances in Three Electrode, Field Enhanced Triggered Gas Switches," presented at Conference on Energy Storage, Compression, and Switching, Torino, Italy, November 1974.

2. C. D. Child, Phys. Rev. 32, 492, (1911).

3. P. Spence, Physics International Company, San Leandro, California, private communication.

PULSED ELECTRON BEAM GENERATORS

OPERATING IN C.E.A.

J. Chevallier , J. Cortella , J. C. Jouys , G. Raboisson
CEA - 33, rue de la Fédération - PARIS - France

I - INTRODUCTION

Since 1965, in order to make pulsed irradiation experiments with electrons or X-rays, the French Atomic Energy Commission (CEA) has developped 3 generators whose instantaneous electrical power ranges between 10^{11} and 2.10^{12} watts. These generators are being used in VALDUC Center, near DIJON.

Two of them, THALIE and EUPHROSYNE are more specially designed as X-ray generators, this being simply achieved by slowing down the electrons in a high Z material such as tantalum. The irradiations with the third generator AGLAE, directly use the electron beam.

Operating this kind of generator requires in the first place an electric pulse : its instaneous power must be great and last about 100 nanosecond. This pulse is powering a "cold-cathode", so that the electric fields on the cathode surface may reach up to 10^{10} V.m^{-1} and release electrons by Fowler-Nordheim effect. The same principle has been applied to the 3 generators with, however, functions matching different voltage, current and energy range. Each generator consists of :

a - An electron tube or diode on whose cathode is applied the electric pulse. Each diode is in fact a vacuum chamber with dielectric walls to withstand anode to cathode voltage.

b - A coaxial line whose discharge creates the pulse powering the diode. The line discharge is initiated by an adequate switch.

c - A Marx generator to charge the line electrodes.

We shall now briefly study the main specifications of the three generators.

II - THALIE GENERATOR

THALIE generator has been designed to produce X-Ray pulses lasting 70 nanoseconds. The "test volume" is a 30 cm diameter, 20 cm high cylinder. The radiation level in this volume is higher than 10^{11} rad (Si). s^{-1}, allowing for a 12 % inhomogeneity. The specifications for the electron beam are as follows :

- 15 MeV kinetic energy
- 200 kA current

i. e. a 2.10^{12} watts peak power, and a 140 K. joules total beam energy.

The diode insulating walls are made of 22 "Araldite" (epoxy resin) rings, 2 meters in diameter and 10 cm thick each. Between the rings, metal discs distribute the potential. The 40 cm hemispherical cathode gives to the diode an impedance of about 50 Ω.

The epoxy resin used for insulation is responsible for electrical breakdown of the rings and we intend to change this material.

The triaxial line, or Blumlein, is dissymetric. The outer and intermediate electrodes constitute a 12 Ω impedance transmission line, the intermediate and inner electrodes a 20 Ω impedance line, the total impedance being therefore 32 Ω. The electric fields on the surface of the inner and intermediate electrodes are almost the same. The insulation of the Blumlein electrodes is achieved by some mineral oil (Esso Univolt 64), with a breakdown field higher than 300 kV. cm^{-1}.

The Marx generator charges the Blumlein in 0. 8 μs and the Blumlein discharge produces a 100 ns FWHM pulse This discharge is initiated by a self-firing axial switch, using oil as a dielectric.

Its shape is such that a minimum self-inductance may be obtained at the time it is operating, by creating several conducting plasma channels. The Marx generator was designed according to a simple, asymetric scheme. It consists of 46 stages, including each :

- 1 x 0, 347 nF capacitor (made by Haefely) with a maximum charging voltage of 240 kV.

- 2 liquid resistors made of water saturated with copper sulfite. The values are 50 kΩ and 5 kΩ .

- 1 SF$_6$ switch. The 46 switches are included in a lucite column, where the gas pressure varies from o to 4 kg. cm^{-2}. The distance between the poles is 20 mm. Only the first switch on the ground side is triggered, the others being self-firing by overvoltage.

The main output characteristics of the Marx generator are as follow :

- Energy   : from 90 to 500 kJ

- Peak voltage (no lead) : 11 MV

- Output capacity   : 7, 4 nF

- Serie-inductance : 28 μH

- Serie-resistor   : 5. 4 Ω

- Parallel-resistor : 209 kΩ

This generator is quite reliable since after 2000 shots in usual conditions no element failure has been observed.

Besides dose measurements obtained from thermoluminescent FL$_i$ powder calibrated by means of an adiabatic calorimeter, some electric diagnostic at different places in the generator give :

- the Blumlein charging voltage
- the diode current and voltage

This is done with capacitive or resistive dividers for voltage measurements and with a loop for the diode current.

THALIE generator has been in routine operation since january 1974.

III - EUPHROSYNE GENERATOR

EUPHROSYNE generator has been built in 1968. With this generator, experiments with photons at a radiation level in the 10^{10} to 10^{11} rad. s^{-1} range may be conducted within a test volume limited to about 200 cm^3.

Thanks to the knowledge obtained from the EUPHROSYNE generator project, we were able to estimate the THALIE specifications.

The electrical characteristics of the EUPHROSYNE diode are : 60-kA beam current and 3-MV voltage. The insulating walls of the diode consist of 11 lucite rings. With this material, we can reach 1000 shots without lowering the dose by more than 20 % . Contrary to what happens with the araldite used for the rings of THALIE generator, there is no risk of electrical breakdown.

The diode is powered by the discharge pulse of a 36 Ω symetrical Blumlein. The pulse width is 50 nanoseconds. Mineral oil is used to insulate the electrodes. The switch firing the pulse is an oil spark-gap, which shorts the intermediate and outer electrodes of the Blumlein. One must notice that mineral oil for the three generators is permanently filtered and cleaned by outgasing to maintain its necessary qualities.

The Marx generator connected to the intermediate electrode of the Blumlein is designed after the same scheme as that of THALIE. It consists of 15 stages, with 0. 15 μF capacitors and 200 kV maximum charging voltage. The output characteristics are as follows :

- output voltage (no load) : 3 MV
- output capacity   : 10 nF
- energy   : 20 to
                                    45 kJ
- serie inductance   : 22 μH
- serie resistor   : 4 Ω

The EUPHROSYNE generator has been in routine operation since 1969, at a rate of approximatly 3000 shots a year.

IV - AGLAE GENERATOR

For the two previous generators, the characteristics optimisation was calculated by the criterion of the maximum X-ray dose. The dose is given by :

$$\overset{\circ}{D} = k\, I\, V^{\alpha} \quad \text{with } 1 < \alpha < 3$$

So, we must try to increase voltage, and the diode impedances can be relatively high. On the other hand, for AGLAE generator which is used for irradiation experiments with electrons of limited kinetic energy, the parameter to be increased is :

$$W = V \cdot I \cdot T$$

and we have to obtain the lowest impedance at the diode and coaxial line.

This involves two operating modes, using two different coaxial lines and diodes which may be connected to the same Marx generator :

a - The first line is a water insulated Blumlein. The water resistivity is about 10 MΩ-cm^{-1}, and the high dielectric constant allows a 10 Ω total impedance. This Blumlein operates with a 1 MeV maximum voltage, and it is connected to a matched diode. Consequently, the beam characteristics are the following :

- 100 kA current
- 1 MV kinetic energy

b - The second line is a single water line, with a 2 Ω characteristic impedance. So, in the same conditions, we get a pulse voltage of 500 kV, and the available current in a matched load is 250 kA.

In these two cases, the discharge is triggered by an axial, SF$_6$ spark gap.

The same Marx generator is coupled either to the first or to the second line. It consists of 9 stages, with 0.15 μF capacitors.

Besides the usual diagnostics, we can mesure energy by a calorimetric method, and the beam current by a Faraday cup.

V - CONCLUSION

The characteristics of the 3 pulsed electron beam generators operating in CEA that we have here briefly surveyed, show us how to deal with the energy storage and switching problems in the 10^4 to 5.10^5 J energy range.

The advanced capabilities, of these generators and their routine operation, sometimes for years, demonstrate the validity of the process.

FAST MARX GENERATOR [*]

H. Aslin
Physics International Company
2700 Merced Street
San Leandro, California 94577

ABSTRACT

A compact modular Marx generator of a unique low inductance design for application in oil or gas insulated systems is described and test results presented.

Marx generator dc charge voltage is 100 kV, balanced with respect to ground. Pressurized gas (sulfur hexafluoride-SF_6) switches are employed and each is equipped with a midplane trigger electrode. Resistive trigger coupling of the Marx switches is employed with the first four switches triggered from an external pulse source.

Low Marx inductance is achieved principally as a result of the energy storage capacitors which comprise the Marx stages. Inductance ranges from about 0.6 μH/MV for low voltage reversal Marx applications to about 1 μH/MV for high reversal applications.

Thirty-five stage Marx generators of the described design have been fabricated and used in both gas and oil insulated systems.

In gas, two such Marxes have been used in series in a system developing a peak output voltage of 6.7 MV. In oil, a Marx of the described design, tested into resistive loads, has produced peak voltage greater than 3 MV.

Marx erection timing jitter ranges between 2 and 4 nsec rms for both the gas and oil insulated designs depending upon the percentage of self-fire voltage at which the Marxes are operated.

INTRODUCTION

The Marx generator design described here was prompted initially by the need for a low inductance gas insulated Marx for TEMPS (Transportable EMP Simulator) designed and built by Physics International Company for the Harry Diamond Laboratories under the fiscal sponsorship of DNA (Defense Nuclear Agency). TEMPS is a Marx generator--peaking capacitor electrical circuit and peaker size is directly proportional to Marx inductance. And thus, a low inductance Marx design was sought.

Additional requirements for low inductance Marx generators have arisen since TEMPS for applications where the Marx pulse charges a water dielectric store. To effectively utilize the time dependent breakdown field strength of water requires a short duration pulse operation which in turn requires a low inductance initial energy store.

TECHNICAL DISCUSSION

Compact low inductance Marxes for operation in atmospheric pressure gas (SF_6) and oil ambient insulation environments are illustrated in Figs. 1 and 2 respectively and both contain 35 capacitor stages dc charged to 100 kV maximum. Low overall Marx inductance is achieved principally as a result of the energy storage capacitors which comprise the Marx stages and their physical configuration.

The Marx stages are built up from tubular 50 kV capacitors connected in parallel. In the gas immersed Marx, tubular capacitor

Fig. 1   TEMPS Marx generator.

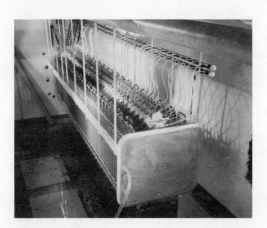

Fig. 2   Sandia Marx in test tank.

dimensions are 1-1/8 inch in diameter by about 16 inches long not including the 5/8-inch long axial studs at each end of the tube to which electrical connection is made. Tube capacity in this case is approximately 23 nF and stores about 29 joules at 50 kV. The volume of the capacitor, including the case, is 261 cm^3 yielding a volumetric energy density of 0.11 J/cc. Tube weight is approximately 0.36 kgm, yielding an energy density of 80.5 mJ/gm.

In the oil-immersed Marx, tube dimensions of 1-5/8-inch diameter by 22-3/8 inches long have been employed. Tube capacity in this case is about 32 nF storing 40 joules at 50 kV. Volumetric energy density and energy density is 0.53 J/cc and 38 mJ/gm respectively.

Fabricated in this way, from parallel connected tubes, the Marx stages are extensible in capacity by connecting additional tubular capacitors in parallel.

Each Marx generator stage consisting of two series connected groups of parallel connected tubes is "folded" and supported on an acrylic plastic sheet (Fig. 3). The support sheet in turn is supported on horizontal support members which run the full length of the Marx at its sides (Figs. 1 and 2). Designed in this way, each Marx stage may be easily removed without disassembly of the Marx as a whole.

Fig. 3   Marx module.

The plastic support sheets also provide dc insulation between stage halves. Dielectric barriers installed between stages provide both dc and pulse insulation.

The Marx switches are pressurized SF_6 gas switches (Fig. 4) and all are equipped with midplane trigger electrodes. Gap spacing is nominally 0.75 cm and single switch self breakdown as a function of SF_6 gas pressure for low discharge current conditions is shown in Fig. 5. The solid line in Fig. 5 is the ideal self-breakdown curve fitted to the measured data at 0 psig. The values listed at each data point are the standard deviation of breakdown voltage for ten samples taken at each point.

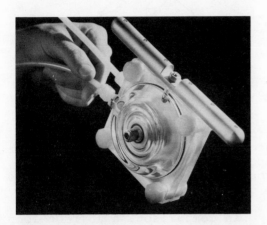

Fig. 4   Switch model 6-70.

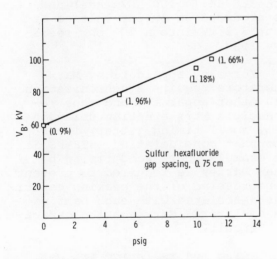

Fig. 5   Single switch self-breakdown

Marx resistors, which form the charging and triggering chains are low inductance wire wound resistors connected in series and potted in epoxy. Epoxy potting is used principally to provide mechanical rigidity. A typical resistor assembly is illustrated in Fig. 6.

Charging resistors consist of five 200 ohm wirewound resistors

connected in series, but potted in epoxy in three sections. A typical assembly is 1 kΩ.

Fig. 6   TEMPS trigger coupling resistor.

Triggering resistors for an m=4 coupled Marx consist typically of ten 100 ohm wirewound resistors potted in epoxy. In general, the triggering resistors form the greatest shunt load upon the erected Marx generator. Each resistor bridges m stages and thus the energy deposited in each resistor is,

$$U_R = \frac{(m \, V_{dc})^2}{R_T} \, \tau_{eff}$$

where R_T is the resistance of each trigger resistor and τ_{eff} is the effective output pulse duration. For Marxes of more than a few stages the total number of trigger resistors is n, the number of stages. Therefore, the total energy dissipated in the trigger resistors is

$$n \, U_R = \frac{n(m \, V_{dc})^2}{R_T} \, \tau_{eff}$$

and this may be equated to a single effective shunt load resistor R_{TEQ}, i.e.,

$$n \, U_R = \frac{n(m \, V_{dc})^2}{R_T} \tau_{eff}$$

$$= \frac{(n \, V_{dc})^2}{R_{TEQ}} \, \tau_{eff}$$

Solving for R_{TEQ} yields

$$R_{TEQ} = R_T \frac{n}{m^2}$$

For a 35 stage Marx, m=4 and
R_T=1 kΩ

$$R_{TEQ} = 1 \text{ k}\Omega \frac{35}{16} \simeq 2.2 \text{ k}\Omega$$

A simplified equivalent circuit of the erected Marx generator
is illustrated in Fig. 7 and is an
adequate representation for analyses where the load placed on the
Marx is capacitive with a value
much greater than the total stray
capacity of the Marx to ground.
For a 35 stage Marx in oil, this
minimum value of load capacity is
about 1 nF. For smaller capacitive
loads, or resistive loads, the distributed nature of the Marx strays
must be considered to adequately
evaluate Marx generator output voltage as a function of time, but
this is easily done with the aid of
a computer circuit analysis program.

For the oil insulated Marx
illustrated in Fig. 2, the approximate value of stray capacity to
ground is 3 pF per stage.

The series inductance L shown
in Fig. 7 is about 60 nH/stage for
the gas insulated Marx and about
100 nH/stage for the oil immersed
Marx.

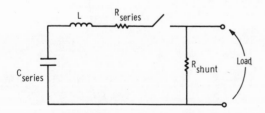

Fig. 7 Simplified equivalent circuit of the Marx during discharge.

TEMPS, a system which employs
two 9 feet long, 35 stage, gas insulated low reversal Marxes contained within tapered conducting enclo-
sures (Fig. 8) possesses an inductance (measured) of 2.15 µH per Marx
or 61 nH/stage. In TEMPS, the
Marxes are connected in series by
means of a high pressure gas insulated output switch and have produced

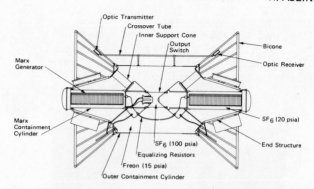

Fig. 8 TEMPS pulser assembly.

an output peak voltage of 6.7 MV
into a load which is essentially a
dipole antenna spaced above earth
ground, and resistively terminated
to ground at its ends.

In comparison, a 13 foot long
35 stage high reversal oil insulated Marx contained within a rectangular cross-section tank has a measured inductance of a little less
than 3.6 µH (≃ 100 nH/stage) and
has produced an output voltage of
about 3.2 MV into a 200 ohm resistive load.

For many applications Marx
generators require synchronization
with other apparatus and thus reproducible Marx erection delay,
i.e., small timing jitter, is an
important consideration. TEMPS is
one example where good timing between Marxes is required to prevent
overstressing of the peaking capacitors associated with each half of
the bi-lateral Marx peaking capacitor circuit.

Using m=4 resistive trigger
coupling, Marx timing jitter is
typically 2-3 nsec rms throughout
a 2:1 operating voltage range for
conditions where the Marx is operated ≥70% of self-fire and for a
trigger input voltage >100 kV rising
to this value in ≤10 nsec (10-90%).

* This work supported in part by
the Defense Nuclear Agency.

HIGH-POWER PULSE GENERATION USING EXPLODING FUSES[*]

J. Benford, H. Calvin, I. Smith & H. Aslin
Physics International Company
2700 Merced Street
San Leandro, California 94577

ABSTRACT

Exploding foils provide an opening switch technique that can deliver energy from an inductive store to a resistive load. The work described in this paper derives from that of DiMarco (Los Alamos Scientific Laboratory) who used copper foils to generate ~100 nsec pulses of up to 100 kV. Our objective was to establish the feasibility of generating still shorter pulses (in the 10 to 100 nsec range) with voltages of up to 1 MV. Initial experiments used a 25 kV capacitor bank energy source, similar to that used by DiMarco, charging a stripline inductive energy store provided with distributed capacitance to permit pulse shaping. Subsequent experiments used a 60 kV bank with a 1 μsec risetime, and some tests were conducted using a water capacitor energy store charged to more than 100 kV. In all cases the foils were immersed in liquid to discourage re-strike. A number of scaling laws were postulated and were found to be in general agreement with experiments. Voltages of more than 200 kV were generated in 50 nsec pulses and voltages in excess of 400 kV were generated in 25 nsec pulses. Of the foil materials tested, copper was found to give the best performance. This result is believed to be associated with its low resistivity at room temperatures which suggests that low temperature foils may give still better results. In the time regimes studied, the limits of the technique appear to be maximum voltage gains between five and 6 and pulse duration "compressions" of the order of twenty.

INTRODUCTION

Inductive energy storage offers advantages over electrostatic storage in high energy applications. The primary advantage is that magnetic energy density can be very high, reducing the size and cost of the store. Effective utilization of the magnetic energy in most cases requires an opening switch which can interrupt large currents. Switching diverts currents from a low impedance circuit into a circuit of higher impedance. Thus energy is delivered at multiplied voltage in a shorter time scale (pulse compression).

One type of switch operating at ambient pressure is investigated here: the exploding metal foil. A number of workers have developed this switch.[1-3] In particular, it has been in use at LASL[*] for several years in a generator that is routinely used to drive a Z-pinch experiment with 200 nsec, 0.7 MA pulses at up to about 100 kV.

_____
*Los Alamos Scientific Laboratory

Operation of the switches in parallel has been demonstrated, and adequate synchronization appears possible. The energy source is a comparatively slow, low voltage (20 kV) capacitor bank. Thus the voltage and pulse duration are close to or within the range of interest for diodes to produce intense relativistic electron beams.

The purpose of the empirical study presented in this paper is to identify and experimentally assess scaling relationships that permit extrapolation from present exploding foil switch electrical circuits and foils to other circuits, particularly to circuits operating on much faster time scales. To accomplish this, three capacitive energy storage and discharge systems are employed in experiments which extend foil energy deposition time from the few microsecond time regime down to about 100 nsec.

In a very simplified sense, the

circuits in this study take the form illustrated in Fig. 1. A capacitive energy store is dc charged to voltage V_i and is discharged into the series combination of inductance L_o and time varying resistance R_f which represents the metallic foil opening switch.

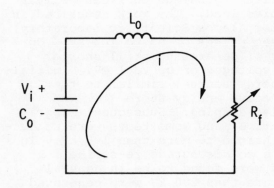

Fig. 1 Simplified electrical circuit--opening switch study.

In the very idealized case where R_f is assumed to undergo a step resistance change from 0 to R_o ohms, opening switch performance can be characterized in the following way.

Initial energy stored in capacitance C_o (Fig. 1) is $U_i = 1/2\ C_o V_i^2$. For $R_f = 0$, energy is transferred to the inductance L_o, $U_t = 1/2\ L_o i^2$, where i is instantaneous circuit current. The energy transferred can be equated to some fraction f of the initial energy, i.e.,

$$U_t = 1/2\ L_o i^2 = fU_i = f^{1/2} C_o V_i^2$$
$$L_o i^2 = f\ C_o V_i^2$$
$$i = \left(f\ \frac{C_o}{L_o} \right)^{1/2} V_i$$

If, at some arbitrary time during energy transfer, R_f increases instantaneously to R_o ohms, then the output voltage across R_o is

$$V_o = R_o i = R_o V_i \left(f\ \frac{C_o}{L_o} \right)^{1/2}, \quad \text{or}$$
$$\frac{V_o}{V_i} = R_o \left(f\ \frac{C_o}{L_o} \right)^{1/2} \qquad (1)$$

The quantity V_o/V_i is termed the "multiplication" and is seen, for this very idealized case, to depend upon the value R_o and the square root of the product $f\ C_o/L_o$.

In Eq. (1), the multiplication for fixed resistance change, R_o, and for fixed C_o and L_o, is maximum for the condition where all the energy initially stored in C_o is transferred to L_o at the instant in time when R_f increases from 0 to R_o ohms.

Energy transfer time τ_t for the circuit in Fig. 1 is

$$\tau_t = \frac{\pi}{2} \left(L_o C_o \right)^{1/2}, \quad \text{and the}$$

e-folding output pulse duration τ_o is

$$\tau_o = \frac{L_o}{R_o}.$$

Expressing Eq. (1) in terms of τ_t and τ_o yields

$$\frac{V_o}{V_i} = \frac{2}{\pi}\ \frac{\tau_t}{\tau_o} \qquad (2)$$

the quantity τ_t/τ_o is termed the "time compression" and for this very idealized case, is directly proportional to the multiplication.

In practice, energy is supplied to the opening switch material (metallic foils) from the circuit in the form of $i^2 R$ heating, thereby, in general, increasing fuse resistivity. If it is assumed for a particular material that a specific energy per unit mass is required to affect a resistance change R_o, then it is clear that, for a given electrical circuit, there is an upper bound upon the mass of opening switch material; for if the mass of material is too large, then the energy stored initially in the circuit is insufficient to produce a significant resistance change.

In some cases, a specific criterion has been used to determine the required mass of material for a given electrical circuit and stored energy. From Ref. 2, this criterion, briefly stated, is that the mass of material is adjusted to just result in vaporization of the material under the action of circuit energy deposition. If the mass is too large, then complete vaporization will not occur. If the mass is too small then the vaporized material, from Ref. 2, is heated to ionization which is apparently to be avoided from a voltage restrike or breakdown standpoint.

This criterion, which appears to be without adequate basis and unnecessarily restrictive, has not been applied to the empirical study described in this paper.

SCALING

Circuits containing exploding foils are hard to analyze because the foil is a time-varying, and perhaps nonlinear element. The purpose of the following paragraphs is to identify simple physical scaling laws that at least enable experimental results to be used to predict the behavior of other circuits, especially ones in which the time scale is different.

Scaling is based upon the circuit shown in Fig. 1 where fuse volume is adjusted along with the circuit to maintain fuse energy density constant. Moreover, the fuse volume is changed in a way which yields equal relative circuit performance.

The simplest scaling laws apply when the time scale does not change. For example, suppose that the circuit in Fig. 1 is modified by decreasing L_o by a factor n. To maintain the time scale constant requires that C_o be increased by a like factor. As a result, circuit energy increases by n and so does the current since it is proportional to $(C_o/L_o)^{1/2}$. Circuit "impedance" has decreased by $1/n$ since "impedance" is proportional to $(L_o/C_o)^{1/2}$. Thus, to maintain equal relative circuit performance requires that the fuse cross section be increased by n with length unchanged.

Other examples of scaling with constant time scale are summarized in Table 1. In these cases (Examples 1-3 in Table 1), current density and electric field in the fuse are never changed. The scaling rules here are hence independent of possible time dependence of fuse resistivity, and nonlinearity of conduction in the foil, and only assume the absence of geometry effects, e.g., edge and boundary effects.

If we wish to find scaling rules that involve changing the time scale, we must assume both time independence and linearity. For example, if L_o alone is decreased by n, the time scale decreases by $n^{1/2}$ and current increases by $n^{1/2}$. Energy is constant, but circuit impedance decreases by $n^{1/2}$. Decreasing fuse length and increasing cross section area by $n^{1/4}$ preserves fuse mass and decreases resistance by $n^{1/2}$ as required. Note that current density and the electric field in the fuse both increase by $n^{1/4}$. Two other examples where the time scale is not constant are summarized in Table 1.

The scaling rules exemplified by Examples 4-6 in Table 1 hold whatever the dependence of resistivity on deposited energy, because the energy density is always the same in the final state and at corresponding scaled times in between. It is seen that scaling in time always implies a change in electric fields and current densities. (This is the reason that linearity must be assumed when the time scale changes.) Moreover, the electric field goes like the inverse square root of the time scale. Also note that the voltage multiplication is unchanged in any scaling considered. To increase the voltage gain, the relative increase in energy density must be changed.

TABLE 1 FUSE SCALING

Variable Parameter(s)	Time Scale	Energy	Current	Fuse		Fuse	
				Length	Cross-Section	Current Density	Field
1. { Increase C_o by n / Decrease L_o by n	Unchanged	Increased by n	Increased by n	Unchanged	Increased by n	Unchanged	Unchanged
2. { Decrease C_o by n / Increase L_o by n	Unchanged	Decreased by n	Decreased by n	Unchanged	Decreased by n	Unchanged	Unchanged
3. Increase V_i by n	Unchanged	Increased by n^2	Increased by n	Increased by n	Increased by n	Unchanged	Unchanged
4. Decrease L_o by n	Decreased by $n^{1/2}$	Unchanged	Increased by $n^{1/2}$	Decreased by $n^{1/4}$	Increased by $n^{1/4}$	Increased by $n^{1/4}$	Increased by $n^{1/4}$
5. Increase C_o by n	Increased by $n^{1/2}$	Increased by n	Increased by $n^{1/2}$	Increased by $n^{1/4}$	Increased by $n^{3/4}$	Decreased by $n^{1/4}$	Decreased by $n^{1/4}$
6. Increase V_i by n and decrease C_o by n^2	Decreased by n	Unchanged	Unchanged	Increased by $n^{1/2}$	Decreased by $n^{1/2}$	Increased by $n^{1/2}$	Increased by $n^{1/2}$

EXPERIMENTAL APPARATUS

Two sets of apparatus were employed in the described study and are shown in Figs. 2 and 3, respectively. Facility 1 (Fig. 2) uses a dc charged capacitor bank as the initial store whereas Facility 2 (Fig. 3) uses a pulse charged water dielectric capacitor. Facility 2 operates at a higher voltage than Facility 1 and was designed specifically to reduce fuse energy deposition time to about 100 nsec. Each of the facilities are described in the following paragraphs.

FACILITY 1

Facility 1 consists of a dc charged capacitor bank connected by way of a low inductance triggered gas switch into a strip-line transmission line immersed in a water-filled tank (Fig. 2). Insulation between the strip line conductors is 1/2-inch-thick polyethylene sheet (actually two 1/4-inch-thick sheets). The purpose of the water medium is to exclude air and, because of the high dielectric constant of water, to place the electric fields in the electrically stronger solid dielectric material.

The fuse material, installed at the output end of the transmission line also resides in the ambient water medium, except isolated from the main tank to facilitate removal of fuse debris.

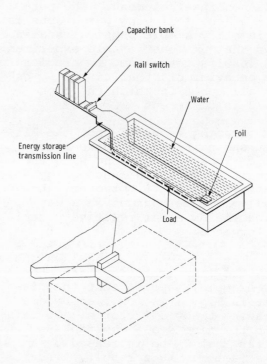

Fig. 2 Circuit of the experiment and configuration of the foil.

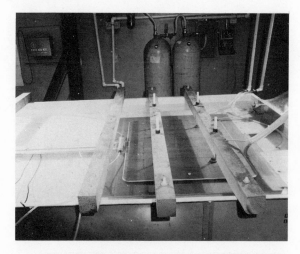

Fig. 3 Opening switch test facility.

The fuse material for the described experiments was generally copper, although aluminum, tungsten, and tantalum were tried also. Figure 2 illustrates the configuration of the fuse installed at the end of the transmission line. In Fig. 2, a small piece of folded Mylar holds the fuse in position.

Two capacitor banks are used to drive Facility 1 and their electrical characteristics are listed in Table 2.

Facility 1 transmission line inductance and stray shunt capacity are about 100 nH and 3 nF, respectively. Table 2 also summarizes the maximum peak current and time to peak current that can be achieved with each capacitor bank.

Fuse voltage and current probe locations are illustrated in Fig. 2. The voltage probe is constructed from 2 watt carbon composition resistors potted in epoxy. The measured risetime of the voltage probe is less than 10 nsec (10 to 90%).

The current measurement probe is a Rogowski coil consisting of 5 turns of #34 AWG copper wire wound on a 3/8-in. form. The probe output is integrated with a 1 msec passive integrator to yield an output signal proportional to current rather than its derivative.

Both probes were frequently calibrated during the experiments.

FACILITY 2

Facility 2 consists of an oil-insulated, 3-stage Marx generator which pulse charges a parallel plate water capacitor as illustrated in Fig. 3. Maximum pulse charge voltage is 300 kV, and charge time is about 1 μsec. When charged, the water dielectric store is switched, by means of a pressurized SF_6 (sulfur hexafluoride) gas spark gap, into a transmission line (the inductive store). The transmission line is 1-1/4 in. wide by 3 feet long, spaced 1 cm over an extensive ground plane and insulated with polyethylene. The transmission line and fuse, installed at the end of the line is immersed in water as the Facility 1 components.

TABLE 2   FACILITY 1 ELECTRICAL PARAMETERS

Capacitor Bank	1	2
Capacity (μF)	82.9	3.7
Max. dc charge voltage (kV)	20	60
Bank and switch inductance (nH)	30	28
Load inductance (nH)*	100	100
Maximum peak current (MA)	0.51	0.32
Max. stored energy (kJ)	16.8	6.7
Time to peak current (μsec)	5.2	1.1

*Not including stray inductance of fuse

Table 3 summarizes Facility 2 electrical parameters.

Output pulse monitors are contained with a removable section of the output transmission line ground plane near the end of the line. Two probes are used, one is an i dot monitor integrated with a 1 μsec integrator at the oscilloscope to yield an overall sensitivity of

2.36 kA/V. The second probe is a
flush plate parallel plate dipole,
a capacitive divider, also integrated at an oscilloscope with a 1 μsec
integrator to yield an overall sensitivity of 59.4 kV/V. The capacitive
probe is located 8 inches from the
end of the transmission line.

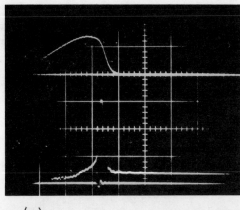

(a)

TABLE 3   FACILITY 2 ELECTRICAL
          PARAMETERS

1. Marx Generator

    Number of stages: 3

    dc charge voltage: 100 kV max.

    Erected series capacity: 135 nF

    Stray series inductance   4.6 μH
    (inc. connections to water
    capacitor)

    Stored energy: 2 kJ max.

2. Water Capacitor

    Length: 3 feet

    Width: 4 feet

    Separation: 1 inch

    Capacity: 38 nF

3. Transmission Line

    Length: 3 feet

    Width: 1-1/4 inch

    Separation: 1 cm
                     (polyethylene)

    Inductance (inc.
    switch):   219 nH

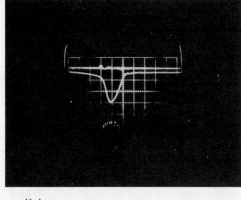

(b)

Fig. 4 Typical exploding foil voltage and current traces for the
low voltage bank.
(a) 1 μsec/cm
     I_{peak} = 215 kA; V_{peak} = 83 kV
(b) 200 nsec/cm
     V_{peak} = 83 kV

EXPERIMENTAL RESULTS

    Copper, aluminum, and tantalum
foils have been exploded using the
Facility 1 low voltage bank from
10 kV to 22 kV. Copper alone has
been used on the high voltage Facility 1 bank and the Facility 2 bank
at voltages of 30 to 60 kV, and up
to 150 kV, respectively. A typical
shot from the Facility 1 low voltage
bank is illustrated in Fig. 4. The
residual voltage in Fig. 4 is that
of the capacitor bank which has not
been fully discharged.

    A number of variations of fuse
parameters were experimentally evaluated in the course of this study
and are summarized in the following
paragraphs.

Constant Fuse Mass. Fuse cross-
section for fixed fuse length must
be chosen so that the output
occurs at the desired time approximating peak current. If the fuse
cross-section and length are held
constant while the width and thickness are adjusted, the initial

(cold) fuse resistance is unaltered. Figure 5 illustrates fuse performance as a function of fuse thickness for constant cross-section and length. Voltage gain varies from 2-1/2 to 5-1/2 depending upon the thickness of the copper fuse and thus it must be concluded that fuse resistivity is not just a function of energy density. We interpret this as related to the ability of the heated material to expand into the water. Optimum voltage multiplication is achieved for 0.2 mil thickness.

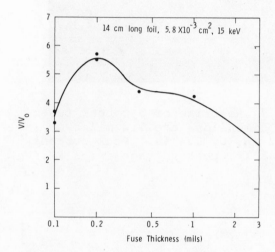

Fig. 5 Voltage multiplication vs fuse thickness for fixed cross-sectional.

Variable Fuse Mass. Since the voltage gain is directly related to the resistance change that the fuse undergoes during the heating, and since, for the materials examined here, resistivity is an increasing function of the energy density, improved performance of a fuse is likely to be achieved by reducing the amount of fuse material used provided that significant energy is transferred to L_O. Figure 6 illustrates the voltage muliplication as a function of fuse length for two fuse cross-section.

The multiplication associated with the large cross sectional area fuse decreases with increasing fuse length greater than about 13 cm. The interpretation here is that the fuses are too massive, and while the energy transferred to L_O is large, the fuse resistance change

is inadequate to yield good multiplication.

Multiplication (for the large cross-sectional area fuse) increases with decreasing fuse length reaching a maximum of about 6 at 13 cm which corresponds to a mass of about 0.75 grams. A similar multiplication is obtained with the 1.3×10^{-3} cm^2 area fuse at a fuse length of about 6 cm and mass equal to about 0.07 grams.

The small cross-section fuses vaporize much earlier in time than the large ones, and, as a result, less energy is transferred to the circuit inductance. However, since the maximum multiplications are equal in both cases, the resistance change undergone by the small cross-section materials is clearly greater than that associated with the more massive fuses.

Multiplication decreases from peak value with decreasing fuse length in both cases, although not markedly, until a fuse length of 3-1/2 cm is reached at which point multiplication decreases rapidly with decreasing fuse length. The gradual reduction in multiplication down to a length of about 3-1/2 cm is interpreted as being due to the relatively early vaporization of the fuse material. Although the resistance change undergone by the fuse may increase with decreasing fuse length (and therefore mass), this is more than offset by decreased energy transfer to the circuit inductance.

For fuse lengths equal to or less than about 3-1/2 cm, multiplication is limited by restrike or breakdown.

Restrike is evidenced by failure of the measured current to decrease completely to zero as the fuse material vaporizes. A typical measurement where restrike occurred is illustrated in Fig. 7. The maximum re-strike field that has been obtained in these experiments is about 18 kV for the dc charged bank measurements. Limited measurements using Facility 2 suggest a somewhat higher restrike field, perhaps

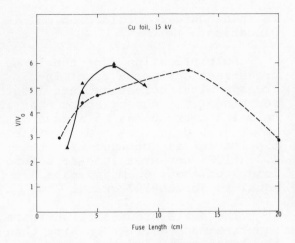

Fig. 6 Voltage multiplication vs fuse length for two cross-sectional areas. Dashed line = 5.8×10^{-3} cm^2, solid line = 1.3×10^{-3} cm^2.

22 kV/cm, but recorded current waveforms are not carried much beyond 300 nsec following foil vaporization, and thus the late time behavior of circuit current is uncertain.

Scaling. The validity of the scaling rules postulated in a previous paragraph was tested by experiment. First a check was made of the first type of scaling, in which the time scale is unchanged. The third example given was chosen as the easiest version of constant time scaling to test in practice. The bank charging voltage, foil length, and foil width were all changed by an accurate factor of 1.5. It was found that consistent with prediction the output voltage changed by the same factor, and the time history of the waveforms was accurately preserved.

The type of scaling in which the time is changed is represented by a comparison of results obtained with the two banks. This in effect combines examples 3 and 5 of Sec. 3. Figure 8 presents Facility 1 low voltage bank data in the form of fuse resistivity as a function of energy density. The plots here are derived from oscilloscope photographs of fuse voltage and current time histories. In all cases here, voltage multiplications are approximately equal. The peak resistivity

attained is between 200 and 300 μohm-cm in all cases with corresponding energy densities (at peak voltage) in the range of 5-8 kJ/gm.

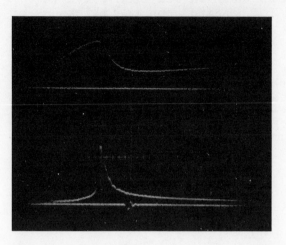

Fig. 7 Restrike of a copper fuse. Fuse dimensions are 0.2 mil by 4-1/2 in. by 3.4 in. Peak current is 246 kA; peak voltage is 51 kV.

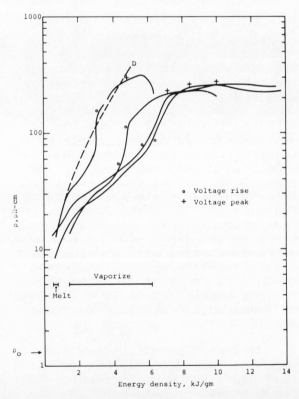

Fig. 8 Foil resistivity as a function of foil energy density using low voltage bank at 15 k-V.

The results of DiMarco and Burkhardt[1] are also shown in Fig. 8 and it is clear that the rate of change of resistivity with energy density in the present experiments is, in general, less than was found by DiMarco and Burkhardt.

Figure 9 shows a similar plot for the Facility 1–3.7 µF bank data. Here also, the peak resistivity is 200 to 300 µohm-cm, but the energy density (measured at peak voltage) is even greater, i.e., 10 to 12 kJ/gm.

Facility 2 data, where energy is deposited in the fuse material in about 100 nsec, yields peak resistivities in the range, 300 to 400 µohm-cm, but with energy densities that lay between the Facility 1 capacitor band data, i.e., energy density at peak voltage is about 9 kJ/gm.

From these results, it may be concluded that the resistivity of fuse materials is not a function of energy density alone. For a given energy density, resistivity decreases as we progress from DiMarco's experiments to Facility 1 low voltage bank experiments and to Facility 1 high voltage bank experiments. However, this trend is not continued by the experimental results from Facility 2 with its very rapid energy deposition times (~100 nsec). In this case, material resistivity as a function of energy density is intermediate to facility 1 low and high voltage bank data.

Figure 10 summarizes fuse energy density to obtain roughly equal resistivity as a function of current density. While there is a fair amount of scatter, the data presented in this way suggests that the energy density required to achieve a given resistivity is directly proportional to the square root of current density.

In earlier work[1], it was also suggested that fuse current density is an important quantity in determining voltage multiplication. Figure 11 illustrates this; for higher multiplication, one must have densities less than a few times 10^7 A/cm^2.

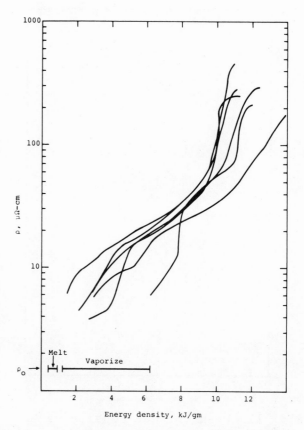

Fig. 9 Fuse resistivity vs fuse energy density for 60 kV bank.

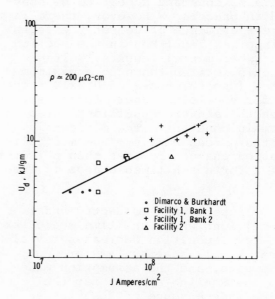

Fig. 10 Fuse energy density as a function of current density for constant resistivity change.

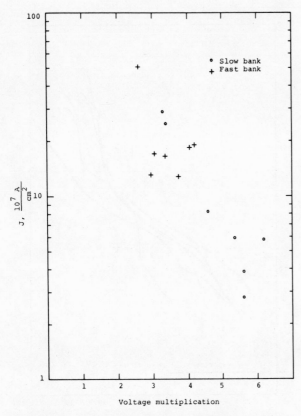

Fig. 11 Peak current density vs voltage multiplication.

SUMMARY AND CONCLUSIONS

Scaling relationships, postulated in the section on Scaling, for the condition where the time scale remains constant appear to be obeyed in practice. However, there is clear evidence of geometry effects since, as described in the section on Experimental Results, voltage multiplication tends to a maximum at a fuse thickness of 0.2×10^{-3} in. under conditions where fuse mass, length, and cross-sectional area remain constant. The maximum is rather broad however; the multiplication changes by less than a factor of 2 for a ten times change in fuse thickness.

Experiments conducted under conditions where the time scale for energy deposition decreases reveal that simple scaling laws are not adequate, and the assumption made in the Scaling section for the time independence of fuse resistivity upon energy density is clearly not valid.

There is firm evidence that the energy density required to yield a given resistivity increases with increasing current density. The best fit for the data suggests, although crudely, that the energy density for this condition is proportional to the square root of current density.

Other more general conclusions from these experiments are summarized as follows:

1. Voltage multiplications of up to 6/1 are readily achieved in practice for current densities in the fuse of a few times 10^7 A/cm^2 or less.

2. Peak opening voltages of up to about 450 kV have been achieved using Facility 2.

3. Output pulse durations about 15% of fuse energy deposition times are readily achieved. Pulses as narrow as 25 nsec (FWHM) have been produced in these experiments.

4. The maximum restrike field obtained in these experiments is in the range of 18 to 22 kV/cm.

REFERENCES

1. J. DiMarco and L. Burkhardt, J. Appl. Phys. 41, 3894 (1970).

2. C. Maisonnier, J. Linhart and C. Gourlon, Rev. Sci. Instr. 37, 1380 (1966).

3. H. Early and F. Martin, Rev. Sci. Instr. 36, 1000 (1965).

QUESTIONS AND ANSWERS

1. Question from W. H. Bostick

At AVCO-Everett, several years ago, multiple wire fuses held with masking tape were used in air. Could you compare the operation of their fuses with yours?

A: Arrays of wires have not been employed in our empirical study and thus, a strict comparison under identical electrical conditions cannot be made.

2. Comment from J. C. Martin

Experiments we did suggested that exploding foils under water and oil did not sustain much higher gradients than in air. The observed F_{max} is about 20 kV/cm. Mesyats also obtains about 20 kV/cm for exploding wires in air.

3. Question from W. F. Weldon

What is the peak current density in the foil?

A: Typically about 2×10^8 A/cm^2.

4. Question from C. Aaland

How does voltage multiplication relate to current or energy transfer to the load? May I suggest that your decrease of voltage multiplication with thinning of the foil is merely a circuit parameter phenomenon.

A: In the ideal case where energy is supplied to a time varying resistance (the opening switch) from a capacitor bank, C_O, by way of a series inductance L_O, it can be shown that voltage multiplication is given by

$$\frac{V_O}{V_i} = R_O \left(\frac{f\, C_O}{L_O} \right)^{1/2}$$

where f is the fraction of energy, initially stored in C_O, which is transferred to L_O and R_O, in this idealized case, is a step resistance change from 0 to R_O ohms.

Thus, voltage multiplication is directly proportional to the resistance change, and is proportional to the square root of the fraction of energy transferred from C_O to L_O just prior to the assumed step change of circuit resistance (the opening switch).

With regard to voltage multiplication as a function of foil thinning, recall that although the foils were thinned, foil width was correspondingly increased resulting in constant cross sectional area, and thus, constant initial resistance (foil length was fixed). Therefore, the circuits were identical initially, and any differences in performance can only be attributed to "geometry" effects.

DESIGN OF VERY FAST RISE AND FALL TIME, LOW IMPEDANCE MEGAVOLT PULSE GENERATORS FOR LASER EXCITATION

J. Harrison, R. Miller, J. Shannon and J.B. Smith
Maxwell Laboratories, Inc.
9244 Balboa Avenue
San Diego, California 92123

ABSTRACT

Recent developments in e-beam excited laser systems have led to a requirement for very fast rise and fall time, short pulse, low impedance e-beam generators. Advances in low impedance e-beam generator technology to meet this requirement are discussed. Two advanced design machines have been built - a 2 MeV, 200 kA generator with 10-90% rise and fall times of 10 ns, and a 1 MeV, 100 kA generator with rise and fall times of less than 5 ns. The design of these machines and their measured output characteristics are presented in this paper.

INTRODUCTION

This paper describes the latest developments in pulse generator design for fast rise-time, fast falltime, low impedance, megavolt electron beams. Earlier designs for these machines were discussed at Princeton.[1] These designs have now been upgraded for faster current risetimes by adding a peaking section to the pulseforming line. Two fast risetime machines have been built and operated with the following output parameters:

1. A 1 MeV, 100 kA, 5 ns risetime generator shown in Fig. 1.

2. A 2 MeV, 200 kA, 10 ns risetime generator shown in Fig. 2.

DESCRIPTION OF MACHINES DESIGN

Electron beam machines with fast rise-times are produced by use of an output peaking system as indicated in the equivalent circuit shown in Fig. 3. Figure 3(a) is a simplified schematic diagram of the conventional type generator which was described earlier;[1] Fig. 3(b) is a diagram for the fast risetime generators showing the added peaking system. Figure 4 is a cross section of the pulse line showing the basic arrangement of the system.

Fig. 1. 1 MeV, 100 kA e-beam generator 3-4 ns risetime

Fig. 2. 2 Mev, 200 kA e-beam generator 10 ns risetime

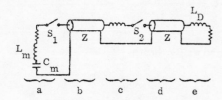

(a) Generator without peaking system

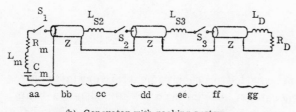

(b) Generator with peaking system

a. Marx Generator   aa. Marx Generator
b. Pulseformer Line   bb. Pulseforming Line
c. Output Switch   cc. Output Switch
d. Output Line   dd. Peaking Line
e. Diode   ee. Peaking Switch
                      ff. Output Line
                      gg. Diode

Fig. 3. Simplified circuit diagram of pulse
generators with and without peaking
systems

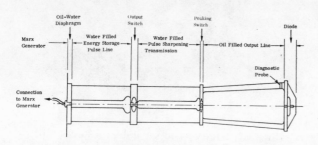

Fig. 4. POCOBEAM electron-beam
transmission line system
with both output and peaking
switches

The major components of the pulseforming
system are:

● A 10 Ohm water dielectric pulseforming
line which stores the energy

● A high pressure SF$_6$ overvoltage triggered output switch

● A water dielectric pulse sharpening
section

● A high pressure SF$_6$ multichannel
overvoltage triggered peaking switch

● An oil dielectric output line

● A vacuum enclosure and diode

The energy is stored initially in a Marx
generator. This energy is then resonantly
transferred to the pulseforming line when the
Marx generator erects (i.e., when S$_1$,
Fig. 1(b) closes).

Switch S$_2$ closes as the voltage on the
pulseforming line approaches 90-95% of its
peak value. Closure of S$_2$ generates a voltage pulse with a 10-90% risetime of 10-12 ns
per megavolt of output voltage in the intermediate line. The early part of this wave is
reflected as it reaches switch S$_3$ prior to its
closure near peak voltage. The wave that is
subsequently transmitted down the output line
by the closure of Switch S$_3$ has a 10-90%
risetime of approximately 3 ns per megavolt
of output voltage. This wave travels down
the output line to the diode load. Due to the
inductance of the diode envelope and the temporal behavior of the diode impedance, the
risetime of the output wave increases approximately 2 ns per megavolt of output voltage.

To accomplish the desired fast rise output waveforms, extensive computer analysis
was utilized in the optimization of switch
inductance, transmission line impedance profile, and the timing between the closure of the
output and peaking switches.

SWITCH DESIGN

These pulse generators are designed to
drive gas laser cavities. It is important that
machines used in these systems generate as
low a shock as possible so that they do not
disturb the optical alignment of the laser system. This constraint led to the use of gas
dielectric (rather than water) for all the
switches in the system. Thus, both the high
voltage switches use high pressure SF$_6$ as a

dielectric. The choice of high pressure SF_6 was made to reduce the switch inductance and resistive risetime to a minimum.

The peaking Switch S_3 is designed to operate in a multichannel mode to obtain the low inductance required for the fast output wave risetime. The number of effective channels in the peaking switch is inversely proportional to the risetime of the voltage applied to it.[2] This applied voltage waveform is generated by the output switch. Thus, both switches must be designed to have low inductance.

The switches are also designed to handle the mechanical stress imposed by the high pressure gas as well as the electric field stress. The mechanical design of each switch is not trivial since the mechanical stress applied to the relatively weak dielectric walls of the switch housing is substantial. The design approach chosen was to design the high pressure enclosure first, and then fit electrodes into the enclosure. The electrode design and the shape of the conductors adjacent to the switch vessel are optimized by using the JASON[3] code for solution of Laplace's equation in the switch envelope.

A cross section of the single-channel output switch design is shown in Fig. 5. The body of the switch is machined from a solid disk of acrylic which fills the diameter of the line. The switch cavity is shaped to provide a near optimum pressure vessel, and the electrodes are fitted into the cavity on the centerline.

The electrodes and their connection to the pulse line inner conductor are designed to provide a nearly uniform field across the gap, thus minimizing the gap length and, therefore, the inductance.

The peaking switch shown in Fig. 6 is a multi-channel triggered gas switch. The radial impedance at the electrodes is approximately equal to the sum of the impedances of the intermediate line and the output line. Thus, the geometric inductance is virtually zero, and the only significant contribution to

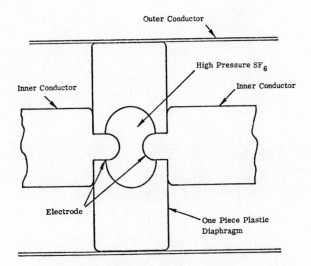

Fig. 5. High pressure gas output switch

switch inductance is that due to the field enhancement electrodes. The calculated inductance of the switch is ~ 18 nH/MV.

TRANSMISSION LINE DIELECTRIC

The pulseforming and peaking lines use water as a dielectric. Water is the most efficient energy storage and transmission media for low impedance (<10 Ohm) short pulse systems with coaxial geometry. However, oil was chosen as the dielectric for the output section to minimize the dielectric discontinuities experienced by the pulse in passing from the peaking switch to the vacuum diode. Large changes in dielectric constant along the transmission path would degrade the wave risetime unless the wave passes through a Brewster angle interface where the transit times of all portions of the wave front from the source to the load only varied by a time significantly less than the required output risetime. These requirements are difficult to meet at the diode interface where the elimination of vacuum breakdown requires interface lengths that limit options in the design of the geometric arrangement. The large dielectric discontinuities are avoided by the use of oil dielectric in the output line.

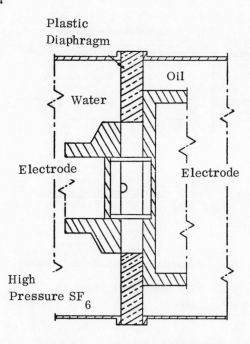

Fig. 6. High pressure gas peaking switch

Design optimization of the impedance profile and the length of the lines is accomplished by computer analysis. The computer program used in this process is a transmission line code which models the varying impedance of the transmission lines. The code also permits lumped circuit transfer functions to be inserted between the lines. This is an important part of the design process in that incorrect design of the transmission line system can lead to severe waveform distortion. An example of the penalties of incorrect design parameters is shown in Fig. 7 which shows three superimposed waveforms: one for a system with close to the optimum relationship between the pulseforming and peaking lines; one with the peaking line shortened, where the nonoptimum configuration produces a severe spike at the end of the pulse; and one with the peaking line lengthened. In this case, the pulse shortens and the fall time is substantially degraded.

IMPEDANCE PROFILE OF TRANSMISSION LINES

The nominal impedance of the two systems discussed here is 10 Ohms. The diameter of the inner conductor of a 10 Ohm water line is approximately 23% that of the outer line, but this is too small to provide good field grading in the switches. Therefore, the inner line diameter adjacent to the switches was increased resulting in reduced line impedance at the switch. This reduced line impedance provides a current "overdrive" through the switches during the risetime of the output pulse. The impedance profile is carefully controlled so that the overdrive provides close to maximum risetime improvement without causing a significant overvoltage spike or oscillations in the output wave.

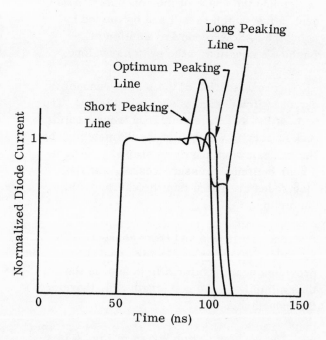

Fig. 7. Computed output current waveforms for variation in peaking line lengths

TIMING BETWEEN CLOSURE OF THE OUTPUT AND PEAKING SWITCH

The time delay between closures of the output and the peaking switches is accurately adjusted to obtain the optimum output rise-time and fall time. The effect of changes in this time delay are shown in Fig. 8 which shows superimposed traces for optimum, early and late closure of the peaking switch. This timing can be adjusted by changing the gas pressure in the peaking switch.

OUTPUT WAVEFORMS

Figure 9 shows traces of the output voltage and current from the 1 MeV, 100 kA generator, and Fig. 10 shows similar traces from the 2 MeV, 200 kA generator. The output voltage in both cases is measured in the output line by a probe placed close to the diode insulator. The measured voltage is the sum of the diode load and diode inductive voltages, and consequently its risetime is faster than that of the load voltage. The measured voltage risetime is consistent with the <20 nH/MV inductance figure calculated for the peaking switch.

The voltage waveform shows a small precursor pulse starting 10 ns/MV before the main pulse. This precursor is generated during the charging of the peaking switch. The amplitude of the prepulse transmitted to the diode during charging of the pulseforming line is less than the resolution of the measuring system, which is less than 0.5% of the output pulse amplitude.

Figure 11 shows the measured Marx and diode voltage waveforms for five super-imposed shots demonstrating the high degree of reproducibility and low jitter of the generator. The oscilloscope trigger signal for these traces is the command signal initiating the system.

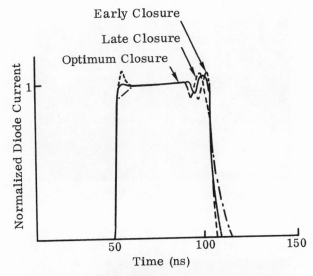

Fig. 8. Computed output current waveforms for variations in peaking switch timing

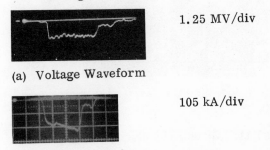

1.25 MV/div

(a) Voltage Waveform

105 kA/div

(b) Current Waveform

Fig. 9. Measured voltage and current waveforms from 1 MeV, 100 kA generator. 20 ns/div.

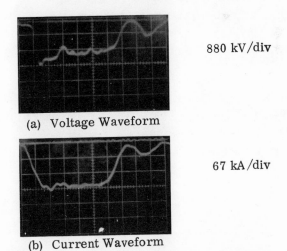

880 kV/div

(a) Voltage Waveform

67 kA/div

(b) Current Waveform

Fig. 10. Measured voltage and current waveforms from 2 MeV, 200 kA generator. 10 ns/div.

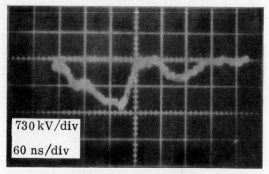

730 kV/div

60 ns/div

Marx Generator Output Voltage

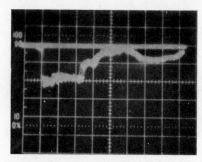

390 kV/div

20 ns/div

Diode Voltage

Fig. 11. Measured voltage and current waveforms showing 5 superimposed traces with the oscilloscope triggered by the command signal

CONCLUSION

Short risetime, high power electron beam machines are now being utilized in electron beam excited laser studies. They have demonstrated rise and fall times of < 5 ns/MV, delivering peak powers of up to ~0.4 TW which corresponds to beam energies of 20 kJ. Mechanical shock associated with machine operation has been eliminated by the use of low inductance pressurized gas switches, thereby avoiding damage misalignment of delicate laser optics and diagnostic equipment.

The performance of the 1 and 2 MeV machines described above confirm the validity of the computerized design techniques used in the design of multimegavolt, very fast rise and fall time pulses to produce e-beams for laser excitation.

[1] J. Harrison et al., "Compact Electron Beam Generators for Laser and Fusion Research", 5th Symposium on Engineering Problems of Fusion Research, Princeton University, November 5th through 9th, 1973, p. 640-651.

[2] J.C. Martin, "Multichannel Gaps", Aldermaster Report SSWA/JCM/703/27, March 1970.

[3] S. Sackett and R. Healey, "JASON - A Digital Computer Program for the Numerical Solutions of the Linear P Equation", University of California Report UCRL-18721, February 1969.

EBFA, A TWENTY TERAWATT ELECTRON BEAM ACCELERATOR*

T. H. Martin and K. R. Prestwich
Sandia Laboratories
Albuquerque, New Mexico 87115

ABSTRACT

The design for an accelerator (EBFA) for studying electron beam driven, inertially confined fusion is described. The EBFA will generate two 3 MV, 20 nsec electron beams in back-to-back diodes with a total beam energy of 200 kJ and a peak power of 2×10^{13} W

The major components of the accelerator are Marx generators, intermediate storage capacitors, pulse forming lines, transmission lines and diodes. Six to eight, 3 MV Marx generators operated in parallel will charge 24 intermediate-storage, water-dielectric capacitors in approximately 1 μsec. The pulse forming lines consisting of two 24-meter diameter, disc-shaped oil-dielectric Blumleins, are charged in 100 nsec by the simultaneous discharge of the intermediate-storage capacitors through triggered gas switches. Several hundred switch channels closing in < 3 nsec will be required to produce a 10 nsec risetime output pulse. The voltage pulses are then transmitted through the disc-shaped transmission lines to the back-to-back diodes where the electron beams are generated and focused onto a spherical target.

INTRODUCTION

An approach to pulsed fusion has recently been proposed by Yonas et al[1] and Rudakov et al,[2] which employs tightly focused high power electron beams (REB's) to spherically implode dense metal shells filled with DT gas. Pellet calculations at Sandia,[3] using classical electron energy disposition in dense metal shells indicates that 1-10 MJ of ~ 1 MeV electrons with powers of $10^{14} - 10^{15}$ W will be required for breakeven. In order to make significant progress in studying the physics of the electron beam deposition and the implosion process, an accelerator with peak power greater than 10^{13} W is needed. In this paper some of the technological problems associated with developing a 10^{13} W accelerator are outlined and a specific design for a 2×10^{13} accelerator is given.

The energy and power needed for breakeven indicate that the beam pulse duration should be ~ 10 nsec. Recent success in pinching beams within the diode leads us to the approach of two-sided irradiation of the target from two diodes with a common anode.[4] A compromise between technology for short pulse accelerators and considerations of classical energy deposition of beams indicate that useful electron energies are in the 1-3 MV range. Thus, the EBFA is designed to produce two 3 MV, 20 nsec beams, each with currents greater than 3 MA.

TECHNOLOGICAL CONSIDERATIONS

Figure 1 shows the basic components of an electron beam accelerator and the approximate power densities in working accelerators with some commonly used dielectrics. The peak voltage and, subsequently, the energy storage capability of the pulse forming line (PFL) is determined by the Marx charge time, typically 0.5-1.0 μsec, and the electrode area. Utilizing the wave propagation velocity in the particular dielectric media when the line discharges into a matched load or transmission line, a power density can be calculated as indicated for oil, water and Mylar. The instantaneous power flow through the transmission line is considerably increased because the voltage is only on from 0.01-0.1 μsec and only one-half of the energy is stored in the electrostatic field. Fig.1 indicates that Mylar and water are superior dielectrics from a power density point of view. Extensive experiments at Sandia [5] and Physics International[6] have indicated that it is extremely difficult

---

*This work was supported by the U.S. Atomic Energy Commission.

to utilize large area Mylar with voltages above 0.5 MV. With 3 MV, 20 nsec pulse and using water dielectric, the separation of the electrodes in the PFL is comparable to the transmission line length resulting in possible wave distortion. For Blumleins, impedance mismatching at the end could be a problem; therefore, transformer oil was selected as the dielectric for the PFL and transmission line for this study. On the other hand, if the above problems with water can be solved, it will probably be a more desirable dielectric to use because the system size decreases with increased dielectric constant and relatively high dielectric strength.

From a power density point of view, the weakest point in the system is the insulator vacuum flashover field which is a function of time and area. The lower power density in the insulator area forces a larger line spacing and necessitates large diameter diode envelopes for the low inductance that is needed to produce fast risetime electron beam pulses.

Inside the diode magnetic confinement of the electrons and beam pinching allow very high power densities.[4]

The other limitation on the output pulse risetime is the PFL switch current risetime. Analysis of present high voltage switch data indicates that rate of current rise in a single channel is limited to 10^{12} - 10^{13} A/sec. Power outputs and risetimes necessary for fusion experiments therefore will require many switches or channels to close simultaneously.

EBFA DESIGN

EBFA has been designed to simultaneously produce two 3 MV, 3.5 MA, 20 nsec pulses for two-sided irradiation of pellets. Figure 2 is a schematic of the accelerator. It consists of two low-inductance high current diodes, disc-shaped transmission lines, two back-to-back, oil-insulated Blumleins with multichannel switching around the periphery of the disc, 24 low-inductance intermediate storage water capacitors and 8 Marx generators. The Marx generators are located in an annulus below the accelerator with the tankage providing support and radiation shielding. Design considerations for each of the components

will now be reviewed starting with the diode.

DIODE

With two diodes, if the output impedance of each transmission line is 0.8 Ω into a matched load, the effective impedance for inductance risetime considerations is 1.6 Ω. If an e-folding risetime < 3 nsec is needed, the diode inductance for each side must be ≤ 5 nH. Empirical relationships for vacuum insulator flashover indicate that a well-graded diode with a 20 nsec pulse can be operated at 150 kV/cm.[7,8] Figure 3 is a sketch of a diode. The diode inductance is roughly:

$$L = \frac{2A}{R} \ [nH] \qquad (1)$$

where A equals the cross sectional area in cm^2 of the insulator volume, and R equals the mean radius in cm of the insulator volume.

For 8 small insulators (~ 2.5 cm x 2.5 cm), the increased area associated with the insulators is ~ 260 cm^2 and the required R would be 100 cm. We have assumed that the cathode can be located near the inner surface of the insulator, and the beam will self-pinch to the axis with electron trajectories that minimize the inductance which is associated with the pinching. Two dimensional computer code calculations of electron motion in diodes give support to this hypothesis,[9] and experiments with large area diodes are now underway.

TRANSMISSION LINES

Since constant impedance 0.8 Ω transmission lines will be needed to connect the Blumleins to the diode, the line spacing changes linearly with the radius. As the voltage pulse is only 20 nsec in duration, significantly higher electric fields can be utilized in this region, allowing the constant impedance to be maintained almost to the diode.

PULSE SHARPENING SWITCHES

Pulse sharpening switches are shown in this transmission line. This switch will be included as a means of decreasing the pulse risetime if the risetime produced by the Blumlein switches is not satisfactory. Figure 4 shows wave shape sketches indicating the pulse sharpening principle.

With the fast rising voltage, breakdown will occur in a few nanoseconds with a large number of channels across a small, highly stressed gap, and an improved risetime will result. The major problem with this type of switch in liquid dielectrics is voltage feed-through caused by the switch capacitance (C) and the transmission line impedance (Z_0). The switch capacity must be charged for voltage to appear across the switch. It charges through the transmission line impedance (Z_0) and, thus, generates a pulse in the line. To minimize the prepulse voltage on the line, $Z_0 C$ must be considerably smaller than the incoming pulse risetime. Since the electrode spacing will be small to permit breakdown in 10 nsec or less, the switch capacity will be high unless care is taken in designing the electrodes.

BLUMLEINS

The spacing between the plates of a conical or strip transmission line is determined by the breakdown electrical field which varies as the inverse cube root of the effective charging time. With oil dielectric and 1 μsec charge time, the diameter of the Blumleins would be ~ 42 meters, but at 100 nsec charge time the diameter becomes ~ 20 meters. A Blumlein was selected which allows the use of lower voltage switches located in an accessible region. The spacing between lines is about 10 cm. Three types of switches are suitable for the system. All three are described in other papers in these proceedings: (1) triggered gas switch[10], (2) triggered oil dielectric[11] and (3) untriggered oil dielectric switches.[12]

SWITCHING

Figure 5 shows estimated switch current risetime (e-folding) versus the number of switch channels that must close simultaneously (within 2 nsec) for both SF_6 and oil switches. The risetime estimates consist of inductive and resistive phases. Although the gas switch is multichannel, the risetime is determined mainly by its electrode and spark inductance. To further reduce the gas switch risetime, the flashover electric field strength of the housing should be

increased. The housing is approximately 30 cm long and the lines are only separated by 10 cm, making low inductance mounting difficult. The curve ends at 180 switches in Fig. 5 because that is the maximum number that could be located around the periphery of the Blumlein. The triggered multichannel oil switching experiments were performed with 0.31 MV/cm electric field (E) between the two main electrodes and the charge time was > 60 nsec. The calculation in Fig. 5 assumes that the switch can be operated with 0.6 MV/cm electric field since the charge time will be 100 nsec. The resistive risetime has been shown to be proportional to $E^{-4/3}$ and is the major contribution to the risetime for large numbers of channels.[13] Calculations indicate that the estimated EBFA risetimes would be 1.7-2.2 times faster than risetimes for the experiment. The experiment[11] produced 12 channels with an e-folding risetime of 3.6 nsec. With this number of channels, EBFA will have a 2.0 nsec risetime. Six channels on the test set up gave a risetime of 5.2 nsec and would correspond to 3 nsec for EBFA. A triggered oil switch operating at 0.6 MV/cm remains to be demonstrated, but even if it must be operated at 0.31 MV/cm, the resulting 3.6 nsec risetime would be acceptable.

The untriggered multichannel oil switch[12] was operated at 0.4 MV/cm with an 80 nsec charge time, and 12 channels closed producing an e-folding risetime of 2.6 nsec. Thus, with similar electric fields and 100 nsec charge time, a 3 nsec risetime will be achievable.

Fast rising trigger voltage pulses of 300 kV or 1.5 MV are needed for both gas and oil-triggered switches. A trigger pulse is needed for each envelope of a gas switch, but only one pulse is required for 6-8 channels with oil switches. One master gap could serve 6-8 switches in either case. For a 3-4 nsec risetime, 16 to 30 master gaps at 300 kV and 2 to 5 sub-master gaps at 150 kV are required for the SF_6 system, and 3 to 8 master gaps at 1.5 MV and one sub-master gap of 300 kV for oil. In the oil system, jitter of the master gaps contributes to the output risetime while jitter of submaster and master gaps would add into the output pulse risetime of the gas system.

INTERMEDIATE STORAGE CAPACITORS

Intermediate storage capacitors are used to provide the 100 nsec Blumlein charge time. Coaxial cylinder, water-dielectric capacitors and gas dielectric switches are located such that the total loop inductance is about 0.5 μH. Since the isolation inductor connecting the high voltage Blumlein electrode to ground would have to be significantly larger than 0.5 μH, plus-minus charging is proposed to allow the high voltage electrode to capacitively float at ground. Twenty-four, 5 μF, intermediate storage capacitors are thus required with alternate plus-minus charging around the periphery. The 24 switches will require low jitter. There are two switches presently available for this use, a Sandia-developed, 3 MV trigatron switch[14] and a Physics International-Sandia developed V/n switch[10] each with jitter < 2 nsec. To minimize the water capacitor size, it is necessary to use several Marx generators on a large diameter circle and charge at a reasonable rate.

The Marx generators are similar to the one developed for the Hydra electron beam generator.[14] Each Marx would have either thirty 0.7 μF, 100 kV capacitors or forty 1.85 μF, 60 kV capacitors, depending on the reliability and adaptability of the 0.7 μF capacitor in this type of Marx. The approximate inductance is 6 μH and the intermediate storage charge time is 700 nsec. The erection time jitter for this type of Marx has been measured to be < 20 nsec and is acceptable for operating 8 generators in parallel.

SUMMARY

Technological problems and design details of a 2×10^{13} W electron beam accelerator have been presented. The design is based on optimum utilization of oil dielectric in the PFL and transmission lines by rapidly plus-minus charging from 24 intermediate storage capacitors. It also utilizes recent advances in low-jitter (< 2 nsec) switching and low-inductance diode designs (3 MV, 5 nH) projected from results of diode pinching experiments and computer beam trajectory calculations.

REFERENCES

1. G. Yonas, J. W. Poukey, J. R. Freeman, K. R. Prestwich, A. J. Toepfer, M. J. Clauser and E. H. Beckner, Sixth European Conference on Controlled Fusion and Plasma Physics, Moscow, p. 483 (1973).

2. L. I. Rudakov, A. A. Samarsky, Sixth European Conference on Controlled Fusion and Plasma Physics, Moscow, p. 487 (1973).

3. M. J. Clauser, 16th Annual Meeting of the Division of Plasma Physics, Albuquerque, New Mexico, 28-31 Oct. 1974.

4. J. Chang, M. J. Clauser, J. R. Freeman D. L. Johnson, J. G. Kelly, G. W. Kuswa, T. H. Martin, P. A. Miller, L. P. Mix, J. W. Poukey K. R. Prestwich, D. W. Swain, A. J. Toepfer, M. M. Widner, T. P. Wright, G. Yonas, Fifth Conference on Plasma Physics and Controlled Nuclear Fusion Research, Tokyo (1974)

5. G. Yonas, K. R. Prestwich, J. W. Poukey, and J. R. Freeman, Phys. Rev. Letters, 30, No. 5, p. 164 (1973).

6. G. Yonas, P. Spence, S. Putnam and P. Champney, 11th Symposium on Electron, Ion and Laser Beam Technology, Boulder, Colorado, p. 421 (1971).

7. T. H. Martin, First IEEE Conference on Plasma Science, Knoxville, Tennessee (1974).

8. J. C. Martin, "Fast Pulse Vacuum Flashover," Internal Report SSWA/JCM/713/157, AWRE, Aldermaston, England (1971).

9. J. W. Poukey, J. R. Freeman, G. Yonas, Vac. Sci. Technol., 10, p. 954 (1973)

10. I. Smith, S. Mercer, T. H. Martin, International Conference on Energy Storage, Compression and Switching, Asti, Italy (1974).

11. K. R. Prestwich, International Conference on Energy Storage, Compression and Switching, Asti, Italy (1974).

12. D. L. Johnson, International
     Conference on Energy Storage,
     Compression and Switching,
     Asti, Italy (1974).

13. J. C. Martin, "Nanosecond Pulse
     Techniques," Internal Report
     SSWA/JCM/704/49, AWRE, Aldermaston,
     England (1970).

14. T. H. Martin, 1973 Particle Accel-
     erator Conference on Accelerator
     Engineering and Technology, San
     Francisco, California, IEEE Trans
     on Nucl. Sci., NS-20, No. 3, p. 289
     (1973).

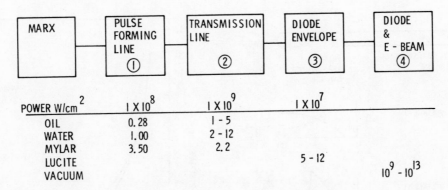

Fig. 1. Power density versus components
and materials in working electron
beam accelerator.

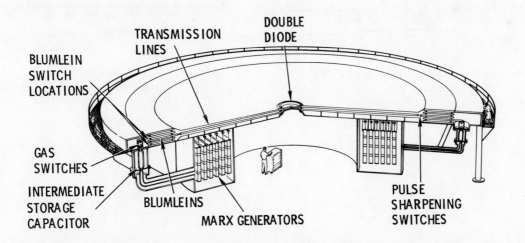

Fig. 2. Artist's drawing of EBFA.

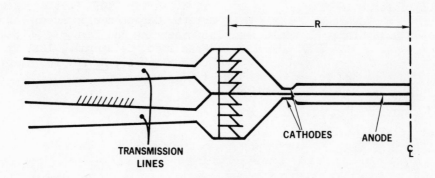

Fig. 3. Diode sketch.

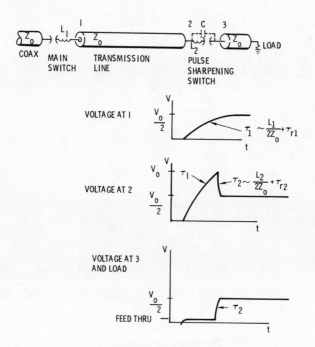

Fig. 4. Pulse sharpening switch voltage
waveshapes.

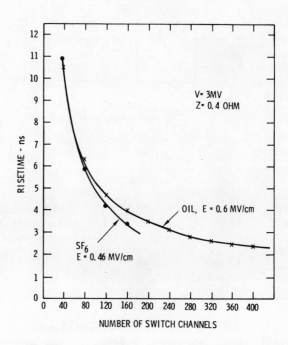

Fig. 5. Estimated EBFA switch
risetime versus number
of channels.

2MV COAXIAL MARX GENERATOR
FOR PRODUCING INTENSE RELATIVISTIC ELECTRON BEAMS

Y. Kubota, A. Miyahara and S. Kawasaki*

Institute of Plasma Physics, Nagoya University, Nagoya, Japan

ABSTRACT

A 2MV Marx generator installed in a completely co-axial form is developed. The structure reduces the internal inductance down to about 150 nH hence makes it possible to produce high intensity electron beams of mare than 20 kA when applied to a field emission type electron gun. The rise time of the output voltage pulse is less than 5 nsec.

INTRODUCTION

The paper concerns the electrical design and the performance of a single pulse generator intended to obtain a high intensity relativistic electron beam (REB) required for the ERA program of IPP-Japan[1]. The high voltage pulse generator has been developed in a way to have a structure minimizing the internal inductance so that it can give an REB of a sufficient intensity (∿20 kA), of a high value of γ (not less than 3) and with a short duration (∿20 nsec), by choosing a completely coaxial structure including the capacitors and the discharge elements. As the result the output pulse has a fast rising part shorter than 5 nsec. With this configuration, the pulse forming line such as a coaxial line or a Blumlim might be got rid of from the standard lineup of the pulser used in the usual high intensity beam facilities[2] and then the loss accompanied by the energy transfer from the Marx to the PFN would be avoided. A prototype of the Marx (600 kV) was built already and reported to show a satisfactory operation[3]. The machine described here is a scaled up model based on essentially the same principle as the previous one.

2MV MODEL

DESCRIPTION OF THE FACILITIES

A cross sectional view of the apparatus is shown in Fig.1. In many respects it is quite similar to the 600 kV model. The whole elements are immersed in the insulation oil contained in a stainless steel tank of 600 mm in diameter and of 2500 mm in length, the outer wall of which

*Present address: Faculty of Science, Kanazawa Univ., Kanazawa, Japan.

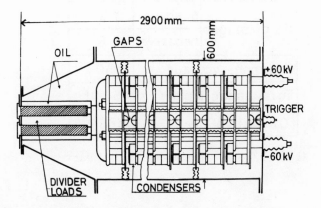

Fig.1. Cross section of 2MV Marx.

serves as the earth side of the co-axial line. The tank is settled horizontally on a supporting system. The generator is composed of multi-stages stacked in series and charged positively and negatively in turn. Total number of the stages is 33. Two sets of 15 barium-titanate capacitors connected in parallel and arranged in a circle on one plane are set in series and consist of unit stage of the capacitor module. Each block has thus 30 ceramic capacitors of 2000 pF with the rated voltage of 30 kV. The equivalent output capacitance is 450 pF. The charging reristors in case of 600 kV model are replaced by the charging inductors of 10 μH each, thus the power loss and the heat production in oil are eliminated. The unit spark gap having two semi-spherical electrodes made of brass is housed in a epoxy of 100 mm in diameter and 70 mm in length. The housings are stacked in series and sealed against the insulation oil outside and the pressurized gas inside as in the prototype. Various kinds of gases such as N_2, CO_2, pure or dry air etc. are tested to improve the discharge characters. The results are discussed later.

The equivalent output capacitance is 450 pF. The Marx is triggered by firing a three electrodes gap at the first stage on the bottom of the apparatus although the triggering at several stages is possible and expected to reduce the jitter of the discharge. The parameters of the 2MV Marx are listed in the Table I and the schematic figure of the electrical circuit is shown in Fig.2.

No.of stages	positive	1 7
	negative	1 6
Charging voltage		± 60 kV
Total capacitance		450 pF
Output voltage		1980 kV

Table I. Main parameters of 2MV Marx.

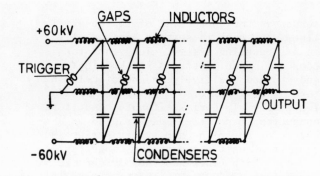

Fig.2. Electrical circuit of 2MV Marx.

EXPERIMENTAL DATA

The Marx has been tested for various working parameters such as the charging voltage, the resistance of the load, the pressure of the gas and so on. A coaxial resistor filled up with copper-sulphate solution is attatched to the output electrode as the load and the voltage divider. The pulse response of the divider is caliprated with a test pulse of square form with variable duration, delivered by the mercury pulse generator, and shows no significant distortion of the pulse at least to 1 nsec. The peak voltages of

the output pulse vs. the load resistances are shown in Fig.3a), from which the output inpedance of the generator is calculated to be about 28 Ω. Fig.3b) gives the outputs vs. the charging voltages with the load of 50 Ω.

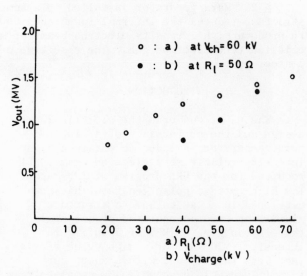

Fig.3. The peak voltage of the output vs.
a) the load resistance
b) the charging voltage.

An example of the wave form of the pulse is shown in Fig.4, where a fast leading head of 5 nsec rise time (10% - 90% of the peak) is followed by a much longer tail with small oscillations superposed. The jitter of the start of the discharge depends on remarkably the working voltage and the pressurized gas. At ± 60 kV charging, the average values of the jitter for various gases are listed in Table 2. Dry air seems to have the best quality in minimizing the jitter. The triggering at multiple gaps is being planed and expected to get a better result. The replacement of the insulation gas in every shot is indispensable for the stable operation. The rate of the discharge is limited by the power of the high voltage source, at present, to one shot per minute.

ANALYSIS

The equivalent circuit of the Marx is shown in Fig.5. After the discharge paths in all the gaps are completed, the transient response of the impulse generator can be analysed with the well known operational technique used in the electronic

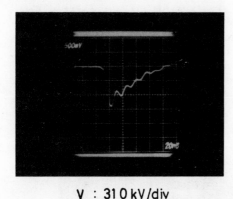

V : 310 kV/div

Fig.4. Output pulse of the Marx, with the load of 25 Ω (310 kV/div.).

Gases name	Jitter (ns)
CO_2	100
N_2	~1 μs
$O_2 + N_2$	100
Dry air	20

Table 2. The jitter of the output for various gases.

engineering[4]. The rise time of the output pulse is approximately determined by the time constant of the unit L-C circuit when the pure resistance component of the discharges in the gaps is negligible compared with the load. 5 nsec rise time for the matched load impedance, e.g. 25 Ω, leads to the unit equivalent inductance L_u of about 10 nH and then the total inductance L is ~150 nH. This value of the equivalent total inductance gives the impedance $\sqrt{L/C}$=18.7 Ω with the total capacitance of 450 pF, which is well consistent with the

experimental result stated above. The decaying part following the sharp peak has a time constnat of ~40 nsec which is not in contradiction with the value expected. The small oscillative structure superposed is attributed to the stray capacities between the circuit elements contained in the tank and the outer wall and/or between the elements of the neighboring stages. With a rough approximation, the frequency f_p of the parasitic oscillation due to the stray capacity c_s is[5]

$$f_p \sim \frac{1}{4\sqrt{Lc_s}\,n} \qquad (1)$$

where n is the number of the stage. The observed frequency of the parasitic oscillation is about 100 MHz which corresponds to c_s of 250 pF/stage and reasonable in our geometry. The value of c_s is much smaller than the capacitance per unit of the discharging condensers, therefore the formula (1) might be justified. At the operation of lower charging voltages with a pressurized gas instead of the insulation oil, a remarkable reduction of the amplitude of the parasitic oscillation can be seen, therefore the explanation given above is verified.

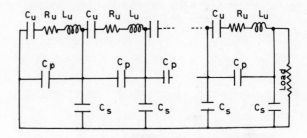

Fig.5. Equivalent circuit of the Marx.

REB PRODUCTION

The high voltage impulse generated by the Marx is applied to a cold emission type electron gun very similar to the one described previously[3]. The schematic view of the gun is shown in Fig.6. The intense REB is emitted from one hundred steel needles fixed to the head of the long shank connected to the output electrode of the Marx. The voltage between the cathode and the annular anode is monitored with a capacitive voltage divider. The gun and the drift tube are immersed in a nearly homogeneous axial guide field of 3000 gauss.

The intensity of the produced REB is measured with a Faraday cup placed on the other end of the drift tube 60 cm apart from the anode.

The Faraday cup is also designed to insure a very high frequency response. One example of the beam current pulse is shown in Fig.7a) and b) with and without the guide field B, respectively.

It should be noted that the nonlinear behavior of the gun impedance distorted greatly the wave form, compared with the case of the pure resistance dummy load. The total electron current emitted from the gun reaches over 20 kA (Fig.8) and lasts during 20 - 40 nsec. The analysis of the beam loss in the drift tube is being carried out and will be reported elsewhere, with the complete description of the electron gun.

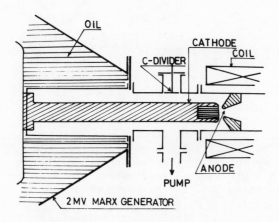

Fig.6. Schematic view of the cold emission gun and the drift tube.

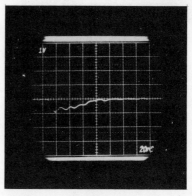

I_e: 20kA/div

Fig.8. Total electron current (20 kA/1 div.).

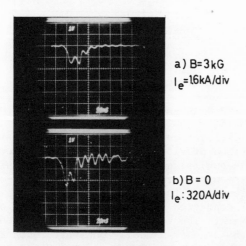

a) B=3kG
I_e=1.6kA/div

b) B = 0
I_e: 320A/div

Fig.7. REB output a) B = 3 kG (1.6 kA/div)
b) B = 0   (320 A/div)

DISCUSSION

We have some motivations to get a Marx having the impedance as small as possible. At first any PFN inserted between a charging power source (Marx) and an electron gun may cause a considerable loss in the energy transfer to REB. To obtain a good efficiency rather severe requirements are imposed on the operation of the Marx. The rise time of the Marx should be of the same order with the characteristic time of the PFN. Secondly the nonlinear impedance of the cold emission cathode makes the pulse shaping meaningless: even if the impulse of a completely square form is applied to the gun, the output beam intensity is much distorted and far from the square pulse. A Marx with a very fast rise time and a short **duration only is needed. The maximum** energy emitted as REB in our case is 600 J

and nearly 70% of the stored energy. A
combination of the Marx with the least
impedance and a post acceleration system
may be best for the ion acceleration
project by REB.

ACKNOWLEDGEMENT

The authors would like to express
their thanks to Mr. Kodaira for his
assistance in preparing the experiments.

REFERENCES

1. S. Kawasaki, A. Miyahara, K. Huke, H.
   Ishizuka, G. Horikoshi, H. M. Saad and
   Y. Kubota: IEEE Trans. on Nucl. Sci.
   NS-20 (1973) 280.
2. See for example G. Cooperstein, J. J.
   Condon and J. R. Boller: J. Vac. Sci.
   Technol. 10 (1973) 961.
3. Y. Kubota, S. Kawasaki, A. Miyahara
   and H. M. Saad: Japan. J. Appl. Phys.
   13 (1974) 260.
4. G. N. G. Glasoe and J. V. Lebacqz ed.,
   Pulse Generators, MIT Rad. Lab. Ser.
   Vol.5(McGraw-Hill, N.Y., 1948).
5. C. Elsner: Arch. Elektrotech. 30
   (1936) 445.

Imploding Systems

THEORETICAL AND PRACTICAL ASPECTS

OF ENERGY STORAGE AND COMPRESSION[*]

O.S.F. Zucker
W.H. Bostick[**]
Lawrence Livermore Laboratory
Livermore, California 94550, U.S.A.

ABSTRACT

The theoretical and experimental aspects of energy storage and compression are reviewed. The role of complementary energy modes combined with nonlinearities in energy compression is described. The need for "coherent nonlinearity" for efficient compression is emphasized. Various fundamental methods for compressions are described. Quantum mechanical limitations are discussed. Energy storage is reviewed from a theoretical and practical point of view with emphasis on new approaches.

INTRODUCTION

Energy compression is the process of increasing the energy density in space-time. Energy comes in the form of static modes (electric, magnetic and gravitational), and dynamic modes (kinetic and electromagnetic). The kinetic, magnetic and electromagnetic modes are repulsive; the gravitational mode is attractive[†]; electric modes can be attractive or repulsive. Energy compression is a dynamic act and thus the kinetic or electromagnetic modes are always involved in energy compression.

Energy storage is an equilibrium situation and involves complementary modes. For example, the kinetic energy mode of the molecules in a gas confined in a pressure vessel is contained by the complementary bonding force between the molecules in the pressure vessel wall. This force is macroscopically referred to as an elastic force but it is really electrostatic in nature. Similarly, a charged capacitor stores electrostatic energy so long as we keep the plates separated. The force that holds the plates apart and prevents the dielectric from collapsing is kinetic in nature: It is the Heisenberg-uncertainty-principle force which causes the electrons to attain higher kinetic energy upon compression, thus preventing them from collapsing into the nucleus. In planetary motion, kinetic energy is confined by gravitational forces. The energy in the complementary mode within one system is sometimes additive and sometime substractive, and sometimes more than two modes of energy are involved. This subject is discussed further in the section on Energy Storage.

This complementarity in energy modes is also the key to energy compression. For example, compressing magnetic energy involves the transfer of magnetic flux from a large to a small volume. Energy can be directly transferred between inductors by the operation of open- and close-circuiting switches. Such an operation, however, always spreads the flux over a larger volume

---

[*] This work was prepared under the auspices of the U. S. Energy Research & Development Administration.

[**] Permanent address: Department of Physics, Stevens Institute of Technology, Hoboken, New Jersey, 07030. Professor Bostick contributed to this work during his appointment as a Summer Employee and later as a consultant to Lawrence Livermore Laboratory.

[†] By "attractive" we mean that the forces tend to reduce the occupied volume and by "repulsive" we mean that they tend to increase it.

(causing decompression). Further-more, it increases the entropy and thus the energy loss. Similarly, transferring electrostatic energy from one capacitor to another or kinetic energy from one mass to

another by inelastic impact) cannot be performed with greater than 25% efficiency, and the energy is always decompressed <u>unless</u> complementary modes are employed. Complementary forms of energy can be always matched to one another: We can convert magnetic energy to electrostatic or kinetic or gravitational or electromagnetic energy with 100% theoretical efficiency. This is likewise true for converting elec-trostatic or kinetic or gravitational energy to all the other forms.

Thus if we wish to compress kinetic energy, for example, we can use elastic energy as an intermediate site between the initial and final kinetic states. Imagine a mass m_1 (Fig. 1a) moving at speed v_1. The objective is to transfer all its energy, i.e., $1/2\ m_1 v_1^2$ to a smaller mass m_2 thus attaining a velocity

$v_2 = v_1 \sqrt{m_1/m_2}$. This can be accom-plished by slowing mass m_1 to zero with a spring; the kinetic energy is momentarily totally converted to elastic energy. At this moment the spring is latched, m_1 is replaced by m_2 and the spring is unlatched. Since the conversion to m_2 is again 100% efficient, the desired com-pression has been achieved. Al-ternatively, a magnetic field can be employed as the intermediate site intead of the spring (Fig. 1b). In Fig. 1c is shown the analogous trans-fer between two capacitors with an intermediate magnetic site.

We have shown the need for complementary modes of energy for both compression and storage. Energy compression is covered in the first section, energy storage in the second section.

ENERGY COMPRESSION

Energy compression in space is directly linked in two distinct ways to energy compression in time. In the quasistatic case, the reduction

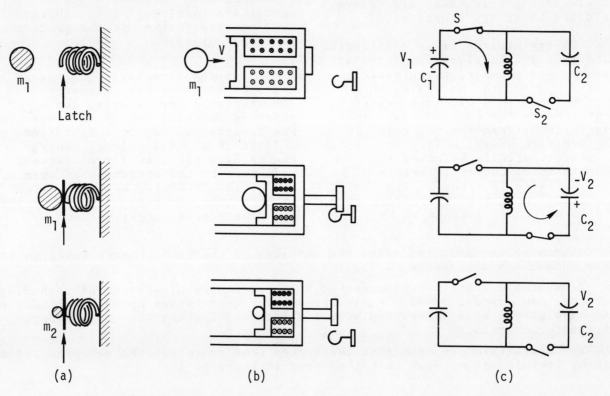

Fig. 1. Complementary transfer between unequal elements.

in volume of an oscillator is linked to the resonant frequency of the elements, and in the dynamic case the time duration and physical dimension of an energy pulse are linked by the group velocity. An added complication in the dynamic case is the difference between transverse and longitudinal compression. In discussing transverse compression, first consider a TEM electromagnetic wave travelling on a transmission line (see Fig. 2a) which undergoes a gradual reduction in its transverse dimension while keeping its impedance constant. The energy in the wave will experience transverse compression. However, the voltage V and the current I, and thus the power, remain the same. Likewise, a pulsed optical plane wave going through a focusing lens (long focal length) (Fig. 2b) suffers a reduction in its wavefront dimensions but the duration of its pulse is unaffected.

The transverse compression process increases the number of photons transversing a given cross-sectional area. Since photons are Bose particles, there is no limit on how many are crowded into a given area. In fact, quantum methanics[1] teaches us that the probability that a photon will occupy a given state (direction and energy) shared by n other photons is n + 1 times greater than the probability that the photon will occupy the state by itself. In other words transverse compression involves the packing of photons, a process that cooperates with nature. We do not need a third body for such compression since it is a linear process and therefore can take place in vacuum.

On the other hand, longitudinal compression is a nonlinear process in both space and time. Consider an electromagnetic lump of energy with dimensions x_0 and duration t_0, as shown in Fig. 3a, moving in a medium. Longitudinal compression causes x_0 and t_0 to decrease. Figure 3b shows the space and time Fourier transforms of our lump of energy. As x_0 and t_0 approach zero, the plot of A(x) and A(t) approaches

the Dirac δ function and the Fourier transforms F(1/x) and F(1/t) approach a constant in wave number and fequency space as in Fig. 3c. Thus we see that longitudinal energy compression involves the transfer of energy from low frequencies to higher frequencies without altering the total amount of energy. In other words the compression process converts many photons (or phonons in the elastic-kinetic case) into fewer but more energetic photons (or phonons). This is a nonlinear process which involves matter to provide momentum and energy conservation. To generate these higher frequency components, we need a nonlinear process. In the case of the energy transfer between the two capacitors (Fig. 1c) the switching action provided the nonlinearity, and in the case of the kinetic energy transfer between two masses (Fig. 1a) it was the latching process. A third body is always necessary to provide the nonlinearity. Two photons in the same state, although they possess a combined energy and momentum of a phonon of twice the frequency, do not undergo such a reaction in vacuum. Electrical

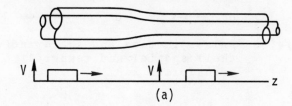

(a)

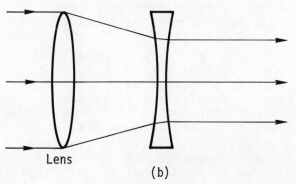

Lens

(b)

Fig. 2. Transverse compression (a) in a transmission line and (b) in a lens system.

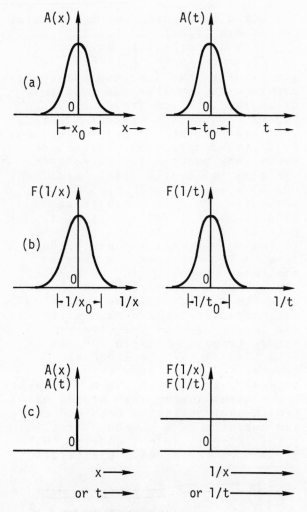

Fig. 3. Pulses in time and space and
their spatial and temporal
transforms.

engineers would prefer to work with
simple resonators such as L-C
resonators rather than photons; in
such cases the role of matter in
providing the nonlinear process
would be performed by ideal switches.

It is possible to picture the
compression process as the total
transfer of energy from a low frequency oscillator through successive stages to a high frequency
oscillator, as shown in Fig. 4.
Here the nonlinearity is provided
by interconnecting switches. It
can now be shown that for efficient
compression (or upconversion),
nonlinearity alone is not sufficient.
What is needed is "coherent nonlinearity." If these oscillators

are interconnected with self-running
switches to provide the nonlinearity,
only a small fraction of the total
energy will arrive at the nth oscillator. However, if the nonlinearities
in the different nonlinear elements
are programmed, then this complete
transfer is possible. In other
words, there is now a definite time
(or phase) relationship between
the onset of the nonlinearities in
the different oscillators. There
must be intelligence in the process.

Even a piece of legislation as
binding as the second law of thermodynamics cannot prevent the energy
compression from d.c. to cosmic rays
at 100% (theoretical) efficiency
from taking place if the process is
carried out intelligently with coherence and sufficient nonlinearity.

Shown in Fig. 5 is one form of
interconnection of harmonic oscillators by nonlinear elements. Ideal
switches are used as the nonlinear
elements, and constant capacitors
with monotonically decreasing inductors in the harmonic oscillators
are used to provide for the increase
in resonant frequency.

The operation of this circuit
is based on resonant transfer characterized by elastic collisions

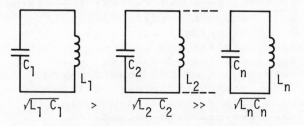

Fig. 4. Successive energy transfer
from low- to high-frequency
resonators.

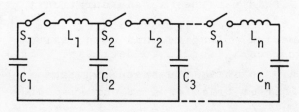

Fig. 5. Energy transfer with coherent
nonlinearity.

between equal masses. Figure 6 shows three kinds of elastic collision transfer systems: capacitive, inductive, and kinetic. The waveforms apply to all. The important point here is that if C_1 is equal to C_2, then the total energy of the circuit will alternate at π-radian intervals from C_1 to C_2; the inductor L never stores more than one-half the total energy. Thus, in Fig. 5a, if $C_1 = C_2 = C_3 \ldots = C_n$, switches S_1 to S_n can be operated sequentially in such fashion that (a) only one switch is closed at a time, and (b) the "closed" time for each switch corresponds to π radians in its corresponding resonant oscillator. With the introduction of such "intelligence," energy can be transferred from C_1 to C_n with 100% theoretical efficiency, assuming lumped parameter idealization.

Outside intelligence is not always necessary for coherence: The flow of energy along the system can provide the timing and thus the coherence.

Let us now consider three kinds of conpressors with such characteristics. The first, an elegant example of this process, is the Melville line.[2] Melville noted that the nonlinearity of ferromagnetic materils can provide the inductance, the switch, and the timing necessary for the Automotization of the compressing network of Fig. 5. Figure 7 shows a typical $\overline{B}$ vs $\overline{H}$ curve for "switching materials" such as 50-50 nickel-iron or ferrite. The salient features are as follows:

1. The abrupt change in slope provides the abrupt impedance change. However, this change in impedance is not from infinity to zero; the impedance ratio is about 10^4 to 10^5 in the 60-Hz range and about 10^2 in the 10^7-Hz range.

2. The moment of saturation is governed by the initial point on the curve and by the flux rate of change. More explicitly the saturation time τ is governed by the equation

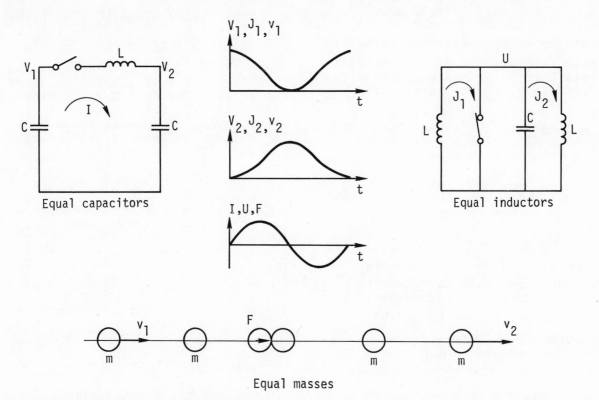

Fig. 6. Complementary transfer between equal elements (elastic collision).

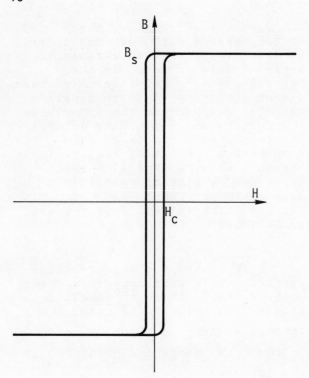

Fig. 7. Hysteresis loop for ferro-
        magnetic material.

$$B_s - B_1 = \int_{t=0} \frac{V(t) \ dt}{nA},$$

where B_s is the material characteristic saturation flux density (1.5
teslas for Ni-Fe and 0.3 teslas for
ferrite), B_1 is the initial flux
density, n is the number of turns,
A is the cross-sectional area, and
V(t) is the voltage function of
time seen by such a saturable inductor.

3. The hysteretic area accounts
for unrecoverable losses.

Figure 8 shows a Melville line.
A typical inductor L_2 is designed
in such a way that the charging
voltage across C_2 will cause it to
saturate the moment C_2 is fully
charged. Upon saturation it attains an inductance L_{2s} which is
small relative to L_1, so that now
the resonant period from C_2 to C_3

is faster than that from C_1 to C_2.
Indeed such lines compress 60 Hz
to 1 MHz in about four stages with
better than 75% efficiency.

There are limitations on such
a resonant transfer compressor: although the frequency increases, the
voltage remains constant; only the
current has increased, thus increasing the power VI. The impedance
V/I has decreased in the compression
process. No impedance reduction
can go on forever. As the frequency
goes up it becomes increasingly
difficult to diverge far from the
free-space impedance. It is impractical to consider impedances
much smaller than one ohm. In
practice this is circumvented by a
more complicated switching arrangement as used in Marx generators, in
Blumleins, and in stacked lines.
These devices raise the voltage and
thus the impedance by the strategem
of "charging in parallel and discharging in series" — an engineering
term for rearranging the circuit.
Such an action does not involve
compression but merely starts the
compression process at a higher
initial impedance and voltage.

Our second example, a compressor
that raises the voltage and thus the
impedance is that proposed by P.R.
Johannessen.[3] Johannessen observed
that the analog of a Melville line
can be built using saturable capaci-
tors (Fig. 9); its $\overline{D}$ vs $\overline{E}$ curve is
shown in Fig. 10. Here the trans-
ition from a very large to a very
small capacitance provides the

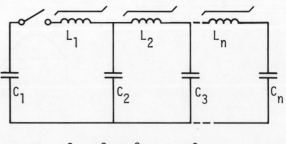

$$C_1 = C_2 = C_3 \ \text{---} \ = C_n$$

$$L_{1s} > L_{2s} > L_{3s} \ \text{---} \ \gg L_{ns}$$

Fig. 8. The Melville line.

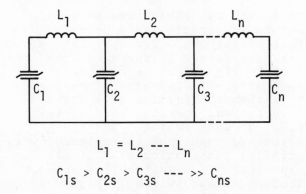

$$L_1 = L_2 \text{ --- } L_n$$

$$C_{1s} > C_{2s} > C_{3s} \text{ --- } \gg C_{ns}$$

Fig. 9. The capacitive (Johannessen) Melville line.

characteristics of a switch that opens. Since Johannessen was not interested in high voltages, he hit upon a most unlikely candidate, the storage diode, whose $\overline{D}$ vs $\overline{E}$ curve is shown in Fig. 11. To date the storage diode has poor voltage-holding properties and its saturation flux density is miniscule, but the last chapter on these diodes as "break switches" has not yet been written.

Lead zirconate, a ferroelectric, has characteristics similar to the desired ones shown in Fig. 10. How-ever, the hysteretic losses are fairly large and, most important, acoustic resonance dictates fine lamination of the ferroelectric material, with its attendant cost, to provide for the desired high-frequency switching.

It is an inherent property of the compression process that switches that close increase the current and decrease the impedance, and switches that open increase the voltage and the impdeance. Thus the use of both kind of switches allows us to main-tain some average impedance regard-less of the number of stages, with the flexibility of moving from one impdedance level to another, up or down at will. For this reason alone, it is possible to justify research on all kinds of opening switches such as exploding wires, superconducting switches, ferro-electrics, and the various pinches in plasmas such as the plasma focus and the electronic ram.

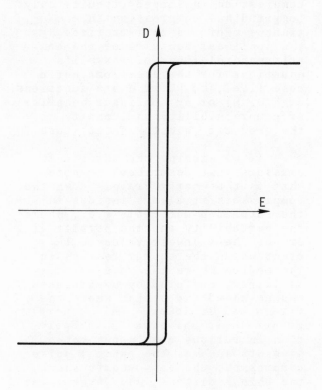

Fig. 10. Ideal D vs E curve for sat-urable capacitors.

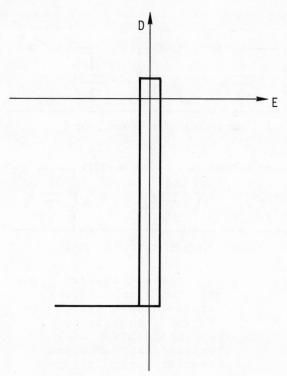

Fig. 11. D vs E curve for the storage diode.

Our third example, a compressor proposed some time ago by one of the authors (O.Z.), utilizes both types of switches (Fig. 12a). We

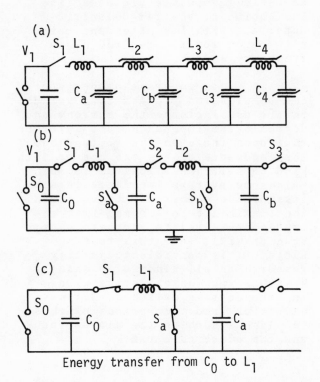

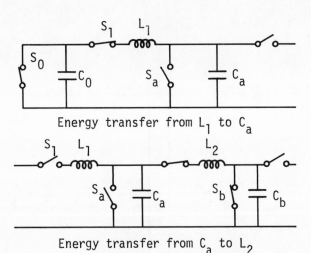

Energy transfer from C_a to L_2

Fig. 12. (a) Constant-impedance compression line utilizing saturable capacitors and saturable inductors. (b) Idealization of a constant-impedance compression line. (c) Step-by-step switch operation for the transfer from C_0 to L_2.

have both saturable inductors and saturable capacitors to perform these functions. The operation of this compressor is shown in idealized form in Fig. 12b. Here as before each saturable inductor is replaced by series combination of a linear inductor and a switch that closes, i.e., a "make switch," and the saturable capacitor is replaced by a parallel combination of a linear capacitor and a switch that opens, i.e., a "break switch". Figure 12c shows the sequential switch operations.

This compressor differs from the first two in two ways. First, the compression rate is faster: Each stage requires one-quarter cycle instead of one-half cycle for compression. Second, since both L and C decrease, the impedance (equal to $\sqrt{L/C}$) can increase or decrease or remain on the average constant. This property is quite useful in matching the compressor on either end to source and load. In other words, both current and voltage increase in the compression process.

Up to now we have examined compression in lumped circuits only. Longitudinal compression in continuous media is characterized by the nonlinear behavior of the constituent relations of Maxwell's equations for the electromagnetic case (i.e., ε, μ, or σ are functions of $\overline{E}$ or $\overline{H}$) or by nonlinear behavior of compressibility and density in the acoustic (elastokinetic) case.

If we examine the three compressors just described, we note that as the energy travels down the compressors it always resides in the saturated material, i.e., where the permability and the permittivity are at their lowest values. In other words the energy resides in the medium where the wave velocity is larger, buffered by unsaturated medium ahead and behind where the wave speed is lower. This general principle is universally in operation in various continuous compressors such as the optical pulse compressor, the gas-dynamic shock, the whip, solitons, and water waves breaking on a beach. The subject of optical pulse compression is covered by R. Fisher's and A. Glass's papers at this conference. So it will not be covered here. In a gas-dynamic shock the large-amplitude, nonlinear, pulsed wave travels largely in the wave-heated gas directly behind the leading edge; the wave speed in this higher-temperature gas is higher than in the gas ahead of the wave. Thus, the wave steepens into a shock with an abrupt leading edge where the energy is concentrated and thus compressed.

The whip (the "bullwhip" especially) provides an excellent mechanical example of continuous energy compression. In a bullwhip the wave propagation velocity (for a transverse wave, like a wave on a guitar string) decreases toward the tip because the tension on the whip decreases toward the tip. Thus, a pulsed wave launched at the handle steepens and shortens as it proceeds toward the tip. The kinetic energy initially stored in the fairly large mass of most of the whip eventually is transferred to just a few centimeters at the tip with the excess momentum extracted by the hand holding the handle. Hence, the velocity of the tip finally reaches a value exceeding that of sound in air. An analysis of the whip is given in Appendix I.

Energy compression by the bullwhip is similar to that in the steepening of a water wave as it approaches a sloping beach, where the wave velocity decreases as the water becomes shallower. It is interesting that the ultimate in water-wave energy compression at a beach is for the wave to become a green water vortex (the heavenly dream of ardent surfers). The stylized design of water waves seen on ancient Grecian urns is that of the profile of a wave becoming a vortex.

For a wave traveling along a string where the wave velocity is being progressively reduced, the ultimate in energy compression is for the string to form a loop (analogous to the vortex). A large-amplitude nonlinear Alfven wave executes energy compression by becoming a traveling force-free vortex

filament. An MHD treatment of Lorentz- and Magnus-force-free vortices is given in the paper by D. Wells in this Proceedings.

Extensive work in shock-like phenomena in electric circuits was performed in the Soviet Union; the work of Kateyev[4] is an outstanding example.

At the beginning of this paper, the prerequisites for longitudinal compression were shown to be energy storage, nonlinearity and coherence. As energy is compressed in space and time, eventually the scene of action is transformed into a plasma. If we wish to continue the compression process, we must provide the above prerequisites in the plasma; indeed, this is done in some fusion devices such as the plasma focus, laser fusion, and electron beam fusion, to name just a few. Here we will mention just a few unique concepts. R. Kidder's paper in this Proceedings discusses laser fusion, while Wintergerg's paper, also in this Proceedings, discusses a scheme for combining laser and electron beam fusion.

THE PLASMA FOCUS

The plasma focus first takes energy stored in a capacitor and transfers it to the inductance of the annular space between the co-axial electrodes. The current sheath driven forward by the Lorentz force of its current flowing across this magnetic field, sweeps up and ionizes the gas in front of it. The resulting plasma, instead of being heated by an inelastic (snow-plow) impact with the current sheath, stores energy in the magnetic field due to the current in the plasma. In the complex structure of the magnetic field, an important role is played by pairs of vortex filaments[5] which form in the corrugations that naturally occur in the current sheath. The energy is stored in complementary modes in the form of helical mass flow in the filaments and as local magnetic fields due to the paramagnetic behavior of the plasma in the filaments. The stronger vortex filaments absorb the weaker ones, leading to concentration of the energy in a smaller number of filaments. The scallops seen in the profile photos of some current sheaths are analogs of the hydraulic jump which releases some of this vortex filament energy at periodically selected radii. Such scallops appear as a circle in the azimuthal direction. They can be considered to some extent the loci where an energy-dissipating phenomenon is occurring. This effect is treated by a paper by Fausto Gratton in these proceedings for a single plasma filament.

As the current sheath approaches the pinch stage at the end of the center conductor, some of the pairs of vortex filaments annihilate each other, releasing energies to the plasma with the emission of soft x-rays and some neutrons (in a deuterium plasma). Other vortex filaments appear to form toroidal solenoids where the magnetic fields (B_z and B_θ) in the central channels become extremely large (~20 000 tesla). The dimensions of the plasma concentrations in these central channels can be spatially very small (0.1 mm in diameter and 0.3 mm long) as measured by x-ray pinhole photos. The rapid decay of these 20 000-tesla fields (in ~10 ns) leads to a production of high-energy electrons with x-rays (up to 0.3 MeV) and high-energy deuterons (up to 0.5 MeV) with both D-D and D-T neutrons, with only deuterium gas filling. Visible light, infrared, and microwave emission are observed. These effects are discussed in a paper by Nardi and Bostick in these proceedings.

In the plasma focus, only a fraction (<10%) of the energy orginally stored in the capacitor is contained in these 20 000 tesla magnetic fields. However, the degree of energy density compression by the plasma focus is high: For the capacitor dielectric stressed to about 2000 V/mil = 8×10^5 V/cm = 8×10^7 V/m with the dielectric constant $\cong$ 5, the energy density is ~0.14 J/cm^3. The energy density of 20 000 tesla is 1.6×10^{15} ergs/cm^3

= 1.6×10^8 J/cm^3. Thus energy density has increased by a factor of ~10^9.

SOLITONS

Solitons[6] are examples of high-amplitude, nonlinear waves that compress energy. In this respect they resemble shock waves, and the "hydraulic jump," self-induced transparency, and the three compressors discussed earlier, in that because of their large amplitude they modify the environment for other small neighboring waves and thus feed on them by attracting them into their vast potential wells. Theoreticians believe that solitons exist in the form of large-amplitude plasma waves. Hasimoto[7] has shown theoretically that solitons can be expected to travel along ideal vortex filaments.

Solitons are solutions to the Korteweg-de Vries equation. These solutions have the property that their asymptotic behavior is localized. In other words, given a train of solitary waves, the velocity of the train and the wave spacings are continuously variable. Solitons can undergo "collisions" in which their identity and coherence are preserved.

Theoreticians are now excited about the identity- and coherence-preserving properties of solitons in a plasma. It was experimentally demonstrated twenty years ago[8] that diamagnetic plasma vortices (plasmoids) can bounce off each other like billiard balls. The plasma vortex is a tightly compressed coherent package of compressed energy in a plasma.

THE ELECTRON BEAM

The electron beam discharge is in itself a highly compressed form of energy (covered briefly in Appendix II); however, a number of schemes are available in which the beam is caused to enhance this compression even further. The most commonly known phenomena are the two-stream instability[9] and the

Bennett pinch.[10] Again, these processes are extensively covered in the literature; they are not discussed here.

One elegant and relatively unknown approach is Raudorf's electronic ram (see his paper in these proceedings). Here a magnetic mirror field slows the electrons from a conventional E-beam gun. The space-charge buildup creates a potential hill which is two orders of magnitude higher than the accelerating potential of the gun. This potential hill acts as a virtual cathode which now emits a corresponding but smaller number of electrons with the attendant higher energy.

Another scheme for compressing electron beam energy is the relativistic beam compressor of L. Kazanskii et al.[11] and M. Friedman[12] Here a propagating electron beam interacts with a cavity in such a way that the first half of the beam pumps the cavity with electromagnetic energy, thus losing energy while the second half drains the cavity, thus gaining energy and spatially compressing the beam energy to half its original size.

There are many similar schemes[13-15] which utilize the interruption of electron beams to accelerate particles to higher energy. The collective accelerator described by Luce,[16] which we are associated with, is a more complicated example (Fig. 13). Here a relativistic electron beam aimed at a hole in a dielectric disk interacts with the disk in a way similar to the way it would interact with a resonant cavity.[17] This is easily seen when one realizes that the surfaces of the dielectric disk are easily ionized, thus becoming conductive. Now the disk, being a dielectric, can support high electric and magnetic fields internally where a disk made entirely of conductive material could not. This interaction produces an intense oscillating electric field in the direction of the beam. In turn

this oscillating electric field generates a (longitudinal) plasma wave traveling in the down-hole direction. The group velocity of these waves is dictated by the slope of the dispersion curve, which is a function of both plasma density and temperature. The conditions down-hole are such that as we travel along the axial direction the density decreases and the temperature increases, both leading to a decrease in the phase velocity v_p and increase in the group velocity v_g, thus causing group acceleration. Slow ions in the vicinity of the hole are gently picked up by the group modes and accelerated in a way similar to the way a wave picks up a surfer and accelerates him toward the beach. In this particular case, accelerated

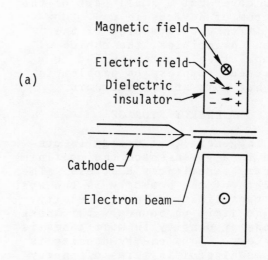

(a)

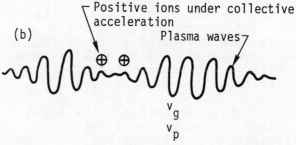

(b)

Fig. 13. (a) Collective accelerator utilizing the resonance of an electron beam with an apertured dielectric disk. (b) Plasma oscillations by the established resonance accelerate ions carried by the group nodes of the plasma wave.

deuterium ions were accelerated to 10 MeV by an electron beam of 1.8 MeV.

The excitation of the hydrogen atom by the absorption of a photon is an example of energy compression. The photon of wave length about 6000 nm is absorbed by an electron of wave length about 6 nm without the aid of an antenna. This can be considered to be a spatial energy compression about 1000-fold in linear dimension. Likewise, the photoelectric effect, Compton scattering, and pair annihilation of electrons and positrons can be considered to be examples of energy compression.

In this section the general principles of energy compression have been discussed. The examples cited cover but a small part of the spectrum of devices in use. However, with understanding of the general principles, the choice of appropriate elements applied to a particular compression problem is somewhat more straightforward.

ENERGY STORAGE

In general, if energy is confined in a given space the confining boundary experiences a stress. The strength of the boundary is the most important limit on energy density. With no limit on boundary thickness, the energy density in most cases is limitless. Thus energy density is most meaningful in terms of energy per unit boundary volume. In this section we investigate the limitations of various energy storage media such as electric, magnetic, electromagnetic and kinetic in conventional and unconventional configurations and combinations.

First, consider a thin-walled cylinder of very high tensile strength material with failure stress σ and with radius r and wall thickness t. By elementary stress calculation we find that the maximum pressure the cylinder can sustain is given by $P = (\sigma t)/r$. We can produce this pressure by filling the cylinder with a gas or a magnetic field, or we can spin the cylinder until we reach the same hoop stress

σ. We find as we shall show presently, that the energy stored in the volume of the cylinder per unit volume of cylinder wall material is always equal to a fraction of σ close to one.

For n moles of a pressurized perfect gas, the energy E stored in creating a volume V_T at a constant pressure P is $PV_T = nRT$ and the internal thermal energy of the gas is $(3/2)nRT = PV_T/(\gamma - 1)$, where $\gamma = 5/3$. For a pressurized gas, the energy generally available for useful work could be roughly approximated by $\sim PV_T/2$ (as in the cylinder of a diesel engine). If V_T is the cylinder volume πr^2 per unit length,

$$E = \frac{1}{2}PV_T = \frac{P\pi r^2}{2} .$$

With this equation we have

$$E \cong \frac{\sigma t}{2r}\pi r^2 = \frac{1}{4}\sigma(2\pi rt) = \frac{1}{4}\sigma V_W,$$

where V_W is the volume per unit length of the confining wall material. Thus, the stored energy per unit volume of wall material is $\cong \sigma/4$.

For a homogeneous magnetic field H, the magnetic pressure is given by

$$P = \mu_0 H^2 = \frac{B^2}{\mu_0} ,$$

equal to twice the magnetic energy density. The stored magnetic energy

$$E = \frac{1}{2}\mu_0 H^2 V_T = \frac{1}{2}PV_T,$$

and

$$E/V_W = \frac{P}{2}\frac{V_T}{V_W} \cong \frac{P}{2}\frac{\pi r^2}{2\pi rt} = \frac{1}{4}\frac{Pr}{t} \cong \frac{\sigma}{4} .$$

If we spin the thin-walled cylinder and call it a flywheel we can show that the energy stored per unit volume of flywheel matter is $\sigma/2$. The limiting stress thus constitutes a universal limitation.

Any equilibrium system involves the balance of forces. The outward (or explosive) force of an inductor a flywheel or a pressure vessel is a repulsive force. There are only two attractive forces available to us to counterbalance and contain these repulsive forces, namely electrostatic and gravitational.

Since the "glue" of matter (between atoms and molecules) is essentially electrostatic, all the above mentioned forms of energy are balanced electrostatically, and σ in the limit equals the strength of the molecular bond. In practice, only single-crystal whiskers come close to this limit; usually imperfections in the form of dislocation and interstitials cause early failure.

Recognizing this universal limitation we now discuss ways to ameliorate it. For the magnetic case it is known that the magnetic field need not be uniform in the container, but rather can rise toward the center without affecting the magnetic pressure on the boundary. The simplest case is given by the coaxial geometry shown in Fig. 14. Here the total magnetic energy is given as

$$E = \mu_0 H_o^2 \ (\pi r_o^2 \ell) \ \ln \frac{r_0}{r_i} \ .$$

If $r_i^2 << r_o^2$ then $V_T \cong \pi r_o^2 \ell$, and

$$E = \mu_0 H^2 \cdot V_T \ \ln \frac{r_0}{r_i} \ .$$

In the uniform case we had $E = (P/2)/V_T$. Thus, there is a gain

of a factor of $2 \ln r_o/r_i$ in energy density, which can be appreciable. It is assumed that with sufficiently reinforced ends, the inner conductor is under three-dimensional compressional stress and thus is able to take a much higher stress than the outer wall under two-dimensional tensional stress.

Another approach takes advantage of the complementarity of the confining stress with the confining field: Imagine a pendulum which, instead of swinging back and forth, spins in a circle around its center as shown in Fig. 15. This pendulum does not exchange its energy between kinetic and potential but rather both energies remain constant; the kinetic energy equals $1/2mv^2$ and the potential equals mgh, where h is the distance that m is elevated from its rest position. As we add energy to the system, the energy distributes itself between the two sites so that for a given centripetal confining force the total stored energy is the kinetic plus potential energy. In this case the ratio of the two is given by

$$\frac{E_P}{E_K} = 2 \ \frac{\cos\theta}{1 + \cos\theta} \ .$$

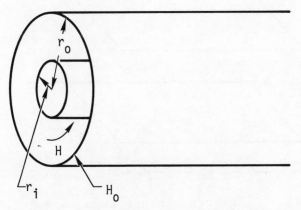

Fig. 14. Magnetic field distribution in a coaxial geometry.

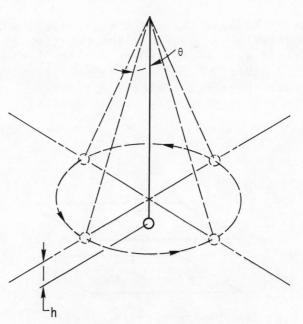

Fig. 15. Energy storage in a spinning pendulum.

For small θ we have $(EP/EK) = 1$ so that the total stored energy is doubled. For small θ, the virial theorem would predict that this is so.

There are other cases where the energy can be simultaneously stored in two complementary modes:

ENERGY STORAGE IN A COMPRESSIBLE-DIELECTRIC CAPACITOR

Consider two capacitor plates held apart by an elastic dielectric, depicted in Fig. 16 as a spring. The elastic restoring force is $-k(d_0 - d)$, and the stored elastic potential energy is $\frac{1}{2}k(d_0 - d)^2$ the electrostatic potential energy is $\varepsilon E^2/2 \, Ad$. Equilibrium occurs when $k(d_0 - d) = qE = CVE = (\varepsilon A/d) \, EEd = \varepsilon AE^2$. Then

Elastic potential energy
Electrostatic energy

$$= \frac{\frac{1}{2}\varepsilon AE^2 (d_0 - d)}{\frac{\varepsilon E^2 A}{2} d}$$

$$= \frac{d_0 - d}{d} = \frac{d_0}{d} - 1.$$

To make the elastic potential energy equal to the electrostatic energy, it is necessary to have the equilibrium d equal to $d_0/2$. In most practical dielectrics, there is essentially very little spatial

compression, that is $(d_0 - d)/d \to 0$. However in the titanates and zirconates, which have extremely large dielectric constants, it is the crystal deformation energy which provides the additional storage ability.

ENERGY STORAGE IN A CAPACITOR WITH ELASTIC (STRETCHABLE) PLATES

The pair of capacitor plates shown in Fig. 17 stretch in the x direction with a restoring force $-kx$ and an elastic potential energy storage of $(kx^2)/2$. They stretch independently in the y direction with a restoring force $-ky$ and an elastic potential energy storage of $(ky^2)/2$. Also when the capacitor is charged, $x = y$. The electrostatic energy stored in the capacitor is $(\varepsilon E^2)/2 \, x^2d$. Equilibrium between the expanding electrostatic pressure εE^2 and the contracting forces $-kx$ and $-ky$ $(= -kx)$ gives $kx = ky = \varepsilon E^2 \, xd$. Hence,

Elastic potential energy
Electrostatic energy

$$= \frac{\varepsilon E^2 \, x^2d}{\frac{\varepsilon E^2}{2} \, x^2d} = 2.$$

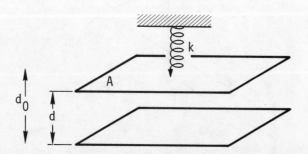

Fig. 16. Energy storage in a compressible dielectric capacitor.

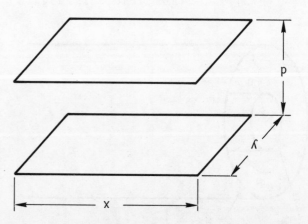

Fig. 17. Energy storage in a capacitor with elastic (stretchable) plates.

A FLUID MECHANICAL ENERGY STORAGE DEVICE IS THE VORTEX

Consider a vortex produced between the two solid plates as shown in the idealized velocity profile in Fig. 18, where the undisturbed hydrostatic pressure in the liquid is p_0. Most of the kinetic energy is contained in the irrotational region where $r_1 < r < \infty$ and

$$v = \frac{v_1 r_1}{r} = \frac{\Gamma}{2\pi r} .$$

The vortex circulation is $\Gamma = 2\pi v_1 r_1$. The kinetic energy per unit length of the vortex is

$$KE \cong \int_{r_1}^{r_2} 2\pi r \, \rho v^2 \, dr$$

$$= \rho 2\pi \frac{\Gamma^2}{4\pi^2} \int_{r_1}^{r_2} \frac{dr}{r}$$

$$= \frac{\rho \Gamma^2}{2\pi} \ln \frac{r_2}{r_1} .$$

The vessel must be terminated at a radius $r_2 < \infty$ if the kinetic energy is to remain finite. The pressure is given by

$$dp = -\frac{\rho v^2}{r} \, dr = \rho v_1^2 r_1^2 \frac{dr}{r^3} .$$

Then,

$$p = p_0 - \int_{r_1}^{\infty} \rho v_1^2 r_1^2 \frac{dr}{r^3}$$

$$= p_0 - \frac{\rho v_1^2 r_1^2}{2\pi r^2} = p_0 - \frac{\rho v^2}{2} ,$$

which gives the pressure profile shown in Fig. 18. There is a

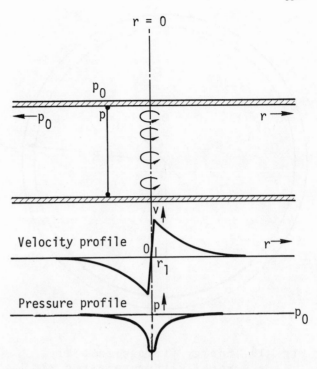

Fig. 18. Energy storage in a fluid mechanical vortex.

pressure differential $p_0 - p = (\rho v^2/2)$ across each metal plate, which one can consider to be produced by elastic bands attached to the inner surfaces of the plates. The elastic potential energy in these stretched bands per unit volume is $\rho v^2/2$ which is locally equal to the kinetic energy.

If a vortex is excited in water in a lake by the oar of a rowboat, hydrostatic pressure is produced by gravity and the vortex produces a cavity along its axis. Potential energy is stored in the earth's gravitational field by "pumping" that water to the surface of the lake and raising the height of the lake. The gravitational potential energy for this vortex plays approximately the same role as the elastic potential energy in the rim of an elastic flywheel, next to be considered:

ENERGY STORAGE IN A FLYWHEEL WITH AN ELASTIC RIM

The mass m of the flywheel diagrammed in Fig. 19 is concentrated

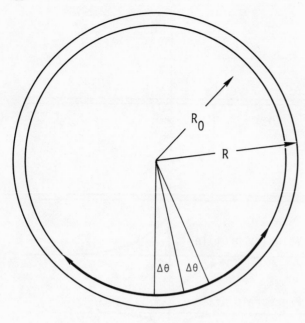

Fig. 19. Energy storage in a fly-
      wheel with an elastic rim.

in the rim of initial radius r_0 and
final radius r. The rim has an
elastic constant k. The tension in
the hoop is $k(r - r_0)2\pi$. The rim
rotates at an equilibrium radius
when

$$2 \left(\frac{m}{2\pi r}\right) r\Delta\theta \frac{v^2}{r} = 2k (r - r_0) 2\pi\Delta\theta$$

or

$$\frac{mv^2}{2} = 2\pi^2 k (r - r_0) r.$$

Then

Elastic potential energy
Rotational kinetic energy

$$= \frac{\frac{1}{2} k 4\pi^2 (r - r_0)^2}{\frac{1}{2} mv^2}$$

$$= \frac{\frac{1}{2} k 4\pi^2 (r - r_0)^2}{2\pi^2 k (r - r_0) r}$$

$$= \frac{r - r_0}{r} = 1 - \frac{r_0}{r}.$$

The elastic potential energy can
equal the rotational kinetic energy
only when $r_0 = 0$. The virial
theorem also prescribes that the
ratio of the energies should be 1
if $r_0 = 0$.

    This system is twice as advan-
tageous for energy storage as the
conventional flywheel where the
energy storage is equal only to the
kinetic energy. Obviously, however,
it is not practical to make large
flywheels out of such elastic
material.

APPENDIX I: THE WHIP

    Perhaps the most ingenious and
effective energy compressor invented
by man is the whip. It is a pity
that such an invention should have
been used so often for punitive
purposes. A description of the whip
by the wave equation and high speed
photography has been given by
Bernstein et al.[18] To consider the
energy-compression aspects of the
whip we first assume that the whip
(Fig. A-I-1) is not tapered and has
a mass per unit length ρ. The whip
master puts the long length of the
whip (ℓ_0 in length) in motion with
an initial velocity v_0 and then
pulls back, either holding the grip
stationary in the earth's frame of
reference or moving it back with a
constant velocity. In either case
he exerts a force on the grip end
pulling with a tension T in the
positive x direction. It is as if
the whip cord in the figure moves
to the left by rolling over an
imaginary rolling wheel and is
progressively laid to rest on the
y = 0 axis. At any point in the
travel, the remaining whip length ℓ
moving at a velocity v has a
kinetic energy $(\rho\ell v^2)/2$ which we
will approximately equate to the
initial kinetic energy $(\rho\ell_0 v_0^2)/2$
imparted to the whip length ℓ_0. The
$v = v_0\sqrt{\ell_0/\ell}$ increases without limit
as the remainder of the whip length
ℓ becomes smaller. We can find the

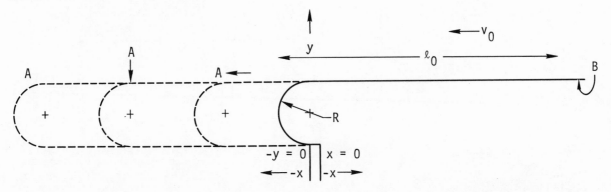

Fig. A-I-1. Idealized bullwhip with constant linear mass density.

expression for the increase of v with time:

$$v = -\frac{d\ell}{dt} = v_0 \left(\frac{\ell_0}{\ell}\right)^{1/2} ;$$

$$-\int_{\ell_0}^{\ell} \ell^{1/2} \, d\ell = \int_0^t v_0 \ell_0^{1/2} \, dt$$

$$-\frac{2}{3} \ell^{3/2} \Big|_{\ell_0}^{\ell} = v_0 \ell_0^{1/2} t \Big|_0^t$$

$$\ell = \left[\ell_0^{1/2} \left(\ell_0 - \frac{3}{2} v_0 t\right)\right]^{2/3}$$

$$v = v_0 \frac{\ell_0^{1/2}}{\left[\ell_0^{1/2}\left(\ell_0 - \frac{3}{2} v_0 t\right)\right]^{1/3}} .$$

The tension at point A is $\rho\ell(dv/dt)$; it decreases linearly along the length ℓ to zero at point B. Hence, the wave velocity $\sqrt{T/\rho}$, where T is the tension, decreases spatially along the whip, thereby developing high-amplitude "shock" in the x direction as the wave travels in the y direction on a string which is moving in the earth system with velocity v. It can be shown that in such a system, where the tension is produced by centripetal acceleration, the wave is always stationary in the earth system, in this case in the y direction. This wave, with gigantic amplitude, is capable of coherently passing the energy

on to smaller and smaller portions of the initial length.

The value of t when $\ell \to 0$ and $v \to \infty$ is given by

$$\ell_0 = \frac{3}{2} v_0 t \quad \text{or} \quad t = \frac{2}{3} \frac{\ell_0}{v_0} .$$

We could expect these relationships to hold quantitatively for a length ℓ as small as R, where R is of the order of the diameter of the cord. If $\ell_0 = 5$ m and $R \cong \ell_{min} = 0.5$ cm $= 5 \times 10^{-3}$ m, the effective

$$v = v_{max} = v \sqrt{\ell_0/R} = v_0 \sqrt{10^3} = 30 \, v_0 .$$

The force F that must be exerted on the handle is given by $\rho v^2 + \rho\ell(dv/dt)$; it increases, theoretically, without limit as $\ell \to 0$.

Suppose that now we taper the whip as shown in Fig. A-I-2 so that its cross sectional radius is given by

$$r = r_0 + (r_1 - r_0) \frac{x}{\ell_0}$$

If ρ' is the mass density per cubic centimeter, the mass of length ℓ is

$$\rho' \int_{x=0}^{x=\ell} \pi r^2 \, dx.$$

If r_0 is taken to be approximately equal to zero,

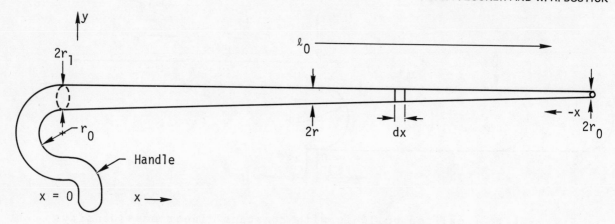

Fig. A-I-2. Tapered bullwhip.

$$r = r_1 \frac{\ell}{\ell_0} .$$

Then the mass of length ℓ is

$$\rho' \int_{x=0}^{x=\ell} r_1^2 \frac{\ell^2}{\ell_0^2} \, dx$$

$$= \frac{\rho' \pi r_1^2}{3} \frac{\ell^3}{\ell_0^2} .$$

For energy conservation we have

$$\frac{1}{2} \frac{\rho' \pi r_1^2}{3} \ell_0 v_0^2 = \frac{1}{2} \frac{\rho' \pi r_1^2}{3} \frac{\ell^3}{\ell_0^2} v^2$$

or

$$v^2 = \frac{\ell_0^3}{\ell^3} v_0^2$$

and $v = \left(\frac{\ell_0}{\ell}\right)^{3/2} v_0 = -\frac{d\ell}{dt} ,$

$$-\int_{\ell_0}^{\ell} \ell^{3/2} \, d\ell = \ell_0^{3/2} v_0 \int_0^t dt$$

$$\frac{2}{5} \ell_0^{5/2} - \frac{2}{5} \ell^{5/2} = \ell_0^{3/2} v_0 t$$

$$\ell = \left[\ell_0^{3/2} \left(\ell_0 - \frac{5}{2} v_0 t,\right) \right]^{5/2} ,$$

and

$$v = v_0 \left(\frac{\ell_0}{\ell}\right)^{3/2}$$

$$= v_0 \frac{\ell_0^{3/2}}{\left\{ \left[\ell_0^{3/2} \left(\ell_0 - \frac{5}{2} v_0 t\right)\right]^{2/5} \right\}^{3/2}}$$

Here $v \to \infty$ and $\ell \to 0$ at $\ell_0 = 5/2 \, v_0 t$, or $t = (2/5)(\ell_0/v_0)$.

Notice that the radius of curvature, which starts out as r_1 in the beginning where the whip is fat, becomes comparable to r_0 toward the end of the travel. If $\ell_0 = 5$ m and the effective final value of is 5×10^{-3} m, $v = v_0 (10^3)^{3/2} \cong 3.3 \times 10^4 v_0$. Thus with fairly modest values of v_0, v can attain values in excess of sound speed and the sonic boom can be heard in the crack of the whip. The increase of kinetic energy per atom of the whip is in the ratio $(v^2/v_0^2) \cong 10^9$, which is a phenomenal theoretical figure. In practice the increase is very likely more modest (probably 10^7 to 10^8). The volume compression of the energy goes as $(\ell_0/\ell)^3 = (10^3)^3 = 10^9$ and the reduction in the de Broglie wavelength $\lambda = h/(mv)$ of the individual atoms of the end of the whip is by a factor of $\sim 3 \times 10^4$.

For a stiff whip like a buggy whip the flexural wave on the whip steepens toward the tip: This situation is a flexural wave on a rod of varying thickness where the wave velocity decreases toward the tip.

APPENDIX II

ENERGY STORAGE IN A RELATIVISTIC ELECTRON BEAM

It is instructive to consider the relationships involved in the energy storage in a relativistic electron beam. (This treatment follows, to a certain extent, that of Linhart.[19]) A relativistic electron beam of radius a is projected at a speed of $v_e = \beta c$ through the center of a conducting cylinder (return path) of radius b as shown in Fig. A-II-1. All velocities, numbers and charge densities are stated in the laboratory frame of reference. The electron density is n_e/m^3 and the positive ion density is n_i/m^3. The extent of the space charge neutralization of the electron beam by the positive ions is characterized by the quantity ϕ where $n_i = \phi n_e$. The net charge density ρ in coulombs/m^3 is

$$-\rho = en_e - en_i = en_e(1 - \phi) = \rho_e - \rho_i.$$

Using Gauss's theorem

$$\int \epsilon_0 E \, dA = \int \rho \, dv$$

$$= \epsilon_0 E \, 2\pi r = \rho \pi r^2 \qquad 0 < r < a$$

$$= \epsilon_0 E \, 2\pi r = \rho \pi a^2 \qquad a < r < b$$

$$E \begin{cases} = \dfrac{\rho r}{2\epsilon_0} & 0 < r < a \\[2mm] = \dfrac{\rho a^2}{2\epsilon_0 r} & a < r < b \end{cases}$$

$$E_a = \frac{\rho a}{2\epsilon_0} = \frac{en_e(1 - \phi)a}{2\epsilon_0}$$

$$E \begin{cases} = E_a \dfrac{r}{a} & 0 < r < a \\[2mm] = E_a \dfrac{a}{r} & a < r < b \end{cases}$$

The electric energy per unit length W_e is

$$W_e = \frac{\epsilon_0}{2} \int E^2 dV = \frac{\epsilon_0}{2} E_a^2 \left[\int_0^a \frac{r^2}{a^2} 2\pi r \, dr + \int_a^b \frac{a^2}{r^2} 2\pi r \, dr \right].$$

$$W_e = \epsilon_0 E_a^2 \pi \left[\frac{r^4}{4a^2} \Big|_0^a + a^2 \ell n \, r \Big|_a^b \right] = \frac{\epsilon_0 E_a^2 \, a^2}{4} \left[1 + 4 \, \ell n \, \frac{b}{a} \right].$$

For the magnetic field associated with the beam,

$$H_\theta \, 2\pi r = \int_A J \, dA \quad \begin{aligned} &= J\pi r^2 \qquad 0 < r < a \\ &= J\pi a^2 \qquad a < r < b \end{aligned}$$

$$H_\theta \quad \begin{aligned} &= \frac{Jr}{2} = H_a \frac{r}{a} \quad 0 < r < a \\ &= \frac{Ja^2}{2r} = H_a \frac{a}{r} \quad a < r < b \end{aligned} \qquad H_a = \frac{Ja}{2} = \frac{\rho v a}{2} = \frac{e n_e \beta c a}{2}$$

$$W_m = \int_V \frac{1}{2} \mu_0 H^2 \, dV = \frac{\mu_0}{2} H_a^2 \int_0^a \left(\frac{r}{a}\right)^2 2\pi r \, dr + \int_a^b \left(\frac{a}{r}\right)^2 2\pi r \, dr$$

$$W_m = \mu_0 \pi H_a^2 \left[\frac{r^4}{4a^2} \bigg|_0^a + a^2 \ln r \bigg|_a^b \right] = \frac{\mu_0 \pi a^2 H_a^2}{4} \left[1 + 4 \ln \frac{b}{a} \right].$$

The power is

$$P_{em} = \int_A (\overline{E} \times \overline{H}) \, dA = E_a H_a \left[\int_0^a \frac{r^2}{a^2} 2\pi r \, dr + \int_a^b \frac{a^2}{r^2} 2\pi r \, dr \right]$$

$$= 2\pi a^2 E_a H_a \left[\frac{r^4}{4a^4} \bigg|_0^a + \ln r \bigg|_a^b \right] = \frac{\pi a^2}{2} E_a H_a \left[1 + 4 \ln \frac{b}{a} \right].$$

In a more compact manner we have

$$W_e = G (1 - \phi)^2 (\pi a^2)$$

$$W_m = G \beta^2 (\pi a^2)$$

$$P_{em} = 2G(1 - \phi)\beta c,$$

where

$$G = \frac{(e n_e a)^2 \left(1 + 4 \ln \frac{b}{a} \right)}{16 \varepsilon_0}.$$

The kinetic energy per unit length is

$$W_k = \left(mc^2 - m_0 c^2 \right) n_e \pi a^2 = m_0 c^2 (\gamma - 1) n_e \pi a^2.$$

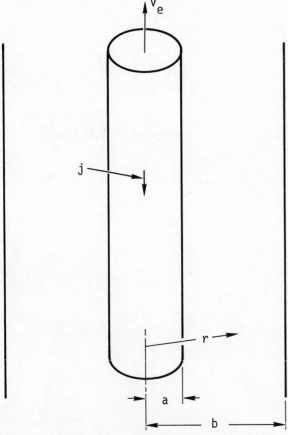

Fig. A-II-1. Electron beam as the center conductor in a coaxial geometry.

The kinetic power flow is

$$P_k = m_0 c^2 (\gamma - 1) n_e \beta c (\pi a^2).$$

The ratio of electromagnetic to kinetic energy in the volume is given by

$$\frac{W_m + W_e}{W_k} = \frac{W_m}{W_k} \left(1 + \frac{W_e}{W_m} \right)$$

$$= \frac{\nu}{\gamma - 1} \left[\beta^2 + (1 - \phi)^2 \right]$$

$$\times \left[\frac{1}{4} + \ln \frac{b}{a} \right]$$

where

$$\nu = \left(n_e \pi a^2 \right) \left(\frac{e^2}{4 \pi \varepsilon_0 m_0 c^2} \right)$$

Here ν is the product of the number of particles per unit beam length by the classical radius of the electron. The ratio $(W_m + W_e)/W_k$ is loosely termed the ν/γ of the electron beam.

The total power is

$$VI = P_{em} + P_k = \pi a^2 \left\{ \frac{E_a H_a}{2} \left[1 + 4 \ln \frac{b}{a} \right] + m_0 c^2 (\gamma - 1) n_e \beta c \right\}$$

The current

$$I = J \pi a^2 = e n_e \beta c \pi a^2$$

$$VI = I^2 R$$

$$R = \frac{\pi a^2 \left\{ 2G (1 - \phi) \beta c \left(1 + 4 \ln \frac{b}{a} \right) + m_0 c^2 (\gamma - 1) n_e \beta c \right\}}{e^2 n_e^2 (\beta c)^2 (\pi a^2)^2}$$

Simplifying, we have

$$R = \frac{60}{\beta} \left[(1 - \phi) \left(\frac{1}{4} + \ln \frac{b}{a} \right) + \frac{\gamma - 1}{2 \nu} \right] \text{ ohms,}$$

where

$$\sqrt{\frac{\mu_0}{\varepsilon_0}} = 120 \ \pi .$$

The first term accounts for the electromagnetic voltage drop and the second accounts for the kinetic voltage drop. If the electromagnetic energy flow is reflected then we can neglect it. Thus,

$$\frac{R}{30} = \frac{\gamma - 1}{\nu\beta} .$$

Let us define a normalized voltage α:

$$\alpha = \frac{V}{V_0} ,$$

where

$$V_0 = \frac{m_0 c^2}{e}$$

so that

$$\alpha = \gamma - 1 \qquad \text{and} \qquad \beta = \frac{\sqrt{\alpha^2 + 2\alpha}}{1 + \alpha} .$$

Thus,

$$\frac{R}{30} = \frac{\alpha (1 + \alpha)}{\nu \sqrt{\alpha^2 + 2d}} = \frac{\gamma}{\nu} \cdot \frac{1}{\sqrt{\frac{\alpha + 2}{\alpha}}}$$

or

$$\frac{\nu}{\gamma} = \frac{30}{R} \sqrt{\frac{\alpha}{2 + \alpha}}$$

For radial equilibrium $F_e = eE$, the electrical force outward on an electron, must equal $F_m = evB$, the containing magnetic force inward at any radius r:

$$F_e = e \ \frac{e \ n_e (1 - \phi) a}{2\varepsilon_0} = - F_m = e \ (\beta c) \ (\mu_0 e \ n_e \beta c a) .$$

Simplifying, we have

$$\phi = 1 - \beta^2 \qquad \text{or} \qquad \frac{n_e}{n_i} = \frac{1}{1 - \beta^2} = \gamma^2$$

Thus if we achieve such neutralization we have

$$\frac{W_m + W_e}{W_k} = \frac{\nu}{\gamma - 1} \left[\beta^2 + \beta^4 \right] \left[\frac{1}{4} + \ln \frac{b}{a} \right]$$

and

$$R = 60 \left[\beta \left(\frac{1}{4} + \ln \frac{b}{a} \right) + \frac{\gamma - 1}{2\nu\beta} \right] \text{ohms} .$$

REFERENCES

1. R.P. Feynman Lectures on Physics, (Addison Wesley, Reading, Mass., 1965) Vol. III, pp. 4-7.

2. W.S. Melville, Proc. IRE 98, 187 (May 1951).

3. P.P. Johannessen, IEEE Trans. Magnetics 3, (3), 256 (1967).

4. I.G. Kateyev, Electromagnetic Shock Waves (Iliffe, London, 1966).

5. W.W. Bostick, W. Nardi, and W. Prior, J. Plasma Phys. 8, 7 (1972).

6. A.C. Scott, F.Y.F. Chu, and D.W. McLaughlin, Proc. IEEE 61, 1443 (1973).

7. H. Hasimoto, J. Fluid Mech. 51, 477 (1973).

8. W.H. Bostick, "Plasmoids," Scientific American October 1957.

9. O. Bunnemann, Phys. Rev. Letters 1, 8 (1968; 10, 285 (1963).

10. W.H. Bennett, Phys. Rev. 45, 890 (1934).

11. L.N. Kazanskii, Atomnaya Energiya 30, 27 (1971).

12. M. Friedman, Phys. Rev. Letters 31, 1107 (1973).

13. A.P. Ishkov, Sov. Phys. - Tech. Phys. 16, 471 (1971).

14. I.A. Grishaev and S.M. Shenderovich, Sov. Phys.-Tech. Phys. 17, 1871 (1973).

15. J.D. Lawson, Particle Accelerators 3, 21 (1972).

16. J.S. Luce "Experiments with collective ion acceleration in relativisitic electron beam deuterated polyethylene target interation," Proc. N.Y. Acad. Sci. Conf. Electrostatic and Electromagnetic Confinement on Plasmas and Phenomenology of Relativistic Electron Beams, in press (1975).

17. O.S.F. Zucker and J. Wyatt, Lawrence Livermore Laboratory, unpublished work (1974).

18. B. Bernstein, D.A. Hall, and H.M. Trent, "On the dynamics of a bull whip," J. Acoust. Soc. 30, 112 (1958).

19. J.G. Linhart, Nucl. Fusion 10, 211 (1970).

A MULTI-MEGAJOULE INERTIAL-INDUCTIVE
ENERGY STORAGE SYSTEM

A.E. Robson, P. Turchi, W. Lupton, M. Ury, W. Warnick
Naval Research Laboratory
Washington, D.C. (U.S.A.) 20375

ABSTRACT

An inertial-inductive energy storage system is described, made up of modules each storing, nominally, ten megajoules. The energy in each module is stored initially in two counter-rotating flywheels which are arranged inside an air-cored inductor to form a self-excited homopolar generator. The storage inductor is also the primary of an air-cored transformer; the primary circuit is broken by a pressurized SF_6 circuit breaker, commutated by a circuit containing a capacitor and a fuse. The function of the fuse is to disconnect the commutating circuit after it has performed its function and thereby minimize the size of the commutating capacitor. The output of the system is taken from the secondary of the transformer, and will be used to drive imploding liners as part of the LINUS program. The rationale of this approach is discussed and the prototype module is described.

INTRODUCTION

It has long been recognized[1,2] that an inductive energy store should be an economical alternative to a capacitor bank for pulsed-power systems in the megajoule range, and where the discharge time is not much shorter than 10^{-3} sec. The economic crossover point depends on the specific application, but probably lies between 1 and 10 megajoules[2]. Experiments in controlled fusion research are now being planned that will require several hundred megajoules of pulse energy[3,4], and although capacitor banks of this energy have been proposed[5,6], and appear technically feasible, their size, complexity and cost make it highly desirable to consider the alternative possibility of using inductive storage for these applications. However, much work needs to be done to bring inductive storage systems to the same level of development as present-day capacitor banks.

The use of superconducting coils as energy storage elements has been extensively discussed and at first sight seems attractive since they can store energy indefinitely with negligible loss. However, their attractiveness is greatly reduced by the difficulty of transferring the energy rapidly in the cryogenic environment, and realistic estimates of the cost of large systems[7] do not show the anticipated economic advantage over capacitive storage. We shall therefore only concern ourselves here with energy storage in normally conducting coils.

The principal problems of inductive storage are in the means of charging and discharging the store. For efficiency, a resistive store needs to be charged in a time short compared to its L/R time, which requires a pulsed DC power supply, and to discharge it involves opening a high-current DC circuit, which has always been a significant electrical engineering problem. The store itself, being simply a coil, presents few problems except that in addition to operating at a high magnetic field level, it must also withstand high voltage upon discharge.

The energy stored in a coil of given shape depends only on the mechanical stress level, and the cube of the linear dimension of the coil. The current level can be chosen independently to suit the characteristics of charging supply and the circuit interrupter, and the number of turns adjusted accordingly. The current requirement of the load is less important, since matching into the load can always be accomplished through a transformer.

The homopolar generator coupled to a flywheel represents an attractive means of charging an inductive store. The generator is rugged and capable of substantial overload, and if run at constant excitation has the electrical characteristics of a capacitor, enabling very efficient transfer of energy to an inductance. The principal disadvantage of the homopolar generator, in its conventional form, is its low output voltage, which necessitates high current levels to satisfy the need for rapid charging of the store. Although very high currents are within the capability of the generator, the necessity of switching the same current to discharge the store may pose serious problems.

An example of the successful application of the homopolar generator to charging an inductive store is to be found in the hot-shot wind tunnel installation at Arnold Air Force Development Center, Tullahoma, Tennessee[8]. Here two large flywheels drive four G.E. homopolar generators to charge a 200 microhenry coil with 100 megajoules at a current of 10^6 amperes. The current is diverted by mechanical switches into a fuse set in the throat of the wind tunnel: the arc formed when the fuse blows dissipates the stored energy in driving the wind tunnel. It is the particular feature of the load being also the current interrupter which simplifies the application of inductive storage in this case, and the scale of the Tullahoma installation is unique. The homopolar generator at the Australian National University, Canberra, Australia, stores 500 megajoules[9], and a proposal[15] has been made to use 20 megajoules to charge an inductive store for a laser flashlamp power supply. Work in progress at Orsay, France, involves the development of a 100MJ homopolar generator[10], which will be used in conjunction with an inductive storage system.

This paper describes a program at the Naval Research Laboratory which has as its objective the development of a versatile, compact inertial-inductive storage system, made up of modules each with a nominal 10MJ storage capacity. Each module consists of a flywheel-homopolar generator, a storage coil, a current interrupter and an output transformer. The system is being developed as part of the LINUS approach to controlled fusion[4], but has more general application to other fusion experiments requiring multi-megajoule pulsed power supplies.

THE NRL APPROACH

The decision to adopt a modular approach to inertial-inductive storage was based on a number of factors. The principal technical considerations in favor of small units are the increased ratio of surface area to volume, which reduces the problems of bearings, brushes

and heat removal, and the ability to work at higher stresses because of the reduced consequences of failure. In application, the modular approach affords considerable versatility through series-parallel combinations of modules, as has been very successfully demonstrated in capacitor banks. Finally, there is the practical consideration that having developed a basic module, very little further development is needed to construct systems of almost arbitrary size, whereas the scaling of electro-mechanical devices from small prototypes to large systems frequently involves qualitatively different problems which require further development work. Against these arguments must be set the consideration that the unit cost of energy is likely to be greater for smaller machines, but the economic equation is a complex one and the optimum size of a module will depend on many factors related to the particular application.

The basis of the NRL design is a module of 10MJ stored energy, and a current level of 100kA. These figures were chosen so that the module could be switched by a single circuit breaker of the largest available capacity. As will become apparent, the characteristics of the switch dominate most of the design decisions, and it is clear that the development of improved switching is the largest single requirement in the whole area of inductive storage.

The restriction on the current that can be switched sets a lower limit to the voltage of the charging system needed to ensure efficient transfer of energy from the inertial store. Unless the store inductor contains an inordinate amount of conductor (in which case its cost will dominate the cost of the system), the L/R time of an inductor, even at the level of 10MJ, is only a few seconds. The charging time must therefore be a fraction of a second, and this can only be achieved if the homopolar generator develops several hundred volts. This voltage is an order of of magnitude greater than developed by homopolar generators in conventional form, which are characteristically low voltage, high current machines.

The c.m.f. developed across a disc of radius r cm rotating at ω radians/sec in a field of B gauss is given by

$$V = \tfrac{1}{2}Br^2\omega \cdot 10^{-8} = \tfrac{1}{2}Brv_p \cdot 10^{-8} \text{ volts}$$

where v_p is the peripheral velocity of the disc. In conventional homopolar generators, v_p is restricted to about 50 meters/sec in order that the brushes may be operated continuously, while B is limited to about 18,000 gauss by the need to use iron yoke magnets for good power transfer efficiency. Thus a disc of 50cm radius develops only about 20 volts. These restrictions can be removed in a homopolar generator designed specifically for pulse-charging an inductive store. The voltage can be increased by an order of magnitude if we spin the rotor to the limit set by its mechanical strength, and a further factor of ~ 2 can be obtained by using a larger excitation field generated by an air-cored coil. It is then a short logical step to make the generator rotor serve also as the inertial energy storage element, and to use the magnetic field of the inductive store as the excitation field, so the whole inertial-inductive system is simply a self-excited homopolar generator.

The problem of making electrical contact at the periphery of high-speed wheels is a significant one, the heating effect of brush friction being the limiting factor. For a pulsed machine however, the brush need only be in contact with the rotor during the discharge of the generator, and by using brushes of advanced design it should be possible to keep the heating within acceptable limits in the 10MJ module.

The use of high-speed wheels and air-cored coils makes it impractical to operate the homopolar generator as its own motor, as is down in the Canberra[9] and Texas[21] designs. However, it is a simple matter to drive the wheels with independent motors, and this is generally a more efficient arrangement.

The design of the NRL module has evolved over the past year and takes into account the many requirements that must be satisfied simultaneously by the motors, wheels, bearings, coils, and brushes. In what follows, we shall describe the system in some detail; we will not go into the manner in which all the design decisions were made, but merely point out that the elements of the system depend on each other in a rather complex fashion and we do not claim that the present design is fully optimized. Experience with the present prototype module should, however, enable us to design a production module which will form the building block for large inertial-inductive storage systems.

THE NRL 10 MEGAJOULE MODULE

The layout of the 10 megajoule module is given in Figure 1. The module has two identical counter-rotating energy storage wheels each driven by a small hydraulic motor. The wheels run in ball bearings and are contained in a massive fiberglass housing, which is evacuated to reduce windage loss. The wheels, motors and housing form a self-contained "inertial package" which is set in the center of the energy storage coil. The latter consists of a single layer of copper strip, wound on edge on a diameter of twice the wheel diameter. The coil is cooled to liquid nitrogen temperature.

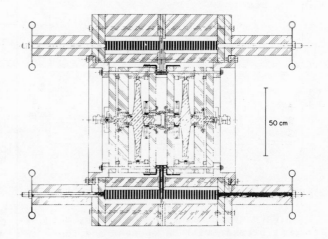

50 cm

Fig. 1. Layout of NRL 10MJ module.

Pneumatically operated brushes make contact with each wheel at the periphery and close to the shaft. The inner brushes connect the two wheels together, while the outer brushes connect to a series feed point at the center of the coil. The outer connections to the coil are taken through high-voltage bushings at each end and thence to the circuit breaker and the load. For the initial tests, the load will simply be a large resistor.

In operation, the wheels will be spun up to full speed over a period of about 15 minutes. A small initial magnetic field will be created by passing a few hundred amperes through the coil from a lead-acid battery. To discharge the generator, the brushes will be lowered, and as the coil current rises and the excitation field increases, the wheels slow down at an increasing rate until the energy originally stored as kinetic energy of the wheels is transformed into magnetic energy of the inductor. The counter-rotating wheels ensure that the angular momentum is always zero and that there is no torque on the frame of the machine during discharge.

The equations describing the discharge phase are given in the Appendix. The current and wheel speed as functions of time for our system are given in Figure 2. Because this is a self-excited generator, the waveform is non-sinusoidal, but the main phase is roughly equivalent to a sinewave of quarter-period 0.4 second.

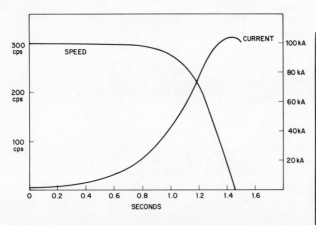

Fig. 2. Wheel speed and current vs. time.

When the current reaches peak, the circuit breaker will be opened, and the commutating circuit fired to bring the current in the breaker to zero. As the breaker recovers, the energy in the inductor will be transferred into the resistive load. The transfer L/R time should be approximately 500 microseconds.

For subsequent application, the output of the machine will be taken from a single turn secondary winding closely coupled to the field of the coil. In this way the current will be transformed from the 100kA of the primary to the megampere level needed to power a LINUS compression coil. However, since the load represented by an imploding liner is effectively resistive (dL/dt), the initial test arrangement using a load resistor connected across the primary will simulate the liner load with fair accuracy.

We shall now go on to discuss the separate elements of the system in greater detail.

THE WHEELS

The wheels are designed to operate at very high speed to maximize the energy stored in the available volume. The specific energy E stored in a wheel is given by

$$E = k\sigma 10^{-7}/\rho \text{ joules/gm } (=\text{megajoules/tonne})$$

where σ is the working stress in the material in dynes/cm^2, ρ the density in gm/cm^3 and k is a factor which depends only upon the shape of the wheel.

Table 1. Energy storage in flywheels.

SPECIFIC ENERGY $E = k\varsigma/\rho$	ς = STRESS ρ = DENSITY	k-FACTOR
THIN RING		0.5
SOLID DISC		0.6
DISC WITH HOLE FOR SHAFT		0.3
CONSTANT STRESS WHEEL $d = d_0 e^{-ar^2}, r \rightarrow \infty$		1.0
MODIFIED CONSTANT STRESS WHEEL (NRL)		0.9

The k-factor for wheels of several shapes is given in Table 1. It can be seen that by optimizing the shape a factor of about 3 can be gained in the specific energy, compared to the simplest case of a disc with a hole for the shaft. The optimum shape is the uniform stress wheel, but since for practical purposes the wheel must have a finite radius, a nearly uniform stress pattern may be obtained by adding a rim to a truncated exponential wheel. This is the principle of the NRL design.

The shape of the NRL wheel was developed with the aid of the 2-dimensional NASTRAN code to achieve uniform stress over the main body of the wheel, and a k factor of 0.9. The resulting design is shown in Figure 3. The wheels are made from beryllium-copper forgings and the shafts are bolted to the main disc to avoid a central hole. The material, Berylco Alloy 25, was chosen on account of its high strength and good electrical conductivity. Alloy steel would have provided greater strength but was rejected because ferromagnetic materials lead to serious problems with end thrust on bearings. The minimum yield strength of the beryllium-copper is 145,000 psi (102kg/mm^2); the working stress is 85,000 psi (60kg/mm^2) at the design speed of 18,000 rpm.

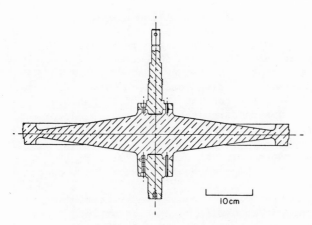

10cm

Fig. 3. NRL 5MJ flywheel.

Each wheel weighs 79kg and stores 5.1 megajoules, a specific energy of 64.5MJ/tonne. This is an order of magnitude greater than is usual in 'conventional' large steel flywheels, such as the Canberra and Tullahoma wheels, where the specific energy is about 7MJ/tonne. The improvement is obtained through the use of small, shaped wheels operated at high stress levels: still higher performance (85MJ/tonne) has been obtained in steel flywheels designed for vehicle propulsion systems[11].

Each of the NRL wheels is driven by a small hydraulic motor, developing 10 hp and occupying a volume of only 350cm^3. This method of driving the wheels was chosen because, being non-electrical, it does not interact with the electrical operation of the system, and being highly compact, it allows for the stacking of several inertial packages inside a single long storage coil, which is one arrangement being considered for large systems.

THE BRUSHES

The peripheral velocity of the wheel is 540 meters/sec, so the brushes represent a significant technological problem. Although successful operation of conventional copper-graphite brushes has been reported under pulsed conditions on wheels with peripheral speeds up to 150 meters/sec[12], these brushes do not seem to be applicable in our case due to the substantially higher rubbing velocity, which will lead to excessive temperature of the brush-wheel interface due to frictional dissipation.

The power generated by friction at the interface is $P = \mu pv \cdot 10^{-7}$ watts/cm^2, where μ is the coefficient of friction, p is the contact pressure in dynes/cm^2 and v the rubbing velocity in cm/sec. For copper-graphite brushes, $p = 3.45 \times 10^6$ dynes/cm^2 (50 psi), $\mu = 0.2$, so at our velocity $P = \sim 3,700$ watts/cm^2.

The heat transfer in a graphite brush is a complex problem because μ and the thermal properties vary with temperature. If we assume, for simplicity, that the heat input is constant during an equivalent slowing down time τ of the wheel, and that we may treat the brush as a semi-infinite solid on this time scale, the maximum surface temperature θ is given by[13] $\theta = 2P(\varkappa\tau/\pi)^{\frac{1}{2}}/4.2K$ where K is the thermal conductivity, and $\varkappa$ the thermal diffusivity. With the assumption that a copper-graphite brush has 90% of the K and $\varkappa$ of solid copper, and that half the frictional dissipation goes into the wheel, we find that for $\tau \sim 1$ second, $\theta \sim 600^\circ$C. Although approximate, this result indicates the seriousness of frictional heating at very high wheel speeds.

Fortunately, recent developments in brush technology offer a solution to this problem. Brushes made of silver-plated carbon fibers have been developed by a British company[15] and have the outstanding characteristic that they can be operated at a pressure of only 0.5 psi compared to 50 psi required for copper-graphite brushes[12]. Other things being equal, this reduces the temperature rise by two orders of magnitude. Accurate data on the thermal properties of the composite material are not available, but approximate calculations indicate that the temperature rise due to friction will be less than 50°C.

The maximum current densities in the inner and outer brushes are 1480 A/cm^2 and 900 A/cm^2 respectively; the equivalent square wave pulse time is 0.28 sec. This duty is significantly less severe than encountered by the solid brushes of the Canberra homopolar generator[12], and we estimate that the temperature rise due to electrical dissipation will be comparable with that due to friction. Although carbon fiber brushes have not yet been tested on very high speed wheels such as ours, there seems to be no fundamental reason why they should not prove satisfactory for our application.

THE COIL

The coil consists of a single layer of 42 turns of copper strip wound on edge. The coil is clamped between fiberglass end plates and surrounded by a plastic foam container filled with liquid nitrogen. The internal diameter is one meter, the size of the conductor is 7.5 x 1.9 cm, and the total weight of copper is 1.7 tonnes. The inductance of the coil is 1.5 millihenries, and the DC resistance, 300 micro-ohms. The magnetic field at the center of the coil is 40 kilogauss for a current of 100kA; the maximum stress in the copper is then 11,500 psi (8kg/mm^2).

The L/R time of the coil is 5.0 seconds, which from the analysis given in the Appendix, should lead to an efficiency of energy transfer from the wheel to the coil of 90%. However, when the skin effect in the coil is taken into account, the effective resistance is about 800 micro-ohms, which reduces the efficiency to 75%. The efficiency could be restored to 90% if the same quantity of copper were wound in a more complex fashion. However, the present coil arrangement was chosen for simplicity of construction, ease of cooling, and uniform voltage grading along its length when the circuit is interrupted. In return for these advantages it was decided that the inefficiency could be tolerated in the prototype module, but a more efficient coil is being designed and will be tested at a later stage in the program.

The efficiency of the system is also affected by eddy currents induced in the beryllium-copper rotors. The characteristic time for magnetic flux penetration into a rotor is estimated to be ~ 35ms; this is short compared to the charging time of the coil but long compared to the discharge time. Thus flux which has penetrated the rotor during charge cannot be extracted during discharge. The amount of flux involved has been determined by measuring the high frequency inductance of a scale-model coil with and without the rotors; the presence of the rotors reduces the inductance by 10%. Taking into account the ratios of the appropriate time scales, we estimate that eddy current losses will absorb ~ 2% of the energy on charge and ~ 10% on discharge.

INTERRUPTION OF THE CIRCUIT

When the current has reached its peak value of 100kA, the circuit will be interrupted and the current diverted into a 2-ohm resistive load. The circuit interrupter consists of a Westinghouse SFV-series circuit breaker and an appropriate commutating circuit. The breaker[20] operates in SF$_6$ at 240 psi and has an AC rating of 63kA rms at 145kV rms; it has a voltage recovery to 200kV, 100 μsec after current zero in a 60Hz circuit. To interrupt a DC circuit, an artificial current zero will be created by the commutating circuit shown in Fig. 4. This consists of a capacitor, a closing switch, an optional crowbar switch, and a series fuse.

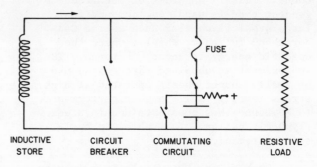

Fig. 4. Energy transfer circuit.

Taking a value of 2 microhenries for the inductance of the breaker circuit, the energy required to bring the current to zero is about 10 kilojoules. The time to bring the current to zero can be varied by varying the size of the capacitor, either directly or effectively through a transformer. For example, for a time of 300μsec a capacitor of 18,000μf at 1,000V would be needed. If however the capacitor remained in the circuit during the transfer of energy from the coil to the load resistor, it would have to withstand the full transfer voltage of 200kV and be capable of storing 2 mega-joules. It is therefore necessary to devise a means of disconnecting the commutating capacitor after it has per-formed its function of bringing the current to zero. This will be accomplished by means of the fuse.

The fuse will consist of a metallic foil exploding under water. Considerable experience has been accumulated on the behavior of fuses under pulsed conditions[16], including work at Braunschweig[17] and NRL[18] on fuses under water. We shall not go into details here, but the fuse we propose to use should conduct for several hundred microseconds and then interrupt in about 100 microseconds, generating a voltage of 200kV. The characteristics of the fuse determine the rate at which voltage appears across the breaker after current zero, and some experimentation will be necessary to match the characteristics of these two elements.

Analysis of circuits without a means of disconnecting the capacitor[19] leads to the requirement that the capacitor should store ~ 20% of the inductive energy, and the cost of this capacitor dominates the cost of the system. The introduction of the fuse leads to a spectacular reduc-tion in the size of the commutating capac-itor, and might even allow it to be elim-inated entirely.

The circuit breaker-fuse combination offers in principle a very cheap method of switching inductively stored energy, but suffers from being very cumbersome and clearly only suitable for low repetition rate systems. Fortunately, many fusion experiments fall into this category and it seems likely that the system described here would be useful for such experiments where minimum capital cost is a dominant requirement. The application of inductive storage to more advanced systems needing greater efficiency and high repetition rates will require the development of improved switching techniques, which will doubtless be based on different principles.

THE EXPERIMENTAL PROGRAM

The objective of the experimental program is to construct and operate the entire system that we have described in this paper. In doing so, we shall verify the assumptions of our design, and obtain experience in the interaction of the different elements of the system upon each other. As we have noted earlier, this is the dominant consideration of the design of systems such as these.

When the system is operating, we shall use it as a test bed for the develop-ment of improved switching methods; the availability of the module will enable new ideas to be tested at the megajoule level.

Finally, we shall design a production module, incorporating any changes suggest-ed by our experience with the prototype, and paying particular attention to improv-ing the efficiency and reducing the unit cost. This module will form the building block of systems involving several hundred megajoules of stored energy.

APPENDIX

DISCHARGE OF SELF-EXCITED HOMOPOLAR GENER-ATOR*

The generator consists of two rotors, each of radius r and moment of inertia $M/2$, rotating with angular velocity ω inside a coil which produces a magnetic flux $\pi\alpha$ through each rotor for unit current in the coil. The total emf gener-ated is then

$$V = \alpha I \omega \qquad (1)$$

If the coil has inductance L and resist-ance R, the electrical circuit equation is

$$L\frac{dI}{dt} + I\,(R-\alpha\omega) = 0 \qquad (2)$$

The torque on a single rotor is

$$T = \alpha I^2/2 \qquad (3)$$

The mechanical equation is then

$$M\frac{d\omega}{dt} = -\alpha I^2 \qquad (4)$$

From Eq. (2) we see that a current rise ($dI/dt > 0$) can be obtained only if $\omega_0 > \omega_c$, where

$$\omega_c = R/\alpha \qquad (5)$$

---

*This treatment follows closely the analysis of Knoepfel[22].

is the critical angular frequency of the system. After eliminating t from Eqs. (2), (4) and integrating, we find

$$I^2 + \frac{M}{L}(\omega - \omega_c)^2 = A; \qquad (6)$$

where A is an integration constant, which is determined by the initial condition $\omega = \omega_0$, $I = I_0$ at $t = 0$. (Note that to start the generator there must be an initial current I_0). From Eqs. (2), (4), (6), we calculate by integration

$$I = \frac{\sqrt{A}}{\cosh\{\alpha(t-t_0)\sqrt{(A/ML)}\}} \qquad (7)$$

with

$$t_0 = \frac{\sqrt{(ML)}}{\alpha\sqrt{A}}\ln\left(\frac{\sqrt{A}}{I_0} + \sqrt{\left\{\frac{A}{I_0^2} - 1\right\}}\right). \qquad (8)$$

Maximum current I_m is obtained at $t = t_0$ when the hyperbolic cosine is equal to one. Then, assuming the initial current to be small, so that $A \approx M(\omega_0 - \omega_c)^2/L$, we find

$$I_m \approx (\omega_0 - \omega_c)\sqrt{\frac{M}{L}}, \qquad (9)$$

$$t_0 \approx \frac{L/R}{(\omega_0/\omega_c)-1}\ln\left(2\frac{I_m}{I_0}\right). \qquad (10)$$

More significant perhaps is the time τ for the current to go from $I_m/2$ to I_m. This is given by

$$\tau = (\cosh^{-1}2)\cdot(ML/A)^{\frac{1}{2}}/\alpha = 1\cdot31\ L/\alpha(\omega_0-\omega_c) \qquad (11)$$

The system efficiency (transformation of initial kinetic energy into magnetic energy in the inductance L) is given by

$$\eta = \frac{LI_m^2}{M\omega_0^2} \approx \left(1 - \frac{\omega_c}{\omega_0}\right)^2. \qquad (12)$$

ACKNOWLEDGEMENT

The authors are pleased to acknowledge the valuable contributions of Messrs. T. J. O'Connell and R. E. Lanham to the mechanical design of the homopolar generator.

REFERENCES

1. H.C. Earley and R.C. Walker. Economics of Multimillion-Joule Inductive Energy Storage. A.I.E.E. Communications and Electronics 31, 320 (1957).

2. R. Carruthers. The Storage and Transfer of Energy. Proc. of the International Conference on High Magnetic Fields, M.I.T., Nov. 1-4, 1961, p. 307.

3. F.L. Ribe, et. al., Proposed Experiments on Heating, Staging and Stabilization of Theta Pinches. LA-5026-P, February 1973.

4. A.E. Robson. LINUS--An Approach to Controlled Fusion Through the Use of Megagauss Magnetic Fields. Report of NRL Progress, June 1973, p. 7.

5. T.W. Hunt and T.E. James. The Feasibility of Large Capacitor Banks Storing about 100MJ. Proc. of the 7th Symposium on Fusion Technology, Grenoble, 24-27 Oct. 1972, p. 461.

6. E.L. Kemp. The Study of Capacitive Energy Storage for a Theta-Pinch PTR, Proc. of the Fifth Symposium on Engineering Problems of Fusion Research, Princeton Univ., Nov. 5-9, 1973, p. 303.

7. R.A. Krakowski, F.L. Ribe, T.A. Coultas and A.J. Hatch. An Engineering Design Study of a Reference Theta Pinch Reactor. LA-5336, March 1974.

8. J.N. Patterson. Inductive Power Supply for a 100-in. Hotshot Wind Tunnel. ARO, Inc. AECD-TR-66-260, March 1967.

9. J.W. Blamey, et. al. The Large Homopolar Generator at Canberra. Nature 195, 113 (1962).

10. N. Chaboseau, R. Guillet, J. Lucidarme and C. Rioux. Design and Realization of 100MJ Pulsed Unipolar Generator. Proc. of the International Conference on Magnet Technology, Hamburg, 1970.

11. R.R. Gilbert, et. al., Flywheel Feasibility Study and Demonstration, Lockheed Missiles and Space Company. LMSC-DO07915, April 1971.

12. R.A. Marshall. Design of Brush Gear for High Current Pulses and High Rubbing Velocities. IEEE Trans. on Power Apparatus and Systems 85, 1177 (1966).

13. H.S. Carslaw and J.C. Jaeger. Conduction of Heat in Solids, p. 75, Oxford 1959.

14. I.R. McNab and G.A. Wilkin. Carbon Fibre Brushes for Superconducting Machines. IEE Journal, Electronics and Power, January 1972, p. 8.

15. E.K. Inall. A Proposal for the Construction and Operation of an Inductive Store for 20MJ. J. Physics E; Sci. Instr., 5, 679 (1972).

16. C. Maisonnier, J.G. Linhart and C. Gourlan. Rapid Transfer of Magnetic Energy by Means of Exploding Foils. Rev. Sci. Instr. 37, 1380 (1966).

17. U. Braunsberger, J. Salge, U. Schwarz. Circuit Breaker for Power Amplification in Poloidal Field Circuits. 8th Symposium on Fusion Technology, Noordwijkerhout, Netherlands, 17-21 June 1974.

18. M. Friedman and M. Ury. High Voltage Pulse Generator using a Novel Opening Switch for the Interruption of Current Flowing through an Inductor. Bull. Am. Phys. Soc. 18, 660 (1973).

19. R.B. McCann, R.E. Rowberg and H.H. Woodson. Engineering Feasibility Studies of Inductive Energy Storage for Controlled Fusion Devices. Proc. of the Fifth Symposium on Engineering Problems of Fusion Research, Princeton Univ., Nov. 5-9, 1973, p. 454.

20. R.E. Kane, R.C. Newsome, C.L. Wagner. New generation of oil-less circuit breakers 115kV-345kV. IEEE Trans. on Power Apparatus and Systems, PAS-90, 628 (1971).

21. H.G. Rylander, R.E. Rowberg, K.M. Tolk, W.F. Weldon and H.H. Woodson. Investigation of the Homopolar Motor-Generator as a Power Supply for Controlled Fusion Experiments. Proc. 5th Symposium on Engineering Problems of Fusion Research, Princeton University, Mar. 5-9, 1973, p. 447.

22. H. Knoepfel. Pulsed High Magnetic Fields, p. 151, N. Holland 1970.

MAGNETICALLY DRIVEN METAL LINERS FOR PLASMA COMPRESSION*

James W. Shearer and William C. Condit
University of California, Lawrence Livermore Laboratory
P. O. Box 808, Livermore, CA. 94550, USA

ABSTRACT

Adiabatic compression of a plasma can be accomplished by the implosion of a metal liner driven by an outer magnetic field. An inner magnetic field insulates the plasma from the metal. Analytic and computer studies have been done in cylindrical geometry for a copper liner. The minimum input energy per pulse E_{min} is found as a function of final plasma density, volume fill factor, liner compressibility, liner thickness, and the desired value of fusion energy multiplication α. We find that at optimum deuterium-tritium reaction temperatures (10-20 keV) $E_{min} \propto \alpha^3/n_f$, where n_f is the final plasma density. For given values of E_{min} and n_f, one can then determine the requirements for the outer magnetic field power supply that drives the liner implosion. Two possible cases are considered: an axial (B_z) magnetic field with azimuthal eddy current in the liner and an azimuthal (B_θ) magnetic field. The latter is shown to be inherently more efficient for transferring energy from the outer magnetic field to the liner. Finally, certain practical problems of liner formation and replacement in a fusion reactor vessel are briefly discussed.

I. INTRODUCTION

A well-known fusion-reactor proposal is the compression of deuterium-tritium plasma to thermonuclear temperatures using an imploding metallic liner.[1-3] Today the largest liner-study programs are located in the USSR[4] and at the Naval Research Laboratory, USA.[5]

In this paper we formulate an approximate analytical model of liner compression in cylindrical geometry and apply the results to reactor applications. The emphasis will be on the imploding metal liner itself as a means of energy compression for fusion.

Figures 1 and 2 show the essential features of the cylindrical liner system under consideration. A high-energy outer magnetic field implodes the metal liner, which surrounds a DT plasma that is insulated from the liner by a magnetic field. The outer magnetic field energy is converted to kinetic energy of the liner and is then converted again to plasma and field energy in the interior. A lithium blanket surrounds the reactor for two purposes: to capture the neutron reaction energy and to regenerate tritium. The overall objective

of the system is to heat the plasma to fusion temperatures and contain it long enough for sufficient reactions to occur to provide a net energy gain.

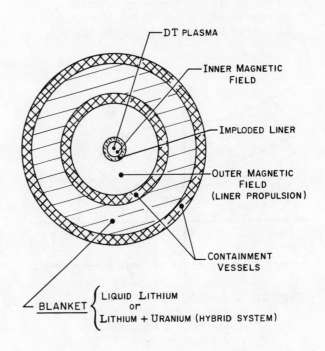

Fig. 1. Contained liner concept.

---

*This work performed under the auspices of the United States Atomic Energy Commission.

In our version of the scheme, the 14 MeV neutrons from the reaction penetrate the liner and the container to reach the lithium blanket, where their energy is absorbed. This is in contrast to other proposals, which assume that the lithium liner is thick enough to stop the neutrons before they reach the wall.[2] When the thick lithium liner is used, it protects the container wall from damage by the 14 MeV neutron flux. Unfortunately, the density of lithium is so low that it provides poor inertial containment for the reacting plasma. It may be possible to overcome this objection by alloying lithium with a heavier metal.[6]

The model developed here is more suitable for our thin, heavy liner concept, although it may be applicable to some composite lithium-heavy metal liners.

II. NONNUCLEAR ENERGIES AT TURNAROUND

At turnaround — when the liner radius reaches its minimum radius r_0 (see Fig. 2) — the energy has three major components: the plasma energy, the axial magnetic field energy, and the compressional energy of the liner. The kinetic energy remaining in the liner is

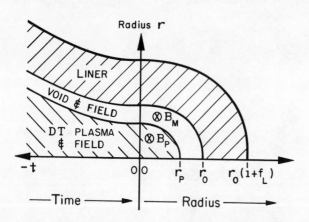

Fig. 2. Time and radius dependence of plasma, magnetic field, and liner near turnaround.

assumed to be zero. These should be good approximations because the sound velocity (or Alfven velocity) of the hot plasma is much greater than the sound velocity of the cold dense liner. In addition, we shall neglect the diffusion of the magnetic field through both the liner

and the plasma; this will be justified in Sec. V.

The total pressure P_0 inside the liner can be written (see Fig. 2)

$$P_0 = \frac{B_M^2}{8\pi} = 2nkT + \frac{B_p^2}{8\pi} , \qquad (1)$$

where n is the ion or electron density of the DT plasma and T is its temperature. Define the pressure ratio β in terms of the external magnetic field:

$$\beta \equiv 16\pi nkT / B_M^2 . \qquad (2)$$

Also, define the geometrical fill factor f_f in terms of the area ratio:

$$f_f \equiv (r_p/r_0)^2 . \qquad (3)$$

Then the plasma energy E_p per unit length is

$$E_p = 3nkTf_f\pi r_0^2 = \frac{3}{2}\beta f_f \pi r_0^2 P_0 . \qquad (4)$$

The magnetic energy E_M per unit length is

$$E_M = \frac{B_M^2(1-f_f) + B_p^2 f_f}{8\pi} \pi r_0^2$$

$$= (1-\beta f_f)\pi r_0^2 P_0 . \qquad (5)$$

To find the compressional energy of the liner, we must make a fit to published compression data and estimate the pressure and density distribution in the liner. The data for many metals[7] can be fitted by an equation of the form

$$\frac{\rho_0}{\rho} = \frac{v}{v_0} = \sum_i a_i e^{-P/P_i} . \qquad (6)$$

For example, for copper two terms suffice to make a close fit over the pressure range 0-4.5 Mbar:

$$\left(\frac{v}{v_0}\right)_{Cu} = 0.3e^{-P/0.65} + 0.7e^{-P/15} . \qquad (7)$$

Let W be defined as the energy of compression per gram. Then one can find

$$W = -\int_0^P P \frac{dv}{dP} dP$$

$$= v_0 \sum_i a_i P_i \left[1 - \left(1 + \frac{P}{P_i} \right) e^{-P/P_i} \right] . \quad (8)$$

To estimate the pressure and energy distribution in the liner, it is convenient to define a mass thickness parameter η in units of g/cm^2:

$$d\eta \equiv \rho dr . \quad (9)$$

At the outside of the liner $[r = r_0(1+f_L)$, see Fig. 2], where the pressure is zero, we define $\eta \equiv 0$. At the inside of the liner $(r=r_0)$, we define $\eta \equiv \sigma_0$, where σ_0 is called the total mass thickness. For the pressure distribution to be consistent with the impulse-momentum theorem, we write

$$P/\eta = P_0/\sigma_0 . \quad (10)$$

This approximate formula for the pressure distribution neglects the inertia of the plasma, convergence effects in the cylindrical geometry, and liner heating.

Next, we need a relation between the total mass thickness σ_0 and the thickness parameter f_L (Fig. 2). This is found by integration:

$$f_L r_0 = \int_{r_0}^{r_0(1+f_L)} \frac{dr}{d\eta} \frac{d\eta}{dP} dP$$

$$= \frac{v_0 \sigma_0}{P_0} \sum_i a_i P_i \left(1 - e^{-P_0/P_i} \right)$$

$$\equiv v_0 \sigma_0 z (P_0) , \quad (11)$$

where we have used Eqs. (6), (9), and (10), and where we have defined the "first compression function" $z(P)$.

Now we are ready to calculate the liner compressional energy per unit length E_L:

$$E_L = \int_0^{\sigma_0} 2\pi r_0 W d\eta . \quad (12)$$

Substituting equations (8) and (10) and

carrying out the integration we obtain

$$E_L = 2\pi r_0 \sigma_0 v_0 P_0 \sum_i a_i \frac{P_i}{P_0} \left[e^{-P_0/P_i} + 1 \right.$$

$$\left. + 2 \frac{P_i}{P_0} \left(e^{-P_0/P_i} - 1 \right) \right] . \quad (13)$$

Finally, we wish to express the energy E_L in a form similar to Eqs. (4) and (5) and as a function of f_L rather than σ_0. Substitute Eq. (11) into Eq. (13):

$$E_L = f_L s(P_0) \pi r_0^2 P_0 , \quad (14)$$

where the "second compression function" $s(P)$ is given by

$$s(P) \equiv \frac{2}{z(P)} \left\{ \sum_i a_i \frac{P_i}{P} \left[e^{-P/P_i} + 1 \right. \right.$$

$$\left. \left. + 2 \frac{P_i}{P} \left(e^{-P/P_i} - 1 \right) \right] \right\} , \quad (15)$$

where $z(P)$ was defined in Eq. (11).

As an example, for copper (Eq. 7) one obtains the compression functions shown in Fig. 3.

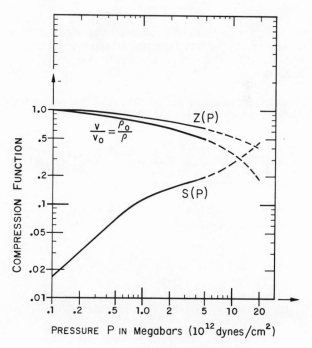

Fig. 3. Copper compression functions versus pressure [see Eqs. (6), (7), (11), and (15).]

To obtain the total nonnuclear energy per unit length E_T, add Eqs. (4), (5), and (14):

$$E_T = E_p + E_M + E_L$$

$$= [1 + \tfrac{1}{2}\beta f_f + f_L s(P_0)] \, \pi r_0^2 \, P_0 \,. \quad (16)$$

Most of this energy must be provided by the external energy supply that implodes the liner; this aspect of the problem will be considered in Sec. IX.

III. TURNAROUND DYNAMICS AND NUCLEAR ENERGY OUTPUT

Consider the time-dependence of the liner motion up to the time of turnaround (see Fig. 2). We ignore the subsequent expansion because Rayleigh-Taylor instabilities on the inner surface of the liner would probably destroy the symmetry at later times.[8] For the thin-liner approximation we have

$$\frac{d^2 r}{dt^2} = P/\sigma = P_0 \left[\left(\frac{r_0}{r}\right)^2 \right]^2 \frac{1}{\sigma_0} \frac{r}{r_0}$$

$$= \frac{P_0}{\sigma_0} \left(\frac{r_0}{r}\right)^{2\gamma-1} \,, \quad (17)$$

where we have approximated the dynamics of the plasma/field combination by adiabatic compression of a simple gamma-law gas.

The first integral of Eq. (17) is obtained by introducing the velocity $u = dr/dt$:

$$\int \frac{d^2 r}{dt^2} \, dt = \int_u^0 u \, du = \frac{1}{\frac{1}{2}\tau_0^2} r_0^{2\gamma} \int_r^{r_0} \frac{dr}{r^{2\gamma-1}} \,, \quad (18)$$

where the hydrodynamic time constant τ_0 is defined by

$$\tau_0 \equiv (\sigma_0 \, r_0 / P_0)^{\frac{1}{2}} \,. \quad (19)$$

Integration of Eq. (18) leads to the following equation for the velocity u:

$$u = \frac{dr}{dt} = -\frac{1}{\tau_0 \sqrt{\gamma-1}} \left[1 - \left(\frac{r_0}{r}\right)^{2(\gamma-1)} \right]^{\frac{1}{2}} \,. \quad (20)$$

For most values of γ (such as $3/2$ or $4/3$), the integration of Eq. (20) leads to transcendental equations for r that are analytically awkward. However, for the value $\gamma = 2$, one finds a simpler result:

$$\left(\frac{r}{r_0}\right)^2 = 1 + \left(\frac{t}{\tau_0}\right)^2 \,. \quad (21)$$

For our approximate analytic model we will use this simple result, although most plasma-field systems would be expected to be "softer" (lower γ). Systems with large magnetic energies would be closest to having an effective gamma of 2. Plasma-dominated systems would be closer to a gamma of $5/3$; an additional heating source, however, such as a laser or an electron beam, could be used to raise the effective gamma during the liner compression.

Now consider the nuclear energy output Y from the DT reaction in the plasma near turnaround (see Fig. 4):

$$Y = E_{DT} \, \pi r_0^2 \, f_f \int_{-\infty}^0 \frac{n^2}{4} \, \overline{\sigma v} \, (T) \, dt \,, \quad (22)$$

where E_{DT} is the useful energy release per reaction, where $\overline{\sigma v}$ (T) is the nuclear cross section averaged over the Maxwellian velocity distribution at temperature

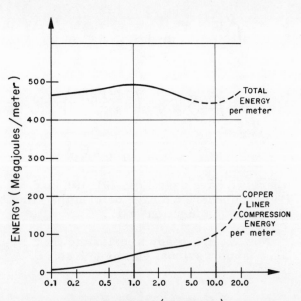

Fig. 4.   Breakeven energies for DT plasma example ($\alpha = f_L = 1$, $\beta = f_f = 0.707$).

T, and where we have assumed a 1:1 mixture of deuterium and tritium ions.

The reaction cross section[9] $\overline{\sigma v}$ (T) will be approximated by a quadratic fit that is accurate to about 20% in the temperature range 7 keV < T < 20 keV:

$$\overline{\sigma v} \ (T) \approx 0.4 \ (kT)^2 \ . \tag{23}$$

Combining Eqs. (1), (2), (17), (21), (22), and (23) we obtain

$$Y = \frac{1}{40} \ E_{DT} \ \pi r_0^2 \ f_f \ \beta^2 P_0^2 \int_{-\infty}^{0} \frac{dt}{\left[1 + \left(\frac{t}{\tau_0}\right)^2\right]^4} \ . \tag{24}$$

Using the substitution $u = (t/\tau_0)^2$ one can rewrite the integral in the following form:

$$\tau \equiv \frac{\tau_0}{2} \int_0^\infty \frac{u^{-\frac{1}{2}} du}{(1+u)^4} = \frac{\tau_0}{2} B\left(\frac{1}{2}, \frac{7}{2}\right)$$

$$= 0.491 \ \tau_0 \ , \tag{25}$$

where $B\left(\frac{1}{2}, \frac{7}{2}\right)$ is a "Beta function".[10] In Eq. (25) τ is the effective nuclear reaction time constant; it is smaller than the hydrodynamic time constant τ_0 because the nuclear reaction rate is a steep function of both plasma density and plasma temperature.

Combining Eqs. (24) and (25), one obtains

$$Y = \left(0.0123 \ E_{DT} \ \beta^2 \ f_f \ P_0 \ \tau_0\right) \pi r_0^2 \ P_0 \ , \tag{26}$$

where the nuclear energy output Y is written in the same form as the total nonnuclear energy E_T [see Eq. (16)]. This result for the output energy Y will be an underestimate when the effective γ of the plasma-field mixture is less than 2. In such a case the total pressure will not fall as rapidly when the radius is increased; consequently, there will be a few more reactions at large radii than have been calculated here. However, this conservatism is offset by the fact that we have not taken radiative losses into account.

IV. ENERGY MULTIPLICATION AND ENERGY PER UNIT LENGTH

We define the energy multiplication α for the total system involving plasma,

field, and liner:

$$\alpha \equiv \frac{Y}{E_T} = 0.0123 \ E_{DT} \ \frac{\beta^2 \ f_f \ P_0 \ \tau_0}{1 + \frac{1}{2}\beta f_f + f_L s(P_0)} \ , \tag{27}$$

where we have used Eqs. (16) and (26). If we set $E_{DT} = 17.6$ MeV and solve for the product $P_0 \tau_0$, we find

$$P_0 \ \tau_0 = 2.89 \times 10^6 \ \frac{\alpha}{\beta^2 \ f_f}$$

$$\times [1 + \frac{1}{2}\beta \ f_f + f_L \ s(P_0)] \ . \tag{28}$$

Equation (28) can be compared with Lawson's $n\tau$ criterion at T = 10 keV using Eqs. (1), (2), and (25):

$$n\tau = \left(\frac{\beta \ P_0}{2kT}\right)\left(0.491\tau_0\right) = 4.44 \times 10^{13} \ \frac{\alpha}{\beta \ f_f}$$

$$\times [1 + \frac{1}{2}\beta \ f_f + f_L \ s(P_0)] \ . \tag{29}$$

As a specific example, choose "break-even" ($\alpha = 1$) for a field-free ($\beta = 1$) plasma contained inside a rigid nonconducting wall [$f_f = 1$, $s(P_0) = 0$]. We find

$$n\tau = 6.66 \times 10^{13} \ . \tag{30}$$

This value is consistent with the Lawson criterion.[11]

To consider specific models, it is interesting to find the final system radius r_0 for a given choice of plasma and liner parameters (α, β, f_f, f_L, and P_0). From Eqs. (11) and (19),

$$r_0^2 = \left(\frac{1}{\rho_0 \ f_L}\right)\left[\frac{z(P_0)}{P_0}\right]\left(P_0 \ \tau_0\right)^2 \ , \tag{31}$$

where we can substitute for $P_0 \tau_0$ from Eq. (28). The total nonnuclear energy E_T [see Eq. (16)] can then be rewritten in terms of this solution:

$$E_T = \frac{\pi}{\rho_0 \ f_L} \left[z(P_0)\right]\left(P_0 \ \tau_0\right)^2$$

$$\times [1 + \frac{1}{2}\beta \ f + f_L \ s(P_0)] \ . \tag{32}$$

Combining Eqs. (28) and (32), we obtain

$$E_T = 2.624 \times 10^{13} \frac{z(P_0)}{\rho_0 f_L}$$

$$\times \frac{[1 + \frac{1}{2}\beta f_f + f_L s(P_0)]^3}{\beta^4 f_f^2} \alpha^2 . \quad (33)$$

Equation (33) is an important result of this approximate analysis; it is the energy per unit length that the liner system must have as a function of the plasma model β and f_f, of the desired energy multiplication α, of the final pressure P_0, of the liner parameters ρ_0 and f_L, and of the liner compressibility functions $z(P)$ and $s(P)$.

Consider again the idealized rigid-wall example of Eq. (30) [$\alpha = \beta = f_f = z(P_0) = 1$, $s(P_0) = 0$]. Then Eq. (33) becomes

$$E_T = (885/\rho_0 f_L) \text{ (in MJ/m)} . \quad (34)$$

Note that this result is independent of the final pressure P_0. The coefficient is a large amount of energy; a dense liner (high ρ_0) is desirable to reduce E_T. A thick liner (high f_L) would also seem desirable; however, it must be remembered that this approximate theory only holds for comparatively thin liners. Choosing a maximum value of $f_L = 1$ and a maximum practical liner density of 10 to 20, we find that the ideal minimum possible value of the nonnuclear energy E_T approaches 50 to 100 MJ/m.

Next, consider a somewhat more realistic breakeven ($\alpha = 1$) case. Choose $\beta = f_f = 0.707$ and consider copper liners, for which $\rho_0 = 8.93$ and $f_L = 1$. Then we find

$$E_T = 460 \, z(P_0) \, [1 + 0.8 \, s(P_0)]^3 \text{ (MJ/m)} .$$
$$(35)$$

In this case the more realistic plasma model causes the minimum value of the nonnuclear energy to be about 5 to 10 times higher than in the previous example. The compressibility function $z(P_0)$ tends to lower this value by increasing the effective liner density, but the corresponding energy function $s(P_0)$ tends to cancel the effect. The overall result is a gentle variation in the value of E_T as a function of P_0, as shown in Fig. 4. Figure 4 also shows the compressional energy in the liner for this case.

Other important parameters of the liner compression are the liner mass m_0 and initial velocity u_L. The mass m_0 per unit length is most easily determined from Eqs. (19) and (28):

$$m_0 = 2\pi r_0 \, \sigma_0 = 2\pi \frac{(P_0 \tau_0)^2}{P_0} . \quad (36)$$

The approximate initial liner velocity u_L is found by neglecting the initial energy of the plasma and field:

$$u_L^2 = \frac{2E_T}{m_0} = \frac{P_0 \, z(P_0)}{\rho_0 f_L}$$

$$\times [1 + \frac{1}{2}\beta f_f + f_L \, s(P_0)] , \quad (37)$$

where we have substituted Eqs. (32) and (36). Note that the required initial velocity u_L is an increasing function of the final pressure P_0. This method of finding the initial velocity u_L is more exact than differentiation of Eq. (21) because it does not assume $\gamma = 2$.

The final radius r_0 and the initial liner velocity u_L are plotted in Fig. 5 for the plasma breakeven case previously considered in Eq. (35) and Fig. 4. From

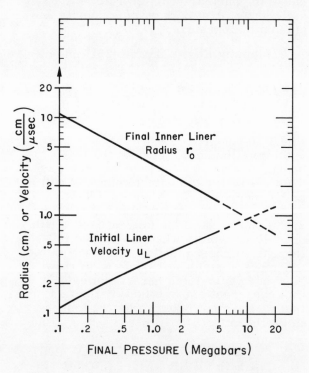

Fig. 5. Radius and velocity for DT plasma example ($\alpha = f_L = 1$, $\beta = f_f = 0.707$).

this figure it appears that the most practical breakeven copper-liner experiment would have a final pressure P_0 near 1 Mbar. Lower pressures imply large radii, and higher pressures imply large liner velocities.

It should be remarked that an energy E_T of 450 MJ/m is comparable to the energy release from 90 kg of TNT per meter. Such large energies imply very large systems; furthermore, $E_T \propto \alpha^2$, so it is of considerable interest to investigate methods of keeping practical liner devices small. Possible ways of doing this will be considered in Sec. VIII.

V. MAGNETIC DIFFUSION IN THE LINER

We have thus far neglected the diffusion of the magnetic field into the metallic liner. Now we shall estimate the size of this effect. To do this we make use of a previously published skin depth (δ) approximation[12]:

$$\frac{d}{dt}(\delta^2) \approx 0.6 \frac{\eta_0}{4\pi} \frac{B^2}{8\pi \rho_0 c_p T_0} , \quad (38)$$

where η_0 is the metal liner resistivity at temperature T_0, ρ_0 is its density, and c_p is its specific heat. For the copper liner of our previous example we obtain

$$\frac{d}{dt}(\delta^2) = 2.62 \times 10^{-10} B^2 , \quad (39)$$

when B is in gauss.

As a first approximation we substitute the flux-conserved value of B (no diffusion) and integrate over time:

$$\delta^2 = 2.62 \times 10^{-10} B_0^2 \int_{-\infty}^{0} \frac{dt}{\left[1 + \left(\frac{t}{\tau_0}\right)^2\right]^2} , (40)$$

where we have substituted Eq. (21). The integral is $\pi/4$, and the skin depth δ becomes

$$\delta = 0.0143 \sqrt{\tau_0} \, B_0 = 0.0719 \, (P_0 \, \tau_0)^{\frac{1}{2}} , (41)$$

where we have used Eq. (1). By comparison with Eqs. (28)-(30) one finds that the skin depth δ is almost independent of the pressure P_0 or the size r_0. For the

example plotted in Figs. 4 and 5, the skin depth is computed to be about 0.25 cm, which is small compared with the final inner liner radius r_0 at all pressures except $P \geq 10$ Mbar, for which the compressional energy is also rising.

This skin depth estimate neglects the decrease in metallic resistivity due to compression, and so it may be an overestimate. On the other hand, the metal vapor cloud observed at high fields[13] may blow across the void gap and contaminate the DT plasma. Such effects require further investigation.

Overall, we find that the comparatively small value of the skin depth in the liner justifies our neglect of magnetic diffusion in these large-liner systems. Magnetic diffusion in the plasma is also neglected; its effects can be roughly taken into account by adjustment of the plasma parameters β and f_f.

VI. NUMERICAL COMPUTATIONS OF LINER SYSTEMS

An independent evaluation of this analytical model can be made by comparing it with numerical computations of cylindrical liner compression. Several such calculations have been done in cylindrical geometry[14] using the multizone, two-temperature hydrodynamic code LASNEX.[15]

To the basic code an axial magnetic field (B_z) has been added, as in the MAGPIE code.[16] The equation of state of the copper liner that was used is more elaborate than the approximation of Eq. (6). When $\rho > \rho_0$ it uses a Grüneisen formulation.[7] When $\rho < \rho_0$ a virial expansion is made that is matched to the estimated critical-point parameters.[17] In addition, the energy of the alpha particles produced in the DT reactions is redeposited in the plasma using approximate formulas for the range and time delay of the alpha particle.

Most of the code problems cannot be compared with this model because they used thick liners; however, one problem had a thin liner for which $f_L = 1.16$, when the liner kinetic energy was minimum. At this time the inner liner radius $r_0 = 1.04$ cm, the total pressure $P_0 = 1.65$ Mbar, the fill factor $f_f = 0.55$, and the plasma $\beta = 0.95$.

Table 1. Comparison of approximate model with a computer calculation — at turn-around: $\beta = 0.95$, $f_f = 0.55$, $f_L = 1.16$, $r_0 = 1.04$ cm, $P_0 = 1.65$ Mbar.

	Computer run	Approximate model
Plasma energy E_P (MJ/m)	37.8	43.9
Field energy E_M (MJ/m)	26.9	26.7
Liner compressional energy E_L (MJ/m)	11.6	9.1
Total nonnuclear energy E_T (MJ/m)	76.3	79.7
Energy multiplication α	0.31	0.59
Nuclear energy output Y (MJ/m)	23.5	47.0

In Table 1 we show a comparison of the energies computed by the code and by the approximate model. The agreement is seen to be fairly good except for the nuclear yield parameters α and Y. The principal reason for this discrepancy is believed to be the fact that diffusion of the magnetic field into the plasma is neglected. Consequently, most of the plasma has an effective β lower than 0.95, which was computed **at** the plasma center. A lower value of β would bring both the plasma energy E_p and the nuclear parameters α and Y into better agreement.

Another important result shown by the computer runs[14] is that the nuclear yield near breakeven does not increase for liners thicker than $f_L \approx 1.0$ if one holds the total problem energy constant. The significance of this result is that our approximate thin-liner model is a good way of estimating the total energy requirements for breakeven given by the more exact computations, even for thicker liners beyond the range of validity of the initial thin-liner assumptions.

Thus, we conclude that the computer runs support the results of this approximate model for liner compression.

VII. LENGTH AND TOTAL ENERGY

In addition to the radial compression, axial flow of the plasma out of the ends of the cylindrical liner must also be considered in any complete system analysis. A complete two-dimensional computation of this problem has not yet been undertaken; therefore, we shall adopt a simple approximate criterion for the length of the liner system that should illustrate the magnitudes of the quantities involved. The criterion is that the re-

action time τ must be at least as short as the time it takes the plasma to escape from the ends:

$$L \geq 2v_a\, \tau \approx 2 \times 10^8\, \tau\,, \qquad (42)$$

where L is the length of the liner-plasma system and v_a is the acoustic velocity in a 10 keV DT plasma.

For the "ideal" plasma of Eq. (30), we find

$$L \geq \frac{1.33 \times 10^{22}}{n}\,. \qquad (43)$$

For a dense theta pinch ($n = 10^{17}$), one obtains a length of 1.33 km, which is within a factor of two of other estimates.[18-20] For the dense liner systems considered here, the length is considerably shorter.

A more general result for L is found by combining Eqs. (25), (28), and (42):

$$L = \frac{2.838 \times 10^{14}}{P_0} \frac{\alpha}{\beta^2 f_f}$$
$$\times [1 + \tfrac{1}{2}\beta f_f + f_L\, s(P_0)]\,. \qquad (44)$$

The total nonnuclear energy E_{min} for the whole length of the liner is then found from Eqs. (33) and (44):

$$E_{min} = L\, E_T = 7.447 \times 10^{27}$$

$$\times \frac{z(P_0)}{P_0 \rho_0 f_L} \frac{[1 + \tfrac{1}{2}\beta f + f_L\, s(P_0)]^4}{\beta^6 f_f^3}\, \alpha^3\,. \qquad (45)$$

Inspection of Eqs. (44) and (45) shows that, to a first approximation, the required total energy is inversely proportional to the final pressure P_0. Detailed calculations of L and L E_T for the more realistic plasma model are plotted in Fig. 6, which confirm the $1/P_0$ relationship except at the highest pressures, at which the liner compressibility becomes important.

Since $P_0 \propto n_f$, we find $E_{min} \propto \alpha^3/n_f$ as stated in the abstract. Note, however, the importance of the parameters β and f_f. Low values of β and f_f require much larger values of E_{min} for breakeven.

It is interesting to compare Eq. (45) with the expression for the minimum energy $E_{L.P.}$ required for ignition of a spherical laser-heated DT pellet[21]:

$$E_{L.P.} \approx 10^{15} \frac{\alpha^3}{\epsilon^4 \eta^2}, \qquad (46)$$

where ϵ is the efficiency of laser light absorption and η is the compression ratio ρ/ρ_0 for solid deuterium-tritium. One sees the same power of α for both inertial systems. The density factor in Eq. (46) is replaced by the product $n_f \rho_0$ in Eq. (45), where n_f is the plasma density and ρ_0 is the liner density. The efficiency factor ϵ^4 is related to the reciprocal of the fourth power of the bracketed term in Eq. (45). Thus, inertial containment follows similar laws in either cylindrical or spherical cases.

VIII. SMALLER-LINER SYSTEMS

Figure 6 shows that breakeven for the DT plasma example will require a total liner energy of the order of a gigajoule. Pulses greater than this ($\alpha > 1$) will require even larger energies, in proportion to α^3, as given by Eq. (45). One gigajoule is approximately equivalent to the energy release from 200 kg of TNT explosive. The application of such very large explosions to electrical power production would require extraordinarily large containers and novel engineering solutions.

In this section we will describe some possible ways in which this large explosion can be reduced to more manageable size. The first of these is a hybrid system in which the lithium blanket is replaced by a composite blanket containing both lithium and fissionable material,

such as uranium. Such blankets have been calculated[22] to be capable both of breeding tritium (from the lithium) and yielding an energy multiplication (from the uranium) of more than 10 times the energy release of the DT reaction.

Thus, breakeven for the overall hybrid system would require $\alpha = 0.1$ in Eq. (45). In that case the energy E_{min} would only be of the order of 1 MJ, or 200 g of TNT equivalent. Containment of such an explosion is quite conceivable within current engineering practice. This radical improvement for the hybrid system is a consequence of the cubic power law for α in Eq. (45). Of course, all the other parameters of the system (radius, length, liner mass, etc.) will be reduced according to the various equations developed above.

Another possible approach to reducing the liner size is to form a "two-component" plasma[23] inside the liner. In this case some of the ions are non-Maxwellian, having energies of 50 to 200 keV. As these ions slow down, they contribute additional in-flight nuclear reactions that would add an additional term to our

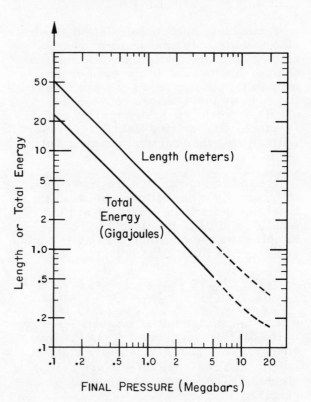

Fig. 6. Length and total energy for DT plasma example ($\alpha = f_L = 1$, $\beta = f_f = 0.707$).

expression for the nuclear energy output Y [see Eq. (26)]. It has been estimated[23] that such a plasma might have an effective $n\tau$ 2 to 3 times lower than for a Maxwellian plasma. Thus, one might conceive of lowering E_{min} by a factor of about ten. This would still be a rather large explosion, however.

A third approach is to lower the required liner length L by changing the design of the ends of the system, where the plasma escapes. End plugs,[18] multiple mirrors,[24,25] cusps,[26,27] and (in the USSR) toroidal plasmas[28] have been suggested for this purpose. These designs would reduce the overall length requirement, but would not affect the cylindrical calculations of Sec. IV.

A fourth possibility — recovering some of the liner energy — is described in the next section. Further attempts to minimize the size of the liner system are desirable; the cubic power law for α offers hope that such an effort can lead to smaller systems.

IX.   MAGNETIC IMPLOSION OF THE LINER

A complete system calculation should also include additional energy losses arising from the inefficiency of the method of liner implosion. In the case of the magnetic field implosion concept (see Fig. 1), we must choose between the usual "θ-pinch" (B_z) driving field and a "z-pinch" (B_θ) driving field.[29,30] Figure 7 illustrates the practical geometry of the two concepts.

It has been shown that the B_θ system is inherently more efficient because the local magnetic field is largest at the smaller radius of the liner.[30] The simplest way to demonstrate this is to consider the liner kinetic energy W:

$$W = E_{m_i} - E_{m_f}, \tag{47}$$

where E_{m_i} is the initial magnetic energy in the driving field, and E_{m_f} is the final energy. (All quantities are per unit length, and resistive effects are neglected.) The magnetic energy E_m is given by

$$E_m = \tfrac{1}{2}\frac{\phi^2}{L}, \tag{48}$$

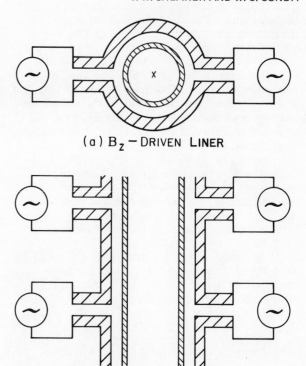

(a) B_z – DRIVEN LINER

(b) B_θ – DRIVEN LINER

Fig. 7.   Schematic arrangements of power supplies for driving the liner.

where ϕ is the flux, and L is the inductance per unit length. But the flux ϕ is a constant; therefore, the driving efficiency η can be written

$$\eta \equiv \frac{W}{E_{m_i}} = 1 - \frac{L_i}{L_f}. \tag{49}$$

Substituting the appropriate inductance formulas for the two cases, one obtains

$$\eta_z = \frac{r_i^2 - r_f^2}{R^2 - r_f^2} \tag{50a}$$

$$\eta_\theta = \frac{\ln(r_i/r_f)}{\ln(R/r_f)}, \tag{50b}$$

where R is the radius of the container. Comparing the two efficiencies, one finds that $\eta_\theta > \eta_z$, as was to be shown.

In addition to its higher efficiency, the B_θ container geometry (Fig. 7) may permit the construction of a higher-pressure vessel due to the possibility of having a higher hoop stress in a cylinder that is unbroken in the azimuthal direction.

In a complete fusion system one must compensate for the inefficiency $1 - \eta$ by specifying a higher α [Eq. (27)]. However, it has been suggested[31] that, if the liner maintains its integrity after the implosion, its outward motion (explosion) will pump energy back into the driving field, thus reducing the required α. It is not presently known whether such stability is possible.

The power supply for the driving magnetic field is the main energy source for the liner implosion. For a radius ratio of 30 (corresponding to an adiabatic temperature ratio of 90 and an initial plasma temperature of 200 eV), the liner implosion time τ_i is roughly $30 r_0 / u_L$, where u_L is given by Eq. (37). Thus to the first order τ_i varies as $1/P_0$. At $P_0 = 6$ Mbar, $\tau_i \approx 50$ μsec, which requires a very fast system. For the larger-energy systems, τ_i is longer. In that case one can consider inherently slow power-supply systems, such as inductive energy storage.

X. LINER FORMATION

Consider a small fusion reactor power requirement of 10^8 W average power. Extrapolating from Fig. 6, one finds that a 1-m breakeven system at 6 Mbar pressure would have $Y = 4 \times 10^8$ J per explosion. If a reactor system operated at the same final pressure (6 Mbar) with a multiplication of $\alpha = \sqrt{5}$, then the yield Y per explosion would be

$$Y = \alpha \, E_{min} = 10^{10} \text{ J/explosion .} \qquad (51)$$

A power of 10^8 W would require a new explosion inside the container every 100 sec (36 explosions per hour). Within this cycle time one would have to pump out all the debris from the previous explosion and form the next liner-plasma system.

We can suggest two possible methods of forming the liner that might be investigated further. The first is the continuous casting of a solid cylinder in place inside the container. As a result of a brief contact with the light metals industry, we estimated that the capability to cast 36 metal liners per hour would cost about 1.2×10^7. If interest plus

payment of principal amounts to 20% per year, then the casting cost would come to about $275 per hour. For a 10^8 W output, this would be $0.00275 per kWh.

A similar cost estimate can be made for the containment shell; it amounts to $0.0010 per kWh. No complete system cost estimates have been made, but the sum of these two costs is less than the market value of $0.01 per kWh.

A second possible method is gravity flow of a liquid liner. This would require that the axis of the liner system be vertical and that the mass flow rate through the annular orifice at the top of the container be varied in time so that the thickness of the liner would be constant (as a function of axial position z) at the time of the magnetic implosion. A simple calculation shows that a linearly decreasing mass flow rate will meet this criterion. Any low-melting-point metal can be used. No cost estimates have been made for this method.

Other practical problems needing further assessment are the pump-out problem and the question whether the exploding liner (at late times after turnaround) will damage the container. This latter problem will be particularly severe if the liner breaks up into chunks of metal shrapnel.

Additional studies are needed to clarify these practical reactor problems.

XII. CONCLUDING REMARKS

This model is a good approximation for thin, cylindrical, compressible liners. Comparison with numerical calculations suggests that its usefulness can be extended to thick compressible liners. Thick incompressible liners (at lower pressures) can best be treated by the method of Robson.[32]

These calculations of cylindrical-metal-liner compressions for fusion purposes have shown that very-large-energy explosions will be needed to surpass breakeven if the usual long-theta-pinch plasma geometry is employed. However, the structure of the equations arouses hope that other plasma configurations may reduce the energy per explosion. If such reductions can be achieved, then metal liners should be taken seriously as a method for achieving fusion.

REFERENCES

1. J. G. Linhart, "Plasma and Mega-gauss Fields," in Proceedings of the Conference on Megagauss Magnetic Field Generation by Explosives and Related Experiments, H. Knoepfel and F. Herlach, Eds., Euratom, Brussels, Rept. EUR 2750.e (1966), pp. 387-396.

2. E. Velikhov, in Proceedings of the Fifth European Conference on Controlled Fusion and Plasma Physics (Grenoble, August, 1972).

3. D. B. Thomson, R. S. Caird, W. B. Garn, C. M. Fowler, "Plasma Compression by Explosively Produced Magnetic Fields," pp. 491-514 of Ref. 1.

4. E. P. Velikhov, A. A. Vedenov, A. D. Bogdanets, V. S. Golubev, E. G. Kasharskii, Zh. Tekh. Fiz. 43, 429 (1973) [English transl. Sov. Phys. Tech. Phys. 18, 274 (1973)].

5. J. P. Boris, R. A. Shanny, and N. K. Winsor, Bull. Amer. Phys. Soc. 17, 1029 (1972).

6. P. J. Turchi, Naval Research Laboratory, Washington, D. C., private communication.

7. R. N. Keeler, "High Pressure Compressibilities," in American Institute of Physics Handbook, 3rd ed., D. E. Gray and M. W. Zemansky, Eds. (McGraw-Hill, New York, 1972), pp. 4-96 to 4-104.

8. A. Barcilon, D. L. Book, and A. L. Cooper, Hydrodynamic Stability of a Rotating Liner, Naval Research Laboratory, Washington, D. C., NRL Memorandum Rept. 2776 (1974).

9. S. L. Greene, Jr., Maxwell-Averaged Cross Sections for Some Thermonuclear Reactions on Light Isotopes, Lawrence Livermore Laboratory, Livermore, Calif., Rept. UCRL-70522 (1967).

10. NBS Handbook of Mathematical Functions, M. Abramowitz and I. A. Stegun, Eds. (Dover Publications, New York, 1968), pp. 258-272.

11. J. D. Lawson, Proc. Phys. Soc. London, Sec. B 70, 6 (1957).

12. J. W. Shearer, F. F. Abraham, C. M. Aplin, B. P. Benham, J. F. Faulkner, F. C. Ford, M. M. Hill, C. A. McDonald, W. H. Stephens, D. J. Steinberg, and J. R. Wilson, J. Appl. Phys. 39, 2102 (1968).

13. J. W. Shearer, J. Appl. Phys. 40, 4490 (1969).

14. W. C. Condit, A Parameter Study of B_z Liner Implosions for Possible CTR Applications, Lawrence Livermore Laboratory, Livermore, Calif., Rept. UCID-16474 (1974).

15. G. B. Zimmerman, Numerical Simulation of the High Density Approach to Laser Fusion, Lawrence Livermore Laboratory, Livermore, Calif., Rept. UCRL-75173 (1973).

16. R. E. Kidder, "Compression of Magnetic Field Inside a Hollow Explosive-Driven Cylindrical Conductor," pp. 37-54 of Ref. 1.

17. D. J. Steinberg, Lawrence Livermore Laboratory, Livermore, Calif., private communication.

18. J. M. Dawson, A. Hertzberg, G. C. Vlases, H. G. Ahlstrom, L. C. Steinhauer, R. E. Kidder, and W. L. Kruer, "Controlled Fusion Using Long Wavelength Laser Heating with Magnetic Confinement," to be published in Fundamental and Applied Laser Physics, M. S. Feld, A. Javan, and N. A. Kurnit, Eds. (Wiley, New York, 1973).

19. R. L. Morse, Phys. Fluids 16, 545 (1973).

20. W. Koppendorfer, W. Schneider, and J. Sommer, "Conclusions on Linear Reactor Devices from Theoretical and Experimental Investigations of End Losses from Theta Pinch Plasmas," 5th European Conference on Controlled Fusion and Plasma Physics, 21-25 August 1972, Grenoble, France.

21. R. E. Kidder, "Some Aspects of Controlled Fusion by Use of Lasers," to be published in Ref. 18.

22. J. D. Lee, Neutronics of Subcritical Fast Fission Blankets for D-T Fusion Reactors, Lawrence Livermore Laboratory, Livermore, Calif., Rept. UCRL-73952 (1972).

23. H. P. Furth and D. L. Jassby, Phys. Rev. Lett. 32, 1176 (1974).

24. V. V. Mirnov and D. D. Ryutov, Nucl. Fusion 12, 627 (1972).

25. B. G. Logan, I. G. Brown, M. A. Lieberman, and A. J. Lichtenberg, Phys. Rev. Lett. 29, 1435 (1972).

26. A. E. Robson, The Flying Cusp: A Compact Pulsed Fusion System, Naval Research Laboratory, Washington, D. C., NRL Memorandum Rept. 2692 (1973).

27. P. J. Turchi, "Spherical Implosion of Thick Liners with Compressibility and Plasma Loss," Naval Research Laboratory, Washington, D. C., NRL Memorandum Rept. 2711 (1974).

28. H. Sahlin, Lawrence Livermore Laboratory, Livermore, Calif., private communication.

29. S. G. Alikhanov, V. G. Belan, A. J. Ivanchenko, V. N. Karasjuk, and G. N. Kichigan, J. Sci. Instrum. (J. Phys. E) Series 2, 1, 543 (1968).

30. J. W. Shearer, The Theta Field Liner Concept, Lawrence Livermore Laboratory, Livermore, Calif., Rept. UCID-16418 (1973).

31. A. E. Robson, Naval Research Laboratory, Washington, D. C., private communication.

32. A. E. Robson, Fundamental Requirements of a Fusion Feasibility Experiment Based on Flux Compression by a Collapsing Liner, Naval Research Laboratory, Washington, D. C., NRL Memorandum Rept. 2616 (1973).

THEORETICAL AND EXPERIMENTAL STUDY OF EXPLOSIVE DRIVEN
MAGNETIC FIELD COMPRESSION GENERATORS

B. Antoni C. Nazet L. Pobé

Commissariat à l'Energie Atomique, Centre d'Etudes de Limeil
B.P. n° 27, 94190-Villeneuve-Saint-Georges, FRANCE

ABSTRACT

We describe generators of electrical energy in which magnetic field compression
is achieved by a solid explosive. The magnetic flux losses have been calculated for generators of various configurations. There is a good agreement between these calculations
and the experimental results for coaxial and "bellows" type generators. In helical generators the magnetic flux losses are higher than those calculated by considering diffusion
only. We show by detailed calculations of the motion of the explosive driven inner
conductor that additional losses come from the jumps encountered by sliding contact
moving along the helix. The jumps are caused by little geometrical defects and the
consequence on losses is strongly dependent on current intensity. The jumps are detrimental to the efficient use of the explosive energy. With helical generators only 5 %
of the energy is transferred into magnetic energy, to be compared to the computed value
of 20 % in the case of bellows generating 15 MA in 40 nH with an e-folding time of
70 μsec.

INTRODUCTION

In the "Centre d'Etudes de Limeil",
we are studying two types of magnetic field
compression generators (helical type and
"bellows" type). The mode of operation of
these generators is based on a same well-
known principle [1].

DESCRIPTION OF THE HELICAL GENERATOR

The helical generator is shown on
Fig. 1. It consists of two copper coaxial
cylinders. The inner one is filled with solid explosive. In the wall of the outer one,
there is cut a variable pitch helix. We have
conducted experiments on two models hereafter referred as helix-1 and helix-2.
Helix-1 has an outer diameter of 140 mm and
an explosive diameter of 50 mm. The total
length of the helix is 800 mm and the explosive energy is about 12 MJ. Helix-2 has an
outer diameter of 200 mm and an explosive
diameter of 69 mm ; its length is also
800 mm and its initial explosive energy is
about 24 MJ. In both cases the initial inductance is about $1.3 \ 10^{-6}$ H. The total
energy efficiency never exceeds 5 % and the
cost is ≠ 500 for helix-1 and ≠ 800 for
Helix-2.

The inductance decrease is obtained
in the following way. After initiation of
the explosive a detonation wave travels from
left to right (see Fig. 1). As a result of
the detonation the inner cylinder, initially
at rest, is lifted radially. The cone thus
formed acts like a moving wall which is displaced towards the outer stationnary

cylinder, hits it and reduces the length of
the current lines and thus the inductance.
The cone displaces the magnetic flux into
the loop made by the load L_u.

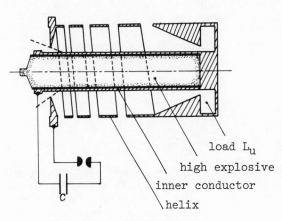

load L_u
high explosive
inner conductor
helix

Fig. 1. Helical generator

DESCRIPTION OF THE BELLOWS TYPE
GENERATOR

The bellows first prototype is
shown in Fig. 2. The inner conductor consists of two copper plates containing the
explosive in between. On both sides of this
sandwich arrangement, two other copper plates form the outer conductor. The conductor width varies exponentially so that the
magnetic field remains constant in the contact zone during the compression. The initial inductance is $4 \ 10^{-7}$ H.

CALCULATION OF THE FLUX LEAKAGE DURING THE OPERATION OF THE EXPLOSION GENERATOR

According to the electric equation

$$(1) \qquad \frac{d}{dt}(LI) + RI = 0$$

the numerical simulation of the operation process requires knowledge of the variation of the inductance versus time and an estimation of the flux leakage. The inductance variation has been determined by using the two following methods :

1) By experimental measurements on laboratory models reproducing the generator configurations at the various times of its operation [2] ;

2) By numerical calculations for the same configurations [3].

The calculations errors are less than 10 % for helical generators and less than 5 % for bellows generators.

With the diffusion equation for a plane conducting half-space of constant electrical conductivity to (these approximations are justified for our devices [4]) Eq. (1) becomes :

$$(2) \qquad \frac{d}{dt}(LI) + \sqrt{\frac{\mu_o}{\pi \sigma_o}} \, G(t) \int_0^t \frac{I'(u)}{\sqrt{t-u}} \, du = 0$$

with :

$$(3) \qquad G(t) = \int_{\ell_1(t)}^{\ell_2(t)} \frac{d\ell}{\ell_a(\ell)}$$

where ℓ is the curvilinear abscissa of a current line and $\ell_a(\ell)$ is the width of the current path. The flux leakage term is the product of two integrals ; one of them is the form factor $G(t)$ only depending on the configuration and the other

$$\int_0^t \frac{I'(u)}{\sqrt{t-u}} \, du$$

represents the influence of the time variation of the current. Flux leakage was also calculated alternatively by means of the skin-layer method. The skin depth was defined as :

$$(4) \qquad \delta_o = \sqrt{\frac{I}{I'} \, \frac{1}{\sigma_o \mu_o}}$$

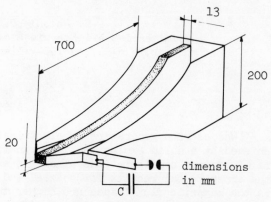

Fig. 2. The bellows first prototype

dimensions in mm

This approximation gives results close to those obtained by the more precise calculation mentioned above. The linearity of Eq. (2) shows that the current amplification at a moment t, $I(t)/I(o)$, must be the same for anyvalue of the initial current $I(o)$. Helix experiments indicate, on the contrary, that the current amplification decreases when $I(o)$ is increased. This behaviour might have come from non linear processes due to the Joule effect. We have calculated them using the skin-layer method. We consider the skin depth variation in the contact zone only, i.e. where magnetic field and therefore heating are maximum. For its calculation, we use the following equations :

$$(5) \qquad \tilde{\delta} = \sqrt{\frac{I}{I'} \, \frac{1}{\sigma_o \mu_o}}$$

$$(6) \qquad \sigma = \sigma_o \, \frac{1}{1 + (B^2/2\mu_o \rho \, cT_o)}$$

where B is the magnetic field in the contact zone, ρ the density of copper, c the specific heat, $T_o = 273$ °K. Taking into account the correction term due to non linear phenomena, Eq. (2) becomes :

$$(7) \qquad \frac{d}{dt}(Li) + \sqrt{\frac{\mu_o}{\sigma_o \pi}} \, G(t) \int_0^t \frac{I'(u)}{\sqrt{t-u}} \, du + 2\mu_o \, (\tilde{\delta} - \delta_o) \, I \, \frac{d}{dt} G(t) = 0$$

It is to be noticed that, on the one hand, for our devices, the flux leakage calculated with this equation has never exceeded more than approximatively 10 % [5] the flux leakage given by Eq. (2). On the other hand, because of the assumptions made for Eq. (5)(6) the non linear effects are overestimated.

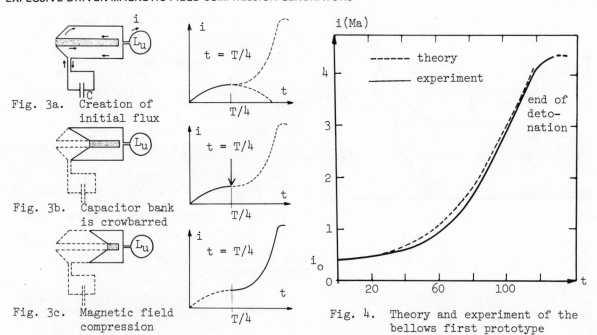

Fig. 3a. Creation of
          initial flux

Fig. 3b. Capacitor bank
          is crowbarred

Fig. 3c. Magnetic field
          compression

Fig. 4.   Theory and experiment of the
          bellows first prototype

EXPERIMENTAL RESULTS

As shown in Fig. 4 the experimental results obtained with bellows are in good agreement with calculations.

In fact, the behaviour remains linear even with a high injection current but only if the magnetic field does not exceed 30 Teslas in the contact zone.

The agreement between experiment and theory is not so good for helical generators.

The current they deliver lies always below the theoretical value, more especially when the initial current is high. In spite of the over estimation of the non linear effects, the discrepancy between calculation and experiment remains important in our devices. The heating of the walls, therefore, is not sufficient to account for the flux leakage.

All this indicates that there are other flux losses than the ones considered in the calculation.

For the interpretation of the whole results, we have to take into account the travel conditions of the sliding contact motion is not always continuous along the helix because of possible constructional defects and asymmetrical distribution of the magnetic field. The sliding contact does not sweep smoothly along the whole helix but may sometimes make jumps from one point to another.

STUDY OF THE SLIDING CONTACT MOTION DISCONTINUITIES

These discontinuities cause as a direct consequence, sudden jumps in the inductance value and therefore in the current derivative dI/dt. The jumps are easily observable on the curve. They form a succession of peaks on the current derivative curve of the helix (Fig. 5a). The discontinuities may be the consequence either of excentricity defects of the two cylinders or of a non uniform wall thickness of the inner cylinder. The defects lying on a generatrix, the discontinuities appear periodically, every time the sliding contact describes a complete revolution along its helicoïdal path. This very curious effect was also observed by other authors who called it "2π clocking effect" [6].

Defect analysis has shown that such discontinuities cannot occurs on bellows. Actually, the experimental current derivative of bellows is always continuous and monotonous (Fig. 5b).

The above mentionned discontinuities in the sliding contact motion explain not only the shape of the current derivative curve but also the additional flux losses appearing during helix operation.

Figure 6 shows schematically the sliding contact motion in the case of certain texture and surface defects.

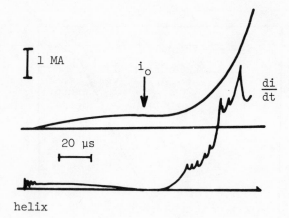

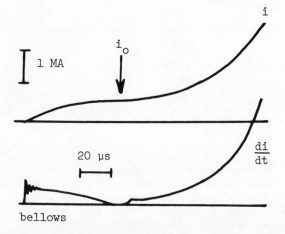

Figure 5

An ultrafast framing camera reveals **these** defects on the lifting cone surface which reminds of an orange peel. The defect size is about 1 mm.

The contact quality depends essentially on the value of the angle between the current line and the contact zone. When there are no geometrical defects, this angle is Ψ_1. Geometrical defects involve an additional angle Ψ_2. The resulting angle can be positive or negative. If Ψ is always positive ($\Psi_1 \gg |\Psi_2|$) there are no additional flux losses due to the sliding contact operation. But if Ψ is sometimes negative ($\Psi_1 \approx |\Psi_2|$) there are flux losses. The magnetic flux in the hachured zone (Fig. 6) is trapped and lost for further compression.

The surface defects are uniformly distributed on the lifting cone. If their mean size is $\Delta E/2$, resulting flux losses $\Delta \Phi$ during the time t are given by :

$$(8) \qquad \frac{\Delta \Phi}{\Delta t} \simeq \mu_o \frac{\Delta E}{2} I \frac{d}{dt} G(t)$$

In Figure 7, the experimental current of helix-one is compared with the current calculated by taking into account the flux losses $\Delta \Phi$. It is easy to see that the helix behaviour is explained when the injection current is low enough (up to 200 kA). In order to explain the behaviour for higher injection current, it is neces-

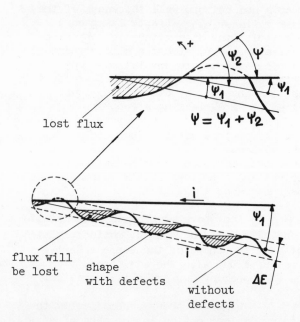

Fig. 6. Flux losses when $\psi < 0$

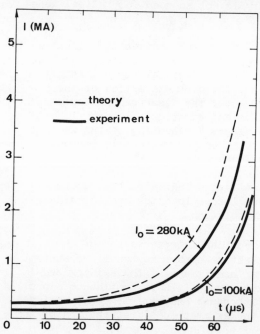

Fig. 7. Theory and experiment of helix 1

sary to take into account the diminution of the angle Ψ_1 and to compare the two angles Ψ_1 and Ψ_2.

We have calculated Ψ_1 for helix-one at different injection current (Fig. 8). We assume that there is no additional flux-leakage due to contact irrgularities. In this helix, angle Ψ_1 lies under 2° and decreases with increasing I_0.

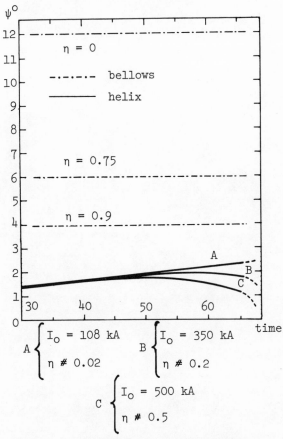

Fig. 8.   Angle ψ for bellows and helix with different value of η

For the helix, excentricity and thickness defects involve angles Ψ_2 of 1-2°. As a result, with growing injection current intensities, the frequency of the jumps and therefore the flux leakage increase.

The efficiency of these devices is the ratio of the kinetic energy transformed into electrical energy and initial kinetic energy of the lifting cone. Theoretically, as reported Fig. 8, η increases with growing I_0 but, as we have seen above in that case the flux leakage grows up, so that the final efficiency is limited at about 5 %.

In the bellows, Ψ_1 is generally higher than Ψ_2 ($\Psi_2 \approx 1°$) Ψ_1 decreases with growing η but remains high even for great values of efficiency ; for instance if $\eta = 0,4$, $\Psi_1 = 9°$.

A final efficiency of 20 % is easily obtainable with bellows.

CONCLUSION

The result of these studies was the construction of a bellows generator with the following characteristics (Fig. 9).

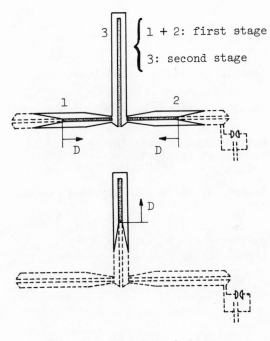

Fig. 9.   Two stage bellows

It delivers 5 megajoules into twenty nano-henries. Its cost, relatively moderate, is of 1 000 $ for one prototype.

It consists in two stages :

- the initial one, with two small high inductive bellows generators, the conductors of which have constant width ;

- the final one, with third bellows, having exponentially varying width.

The three bellows are series connected.

When the magnetic flux is maximum in the whole loop constituted by the three series-connected bellows, the two bellows of the first stage are detonated simultaneously and start moving the magnetic flux into the final stage.

After the detonation of the first stage, the final one is crowbared and pushes the magnetic flux into the inductive load.

We think it is possible to realize generators of this type being able to produce higher magnetic energy.

At the present stage of research there appears no limitation in the development of higher energy generators.

REFERENCES

[1] A.D. Sakharov et al., Sov. Phys. Dokl., 10, 11 (1966) 1045

[2] B. Antoni, C. Nazet, (unpublished)

[3] B. Antoni, C. Nazet, L. Pobe (unpublished)

[4] B. Antoni, Thèse d'Etat (Orsay, 18 juin 1974, CNRS n° AO 8337)

[5] B. Antoni, C. Nazet (CEA Report to be published)

[6] J.W. Shearer et al., J. Appl. Phys., 39, 4, (1968), 2102

HIGH ENERGY PULSED POWER BY INDUCTIVELY DRIVEN
IMPLODING LINER FLUX COMPRESSION

Peter J. Turchi
Plasma Physics Division
Naval Research Laboratory
Code 7760, Washington, D.C. 20375

ABSTRACT

The use of electromagnetically driven imploding liner flux compression techniques coupled with high energy inertial-inductive storage systems is investigated theoretically. Basic flux compression generator operation is reviewed and two techniques are discussed for obtaining high power, high energy electrical pulses. Power levels of several terawatts, involving tens and hundreds of megajoules of electrical energy in a few microseconds appear possible with overall efficiencies of 7-10%.

INTRODUCTION

Very high energy electrical pulses can be generated by the motion of conducting liners in suitable magnetic geometries. This technique is well-known[1] and has been used for many years as the basis of explosive generators. The fundamental requirements for operation are high speed conductors, moving against high magnetic fields, over sufficiently large dimensions, (magnetic Reynolds number $\gg 1$). Typically, speeds on the order of a few mm/μsec, fields up to a few megagauss, and distances of several centimeters are involved, requiring tens of megajoules of initial energy. Such energies must be delivered in short times, so that flux compression techniques for pulsed power applications have generally involved explosives, and the associated difficulties of explosives handling and on-site destructiveness. With the advent of fast inertial-inductive power systems[2], explosives and their disadvantages can be eliminated, allowing flux compression techniques to be employed in complex plasma experiments at very high energies.

REVIEW OF THE FLUX COMPRESSION METHOD

To review briefly the operation of a flux compression generator, we consider an idealized system in which resistance is neglected. The rate of transfer from kinetic energy of the moving conductor to magnetic energy is then simply:

$$P = \tfrac{1}{2}\, \dot{L} J^2$$

where $\dot{L}$ is the rate of change of circuit inductance and J is the circuit current. Let the circuit inductance L be given by $L = L_0 + L_G$, with L_0 a fixed inductance value including that of the load, and $L_G = Gx$, where x is the dimension in the direction of conductor motion and G is a geometric constant depending on the generator configuration. The rate of change of inductance is then $\dot{L} = Gu$, where u is the instantaneous speed of the conductor. Defining a parameter λ by $L_0 = \lambda G x_0$, where x_0 is the particular value of x when λ is evaluated, the generator current at any time is given by:

$$J^2 = 2E/L$$
$$= 2E/Gx_0(\lambda + x/x_0)$$

with E, the total magnetic energy in the circuit. The energy transfer rate is then:

$$P = Eu/x_0(\lambda + x/x_0) \ .$$

The conductor speed at any time can be written as $u = (2E_K/M)^{\frac{1}{2}}$ where M is the effective conductor mass and the instantaneous kinetic energy E_K is given in terms of the initial values of kinetic and magnetic energy, E_{Ko} and E_0, respectively and the instantaneous magnetic energy E:

$$E_K = E_{Ko} + E_0 - E \ ,$$

The power is then:

$$P = \frac{E_{Ko} u_0 \varepsilon (1 + \varepsilon_0 - \varepsilon)^{\frac{1}{2}}}{x_0 (\lambda + x/x_0)}$$

where $\varepsilon = E/E_{Ko}$, $\varepsilon_0 = E_0/E_{Ko}$, and u_0 is the initial conductor speed. To obtain the power waveform, we eliminate x in terms of ε by

$$\frac{\lambda + x_i/x_0}{\lambda + x/x_0} = \frac{\varepsilon}{\varepsilon_0}$$

GENERAL SOLUTION FOR IDEAL FLUX COMPRESSION GENERATOR

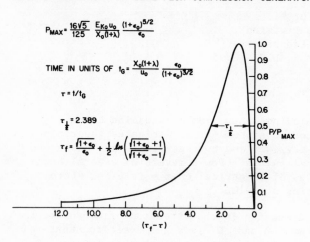

$$P_{MAX} = \frac{16\sqrt{5}}{125} \frac{E_{Ko} u_o}{x_o(1+\lambda)} \frac{(1+\epsilon_o)^{5/2}}{\epsilon_o}$$

TIME IN UNITS OF $t_G = \frac{x_o(1+\lambda)}{u_o} \frac{\epsilon_o}{(1+\epsilon_o)^{3/2}}$

$\tau = t/t_G$

$\tau_{\frac{1}{2}} = 2.389$

$\tau_f = \frac{\sqrt{1+\epsilon_o}}{\epsilon_o} + \frac{1}{2} \ln\left(\frac{\sqrt{1+\epsilon_o}+1}{\sqrt{1+\epsilon_o}-1}\right)$

Figure 1

where x_i is the initial value of x (for which $\epsilon = \epsilon_o$). The elapsed time is then found from:

$$t = t(\epsilon) = \int_{\epsilon_O}^{\epsilon} \frac{E_{Ko}}{P(\epsilon)} d\epsilon .$$

With $x_i = x_o$, it is useful to measure time in units of

$$t_G = \frac{x_o(1+\lambda)}{u_o} \frac{\epsilon_o}{(1+\epsilon_o)^{3/2}}$$

so that $t = \tau t_G$, where τ is a dimensionless time variable. The power waveform obtained in this way is displayed in Fig.1, with values of power normalized by the maximum rate:

$$P_{MAX} = \frac{16\sqrt{5}}{125} \frac{E_{Ko} u_o}{x_o(1+\lambda)} \frac{(1+\epsilon_o)^{5/2}}{\epsilon_o} .$$

Note that time is measured backwards from the time of completion of conductor motion ($E_K = 0$) which is given by

$$\tau_f = \frac{\sqrt{1+\epsilon_O}}{\epsilon_o} + \frac{1}{2} \ln\left(\frac{\sqrt{1+\epsilon_O}+1}{\sqrt{1+\epsilon_O}-1}\right) .$$

Generator action begins at an earlier time determined by comparing the ratio of initial power to maximum power with the curve in Fig.1. This curve thus represents the general solution for the power waveform of a flux compression generator in which resistive effects may be neglected and inductance is eliminated in a linear manner with conductor travel distance. Some interesting features to note are that the time for which $P \geq P_{MAX}/2$ is given by $2.389 t_G$ and that, therefore, more than half the energy ($0.541(1+\epsilon_o)$) is delivered at power levels in excess of $P_{MAX}/2$. The average power level is thus $0.791 P_{MAX}$.

These formulas are subject to the constraint that resistive diffusion during the power pulse time δt is negligible, which may be expressed as $(\eta \delta t/\mu)^{\frac{1}{2}} \ll u \delta t/2$, where η is the effective resistivity of the conductor. With $\eta = 10 \mu\Omega$-cm and $\epsilon_o = 0.1$, to achieve peak powers of 10^{13} watts and microsecond pulses, $x_o > 2.4$ mm, $u_o > 0.564$ mm/μsec and $E_{Ko} > 146$ MJ. It is the last quantity which normally introduces the need for high explosives. Recent development of high energy inertial-inductive storage modules, however, should allow the above values to be attained by nonexplosive means.

INDUCTIVELY DRIVEN LINER FLUX COMPRESSION

Fast-rising high magnetic fields can be used to accelerate appropriately configured conductors up to high speeds and energies, thereby allowing magnetic flux compression without the use of high explosives. This technique has been demonstrated by Cnare[3] and Alikhanov, et al[4], using electromagnetically imploded metallic cylinders or liners in θ-pinch and z-pinch driving configurations, respectively. More recently, at NRL, this work has been extended to very large diameter (28 cm) cylindrical aluminum liners. Fields in excess of 1.3 Mgauss and implosion speeds greater than 1.4 mm/μsec have been obtained on a routine basis using a high strength, single-turn solenoidal driver coil powered by a 540 kJ capacitor bank. To use imploding liner flux compression for pulsed power generation more specialized geometries are needed such as shown schematically in Fig.2 .

**LINER DRIVEN
PULSE GENERATOR**

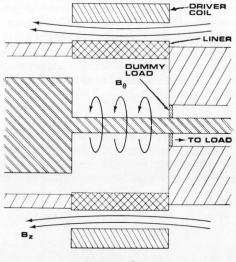

Figure 2

In this approach, a θ-pinch type, solenoidal driver coil is used to implode a cylindrical liner which becomes the outer conductor of a coaxial inductive store. The radial motion of the liner first traps azimuthal magnetic flux in the coaxial chamber and then compresses this flux to achieve high currents and fields. The current flow during most of the implosion can be through a parallel or dummy inductor as indicated in Fig.2 , with the actual load connected in parallel by means of closing switches when the rate of transfer of kinetic energy to magnetic energy has attained the desired power level. Thus, the first portion of the liner implosion is used to provide high magnetic fields against which the continued motion of the liner can work to obtain high powers. (This behavior is indicated in Fig.1). With load inductances much less than that of the dummy inductor, most of the electrical energy generated will be transferred to the load. This has been discussed for explosive generator-foil implosion systems in Refs.5 and 6 , and works equally well for non-explosively-driven flux compression devices.

INERTIAL-INDUCTIVE DRIVING SYSTEMS

For high energy pulsed power generation, involving tens and hundreds of megajoules, convenience and economy of operation requires more compact energy storage than is available with capacitor banks. It is therefore useful to employ high speed flywheels and high strength inductors for energy accumulation and peaking stage energy storage. Such inertial-inductive systems are under development at NRL[2] in the form of self-excited homopolar generators with oversized excitation coils.

At peak current, the field coils of the generator serve as an inductive store and in addition function as the primary windings of an air-core transformer. Interruption of a relatively low primary current (~100 kA) at high voltage(200 kV) is then used to provide the necessary high current for liner implosion. A modular approach is followed, with very high energy systems to be constructed as series/parallel arrays of a basic 10 MJ homopolar generator-inductive storage unit, (in the same spirit as a capacitor bank but with much higher energy elements). The small physical size of the module (approximately 2 m³) is an essential feature for its use in high energy experiments.

To study the coupling of such systems to flux compression generators, the circuit shown in Fig.3 has been analyzed. The first portion of the circuit consists of a self-excited homopolar generator in series with the primary of a transformer (the excitation coils) and an SF_6 breaker, which is initially closed. In parallel with the breaker, upon command closure of a switch, is a capacitor driven circuit which is used to create and maintain an arc current zero. A fuse in this circuit provides the essential dissipative impedance for transferring energy from the primary to the secondary circuit (without requiring significant intermediate energy storage). Interruption of the primary current by the action of the fuse induces current in the secondary (and also in the homopolar discs). This basic circuit behavior occurs simultaneously in a series/parallel array which provides current to the solenoidal driver coil.

The resulting liner implosion is modeled by treating the driver coil and liner as coupled inductors, with the mutual inductance and self-inductance of the liner determined by the liner motion. The liner resistance also changes with time due to joule heating, plastic deformation and simple geometry change. Flux leakage from the driver coil through the liner is thus included in the analysis.

Not shown in Fig.3 is the liner driven generator (of the form in Fig.2), which converts the implosion energy into magnetic energy. This element is introduced in the electromechanical circuit after the liner has acquired most of its maximum kinetic energy. Analysis of the subsequent implosion onto an initial amount of trapped azimuthal magnetic flux includes joule heating of both the liner and the center conducting rod. Transfer of energy to useful loads is also allowed, but is not considered in the present discussions.

SCHEMATIC OF LINUS PULSED POWER CIRCUIT

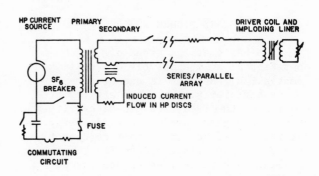

Figure 3

In Figs.4 and 5, the behavior of various portions of the system is displayed as their actions become the controlling features of operation. Note that the time scale along the abscissa changes, with homopolar generator self-excitation requiring tenths of a second, the breaker opening in several milliseconds and current interruption occurring in about a hundred microseconds. In the last regard, a particularly simple model for fuse vaporization was employed which yields the conservative result that the blowing time of the fuse is about equal to its delay or conduction time. Experimentally, fuses tend to blow in shorter times (by a factor of about six) than the conduction time, so longer conduction times for clearing the breaker and shorter interruption times should be possible. The efficiency of system operation should therefore be higher than predicted here since the increase in secondary inductance due to liner motion will be less on the time scale of circuit interruption.

Of the 10 MJ stored initially in the two counterrotating flywheels of each module, about 6.7 MJ of electrical energy is generated in the primary. Using a fifty module array, with ten in parallel and five in series, an aluminum liner (1.5 m wide, 1.5 m initial radius and 1.5 mm initial thickness) is accelerated up to 1.9 mm/μsec and a total kinetic energy of about 92 MJ. The implosion of this liner onto an azimuthal magnetic field with an initial energy of 1 MJ results in a current of about 250 MA in the eighty centimeter diameter center "rod". With a load inductance of one nanohenry to simulate a low impedance foil load, the final load energy is about 37 MJ. The peak electrical power generated in this case is about 7.5 terawatts. While the operation and interaction of the various system components has not yet been completely investigated to provide optimum total performance, the overall efficiency of conversion of flywheel energy to load energy is already quite good (7.4%), with a conversion efficiency from primary electrical energy of 11%.

TOWARD HIGHER POWERS

The basic design trade-off in the operation of an inductively driven flux compression generator is one of power versus efficiency. From the earlier discussion of the idealized generator, it is seen that higher powers at fixed energy and generator size require higher conductor speeds. Higher final speeds for the imploding liner generally mean higher speeds earlier in the liner motion (although some gains are in principle possible with very

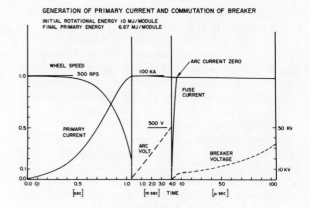

Figure 4

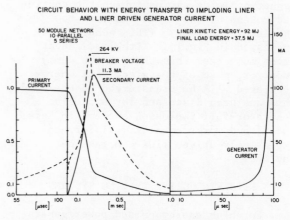

Figure 5

thick liners). The effective inductance of the secondary circuit during the switching process, therefore, increases so that the dissipation in transferring energy from primary to secondary also increases, with an associated loss of efficiency. To achieve tens of terawatts of output power it is necessary to improve the techniques of inductive switching and/or utilize further energy transformation stages to boost power, albeit at some cost in efficiency.

In the latter regard, imploding liner techniques offer the possibility of attaining very high conductor speeds by means of magnetic field gradient acceleration of induced magnetic dipoles. The approach is indicated schematically in Fig.6, in which the magnetic field gradient is due to the nonuniform inside diameter of the imploding liner. As shown, the conducting projectile would be injected at high speed into a target solenoid where its kinetic energy would be converted into electrical energy by the displacement of flux. The concept, labelled here as METEOR, has discussions with M.L. Cowan of Sandia Labs., Albuquerque, N.M., in its ancestry, and follows the work there on the physically larger,

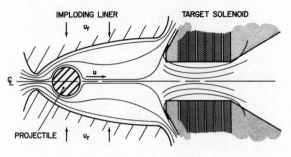

METEOR:
POWER MULTIPLICATION BY
MAGNETICALLY ACCELERATED
HYPERVELOCITY PROJECTILE

Figure 6

but slower PULSAR system[7].

The concept of using imploding liner flux compression to generate high magnetic field gradients for macroparticle acceleration is not new either[8], with the usual result given that the maximum final kinetic energy of the particle will about equal the product of the magnetic energy density in the solenoid and the volume of the particle. A more specialized calculation appropriate to imploding liner flux compression, however, yields the possibility of considerable improvement over this result.

If the projectile is a sphere of sufficiently high conductivity that resistive diffusion can be ignored on the time scale of the final stage of liner implosion (when high field levels are attained), then the axial accelerating force may be written as:

$$F = - \frac{a^3}{2\mu} B \frac{dB}{dz} \quad ,$$

where a is the radius of the sphere and B is the effective magnetic field. By flux conservation inside a liner of nonuniform inner diameter, Br^2 = constant, where $r = r(z)$ is the liner radius at axial station z. As indicated in Fig.6, by proper shaping of the liner surface it should be possible to squeeze the projectile along the axis, continuously accelerating it with a traveling magnetic field gradient of maximum strength.

To estimate the performance of such a system, we consider a shaped liner imploding with uniform radial speed. Upstream of the projectile, the liner radius may be found in terms of the initial energy per unit length of the liner E_1 and the change in the projectile energy E_p :

$$(E_1 - \frac{dE_p}{dz}) \, \delta z = \frac{\pi B_1^2 r_1^2}{2\mu} \delta z \quad ,$$

where B_1 is the field upstream of the projectile, r_1, is the liner radius, and δz is an increment of axial distance. Downstream of the projectile, the magnetic field is B_0 and the liner radius is r_0 , with $B_1 r_1^2 = B_0 r_0^2$. In the annular region adjacent to the projectile surface, the magnetic field will be taken as $B_M = B_0/(1-a^2/r_0^2)$. Substituting in the formula for the axial force, with $dB/dz \approx (B_0 - B_1)/2a$, we obtain, after some manipulation:

$$F = \frac{a^2 B_0^2}{4\mu(1-a^2/r_0^2)} \cdot \frac{(\frac{2\mu}{\pi B_0^2 r_0^2} E_1 - 1)}{(1 + \frac{1}{2\pi} \frac{a^2/r_0^2}{1-a^2/r_0^2})} \quad .$$

Relating E_1 and B_M by:

$$E_1 = K \frac{B_M^2}{2\mu} \cdot \frac{4}{3} \pi a^3/2a = \frac{\pi}{3} \frac{B_M^2}{\mu} a^2 K \quad ,$$

where K is a proportionality constant, the previous equation reduces, in the limit of a closely coupled liner-projectile system $(a = r_0 - \delta$, with $\delta \rightarrow 0)$, to $dE_p/dz = KE_1$. If the liner is shaped so that the "contact point" travels with the accelerating projectile, then $E_p \approx KE_1 z$, and the efficiency of energy transfer is simply K .

To evaluate K it should be noted that the magnetic energy left upstream of the projectile will cause the liner to re-expand. In order to preserve the local field gradient, the time scale for such re-expansion should exceed that for the projectile to move axially a distance about equal to its diameter. It may be shown then that in the same limit of closely coupled elements, the minimum value of K is:

$$K = 1 - \frac{2\delta}{f r_0} (\frac{3}{8} \frac{M}{m})^{\frac{1}{2}} \quad ,$$

where $f = (\frac{\mu}{2\pi} \frac{M}{m} KE_1)^{\frac{1}{2}}/B_1 a \approx 1$ is the ratio of re-expansion time to projectile motion time, and M/m is the ratio of liner mass to projectile mass. The maximum value of projectile speed to liner speed, with K held constant at this minimum value is then $u_p/u_r = 0.577(M/m)^{\frac{1}{2}}$ and is obtained with an efficiency $\varepsilon = K = 1/3$. Thus, with $r_0 \approx a$, it should be possible to increase the output pulsed power by u_p/u_r , at a cost, however, in overall system efficiency.

CONCLUSION

The use of imploding liner flux compression techniques coupled with compact, high energy inertial-inductive storage systems offers the possibility of economical operation of high energy, high power experiments within conventional plasma laboratories. Power levels in excess of several terawatts, involving ten and hundreds of megajoules in pulse times of a few microseconds should be possible, with overall efficiencies of 7-10% into useful loads. Experimental development of this new level of pulsed power technology is proceeding and should have substantial impact on advanced plasma systems in the near future.

REFERENCES

1. H. Knoepfel, Pulsed High Magnetic Fields, North Holland Publishing Co., Amsterdam (1970).

2. A.E. Robson, P. Turchi, W. Lupton, M. Ury, and W. Warnick, Proc. Intern. Conf. on Energy Storage, Compression, and Switching, Nov.5-7,1974, Asti, Italy.

3. E.C. Cnare, J. Appl. Phys., 37, 3812 (1966).

4. S.G. Alikhanov, V.G. Belan, A.I. Ivanchenko, V.N. Karasjuk, and G.N. Kichigin, J.Sci. Instrum.,(Series 2), 1, 543 (1968).

5. P.J. Turchi and R.S. Caird, AFWL Technical Report 74-222, Air Force Weapons Laboratory, Kirtland AFB, N.M. 87117.

6. P.J. Turchi and W.L. Baker, J. Appl. Phys., 44, 4936 (1973).

7. M. Cowan, E.C. Cnare, W.B. Leisher, and R.E. Stinebaugh, Proc. of Intern. Conf. on Energy Storage, Compression and Switching, Nov.5-7, 1974, Asti, Italy.

8. R.L. Chapman, Proc. of Conference on Megagauss Magnetic Field Generation by Explosives and Related Experiments, Sept.21-23, 1965, Frascati, Italy, Euratom, Eur 2750.e, P. 107.

MULTIMEGAJOULE PULSED POWER GENERATION FROM A
REUSABLE COMPRESSED MAGNETIC FIELD DEVICE

M. Cowan, E. C. Cnare, W. K. Tucker, and D. R. Wesenberg
Sandia Laboratories Albuquerque, NM 87115

ABSTRACT

A pulsed power system is described which is potentially capable of producing multi-megajoule pulses with millisecond rise time and repetition rates of more than one per second. A metallic piston is driven by gas combustion through axially aligned coils which are originally energized with relatively small magnetic energy. Experimental results are presented for a pulse generator of this kind which utilized a piston of 20 MJ kinetic energy, and comparison is made with predictions of a numerical model.

INTRODUCTION

For generating pulse power with rise time of about one millisecond, flux compression is an interesting alternative to methods currently being developed which utilize inductive energy stores and switches or rotating machinery. In flux compression, magnetic energy is generated when the inductance of a closed circuit is forcibly reduced with sufficient speed. This has been done previously using high explosive detonations. However, high explosive causes generator destruction which is often unacceptable. In flux compression generators described here, the primary energy source is ultimately intended to be gas combustion; however, experimental results to date have employed artillery shell propellant. The generators do not destroy themselves when they operate and one version which utilizes a superconducting magnet seems particularly adaptable to production of a continuous train of powerful pulses. Within limits, pulse rise time and shape can be varied conveniently without circuit breakers and for some versions without crowbar switches.

This paper will qualitatively describe generator operation and indicate the exploratory program being carried out by examples of numerical simulation and experimental results.

GENERATOR OPERATION

Pulse power is produced by driving a conducting piston (called the armature) through coils which contain an original magnetic flux. The pulse generator identified by the name PULSAR has two versions. One uses normal conductors only, while the other utilizes both normal and superconductors. In the normal conductor version

shown in Fig. 1, the armature is driven through a number of coils sequentially, the first of which is energized by an auxiliary power supply which is shown here as a capacitor bank. After the capacitor energy is discharged into the first coil, the first crowbar switch is closed.

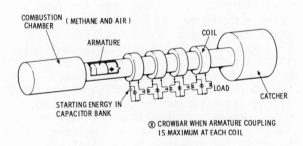

Fig. 1 PULSAR Generator with Normal Conductors

This must be done just before the armature couples strongly to the coil. When the armature has maximum coupling to the first coil the second crowbar switch is closed and so on. If the time for the armature to pass from one coil to the next is short compared to the characteristic time for current decay, magnetic energy can increase from coil to coil at the expense of kinetic energy of the armature until the final pulse is delivered to the load.

The PULSAR which employs a superconducting coil is shown in Fig. 2. In this device, the armature is driven through

having mutual inductance (not indicated in the Figure) with all other coil and armature segments. Current and both

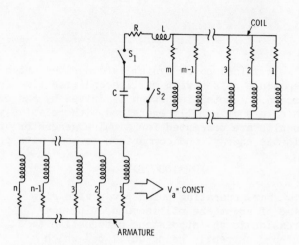

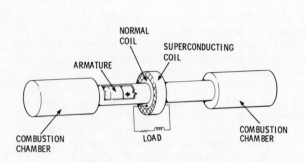

Fig. 2 Superconducting Version of PULSAR

Fig. 3 Circuit to Simulate Normal Coil Operation

two nested coils, the outer one being superconducting and the normal inner one being connected in series with the load. Prior to armature arrival, current in the normal coil is zero while the superconducting coil carries a steady current providing flux in the normal coil by mutual inductance. When the armature passes through the two coils the coupling coefficients involved are so valued that the pulse delivered to the load may have an energy even greater than that originally stored in the superconducting coil and yet the energy in the superconducting coil is restored to its original value soon after the armature has passed. Since no circuit breakers or crowbar switches are required, this design seems well suited to a reciprocating armature arrangement for producing a train of pulses.

PERFORMANCE ANALYSIS

Numerical calculations have been carried out to establish performance estimates for PULSAR. The circuit shown in Fig. 3 has been used to simulate passage of an armature through a single normal coil. Load inductance and resistance are L and R, the capacitor bank has capacitance C, and switches S_1 and S_2 are closed in sequence to first charge and then crowbar. For calculation, both the coil and armature were divided axially into a number of segments, each having inductance and resistance, and each

axial and radial forces were calculated as functions of time for each segment. The coil was assumed to be made of copper sheet wound in a simple spiral with sheet width equal to coil length and sheet thickness the order of skin depth. The armature was assumed to be a shell of aluminum or copper with thickness the order of skin depth. These calculations were not self-consistent in that the armature did not slow down as magnetic force was generated. Also, room temperature resistivities were used with no accounting for increases due to ohmic heating. Examples of axial current distributions computed for the first experimental coil and armature designs are shown in Fig. 4. Distributions are shown for the two armature positions indicated. Values used for these calculations are typical of initial experiments.

Armature Speed - 800 m/s
Coil I.D.   - 300 mm
Pulse Energy   - 100-200 kJ (in the
                 load)

Corresponding radial pressures on the coil are shown in Fig. 5. Notice that pressures are very nonuniform with the peak value exceeding 300 MPa. This shows the disadvantage of the spiral coil which has been used in experiments to date. While it is relatively easy to build and analyze, it is not suitable for high energy operations.

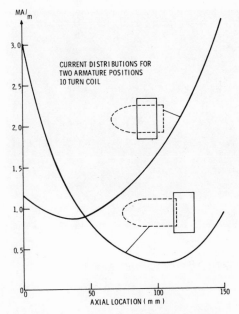

Fig. 4   Current Distribution

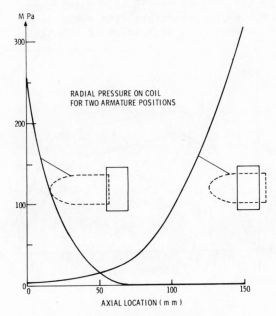

Fig. 5   Pressure Distribution

To estimate performance of the superconducting version of PULSAR, circuits shown in Fig. 6 were numerically analyzed. In this analysis, kinetic energy of the armature, was allowed to change to be consistent with changes in magnetic energy. Mutual inductances and the associated coupling coefficients k_{sa}, k_{sn} and k_{na} were varied with time to simulate passage of the armature through the nested coils L_s and L_n. Currents, armature velocity, and magnetic, kinetic and $\int I^2 R$ dt ener-

gies were calculated as functions of time. One interesting discovery was that a pulse may be generated in the load L without inducing current changes in the superconducting circuit provided the following conditions are satisfied.

$$R_n = 0 \qquad\qquad (1)$$

$$R_a = 0 \qquad\qquad (2)$$

$$k_{sa} = \frac{L_n}{L_n + L}\, k_{sn}\, k_{na} \qquad (3)$$

There would be no change in superconducting current if there were no ohmic heating and if the coupling coefficients met the conditions specified by Eq. 3. Under these conditions, energy conversion would be reversible.

However, of more general interest is the case where the magnetic energy once generated is utilized in some irreversible way, for example, by an R or dL/dt in the load. For this case, current in the superconducting circuit must change. Still, it is possible to return superconductor current to its original value after a small excursion while generating a pulse with energy equal to or greater than that stored. Numerical experiments indicate that this is most effectively done by satisfying the condition expressed by Eq. 3 while also making k_{sn} relatively small and k_{na} as close to one as possible.

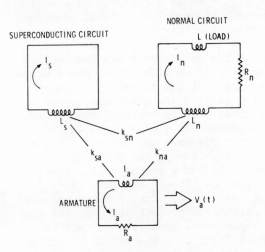

Fig. 6   Circuit to Simulate Operation of Superconducting PULSAR

EXPERIMENTAL PROGRAM

Initial experiments have been carried out using a recoilless cannon to propel an armature through a single normal coil. This cannon, shown in Fig. 7, has a bore of 293 mm and was made by connecting two 10 m long barrels breech-to-breech. Recoil is eliminated by firing the armature and another projectile of equal mass in opposite directions. A nonconducting extension of the barrel incorporates a single coil which is connected in a circuit like the one shown in Fig. 3 including capacitor bank, crowbar switch and load. The experimental arrangement at the end of the barrel and an aluminum armature are also shown.

Fig. 8   Coil and Armature

Fig. 7   Recoilless Cannon

Fig. 8 shows a close-up of a coil structure and armature. These were the first to be constructed. The coil structure was used on the first two experiments. The coil itself was made of 1.6 mm thick, 152 mm wide copper sheet wound in a spiral of 8.8 turns. The nonconducting part of the structure was of fiberglass tape impregnated with epoxy. The armature had a mass of 68 kg, and its outer diameter was 292 mm. It was made bullet-like in shape to reduce peak magnetic pressure on the coil. Armature velocity was 770 m/s and its kinetic energy was 20 MJ. Fig. 9 shows measured and calculated current gain for the second experiment. Measured current increased from 50 kA at crowbar to 125 kA peak about 0.25 ms later. Coil energy at crowbar was 28 kJ and peak energy in

the 4.5 μH load was 35 kJ. Total magnetic energy increased from about 34 kJ at crowbar to a peak value of 100 kJ.

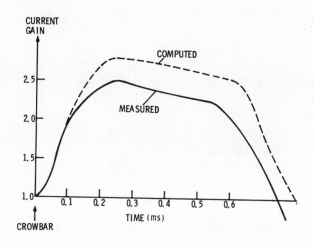

Fig. 9   Measured and Computed Current Gain

The present experimental program is aimed at developing coil and armature designs which will establish limits of performance at the present 290 m diameter. This limit is expected to be a few MJ's in the load with a meter-long coil. More efficient and higher energy operation is possible for larger diameter systems. Development of combustion chamber technology will begin early in 1975.

IMPLOSION AND STAGING SYSTEMS FOR A SCYLLAC FUSION TEST REACTOR

R. F. Gribble, R. K. Linford, and K. I. Thomassen
University of California, Los Alamos Scientific Laboratory
Los Alamos, New Mexico 87544

ABSTRACT

The implosion heating and adiabatic compression processes will be separated in future theta pinch devices. The circuit to achieve the fast implosion heating and power crowbar (staging) for the Scyllac Fusion Test Reactor is described here. The plasma is very tightly coupled to the circuit and presents a varying inductive load. Computer-aided circuit designs which achieve a programmed magnetic field waveform are described. The field approximates a two-step waveform, on-off-on, which is ideal for achieving the large initial plasma radius needed for stability. The components for the circuits have been developed and are being tested in experiments at Los Alamos.

The concepts of implosion and staging in theta-pinch systems were proposed several years ago[1] to meet the disparate requirements of fast rising magnetic fields for initial implosion heating and slow rising, long delay-time magnetic fields for adiabatic compression heating and thermonuclear burning. The staging supply holds the fields at the implosion field value until the compression field comes on. The compression fields are applied to a separate field coil, and are driven by low cost capacitive or magnetic energy supplies. The implosion and staging fields are applied by a low inductance coil nested inside the compression coil.

Experiments aimed at studying the implosion process are underway at Los Alamos[2] and both analytical[3] and computer simulation[4] models of the implosion process are being advanced to aid in understanding the process and in the design of the required circuitry. This paper reports the results of a plasma-circuit design of the implosion system of the Scyllac Fusion Test Reactor (SFTR).

The design of the implosion system depends on the plasma model. A simple bounce is assumed here, with the ions treated as billiard balls in cylindrical geometry that bounce off the magnetic piston with a speed twice that of the piston velocity V. In principle the system efficiency can be increased over that with a unit step field B_o by application of a pulse of strength $B_o/2$ for a time $T = \frac{2}{3} b/V$ followed by a step field B_o at $t = b/V$ where b is the wall radius. Thus the ions move outwards freely during the interval

T<t<1.5 T without losing energy to the field and are reflected by a stationary piston at t = 1.5 T. It is technically difficult to achieve the ideal magnetic field waveform. However, by using a numerical code that includes the simple plasma model and the complete electrical circuit an implosion system was designed based on the free expansion principle that obtains plasma model parameters not very different from those of the ideal waveform.

The circuit, Fig. 1, consists of two or more parallel connected capacitor banks switched sequentially. Varying the capacitor charge and switch timing provides a wide variety of wave shapes. The first 125 kV bank (C_1 of Fig. 1) produces the initial plasma acceleration during most of the first half-cycle of the LC circuit, with a peak field of about 0.6 T. Near

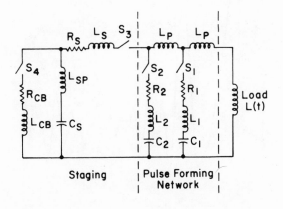

Fig. 1. Implosion Heating Network.

the end of the half-cycle when the field is small, the plasma expands and is reflected near the wall as the second 125 kV bank C_2 is connected, bringing the field up to 1.3 T. Figure 2 presents the calculated values of magnetic field, piston radius and plasma energy. The load coil of Fig. 1 is a 0.11 meter-radius, 0.55 m length Marshall coil of effective turns ratio 0.8. The effect of the compression coil (which surrounds the implosion coil) on the implosion system is minimized by filling the space between the coils with a ferromagnetic material. The values of C_1, C_2, and C_s (per 0.55 meter load coil) are 0.4, 1.2 and 16 µF. Voltages are 125 kV on C_1 and C_2, and 50 kV on C_s. Switch S_1 is closed at t = 0, S_2 at 0.65 µs and S_3 at 0.35 µs.

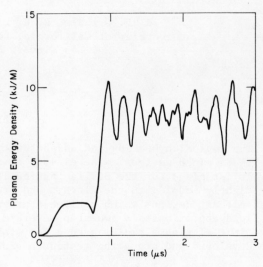

Fig. 2c

Fig. 2.   Computer calculations of a) magnetic field, b) piston reading, and c) plasma energy as a function of time.

Passive clamping of the circuit can provide a field sufficient for containment for several tens of µs, depending on the resistance of the clamp. The circuit of Fig. 1 shows an active crowbar capacitor, C_s, which extends the field sufficiently long for the main compression field rising to 5 T in 1 ms to take over. For the calculated values of Fig. 2, S_3 was actually closed before S_2 to prevent field reversal and to decrease the "catcher" field risetime. Figure 3 shows

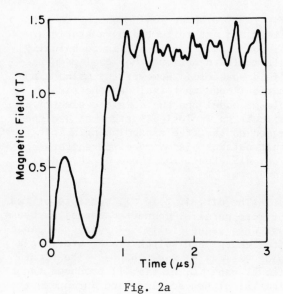

Fig. 2a

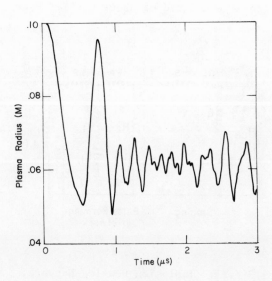

Fig. 2b

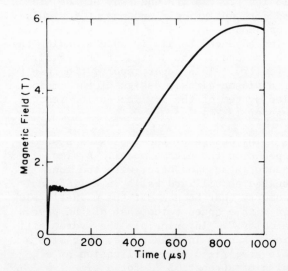

Fig. 3.   Computer calculation showing the sum of the implosion, staging, and compression fields as a function of time.

the field for an extended period. Preliminary results from newly designed crowbar gaps indicate that active clamping may not be necessary.

Although the design is technically feasible (the 125 kV capacitors were designed for Scyllac), development of new spark gap switches was necessary. Requirements for bank 1 switches are reasonable: 125 kV DC holdoff (the capacitors are actually pulse charged), triggering with less than a few ns jitter, and low inductance. Bank 2 gaps must in addition withstand pulse voltages to 300 kV and, for some operating modes, trigger at low voltage. Staging switches must hold 50 kV DC, 200 kV pulse, trigger at low voltage, and exhibit very low inductance and resistance.

Banks 1 and 2 use the same field-distortion type spark gap, and are capacitor mounted in coaxial geometry. The gap air pressure varies from 3 to 10 atmospheres and is adjusted to suit the mode of operation. Low value electrolytic resistors and geometrically matched electrode capacitances provide dynamical biasing of the symmetrical trigger electrode for holdoff of rapidly applied pulses. At 8 atmospheres the holdoff potential is greater than 350 kV. For a rate of rise of 20 kV per ns on the trigger ring, and at 8 atmospheres pressure, 75 kV capacitor charge, these gaps have a delay of 7 ns with jitter of less than one ns. The measured inductance of 16 nH compares well with the estimated value of 15 assuming a continuous current sheet

around the ring electrodes. The staging start and crowbar gaps are linear rail types arranged so that the arc is blown away from the trigger electrode.

These implosion systems represent a departure in technology from present Scyllac systems, and offer significantly improved performance. However, both the technology and physics are evolving, and further work on development and plasma modeling is clearly indicated.

REFERENCES

1. J. P. Freidberg, R. L. Morse, and F. L. Ribe, "Staged Theta Pinches with Implosion Heating," Symp. Tech. of Controlled Thermonuclear Fusion Experiments and Engineering Aspects of Fusion Reactors, Austin, Texas, Nov. 20-22, 1972.

2. K. S. Thomas et al, "Implosion Heating Studies in the Scylla 1B, Implosion Heating, and Staged Theta Pinch Experiments," Fifth Conf. on Plasma Physics and Controlled Nuclear Fusion Research, Tokyo, Japan, Nov. 11-15, 1974. Paper IAEA-CN-33/E 8-4.

3. T. A. Oliphant, "A Mixed Snowplow Bounce Model for Shock Heating in a Staged Theta Pinch," Nuclear Fusion 14 (1974).

4. R. F. Gribble et al, "Shock Heating and Staging Circuit for FTR Test Module," Bull. Am. Phys. Soc. 19, 937 (1974).

ENERGY STORAGE, COMPRESSION, AND SWITCHING IN A
THETA-PINCH FUSION TEST REACTOR

Keith I. Thomassen
University of California
Los Alamos Scientific Laboratory
Los Alamos, New Mexico 87544

ABSTRACT

A new 488 MJ superconducting magnetic energy storage and transfer system is being proposed for a Scyllac Fusion Test Reactor. The 1280 module system uses vacuum interrupters to switch 26 kA storage currents in 0.7 ms through a capacitive transfer circuit at 60 kV to the compression coils in the machine. Many of the components of the system have been built and tested and a prototype section of the machine is planned. Prototype coils with 381 kJ at 26 kA currents will be built by industry using advanced superconducting wire. The wire uses a Cu and Cu-Ni matrix around filaments of Nb-Ti to minimize eddy current losses. These wires are presently used in a 10 kA braided conductor for 300 kJ pre-prototype coils, and can withstand field changes of $\sim 10^7$ gauss/sec without inducing normal transitions. Three such 300 kJ coils are being constructed in industry for the LASL program.

A new magnetic energy transfer and storage system is being developed for the Scyllac Fusion Test Reactor (SFTR), presently in design at Los Alamos. The machine is designed to "breakeven" by producing as much neutron energy in a pulse as is contained kinetically in the plasma.

The plasma is heated and contained in a 20 cm bore toroidal ring, with a 40 m radius. The programmed magnetic field is generated by two different sources. An implosion heating system initially heats the plasma to 2.8 keV by driving a low inductance end-fed implosion coil with 240 kV capacitors switched separately onto the parallel plate feed line. The field is held at the 11 kG level for 200 μsec with crowbars until energy from a 488 MJ superconducting storage ring is transferred (in 0.7 ms) to a separate compression coil, raising the field to 55 kG and bringing the plasma to ignition. The coil is crowbarred to give a 250 ms L/R decay of the field, during which time the neutron energy is delivered.

The 5 MJ implosion heating capacitor bank brings the field up to 11 kG in 200 ns, and the circuit is "tuned" to resonate with the plasma motion. The quarter period determined by the capacitor and inductive load matches the time for the plasma to reach its minimum radius at the first "bounce". This resonant mode, now under investigation,[1] can be very efficient of energy transfer.

The major new system is the 488 MJ superconducting storage and switching system. The system has 1280 modules of 381 kJ coils, each module driving a 20 cm section of the torus. One is forced to have many modules by the constraints imposed by the current interrupters required to initiate the transfer of energy from the storage coil to the compression coil. The constraint is easily understood from the simplified circuit for a single module shown in Fig. 1.

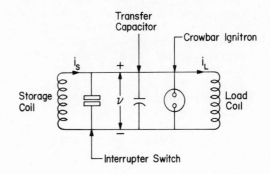

Fig. 1. A circuit for transferring approximately 100% of the stored energy to a load with a transfer capacitor.

To avoid energy dissipation in the switch, a transfer capacitor bank is used to resonantly transfer the energy to the load in the transfer time τ. The bank must handle half the stored energy during the transfer (assuming equal load and storage inductances and, hence, currents), but without it, that energy would be dissipated in the switch. (This follows from flux conservation). Further, a quarter of the energy would be left in the storage coil, giving a maximum of 25% transfer efficiency.

With this transfer circuit, the product of the initial current I and the maximum voltage V (occurring midway through the transfer period) is $\pi E/\tau$, where E is the initial stored energy. The interrupter must break the current I and withstand the voltage V, so its ratings determine the maximum energy E. At present, we envision using vacuum interrupters with I = 26 kA, V = 60 kV. Using a single Westinghouse vacuum breaker we have achieved 13 kA @ 40 kV, limited by the test facility. The higher voltage requirement should be easily achieved, but the current requirement may be at the limit of reliable, long-life operation.

The storage coils are constructed from low-loss a-c superconductors. To achieve extremely low losses one constructs wires of 1/4 - 1/2 mm diameter containing several hundred filaments embedded in a Cu and Cu-Ni matrix. The Cu-Ni surrounds each filament and resistively separates it from the other filaments, preventing eddy currents which could drive the wire into the normal state. The wires are twisted during drawing, and hundreds of these wires (each carrying 50-100 amps) are woven into a fully transposed braid to make a 10-20 kA cable which can withstand rapidly changing magnetic fields without inducing large losses.[2]

To date, a 380 kJ superconducting coil has been constructed and successfully tested at LASL. However, the 10 kA wire was a simple rectangular copper conductor with embedded filaments of Nb-Ti and no Cu-Ni. As a result, the coil goes normal during discharge, resulting in somewhat higher losses than allowed for SFTR. Nonetheless, part of the energy was transferred (to limit V x I) to a dummy load coil in 1 ms @ 40 kV using vacuum interrupters and capacitive transfer. (The full 380 kJ was transferred in several milliseconds at lower voltage, hence all required parameters were tested).

A program of industrial development was initiated to produce the low-loss coils. Three suppliers are building 300 kJ coils using 10 kA low loss cables braided from mixed-matrix multifilamentary wire. These will be tested beginning in late 1975.

The circuit to be used in SFTR[3] is shown in Fig. 2, which contains the linked modules and the necessary switches for energizing the full set of coils, and isolating them into the individual circuits of Fig. 1 prior to the transfer of energy to the machine. A prototype engineering section containing 3 linked modules will be constructed in a few years to test the individual components and controls.

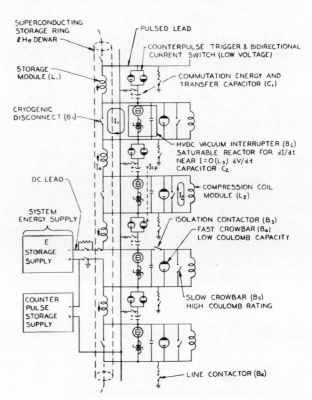

Fig. 2. A circuit for transferring 488 MJ of energy from superconducting storage coils to the compression coils of SFTR, using vacuum interrupters and transfer capacitors.

The coils are stacked ten to a dewar, with 128 ten-packs around the machine. The coils are electrically in series for charging and separated by cryogenic disconnecting switches submerged in the ℓ He. The functions of each component are described in Table 1, but briefly, the circuit operates essentially like the simplified circuit of Fig. 1. Current is transferred through the vacuum interrupters, which are closed following charging, by opening the disconnects B_1. The transfer capacitors are back-biased and discharged through the interrupters after they are opened to give a momentary current zero to extinguish the arc. The counter pulse is applied 5-10 msec after the interrupter contact is broken, which allows sufficient separation of the electrode to withstand the 60 kV voltage. The load current rises like $\sin^2 \pi t/\tau$ and is crowbarred at its peak.

The system described above is at the component development stage, but many of the key features have been tested. Prototype system tests to be made several years from now will determine the viability of this approach in SFTR.

REFERENCES

1. R. F. Gribble, R. K. Linford, G. P. Boicourt, "Shock Heating and Staging Circuit for FTR Test Module, Bull. APS, 19, 937 (1974).

2. H. L. Laquer, Superconductors for Millisecond Pulse Applications, Applied Superconductivity Conference, Oakbrook, Ill. (1974).

3. Swannack, C. E., Blevins, D. J., Harder, C. R., Lindsay, J. D. G., Rogers, J. D., Weldon, D. M., "10 kA, 300 kJ Magnetic Energy Transfer and Storage (METS) Test Facility, Applied Superconductivity Conference, Oakbrook, Ill. (1974) to be published. (Also available as a Los Alamos Scientific Laboratory Report, LA-UR-74-1434, Los Alamos, New Mexico).

TABLE I

METS-FTR CIRCUIT COMPONENTS

Main Power Supply
: 1-2 MW to ramp current to 26 kA in 300 second charging cycle. Energy stored = 1280 x 381 kJ = .488 MJ. Current leads from main supply enter circuit at a point which remains at low voltage at all times.

Counterpulse Power Supply
: 50 kW supply. It charges up the transfer capacitors to several kV. 1-2 MJ needed for counterpulse, total.

Inductors
: L_1 Storage module, superconducting. $L \simeq 1.1$ mH

> E = 381 kJ at 26 kA
> V = 60 kV during 0.7 ms transfer, peak
> B = 2.5 T

L_2 Compression module, multiturn copper. L = 1.1 mH
$\lambda/16$ section, 20 cm of toroidal circumference

> B = 5.5 T
> L/R = $\leq$ 250
> E = 325 kJ when B = 5.5 T inside bore.

L_3 Saturable reactor to allow preferred dI/dt near I = 0 during commutation. Parameters to be determined.

Capacitors
: C_1 Transfer capacitor $\begin{cases} C_1 = 100 \ \mu F \\ V_1 = 60 \ kV \end{cases}$ sets transfer time and peak voltage during transfer. It is reverse charged to $\leq$ 5 kV to supply counterpulse current.

C_2 Transient capacitor to handle the overcurrent during commutation and limit voltage on HVDC interrupter during deionization period, several 10's μsecond. The small series damping resistor eliminates ringing.

TABLE I

(continued)

Breakers, Contactors,
Crowbars

B_1 Cryogenic disconnect within ℓHe cryostat. It carries the charging cycle current up to 26 kA. It is to be designed for a power loss < 24 W at full current. Resistance $\leq 3.8\ (10)^{-8}$ ohms.

B_2 HVDC vacuum interrupter, or equivalent. This closes just prior to B_1 disconnecting, current transfers into the pulsed lead ambient temperature loop to B_2.

B_3 Isolation contactor. Removes E storage voltage from compression coil module during charging cycle.

B_4 Fast crowbar. Crowbars compression coil module at peak current (I = 20 Coulombs/ms)

B_5 Closing contactor. This relieves B_4 after several milliseconds. Voltage drop across each B_5 is insufficient to maintain B_4 conduction, extinguishing all B_4's.

B_6 Electrically operated line contactor. The series resistor provides the proper L/R to discharge the storage coil in a short time if the cryogenic disconnect, B_1, fails to open.

RECENT PROGRESSES IN GAS-EMBEDDED Z-PINCH AND NEUTRON PRODUCTION

Dah Yu Cheng
Plasma Research Laboratory
Santa Clara Research Institute

ABSTRACT

Recent progresses and neutron yield data of an exploding metal tube prefilled with deuterium gas are presented. Some problems related to the stability and control mechanisms of the Z-pinch are discussed.

INTRODUCTION

Alfven first pointed out that gas surrounding a Z-discharge does not bring with it high heat conduction losses. This has helped many of us to understand more about its physical process.

Fusion energy if it is ever to be harnessed by human beings may have to be demonstrated on a small scale experiment so that the physics can be easier to understand, because small scale experiments offer opportunities to perform experiments with a variety of parameters, with finite monetary resources for research. The gas-embedded Z-pinch has the potential to improve the current Lawson condition for fusion in a smaller size experiment, besides it has advantages in simplicity for future reactor designs, such as no vacuum wall contamination, possible direct energy conversion, etc. This also may offer competition in dollars per kw to the current power plants.

Recently, many paralleled efforts both in theory and in experiments[1-6] have been conducted such as the laser triggered Z-pinch, E-beam triggered Z-pinch, Vortex stabilized arc, plasma focus restrike Z-pinch, exploding wires and gas filled exploding tubes, etc. There are vast differences in the conditions a Z-pinch can be formed, and the conditions of most gas-embedded Z-pinches are different only in degree except for the explosion of a metallic tube prefilled with D_2 which has some unique physical process and properties. The discussion in this paper, therefore, will be limited to that process only.

CONDITION FOR THE FUSION EXPERIMENT

To achieve controlled thermonuclear fusion one needs to heat a deuterium plasma to 10^8 °K and to be able to confine the plasma according to Lawson's criteria in $n \tau_c$ to about 10^{14} particles sec/c.c.

Higher plasma density is traded off with a shorter required confinement time. If the plasma number density is 10^{20}/c.c., the required confinement time will be only 1 microsecond. Instability occurs to most of the confined plasmas. The characteristic time for growth τ_g becomes longer with a higher density gas-embedded Z-pinch such that the required confinement time τ_c is shorter at a certain density level. Hence, the troublesome mechanisms to stabilize the plasma can be eliminated. Unfortunately, the plasma density may be too high to reach this ideal condition unless the background gas has a high atomic weight and also a very high density for inertial confinement.

Most gas-embedded Z-pinches are initiated by the nature of their triggering systems such that an arc channel could be formed in a neutral gas having a very small channel diameter; for example, by E-beam, laser, and exploding wires or tubes, etc. The E-beam and laser triggered Z-pinch differ very little from the vacuum type Z-pinch experiment due to the fact that the inertial force around the Z-pinch is too small. The exploding wire produced plasma is not interesting as a fusionable plasma.

A metallic tube prefilled with a high pressure deuterium gas can be exploded by a capacity bank, the magnetic energy stored in the exploding tube is then transferred to the deuterium plasma to provide the energy needs for fast ohmic heating and magnetic confinement. The atomic weight of the metallic vapor can be very high (typically 60) and the number density of the vapor is also very high during the very short period of the desired confinement time. This special condition can prolong the characteristic kink growth period and can provide the inertia for confining the D_2 plasma in the center such that the condition $\tau_g > \tau_c$ can be satisfied. The magnetic energy storage system can also provide exceedingly high voltage peaks with a very short pulse period for heating those high density Z-pinch plasmas without the use of an expensive high voltage capacitor bank. Very preliminary data have indicated the possibility of very high neutron yield, thus making the exploding tube prefilled D_2 experiment worthwhile for further investigation.

From Bennett's relation, the Z-pinch requires the magnetic field B_θ due to I_z in a plasma channel to produce a magnetic pressure larger than the plasma pressure. From the force balance, too much heating would make the plasma pressure high, therefore, the plasma radius to become large. As a result the magnetic pressure is decreasing with the same I_z current.[7] This is especially true in low density rather than high density plasmas. Because the radiation losses from the plasma scale as density $(n)^2$, the amount of energy required to heat a small diameter (.1 cm) plasma is very low; therefore, one can afford to heat a high density plasma channel with the majority of the energy input used to compensate for energy losses. Beside the advantage of shortening the required confinement time τ_c, micro instabilities of the type independent of the plasma density will scale away. The higher density also has the added advantage of efficient ohmic heating.

PHYSICAL PROCESS IN Z-PINCH FORMATION

Illustrative density and pressure profiles can be seen in Fig. 1. If a metal tube with a density ρ_1 is prefilled with D_2 at a pressure higher than the ambient pressure, the electric current will heat the metal quickly to the vapor state, suddenly the vapor becomes a non-conductor for electricity. The originally high density (particles/c.c.) metal has created a pressure over 10^4 atmospheric pressure. This metal vapor cylinder propagates an imploding shock and an exploding shock as can be seen in Fig. 1(b). The shock reflects from the center producing a density well channel for the plasma Z-pinch and adiabatic heating. The prefilled D_2 gas can also be ionized by shock heating even at very high density, the current flowing in the metal tube suddenly is transferred due to the metal vapor becoming a good insulator. This is the Z-pinched plasma having a positive density gradient $\partial\rho/\partial r$; inherently, it can be kink stable.[8] From the pressure profile (a) one can see the original pressure of this experiment is not too high but

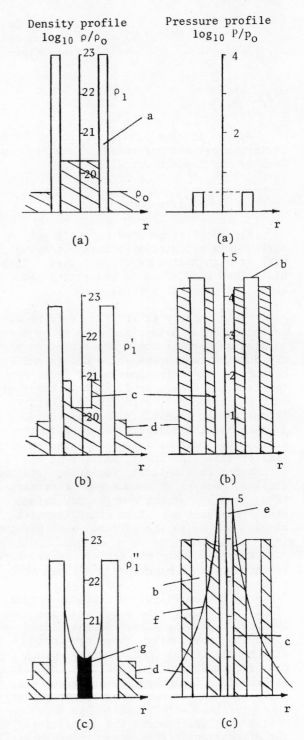

a - metal cylinder   e - pinched plasma
b - metal vapor pressure f - $B_\theta^2/8\pi$
c - imploding shock   g - Z-pinch
d - shock

Fig. 1. Time history of density, pressure and magnetic pressure of exploding tube Z-pinch.

when the tube is exploded it can produce
extremely high pressure which is needed
for stabilizing the Z-pinch, (b) when the
plasma channel is formed, the pinched
plasma is confined by even stronger magnetic pressure.

ENERGY STORAGE AND TRANSFORMATION

Any high density plasma heating
requires a high rise time, and a high
voltage pulse. High voltage capacitor bank
sooner or later runs into technical limitations on insulators and cost. Magnetic
energy storage in conjunction with exploding wires and ribbons, etc., are commonly
used. Their advantages are obvious in
that lower voltage capacitor banks can be
used to generate such a high voltage spike.

In an exploding tube Z-pinch experiment, even the exploding energy deposit in
the tube material is utilized. The size of
the tube is then a critical parameter. The
magnetic field profile can be seen in
Fig. 2. When the current is flowing in
the metal tube, the magnetic flux is completely excluded from the center region.
When the metal solid suddenly becomes
vapor, B_θ flux rush into the center of the
D_2 plasma to form the Z-pinch. Typical
voltage and current traces are shown in
Fig. 3. Some magnetic flux at earlier
stage are annihilated at the center in
order to supply energy for ohmic heating
and others will be excluded when the plasma
becomes hot.

The plasma channel can be formed in
any way that the imploding shock can ionize
the D_2 gas at the center due to the thermal
effect. Such a plasma channel would be
much smaller than the reflection formed
plasma density well. Just how to obtain
the best condition for fusion would be
an on-going research program.

PRELIMINARY EXPERIMENTAL RESULTS

Results of the exploding tube filled
with D_2 have been published (Ref. 2) before
but some high neutron yield results were
never reported for reasons that the phenomena were not understood until recently.
The difference between the two exploding
tube parameters is in the inductance of
their capacitor bank. A low inductance bank
could store magnetic energy better before
the current transfer to the plasma. Before,
the understanding of the plasma formation,
high neutron yields were unexpected at that
time; therefore, neutron counter saturation

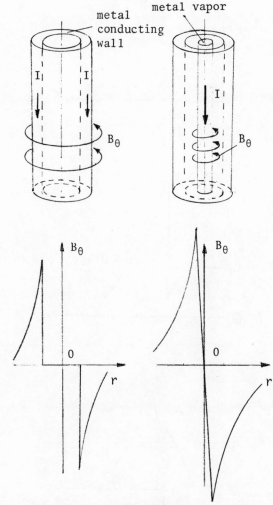

Fig. 2. Magnetic field profile of current
transfer.

and all of the other unexplainable phenomena surrounding the experimental results
have confused us for a long time. The conclusions are only considered to be sketchy
because not enough runs were made to establish a trend. The circumstances of the
experiment is described in the following.

The experiment used a stainless steel
tubing 0.317 in. in diameter with 0.010 in.
wall. The thin wall section was 6 in. long
and prefilled with D_2 up to 60 psig. Neutron detectors used were personal pocket
dosimeter, thermal neutron detector (Ben-
dix Model 609) and silver activated neutron
counter (on loan from LASL). Standard wall
mounted gamma ray detector was used to monitor gamma ray counts. The silver activated counter was 34 in. away from the
experiment and those carrying pocket

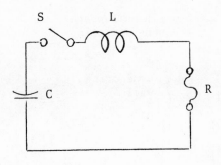

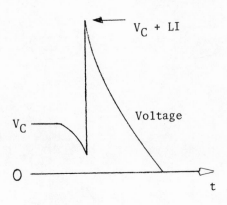

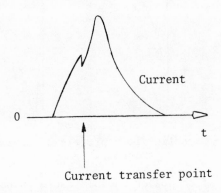

Current transfer point

Fig. 3. Voltage and current traces.

dosimeters were at least 20 ft away from the experiment with a 6-1/2 in. thick concrete wall barrier in between.

The capacitor bank used had 36 cans of 15 µf capacitor with less than 15 nh internal inductance per can charged to about 12 kv. Each capacitor can has an individual ignitron housing of less than 12 nh per housing. Each can has two double coaxial cables with an internal inductance of 28 nh/ft, and a capacitance of 110 mmf/ft. The cable length used was 25 ft. The reason for running at such a low voltage was that ignitrons tended to prefire. Eventually, at a higher voltage the prefiring had

destroyed the low inductance system. When high inductance gas spark gaps were used later, no high neutron yields were recorded again.

During the experimental period, the personal pocket dosimeters typically shifted between 1 to 3 milli-Roentgen after a shot. This was later found to indicate a neutron yield of approximately 10^{14} per shot. The wall gamma ray counter was jumped from 0.07 milli-Roentgen per hour to 0.09 milli-Roentgen per hour immediately. This would take 24 hr to settle back to 0.07 mR/hr. The silver activated counter was given counts below the normal background counts after the first minute. Continuous minute-counts up to 10 min indicated some strange structure. Only at a later date did one find that 10^{14} per shot would cause the counter to be saturated up to 14 minutes.

This high neutron yield was produced in a rapid current transfer mode so that the transfer occurred at above 3 µsec and the current continued to rise. The only explanation of the high neutron yield is possibly due to its extremely high plasma density which was inertially stabilized by an even higher density (approximately, at liquid density of 2×10^{22}/c.c.) heavy molecular weight vapor. The original density of the stainless steel is on the order of 10^{23}/c.c. Such an experimental condition is unique only for the exploding tube initiated gas-embedded Z-pinch.

DISCUSSION

Gas-embedded Z-pinch has long been neglected as a possible controlled fusion device even after it is known for its potential as a clean reactor and its simplicity. There are points leading to the skepticism which should also be cited:

(1) Instability: Many experiments indicated that gas embedded Z-pinches are stable, but very hot pinches also show they can be unstable. A neutral gas surrounding a plasma is not necessarily causing differences in instability characteristics because if the density of a plasma is approximately the same as the surrounding neutral gas then the inertial forces are far from sufficient to stabilize the pinch due to the plasma pressure is far greater than the gas pressure. Previous analysis[4] has shown that unless the surrounding gas can reach a density up to 10^{22}/cc and

if the plasma channel has a density of 10^{20}/cc before it can have a kink growth period of several microseconds which is more than a sufficient condition to reach Lawson criteria. Such a high density in an ordinary Z-pinch requiring high voltage storage systems just for the electrical breakdown would be beyond the current status of the art. However, exploding-tube-initiated Z-pinch is different than all other types of Z-pinches in that those conditions can be reached in simple experimental facilities. When the surrounding gas is an order of magnitude higher in molecular weight, the required surrounding gas density for inertial confinement can also be reduced.

It can be concluded that this experiment can reach fusion with its stability condition just on the boundary of the stability criteria.

(2) <u>End Losses</u>: The propagation of cooling waves from the ends can be estimated by the wave speed which can only be as fast as the speed of sound. Within the required confinement time, therefore, the length of the plasma channel can be scaled accordingly.

(3) <u>High Z-Metal Vapor Reaching the Center</u>: The D_2 gas has to be compressed to a high pressure in equilibrium with the metal vapor before the current-transfer; hence, both the D_2 gas and metal vapor are at very high densities. The metal vapor can penetrate the metal vapor-gas interface only by classical diffusion. Classical diffusion coefficient is inversely proportional to the mean free path and is dependent on the square root of the temperature. At low density it indeed can be a problem but in a high density gas, diffusion is very slow and the required confinement time is short. Again, at a certain high density contamination of high Z-material ought to be negligible.

(4) <u>Overheating</u>: It was shown by Baker and Phillips[7] that the Z-pinch has a tendency to overheat as the plasma would reach a pressure higher than the magnetic pressure. This situation again is alleviated in high density regions that the radiation scales as density squared. In this experiment, it requires more heating than the lower density Z-pinch. This can restrict the experimental condition of the plasma channel radius in two ways:

(1) to produce just enough plasma for fusion with an available pulsed energy input, and (2) to have a high enough magnetic pressure to confine the plasma without excessive current.

At this point, one can only conclude that the knowledge of the gas-embedded Z-pinch is very sketchy but the future potential for controlled fusion certainly is great.

<u>REFERENCES</u>

1. D. Y. Cheng, W. J. Loubsky, V. E. Fousekis, Phys. Fluids, Vol. 14, 11, 2328, (1971).

2. D. Y. Cheng, Nuclear Fusion Letters, Vol. 13, No. 1, 129, (1973).

3. D. A. Tidman, Structure of a gas embedded Z-pinch initiated along a laser produced ionization filament, NRL Tech. Report, by Versar, Inc.

4. W. Manheimer et al., Phys. Fluids 16, 1126, (1973).

5. H. Alfven and E. Smars, Nature (London) 188, 801 (1960).

6. C. W. Hartman, D. Y. Cheng, G. E. Cooper, J. L. Eddleman, and R. H. Munger, IAEA CN-33/H-5-2.

7. D. A. Baker and J. A. Phillips, Phys. Rev. Letters, Vol. 32, pp. 202, 1974.

8. B. B. Kadomtsev, Rev. of Plasma Phys. VI, 165 (1966) Ed. M. A. Leontovich.

HIGH-VOLTAGE LOW-IMPEDANCE ELECTRICAL SYSTEM FOR DRIVING A THETA-PINCH IMPLOSION-HEATING EXPERIMENT

J. Hammel, I. Henins, J. Marshall, and A. Sherwood
Los Alamos Scientific Laboratory
Los Alamos, New Mexico   (U.S.A.)

ABSTRACT

A pulse forming network (PFN) system has been developed having an impedance of ~0.3Ω with an emf of 0.5 MV. It is used to drive a 40-cm-diameter theta pinch, one-meter-long for the purpose of studying the implosion phase of theta pinches. The low impedance is required to match the effective impedance of the imploding plasma which is low because of the relatively high initial density, 10^{21} deuterons/m^3. The coil is energized at four feed slots by PFN's employing fast pulse charging to 125 kV by Marx generators. At firing time the charging current of each PFN is 300 kA, a substantial fraction of the 800 kA design load current. This transfer capacitor operation contributes to fast rise time of the current. The current is designed to be about 800 kA for 500 ns. Techniques have been developed for operating the system in air, which include voltage grading with conducting plastic sheet, "ballooning" of insulation extending beyond edges of strip lines, voltage gasket connections of capacitor to strip lines, field shaping near insulators to prevent air breakdown. Difficulties with capacity of rail gaps to absorb energy has forced development of solid dielectric switches to take their place for the main start switches. More robust rail gaps are being developed. The system has not yet been operated at full energy, but it is hoped that this will take place soon.

EXPLOSIVELY DRIVEN MHD GENERATOR
POWER SYSTEMS FOR PULSE POWER APPLICATIONS

J. Teno and O. K. Sonju*
Maxwell Laboratories, Inc.
4B Henshaw Street
Woburn, Massachusetts 01801

ABSTRACT

Explosively driven MHD generators offer one means of generating very large bursts of energy directly without the requirement for switching. Since switches for systems using conventional techniques for generating very high energy bursts are not presently available, this feature is very significant.

In an explosively driven MHD generator, a high enthalpy plasma is generated by detonating an explosive charge in a containment chamber; this plasma as it is formed is directed toward the channel and expanded at the same time. As the plasma moves through the generator channel cutting a magnetic field at very high speeds, it generates a high energy pulse. This paper reports the results of a preliminary investigation of explosively driven MHD generators. The work performed was both of an analytical and experimental nature including studies of explosives, detonators, expansion of detonation products and evaluation of conductivity of expanded gases resulting from detonation of explosive charges. The results of this work provides a sound foundation for continuing explosively driven MHD generator research and development programs.

INTRODUCTION

The explosively driven MHD (XMHD) generator through direct generation of high energy pulses offers an alternate to conventional methods of generating high energy pulses using energy storage devices which are operated by slowly charging the energy storage device and rapidly discharging the device when the energy is required. Typically these systems are large and must use switches to control the energy storage and discharge. Switches required for very high energy pulse systems are today not available.

Several important features of the XMHD generator system include:

1. With conductivities of 100 MHO/ Meter and velocities of 10 KM/sec, very high power densities -- to 5×10^5 MW/M3 -- are possible leading to generators capable of outputting large energy burst in compact devices.

2. The XMHD generator system is simple in configuration requiring a minimum of hardware and controls.

3. No switches are required in large systems. This is an outstanding advantage.

4. No scheduling with electric utilities is required with instantaneous delivery of energy under completely independent operating conditions possible.

The present program was aimed at the development of compact multi-megajoule XMHD generators with an overall operating efficiency of 5% or larger. With pulse lengths of some 100 microseconds, the power associated with the pulse is some 50 billion watts. Such a system as will be discussed would be simple in terms of hardware and also simple to operate requiring only the detonation of a charge to generate the energy pulse. Such generators would have wide application where there exist requirements for high energy pulses.

*This program was supported by the Advanced Research Projects Agency and the U.S. Air Force. The work was performed by the authors while at the AVCO Everett Research Laboratory.

As conditions relative to XMHD generator technology existed at the time of the study, the feasibility of generating several hundred megawatts of power had been demonstrated several times by different groups. The development of the technology and derivation of scaling laws had barely started.

The present program aimed at upgrading the technology included analytical studies of

1. Explosives
2. Detonation
3. Jet formation using shaped charges
4. Jet expansion
5. XMHD generator of various configurations.
6. XMHD generator and load interfacing

and experimental investigations of

1. Detonation products velocity
2. Expansion of detonation products
3. Containment chambers
4. Conductivity.

The program also included analytical design studies of multi-megajoule, XMHD generators and the development of scaling laws.

This paper presents a very brief review of the program and results obtained.

TEXT

TYPICAL XMHD SYSTEM

Figure 1 shows a schematic of a linear type configuration XMHD generator. Charges seeded with a cesium compound (which were shaped to direct the detonations products), the expansion region, the generator channel section with its electrodes, and the magnet providing the transverse field are identified. Upon detonation of the shape charge a slug of explosive products is formed. The slug expands as it flows through the expansion region and then enters the magnet region. The power pulse is generated as the slug flows through the magnet. The explosive products then are exhausted as shown. To achieve best performance, the generator channel should operate under vacuum conditions of approximately 1 to 10 torr.

Figure 2 shows a more detailed schematic of the magnet and generator channel section. The B field direction and the gas slug which actually is a small fraction, 1 - 10%, of the explosive products, are shown. The velocity of the slug through the channel is 2 to 12 km/sec. The pulse current, I, through the slug is identified. V is the velocity of the slug.

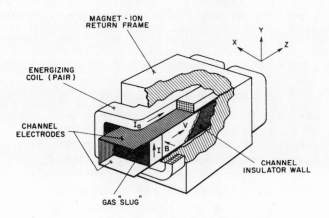

Fig. 2. Detailed Schematic of XMHD Generator System

A detailed analysis of generator performance was developed as part of this work. In particular, inductive coupling, slug configuration, and flow deceleration effects were included.

DISCUSSION OF THE PROGRAM

Energy extraction from the kinetic energy of detonation products is a function of the electrical conductivity of the gas slug, magnetic field flux density, density of the gas slug and the travel time through the generator channel. The energy extracted increases with

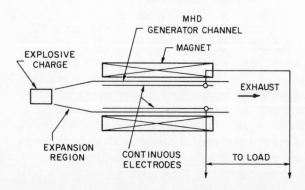

Fig. 1. Schematic of XMHD Generator System

decreasing gas or plasma density and increasing conductivity, magnetic field and travel time.

Curves of constant overall efficiency (output energy/energy of the explosive) are shown in Fig. 3. The top abscissa of the graph shown is scaled to show values of B^2/P (ratio of flux density squared to pressure) for the case where the load is matched to the internal impedance, the conductivity is 10^3 mhos/meter and the travel time is 10^{-4} second. P is the estimated gas pressure in atmospheres.

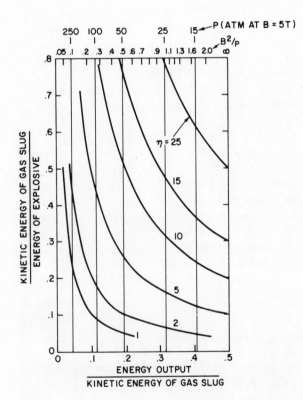

Fig. 3. Performance Curves for XMHD Generator Systems

Based on this simple analysis and the results presented in Fig. 3, certain conclusions can be draws. These are:

1. To obtain an overall efficiency, η, of 5% (B is assumed to be 5 T), the operating pressure must be below 250 atm.

2. At 100 atm, the ratio of the kinetic energy of the gas slug to the total energy of the explosive must be 0.45.

3. To obtain 5% overall efficiency at low energy conversion ratios (kinetic energy to explosive energy) requires operating pressures below 25 atm.

4. Since the detonation pressure developed by an explosive charge during detonation is in excess of 100,000 atm, considerable expansion of the detonation products must occur if reasonable conversion efficiencies are going to be achieved. It is of paramount importance to control the expansion process if fairly uniform flow conditions are to be obtained for satisfactory performance. Further, the expanded gases must have conductivities suitable for generator operation. If it should be required to run at higher pressures to achieve the required conductivities, a larger value of magnetic field would be required to realize the desired performance as a detailed inspection of Fig. 3 reveals.

These conclusions were the guiding and controlling factors in establishing the plan implemented during this program. Specifically, considerable attention, both analytical and experimental, was given to developing an understanding and design data using charges which were shaped to produce high velocities and using the detonation products after a large expansion had occurred, both of which had not been done before, so that the investigation required starting from a position devoid of any background data or experience.

The baseline design that evolved for the explosively driven MHD system consisted of a sperical explosive chamber and a rectangular channel through which the explosive gases would flow.

To verify the feasibility of this design concept, a subscale explosive chamber and channel were fabricated as shown in Fig. 4 and tested with various explosive charge weights. The purpose of the tests in the subscale fixture was to obtain as much information as possible in the following areas:

1. MHD explosive chamber concept feasibility.

2. Structural integrity of the subscale test fixtures.

Fig. 4.   Half-Scale Containment Fixture

3. Design information, particularly limitations to be used in designing the larger explosive chamber.

4. Evaluation of the mylar seal across the exit mouth of the channel.

5. Evaluation of the explosive gas dynamics during the test, and during the subsequent gas evacuation phase.

Forty-two tests were performed in the subscale explosive chamber utilizing explosive charge weights ranging from .01 to 0.75 pound (4.5 to 341 grams). The explosive used in the tests was C4, and RDX composition. The explosive chamber was instrumented during the tests to record chamber and channel gas pressure and to measure resultant stresses in the chamber. Stress data and static stress calculations indicated the chamber should be in the yield range with approximately 0.3 pound (136 grams) explosive. However, no measurable deformation was achieved through tests with 0.6 pound (272 grams) explosive and only slight deformation was observed after the tests at 0.7 pound (318 grams) and 0.75 pound (341 grams).

The subscale explosive chamber was made by welding together two halves of commercially available pressure-fitted end caps. The end caps were approximately 20 inches (51 cm) in diameter with a 0.593 inch (1.5 cm) thick wall with a 38,000 psi yield strength. Attached to the chamber was a four foot (122 cm) long duct with a 8-1/2 cm by 16-1/2 cm opening. Channel iron was used to form the base and cradle for support. Opposite the duct opening in the

chamber was another opening to accommodate a ten inch (25.4 cm) diameter threaded breech. The total weight of the subscale chamber fixture was 928 pounds (422 kg).

The explosive charge geometry was designed as a tubular charge for the purposes of achieving maximum velocity. The charge geometry was scaled for the different charge weights. The charge contained C-4 explosive pressed into plastic cylinders to the desired configuration. The explosive charge was then fit into plastic tubing and inserted into the cavity within the breech. The charge rod length was controlled to place the shaped charge at the center of the explosive chamber.

Initiation of the charge was accomplished by inserting two RDX pellets at the base of the C-4 charge with the pellets in turn being initiated by mild detonating fuze wire.

Experiments were performed using several different test setup. Figure 4 showed the equipment used to study the design aspect of containment chamber. Figure 5 shows a photograph of an assembled model XMHD channel used to study the expansion characteristics of the products of detonation, the velocity of the expanding products and the conductivity of the products.

Fig. 5.   Photograph of Assembled XMHD
Generator Channel

The basic channel structure consisted of one-inch thick mild steel plates reinforced with one-inch thick gussets on the outside. The top and bottom plates were bolted to the side plates which contained an "O" ring groove and seal. The channel assembly was bolted to a support stand, approximately 30 inches high, which was bolted to the concrete pad.

The inside dimensions of the assembled channel were 355 mm wide by 185 mm high by 1000 mm long.

The high explosive (H.E.) charge was located 19.5 inches from the entrance of the channel and supported by a metal guide.

For tests conducted at initial channel pressures of one atmosphere, the metal guide consisted of 1/16 inch thick aluminum sheets welded together in the shape of a pyramidal frustrum.

For tests conducted at initial channel pressures less than one atmosphere, the guide was fabricated from 1/4 inch stiffeners on the outside and a flange was added for attachment of the H.E. charge inside a thin wall aluminum tube with end plate and flange. The detonation initiator was inserted through a rubber stopper which was placed into the end plate of the metal tube. The H.E. charge assembly, was bolted to the guide and the guide bolted to the entrance of the channel. Finally, the exit of the channel was covered by a 3/8-inch thick aluminum plate with provisions for a vacuum hose connection.

Four different high explosive charge compositions were fabricated for the testing program. These were:

1. 95% RDX + 5% WAX
2. 94.9% RDX + 5.0% WAX + 0.1% $CsNO_3$
3. 94% RDX + 5% WAX + 1% $CsNO_3$
4. 90.2% RDX + 4.8% WAX + 5% $CsNO_3$

All of the H.E. charges were of the same configuration. The charge assembly consists of one 2.50-inch O.D. by 1.00-inch long solid cylinder and six hollow cylinders 2.50-inch O.D. by 1.25-inch I.D. by 1.00-inch long. The weight of each charge was approximately 750 grams.

The tip velocity of the slug as it exits from a shaped tubular explosive and propagates away from the explosive was measured both

with a streak camera and a framing camera. Good photographic records of the detonation and the initial phase of the slug propagation into both the atmosphere and into vacuum were obtained. Samples of photographic results obtained both from streaking camera (top photographic record) and framing camera (bottom photographic record) are shown in Fig. 6. In this set of experiments, tests using solid cylinders were conducted for comparison of the motion of the detonation products off to the end of the detonating charge with that of a shaped tubular explosive.

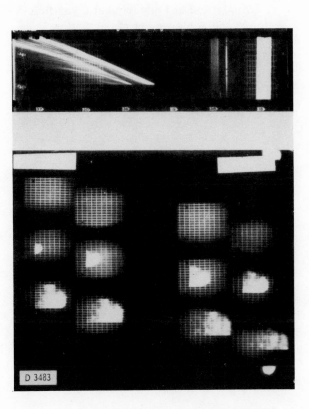

Fig. 6. Streak and Framing Photographs of a 1.5" OD x 1" ID x 4" long Firing in Vacuum

Cameras were used to measure the detonation velocity in the explosive tip velocity of the slug as it exits the tubular esposlive and propagates away from the end of the explosive charge and the expansion rate and the profile of the expanding jet. On an overall basis, the general agreement between the velocities obtained independently by the streak and framing camera indicates that the velocities obtained were generally accurate to 10% or less.

SUMMARY OF RESULTS

During this program, the feasibility of expanding detonation products to low pressures at high conductivities of interest was established thus showing that achievement of high efficiency and large MJ generators was practical. A basic understanding of the performance of XMHD was achieved thus permitting the design of generators capable of operating various energy levels.

The specific results obtained included:

1. Analytical and experimental verification of the achievement of detonation products velocities of 10 KM/sec and higher using shaped charges after expansion in evacuated XMHD generator channels.

2. Experimental verification of the achievement of conductivities of the order of 10^3 MHOS/meter after expansion.

3. Analytical studies of an experimental proof of design of containment chambers for XMHD generator systems.

4. By analytical means, it was verified that interfacing with the load could be accomplished even when requirements exist for waveshaping of the delivered energy pulse.

The main conclusion which was made on the basis of this work was that the XMHD generator power supply is a very attractive alternative for supplying large amounts of energy on demand at full level without the use of switch.

MAGNETIC COMPRESSIVE GENERATORS USING GASEOUS EXPLOSIVE

R. Hahn, B. Antoni, J. Lucidarme, C. Rioux, F. Rioux-Damidau
Laboratoire de Physique des Plasmas
Groupe Electrotechnique et Fusion Controlée
Université Paris-Sud, 91 405 Orsay - France

ABSTRACT

We present a preliminary experimental compression generator for which the normal solid explosive is replaced by mixture $2H_2 + O_2$ at a pressure of 60 bars. A coaxial, conducting structure is formed by an aluminium tube of 92 mm external diameter, 6 mm thick, holding the mixture and placed in a conducting carcass (internal diameter 140 mm). An exploding wire placed axially in the tube initiates the explosion which accelerates the tube radially outwards. The magnetic field initially created by the discharge of a capacitor bank through the coaxial structure is thus compressed by the tube. With an initial current of 0.8 MA, we have obtained 4.75 MA, the energy increasing from 10 to 30 kJ. In this structure, which has the minimum possible dimensions for a conclusive result, flux losses of 50 % have been observed corresponding to theoretical predictions. For larger units a more complex tube structure could be chosen, considerably reducing the losses. With this type of generator, only the tube is destroyed and energies greater than 10 MJ can readily be delivered for a minimum of total cost at a good chemical efficiency. The mixture $2H_2 + O_2$ is produced by electrolysis, avoiding the storage of high energy explosives.

INTRODUCTION

At the present time, production of high magnetic energies (a few MJ in a few 10^{-4}s) is easily done by means of magnetic field compression generators. They all employ solid explosives[1] and they can be used to realise plasma experiments as well as the high energy capacitor banks with short delivery times, but their cost is much lower. Unfortunately, because of its destructive nature, the use of classical explosive driven devices is limited to laboratories having their own explosive facilities.

In this paper, we propose a new type of generator of particularly low cost which can work in a conventional laboratory and we show its feasibility. It uses a gaseous explosive $2H_2 + O_2$ which, for the same chemical energy density, is much less destructive than the solid explosives because the exerted pressure is lower but lasts longer[2,3].

DESCRIPTION OF THE GENERATOR

Figure 1 gives the scheme of the generator.

The liner, a thin cylindrical aluminium tube 1, is coaxial to a cylindrical casing 2. This last one is made of steel and is covered on its inner part by bronze. A gaseous mixture $2H_2 + O_2$ under a pressure of 60 bars is contained in the liner closed by 3. The detonation of the

explosive is initiated along the axis of the system. The liner is then accelerated by the high pressure exerted by the explosion products and is plastically deformed. If an initial magnetic field exists between the inner and the outer conductors, it will be compressed by the

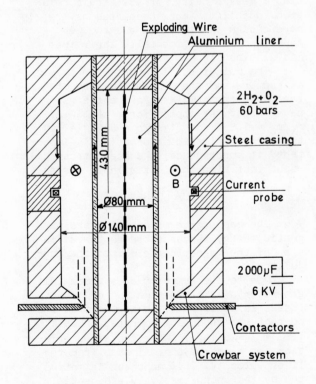

Fig.1 Scheme of the generator.

exploding liner until the magnetic pressure stops it (when flux losses are limited). The chemical energy of the explosive is thus transformed into kinetic energy of the liner and next into magnetic energy.

The initial magnetic field was chosen to be azimuthal because it is easier to create an axial current in the generator. Furthermore, for a given magnetic energy amplification, the liner stops at a greater distance from the casing when the current is axial ; so in that case, the defects in the symmetry of its deformation are of less importance.

MECHANICAL BEHAVIOUR OF THE LINER

In order to achieve the best chemical into kinetic energy transformation, it is obvious that the liner must exhibit a final over initial radius ratio as large as possible, in particular because the polytropic exponent of water vapour is low, about 1.15 [4]. Besides, the highest kinetic energy is obtained when the initial pressure of the explosive gas mixture equals to the yield pressure of the liner. In this last condition, the radial velocity does not depend on the dimension of the liner but only on its mechanical characteristics : mass density, stress-strain relation, ductility. For these reasons, we have chosen aluminium liners and we have studied experimentally their free radial expansion under explosive loading.

The main results concern the strain at rupture and the marked increase of the hoop stress with strain and strain rate. Fractures appear for a strain $r_f/r_0 \simeq 1.5$; for $\dot{\varepsilon} = v/r = 5.10^3 \text{ s}^{-1}$, the work of plastic deformation is important, it corresponds to a mean stress of about four times the initial hoop stress.

FLUX LOSSES

The conversion of kinetic into magnetic energy is connected to the diffusion of the magnetic field into the conductors.

Due to the symmetry of our coaxial generator, a simple estimation of flux losses is obtained if we assume that the current varies during compression as : $I(r) = I_0 \exp(t/\tau)$ where the caracteristic time τ can be approximated from the liner velocity. This approach is motivated by the approximatively exponential current increase observed in flux compression devices.

With this approximation, we can replace the classical field-diffusion equations by the formal equations of a L.R. circuit which we can easily solve. So we obtain literal expressions[2] for the flux coefficient

$$\Phi/\Phi_0 = \exp - \int_o^t dt \, R/L$$

and for the magnetic energy amplification W/W_0.

The dimensions of the system are parameters of these expressions, so they are very useful for the design of a generator. They give the increase of Φ/Φ_0 and W/W_0 with dimensions and show the existence of a smallest dimension to attain an effective magnetic energy amplification.

SIZE OF THE GENERATOR

For the realisation of a first generator we have chosen a sufficiently large system in order to achieve a magnetic energy amplification of three which proves the feasibility of a gaseous explosive driven device, and sufficiently small to be cheap and relatively easy to handle.

This compromise leads to the following dimensions :
casing : Inner radius : 7 cm
liner : Inner radius : 4 cm
         Thickness   : 6 mm
         Length   : 43 cm

PRODUCTION OF THE GASEOUS EXPLOSIVE $2H_2+O_2$

The mixture $2H_2,O_2$ was produced in place by electrolysis of a solution of potash in a special device made of stainless steel. This device can bear a static pressure of 1200 bars so that it can resist an eventual explosion with an initial pressure of 60 bars ; it can be isolated from the volume of the liner by high pressure valves.

In our experiment, the electrical current was furnished by a DC power supply (2000 A ; 0-15 V). With a 100 Amp current, the volume of the liner was filled in approximatively 15 min to the pressure of 60 bars.

This process has revealed to be sure and simple. It needs no storage and manipulation of hydrogen and oxygen under pressure. No untimely explosion has ever been observed.

IGNITION SYSTEM

To initiate the explosion of the mixture $2H_2+O_2$ along the axis of the liner, the most convenient device is an exploding wire operated by a condensor bank. However, a voltage of a few kV per cm of wire is needed. In our generator, which is 40 cm high, we used a special wire formed by the alternating of parts of small and large sections. (see Fig.2). When the condensor

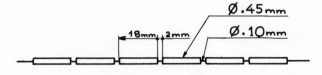

Fig. 2 Exploding wire.

bank is discharged, only the smaller section parts explod and the required voltage is governed by those parts alone. The distance between the efficient parts has been taken to one half of the inner radius of the liner ; 10 kV are then sufficient and experimental results have shown that this process of ignition is quite satisfactory. The manufacture of the wire is very simple : a copper plating is deposited on the non hidden parts of a thin copper wire (diameter : .1 mm).

EXPERIMENTAL RESULTS

The reported currents are measured with inductive probes[5], whereas the displacement-time history of the exploding liner is determined using a set of pin contactors. The pressure of the stoichiometric mixture of hydrogen and oxygen is measured with a piezoelectric pressure gauge.

The capacitor bank system, which provides the initial flux for the compression device, has the following characteristics : energy 36 kJ, voltage 6 kV, series inductance including the connections 15 nH. Its discharge circuit is closed by the liner which, at the start of its expansion, impinge 12 radial contactors. The initial position of these contactors are adjusted so that the discharged current will be maximum when the liner closes the device on itself.

Figure 3 shows the results of a typical shot with the previously described system. The initial conditions were the following :
Pressure of the gaseous mixture
$2H_2+O_2$: 60 bars
Chemical energy   :925 kJ
Initial current   : .8 MA
Initial magnetic energy   : 10 kJ.

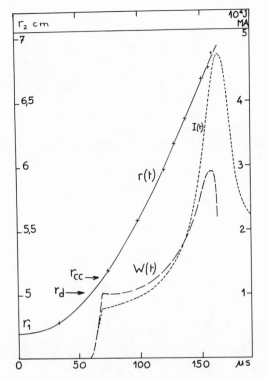

Fig. 3 Experimental results.

The curves r(t), I(t), W(t) are respectively the variations of the external radius of the liner, the current and the magnetic energy. We obtained the following main results :
Work done by the detonation products :

	$\sim$ 120 kJ
Maximum velocity of the liner :	235 m.s^{-1}
Kinetic energy :	55 kJ
Maximum current :	4,75 MA
Maximum magnetic energy :	30 kJ
Flux coefficient at the maximum of energy :	.5

These results agree well with our calculated expectations, the magnetic energy amplification reaches the factor three which is its maximum value as seen previously.

FUTURE DEVELOPMENTS

The experimental results and calculation reported previously prove the feasibility of a gaseous explosive driven generator which can work in a conventional laboratory.

As can be expected, the limitation of the velocity of the liner due to the fact that it contains the initial pressure of the gaseous mixture is the main disadvantage of this device.

In future experiments, the mechanical containment of the initial pressure will

be held by a thin steel cylinder, placed
inside the aluminium liner and which will
open under the detonation pressure. So
the liner, whose thickness is thus determined only by the skin depth, will be
much lighter and will acquire a higher
velocity. This improvement will lead to
very interesting performances. We project
a compact generator to obtain, starting
from about 1 MJ, final magnetic energies
in excess of 15 MJ. The simple mechanical
construction, and the non-destructive
operation (the liner alone is to be
replaced after each shot) make the cost
of such a system very low if compared
with other types of generators.

REFERENCES

1   H. KNOEPFEL, Pulsed High Magnetic
     Fields, North-Holland Pub. Co, 1970.
2   R. HAHN, Thesis (to be published)
3   Brevet ANVAR N° 74 21 181. (18 Juin
     1974).
4   A. BENOIT, Equilibrium Thermodynamic
     Data for the H_2-O_2-H_e system.
     Utias Technical Note n° 128, August
     1968.
5   C. RIOUX, F. RIOUX-DAMIDAU, Revue
     Phys Appl. **vol 7** (1972) 303 and 313.

Plasma Focus and Collective Effects

COLLECTIVE-FIELD ACCELERATION OF HIGH-ENERGY IONS[*]

J. S. Luce

Lawrence Livermore Laboratory, University of California
Livermore, California 94550

ABSTRACT

A collective-field accelerator has evolved from experimental and theoretical research at the Lawrence Livermore Laboratory that uses a high-vacuum diode with an adjustable graphite cathode as well as an insulated anode, and that operates with a relativistic electron beam with ν/γ of ~1. Alternate gradient lenses are used to focus collectively accelerated particles. The gradients are produced by alternate dielectric and grounded lenses. The dielectric lenses are self charged by the electron beam creating a potential difference in reference to the grounded lenses. These lenses focus both electrons and ions by convective processes. Deuterons have been accelerated in pulses of ~10^{14} producing up to 10^{11} D-D neutrons per burst by impingement on suitable targets. Hydrogen, deuterium, carbon, fluorine and chlorine ions have been accelerated to produce both light- and heavy-ion reactions. Analysis of activation data shows that heavy ions with >6 MeV per nucleon and protons with ~15 MeV energy have been produced. Theoretical analysis indicates that the collective ion acceleration mechanisms arise from interactions with plasma-wave trains created by near-resonant, beam-cavity interactions and accelerated (in group velocity) through density and temperature gradients in the secondary plasma beyond the anode.

INTRODUCTION

The first attempt to accelerate positive ions with the self-field generated by high energy electrons was made by Hannes Alfvén and Olle Wernholm in 1952.[1] The potential of this exciting new concept was recognized by many scientists and by 1956 several schemes had been proposed to accelerate ions by this method.[2-6] The promise of economical, compact accelerators that would accelerate particles with powerful self-generated fields created considerable optimism among early workers. This enthusiasm has gradually tempered because of technological difficulties and the need for more comprehensive theory. Funding has also been attenuated by the very successful (and expensive) development of the alternate-gradient strong-focusing principle that is now an essential feature of many particle accelerators throughout the world. It is interesting to note that not until recently has alternate-gradient focusing been considered in collective field accelerators. In 1974 the first preliminary experiments using self-charged lenses were initiated at the Lawrence Livermore Laboratory (LLL).

Many workers have made contributions in the field of collective particle acceleration.[7-47] Linear collective field acceleration of positive ions has up until now been more effective than other methods being investigated. This system is being studied at LLL and several other laboratories in the USA and the USSR. Discussions in this document are limited to linear systems.

The experimental geometry, conditions, and degree of success that have evolved from the LLL program are significantly different from other collective acceleration experiments known to us. Fortunately, both the geometry and the vacuum conditions that optimize the collective acceleration mechanism are particularly well suited for adoption as a production ion accelerator.

THE LLL COLLECTIVE FIELD ACCELERATOR

Research with the LLL collective field accelerator using a Pulserad 422 E-beam machine that delivers 2500 J of energy at 2 MeV in 60 ns has been reported in the literature.[48,49] However, the device has undergone rapid evolutionary changes that are still in progress. Consequently the configuration and results reported herein have not as yet been published.

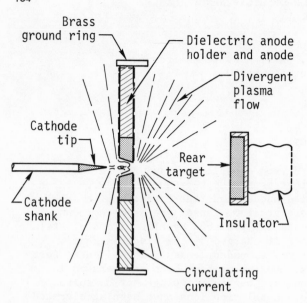

Fig. 1. Original LLL collective field
acceleration geometry.

The first geometry used for collective field acceleration[48] is shown in
Fig. 1. When the E-beam machine is triggered the relativistic electrons move more
or less uniformly along the cathode surface.
Low-energy positive ions produced on the
surface of the cathode prevent space-
charge blowup of the electron beam. The
flow of current along the cathode produces
a strong B_θ field, constraining the electrons to the small diameter cathode and
ejecting a relatively uniform electron
beam into the cathode-anode gap.

No fundamental difficulties have been
encountered with high impedance as a result
of the small diameter cathode, although
major alignment problems and destructive
filaments were encountered until an all-
graphite cathode structure was used. It is
believed that the presence of low-energy
positive ions on the cathode reduces impedance and greatly enhances electron flow.
When the electrons leave the end of the
cathode they enter a hard vacuum and are no
longer space-charge neutralized. Consequently, the electron beam blows up to a
larger diameter determined by the geometry
and pressure in the cathode-anode gap.
The electrons impinge on the inside surface
of an anode containing the proper element
to produce the ions desired for acceleration. When the electrons strike the insulated anode several events occur. The
insulated anode behaves as a resonant
cavity, which modulates the beam. An intense plasma is formed primarily behind the
anode in the space between the anode and

the rear target. This plasma provides the
ions that are collectively accelerated
toward the rear target. Initially part
of the return current flows out of the
cathode side of the insulated anode hole
like the point of a cusp-shaped plume. The
outer periphery of this cusped-current
sheet impinges on grounded surfaces thus
completing the circuit. The sheet contains
both ions and electrons since activation
is detected where the current sheet reaches
ground potential. Ions also impinge on the
cathode where N^{13} is detected when deuterons
derived from a deuterated polyethylene
anode are produced. The ions accelerated
in the cathode-anode region have relatively
low energy compared to the collectively
accelerated ions from the plasma formed
between the insulated anode and the rear
target. The ions that impinge on the rear
target flow in the same direction as the
primary electron beam so it can be safely
stated that these ions are accelerated by
collective fields produced by the primary
electron beam. The ions flowing in the
opposite direction with the return current
are probably also accelerated by a collective process because of the electric
fields in the return current sheet. Ions
reaching the cathode, however, appear to
be accelerated by an electrostatic field
produced by the applied machine voltage. A
sheath is clearly discernable between the
cathode and the anode plasma as shown in
the open shutter photographs in Fig. 2.

The intense plasma formed between the
anode and the rear target produces a widely

Fig. 2. Cathode sheath.

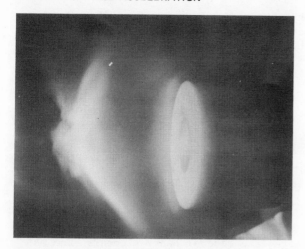

Fig. 3.   Divergent flow of ions and
          electrons.

was developed using alternate gradient
focusing. These gradients were achieved
by using alternate dielectric and conducting lenses. The dielectric lenses
were charged to a high potential by the
electron beam whereas the conducting
lenses were grounded. The last lens was
found to be defocusing when used with a
grounded rear target. This defocusing was
caused by the lack of a gradient between
the (grounded) last lens and the grounded
rear target. Removing the fourth grounded
lens left three lenses and a grounded rear
target acting as a lens. This arrangement
provided the best focus achieved to date
and the highest energy ions. This lens
system was designed by impinging the
collectively accelerated beam on the lenses
and using the resultant radiated areas to
determine various geometrical parameters

divergent flow of ions and electrons as
shown in Fig. 3. This divergence is typical
of linear collective field accelerators.[50]
It was postulated that the convective
nature of the plasma would make it possible
to treat it as a cloud of energetic electrons. If this were true it would be logical to assume that the plasma could be
focused by a suitable lens. This hypothesis
has been shown to be correct and improved
focusing as well as higher energy ions have
been achieved with a self-charged lens.
Figure 4 shows the focusing achieved with
this lens and Fig. 5 shows the geometrical
arrangement used.

        The cathode is a graphite rod 2 mm in
diameter, which is fed through a tapered
graphite transition piece with compound
tangent angles providing an average angle
of about 28°. This parameter, as well as
the length and average diameter of the
transition piece, is determined by the
diameter of the machine cathode shank into
which the transition piece is fitted.

        The anode is a 6-mm-thick piece of
selected plastic containing the necessary
element to produce the ion specie desired.
This anode has a stepped structure, and
is pressed into a plastic disc anode
holder 11 cm in diameter by 6 mm in thickness, providing the insulation required
for efficient operation. A second plastic
plate is used with the anode to prevent
it from being blown out by the beam. The
anode has a hole 9.5 mm in diameter and
the lens is located 6.25 cm behind the anode.

        The improvements resulting from the
use of a single lens led to experiments with
multiple lenses. A four-stage lens system

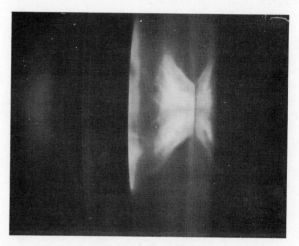

Fig. 4.   Focusing lens. The apparent
          plasma on the right-hand side of
          this photo is a reflection of the
          actual focused plasma.

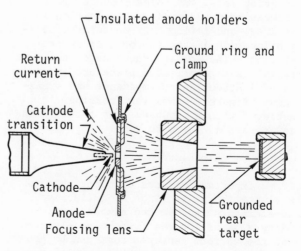

Fig. 5.   New geometry with focusing lens.

Fig. 6. Multiple lenses.

such as length of the lenses, gap distances and hole sizes.

Figure 6 shows the present device in operation. It can be seen that plasma is created in each of the lenses. Thus a series of plasmas are present, isolated by vacuum and high potentials. The electron bunches produced by the oscillating anode pass through these plasmas and probably create additional plasma waves in each isolated plasma. There are strong indications that ions formed in these lens plasmas are also accelerated to the rear target. It is also of interest that the second and third lenses have no effect unless a plasma is formed in the lenses.

These lenses have a lens-to-lens gap of 2.54 cm and a lens length of 3.81 cm. The hole sizes beginning with the first lens are 6.35 cm, 6.98 cm and 7.62 cm respectively. A rear target holder is located 6.25 cm behind the exit of the focusing lens. It holds grounded metal targets when radioisotope production is desired, and teflon[51] or deuterated polyethylene (CD_2) targets for neutron production. The actual grounding of the rear target is a complicated procedure involving careful attention to impedance and breakdown. Figure 7 shows the focusing achieved with a grounded rear target when it is used as a fourth lens.

EXPERIMENTAL RESULTS

Important figures of merit used to evaluate the performance of collective accelerators include the number of elementary particles that can be accelerated and the various isotopes that can be produced. At this time we have accelerated

hydrogen, deuterium, carbon, fluorine and chlorine ions to produce such light- and heavy-ion reactions as

$$^{65}Cu(p,n)^{65}Zn, \ ^{63}Cu(p,n)^{63}Zn,$$
$$^{63}Cu(p,2n)^{62}Zn, \ ^{56}Fe(p,n)^{56}Co,$$
$$^{58}Fe(p,n)^{58}Co, \ ^{12}C(d,n)^{13}N, \ ^{13}C(p,n)^{13}N,$$
$$^{27}Al(^{12}C,He,n)^{34}Cl, \ ^{27}Al(^{19}F,D)^{44m}Sc,$$
$$^{27}Al(^{19}F,D)^{44}Sc, \ ^{27}Al(^{19}F,t)^{43}Sc,$$
$$^{27}Al(^{37}Cl,n)^{63}Zn, \ ^{27}Al(^{12}C,^{4}He,^{3}He)^{32}P,$$
$$\text{and } ^{65}Cu(^{12}C,2n)^{75}Br, \text{ etc.}$$

Unfortunately this list of radioisotopes is necessarily incomplete since the data from numerous runs have not been evaluated. In the coming months we also plan to irradiate several heavy elements such as Ta, Au, Pt, Ho, etc. with Hg ions.

Although incomplete, these data do indicate a high efficiency for ion acceleration that is confirmed by measurements of 10^{14} deuterons per shot with an average energy of >12 MeV. Thus about 190 J of energy is being delivered by these deuteron bursts. Heavier ions with ~6 MeV per nucleon have also been observed. The efficiency of conversion of electron beam energy (2500 J) to ion energy is therefore over 7% and the neutron production efficiency from deuterons incident upon deuterated polyethylene and teflon targets (~10^{11} per shot) is approximately 100 times larger than in other ion acceleration systems.

While it is obvious that much research remains to be done, recent technological advancements suggest that the time has come when it is appropriate to begin considering applications of collective field accelerators.

Fig. 7. Irradiation of grounded copper target.

APPLICATIONS OF COLLECTIVE
FIELD ACCELERATIONS

ACCELERATOR APPLICATIONS

The encouraging results discussed in the previous section naturally lead to speculation regarding the future role of collective acceleration in producing radio-isotopes economically in widely distributed locations and the impact this would have on the related fields of nuclear physics, chemistry and medicine. While it is true that a repetitively pulsed electron-beam machine would be necessary for these and most other applications, it is appropriate to begin investigating these possibilities with present nonrepetitive electron-beam machines.

One of the most impressive features of linear collective field accelerators is their inherently low capital cost. Equally or even more impressive is their surprisingly low operating cost. It can be predicted with considerable confidence that the LLL device will in a reasonable time be able to produce radioisotopes automatically. Automatic operation has already become a reality with certain electron-beam machines used in the so called "flash x-ray" mode. The basic simplicity of these devices assures the success of complete automation including permanent records of machine performance and radiation levels from the target. Human intervention would be needed only in the case of malfunctions or maintenance of equipment. The low cost of these accelerators (probably about half a million dollars for a large machine) would put them within the reach of many scientific organizations that are now unable to participate in advanced accelerator research.

It is probable that if a modest but determined research effort were funded, collective field accelerators could be developed that would open up new fields of research not feasible with present machines. Because of the universality of the collective field principle such machines could theoretically accelerate the ions of all the elements. Since all the elements would also be available as targets, the entire spectrum of possible reactions would, in principle, become available for exploration — limited only by the ultimate energy of the ions and the time required for such an enormous task. When the advances that have been made at LLL in focusing collective beams are considered, the achievement of high-density clashing beams also becomes a real possibility. Since these beams contain both ions and electrons they are space-charged neutralized and their particle density would grossly exceed the density attainable with present space-charge-limited devices. The utilization of laser beams to create highly focused preionized paths for plasma waves is an exciting possibility for further improvements in beam focus. In view of the many surprises that have resulted from high-energy light-ion reactions it appears inevitable that many new discoveries await experimenters using the intense heavy-ion bursts that would be available with a successful high-energy collective accelerator. At a minimum, new radioisotopes and transuranic elements could be anticipated.

Other unique characteristics of collective accelerators that may lead to new research include the use of hybrid beams containing electrons and ionic mixtures. Polarized beams and targets may result from the presence of shock waves produced in the diode region. Metallurgical changes could be expected in irradiated samples exposed to these shock waves and intense bursts of high energy ions and electrons. While these fast pulses have many useful applications, their radiation will swamp out most detection equipment developed for CW operation. This would of necessity eliminate many of the classical experiments now conducted with conventional accelerators. Collective accelerators can be viewed as a different kind of accelerator with characteristics that open up new facets of research not necessarily competitive with present machines.

INJECTOR APPLICATIONS

Collective field accelerators have potential advantages as injectors for frequency modulated accelerators. Even in its present primitive state the LLL device has achieved >6 MeV per nucleon, approaching the performance of tandem Van de Graaffs and Linacs.

Both tandems and Linacs can provide steady-state operation while the collective accelerator is strictly a pulsed device. When these devices are used as injectors, however, steady-state operation is a doubtful advantage since the machines used for final acceleration are usually frequency modulated and require only periodic injection.

Actually Van de Graaffs and Linacs should not be considered injectors since

the injectors they require themselves probably cost at least as much to build and
operate as would a linear collective field
accelerator. Certainly collective accelerators would outperform the injectors now
used for Van de Graaffs and Linacs by a
wide margin.

A little-understood phenomenon makes
it possible to separate high-energy ions
from the primary electron beam in the LLL
device. It has been found, for a given set
of conditions, that the electron beam
"blows up" and disperses at a precise distance from the anode. This dispersion
occurs in a completely reproducible manner
provided operational conditions are not
varied. The dispersion of the electron
beam does not affect ion acceleration,
since the ions are carried by plasma waves.
The dispersion of the primary electron
beam is useful, however, since it reduces
the flow of high energy electrons from the
injector. The ions that enter the main
accelerator would be accelerated as at
present.

The ion injector application of a
collective field accelerator fails to
exploit the full potential of the method.
A much more interesting possibility is to
use several stages of collective acceleration. Empirical data indicate that well-
insulated targets that charge to a high
negative potential reduce the energy of the
accelerated ions whereas grounded targets
increase their energy. This result indicates that the highest energy ions are
achieved when the potentials are arranged
to produce the maximum accelerating fields
for electrons.

This has led to the idea of using an
E-beam machine for initial collective field
acceleration and a reversed charged
(+ potential) Van de Graaff for the second
stage. Ideally this second-stage device
would be involved in the acceleration
process only, thus greatly reducing power
requirements. For the purpose of this
experiment it may be possible to use a
standard Van de Graaff with a target in
the + terminal or a Van de Graaff-type
pulsed E-beam machine that is reversed
charged to produce a + terminal. The advantage of the latter machine is its large
current capacity.

These Van de Graaff possibilities are
suggested as inexpensive methods for
investigating the principle of multistaged
collective accelerators. Basically this
class of ideas is much more appealing than

using a collective field accelerator as an
injector for a conventional machine, but
both applications should be considered.

CTR APPLICATIONS OF COLLECTIVE FIELD ACCELERATION

During the past few years various
methods have been proposed for compressing
D-T pellets to achieve thermonuclear burn.
These systems involve the use of photons,
electrons, ions, collectively accelerated
ions and electrons as well as high energy
neutrals.[52-66] In this section the compression of D-T pellets by collective
acceleration is briefly considered. This
concept differs from previous suggestions
that only considered collectively accelerated ions and electrons.[60] Here we also
present qualitative justification for investigating the compression of D-T pellets
with high velocity colloids that are collectively accelerated by the LLL method.
Two unreported "quick and dirty" experiments conducted at LLL apparently show that
colloid acceleration has occurred in the
system. Because of lack of priority no
diagnostic measurements were made. However,
carbon and copper powders injected into the
anode hole were found plated on the rear
door of the vacuum chamber. This is not
necessarily surprising since high velocity
colloids have been accelerated with other
plasma devices. The simple facts are
basically encouraging: the LLL device
accelerates sharp bursts of heavy ions to
relatively high energies achieving about
>6 MeV per nucleon. It may be feasible to
extrapolate from accelerating heavy ions
to small colloids. It has been observed
that performance of the LLL device depends
upon a large density gradient between the
pulsed plasmas in the anode and lens apertures as well as the vacuum volume surrounding these plasmas. This observation
is consistent with the theory of wave
group trapping and acceleration, which
corresponds in its fundamental process to a
large number of phenomena that are common-
place in physics and in the everyday world.
It arises basically from the fact that a
nonuniform diffusion process will give rise
to a net flow from regions of large diffusion rates into regions of lower diffusion
rates, appearing formally as the effective
pressure (first term) in the expansion of
the diffusion equation for a nonuniform
process:

$$\frac{\partial}{\partial \underset{\sim}{x}} \cdot \left(\underset{\sim}{\partial} (x) \frac{\partial}{\partial \underset{\sim}{x}} f(\underset{\sim}{x}, t) \right) = \left[\frac{\partial}{\partial \underset{\sim}{x}} \cdot \underset{\sim}{\partial} (\underset{\sim}{x}) \right]$$

$$\cdot \frac{\partial}{\partial \underset{\sim}{x}} f(\underset{\sim}{x},t) + \underset{\sim}{\partial}(\underset{\sim}{x}) : \frac{\partial}{\partial \underset{\sim}{x}} \frac{\partial}{\partial \underset{\sim}{x}} f(x) .$$

As such, it is responsible for such diverse phenomena as concentration of suspensions in liquids by ultrasonic vibrations, the Kundt's tube phenomenon, strong focusing in accelerators, runaway electrons in strong electric fields, and the piling up of trash in a freeway divider strip — the important point being that like thermodynamics it is an extremely robust phenomenon.

Short pulses (a few ns duration) of collectively accelerated particles have been observed with the LLL device. These short pulses are both unique and necessary if collective field acceleration is to be used in compressing thermonuclear pellets. It would be logical to study these possibilities theoretically considering both the ions of the elements and appropriate colloids; bearing in mind that momentum transfer as well as ablation can now be considered in pellet compression.

INJECTION AND TRAPPING
OF COLLECTIVE BEAMS

There is some preliminary work under way to study trapping of collectively accelerated ions in gas-filled drift tubes. While these experiments are interesting it would appear fundamental that it is necessary to use only ions and electrons in a vacuum volume because of the untenable electron exchange losses that occur in gases. Experiments conducted in several laboratories show that nonadiabatic effects allow injection and trapping of high-energy beams in appropriate magnetic "bottles." The successful injection and containment of collective beams would lead to important studies of ionic[67] slowing-down nuclear chain reactions[68] heretofore impossible to achieve.

Among the various injection technologies only the LLL system provides both high energy ions and electrons for injection into suitable high vacuum traps. A major loss mechanism in fusion devices is that caused by collisions between high energy ions and slow electrons. Since slow electrons would not be present during the time scale of these experiments it would be possible for the first time to study very hot plasmas in which ion-ion collisions dominate.

ACKNOWLEDGMENTS

It is a pleasure to acknowledge the contributions of W. Bostick, D. Dalgas, H. Sahlin, John Wyatt and O. Zucker. The assistance of the Mechanical Engineering Division in the operation of the Pulserad 422, E-beam machine, the Chemistry Division in the advising and fabrication of critical accelerator components, and the Radiochemistry Division in the identification of radioisotopes has proven invaluable.

The writer is grateful to the LLL Physics Department for its continuing support of this exciting research.

REFERENCES

*This work was performed under the auspices of the U.S. Energy Research & Development Administration.

[1] H. Alfvén and P. Wernholm, Ark. Fys. 5, 175 (1952).

[2] R. B. R-S-Harvie, AERE memorandum G/M87, 1951.

[3] I. J. Billington and W. R. Raudorf, Wireless Engineer 31, 287 (1954).

[4] V. I. Veksler, Proc. CERN Symp. on High Energy Accelerators, 1956, p. 80.

[5] G. I. Budker, Proc. CERN Symp. on High Energy Accelerators, 1956, p. 68.

[6] Ya. B. Fainberg, Proc. CERN Symp. on High Energy Accelerators, 1956, p. 84.

[7] R. M. Johnson, Symposium on Electron Ring Accelerators, Lawrence Livermore Laboratory Report UCRL-18103, 1968.

[8] Ya. B. Fainberg, Sov. Phys.-Usp. 10, 750 (1968).

[9] V. N. Tsytovich, Relativistic Solitons and Non-Linear Waves as Bunches for Coherent Acceleration, Dubna Preprint P9-5090, 1970.

[10] H. Motz and C. J. H. Watson, Advan. Electron. Electron Phys. 23, 153 (1967).

[11] H. Alfvén, Phys. Rev. 55, 425 (1939).

[12] J. D. Lawson, Phys. Lett. 29A, 344 (1969).

[13] V. P. Sarantsev et al., Experiments on α-Particle Acceleration by the Collective Method, Dubna Preprint P9-5558, 1971. English version as CERN translation 71-1.

[14] V. I. Veksler et al., Proc. Sixth Int. Conf. on High Energy Accelerators, Cambridge, Mass., 1967, CEAL-2000, p. 289.

[15] C. Bovet and C. Pellegrini, Particle Accelerators 2, 45 (1971).

[16] Proc. 1971 National Accelerator Conference, Chicago, I.E.E.E. Trans. Nucl. Sci. NS-18, No. 3.

[17] L. J. Laslett and A. M. Sessler, Comments Nucl. Part. Phys. 4, 211 (1970).

[18] D. Koshkarev and P. Zenkevich, Proc. Eighth Int. Conf. on High Energy Accelerators, CERN, Geneva, 1971.

[19] D. Keefe, Particle Accelerators 1, 1 (1970).

[20] R. E. Berg, Hogil Kim, M. P. Reiser, and G. T. Zorn, Phys. Rev. Lett. 22, 419 (1969).

[21] M. Reiser, IEEE Trans. Nucl. Sci. NS-18 3, 460 (1971).

[22] L. A. Ferrari, K. C. Rogers, and R. W. Landau, Phys. Fluids 11, 684 (1968).

[23] L. A. Ferrari and M. S. Zucker, Particle Accelerators 2, 121 (1971).

[24] A. G. Bonch-Osmolovskii, On the Dynamics of Charged Bunch Collisions in Connection with the Impact Mechanism of Acceleration, J.I.N.R. Dubna Preprint P9-5197, 1970. English translation as Lawrence Berkeley Laboratory Report UCRL-Trans.-1421.

[25] W. O. Doggett and W. H. Bennett, Bull. Am. Phys. Soc. 14, 1043 (1969).

[26] W. O. Doggett, Bull. Am. Phys. Soc. 15, 1347 (1970).

[27] V. N. Tsytovich, On Collective Beam Acceleration, Dubna Preprint P9-5091, 1970.

[28] V. B. Krasovitskii, Sov. Phys.-JETP 32, 98 (1971).

[29] A. A. Plyutto et al., JETP Lett. 6, 61 (1967).

[30] A. A. Plyutto et al., Sov. At. Energy 27, 1197 (1969).

[31] E. D. Korop and A. A. Plyutto, Sov. Phys. Tech. Phys. 15, 1986 (1971).

[32] S. E. Graybill and S. V. Nablo, Appl. Phys. Lett. 8, 18 (1966).

[33] S. E. Graybill and S. V. Nablo, IEEE Trans. Nucl. Sci. NS-14 3, 782 (1967).

[34] W. T. Link, IEEE Trans. Nucl. Sci. NS-14 3, 777 (1967).

[35] S. E. Graybill and J. R. Uglum, J. Appl. Phys. 41, 236 (1970).

[36] J. Rander et al, Phys. Rev. Lett. 24, 283 (1970).

[37] S. D. Putnam, Phys. Rev. Lett. 25, 1129 (1970).

[38] N. Rostoker, Proc. Seventh Int. Conf. on High Energy Accelerators, Yerevan, 1969.

[39] J. Rander, Phys. Rev. Lett. 25, 893 (1970).

[40] M. S. Rabinovich, Collective Methods of Acceleration, Plasma Accelerator and Plasma Physics Laboratory of the Lebedev Physical Institute, Moscow, Preprint No. 36, 1969. English translation as Lawrence Berkeley Laboratory Report UCRL-Trans.-1938.

[41] W. B. Lewis, Symposium on Electron Ring Accelerators, Lawrence Livermore Laboratory Report UCRL-18103, 1968.

[42] A. W. Trivelpiece, R. E. Pechacek, and C. A. Kapetanokos, Phys. Rev. Lett. 21, 1436 (1968).

[43] G. S. Janes, R. H. Levy, H. A. Bethe, and B. T. Feld, Phys. Rev. 145, 925 (1956).

[44] J. D. Daugherty, J. E. Eninger, G. S. Janes, and R. H. Levy, IEEE Trans. Nucl. Sci. NS-16 3, 51 (1969).

[45] D. A. Tidman and N. A. Krall, Shock Waves in Collisionless Plasmas (Wiley-Interscience, New York, 1971).

[46] A. M. Sessler, Comments Nucl. Part. Phys. 3, 93 (1969).

[47] A. M. Sessler, Collective-Field Acceleration, Lawrence Livermore Laboratory Report UCRL-19242, and also in Proc. Seventh Conf. on High Energy Accelerators, Yerevan, 1969.

[48] J. S. Luce, H. L. Sahlin, and T. R. Crites, IEEE Trans. Nucl. Sci. NS-20 (1973).

[49] J. S. Luce, Ann. N.Y. Acad. Sci. 251, 217 (1975).

[50] G. W. Kuswa, L. P. Bradley, and G. Yonas, IEEE Trans. Nucl. Sci. NS-20 3 (1973).

[51] Reference to a company or product name does not imply approval or recommendation of the product by the University of California or the U.S. Energy Research and Development Administration to the exclusion of others that may be suitable.

[52] R. E. Kidder, Nucl. Fusion 14, 53 (1974).

[53] J. H. Nuckolls, in Laser Interaction and Related Phenomena, edited by H. J. Schwarz and H. Hora (Plenum Press, New York, 1974) Vol. 3, pp. 399-425.

[54] N. G. Basov, E. G. Gamaly, O. N. Krokhin, Yu. A. Mikhailov, G. V. Sklizkov, and S. I. Fedotov, in Laser Interaction and Related Plasma Phenomena, edited by H. J. Schwarz and H. Hora (Plenum Press, New York, 1974) Vol. 3, pp. 553-590.

[55] J. S. Clarke, H.N. Fisher, and R. J. Mason, Phys. Rev. Lett. 30 (3), 89 (1973).

[56] R. R. Johnson, Bull. Am. Phys. Soc. Ser. II 19 (9) 886 (1974).

[57] G. Yonas, J. W. Poukey, K. R. Prestwitch, J. R. Freeman, A. J. Toepfer, and M. J. Clauser, Nucl. Fusion 14, 731 (1974).

[58] L.I. Rudakov and A. A. Samarsky, in Proc. Sixth Euratom Conf. Controlled Fusion Plasma Phys., Moscow, 1973 (Euratom-CEA, Grenoble, 1973) Vol. I, pp. 486-490.

[59] F. Winterberg, Plasma Phys. 17, 69 (1975).

[60] J. S. Luce, Ann. N.Y. Acad. Sci. (1975).

[61] C. L. Olson and J. W. Poukey, Phys. Rev. A 9 (6), 2631 (1974).

[62] H. Conrads and P. Cloth, in Proc. Fifth Euratom Conf. Controlled Fusion Plasma Phys., Grenoble, 1972 (Euratom-CEA, Grenoble, 1972).

[63] R. Gullickson, A Measurement of the Distribution of Very Energetic Ions in the Plasma Focus Device, Lawrence Livermore Laboratory Report UCRL-76374, 1975.

[64] S. Humphries, J. J. Lee, and R. N. Sudan, Appl. Phys. Lett. 25, (1) 20 (1974).

[65] S. Humphries, J. J. Lee, and R. N. Sudan, Advances in the Efficient Generation of Intense Pulsed Proton Beams, Cornell University Laboratory of Plasma Studies, Ithaco, New York, Report No. LPS-154, 1974.

[66] J. W. Shearer, Ion Beam Compression of Thermonuclear Pellets, Lawrence Livermore Laboratory Report UCRL-76519, 1975.

[67] Sometimes referred to as the "two component" system.

[68] J. L. Wyatt, K. Hinrichs, and W. Day, Properties of Slowing Down Reactions, AFOSR Final Scientific Report No. 70-2101-7R, 1970.

MEGAGAUSS FIELDS BY AUTOMODULATING CURRENTS *

V. NARDI

Istituto Elettrotecnico Nazionale Galileo Ferraris, Torino, Italy
Stevens Institute of Technology, Hoboken, New Jersey, 07030

A B S T R A C T

The magnetic field building-up is related with the filamentation of current channels which is typical of dense plasmas in a variety of experimental conditions. Specifically a sequence of different typical stages is considered for the filamentary current sheath of a coaxial accelerator (plasma focus) in each of which the plasma is in a quasi-stationary state. A description of plasma and field fine structure is given in term of phase-space particle density and electric and vector potentials. The break-up of the filamentary structure generates three-dimensional current loops in the plasma with local amperian currents which can be higher than the accelerator total current. These loops have toroidal and poloidal fields with a maximum of amperian current on the loop axis in the region of poloidal-field peak intensity. Mirror effect on this region creates a steep potential wall with some interruption of the current carrier flow.

Extreme values of current and magnetic field (up to $\sim 10^8$G) are due to the high-energy particles which are produced by the potential wall as in an electronic ram. The time for potential wall and magnetic field building-up is compressed by radiative energy losses for an increasing plasma transparency when the electron cyclotron frequency reaches the plasma frequency and negative-temperature conditions are created.

1. INTRODUCTION

Plasma-focus experiments[1,2] suggest that the building up of self-consistent magnetic fields with intensity $B \gtrsim 10^6 - 10^8$G in the plasma can feed on the fine filamentary structure of the current sheath which is formed starting long before ($\sim 1-2$ μs $\sim T/4$, T = period of discharge circuit) the focus stage of the coaxial-accelerator discharge.[3,4,5]

The emission of bremsstrahlung x-rays (photon energy $\varepsilon > 1$keV–1 MeV) has a maximum of intensity on the current sheath (1.0–0.1 mm thick) consistently in all stages of the discharge[1] which have a high emission intensity. On the other hand the life time $\sim$ 100 ns of the plasma column which is formed along the electrode axis by the convergent motion of the current sheath is also much longer[6,7] than the hydrodynamic time $t \sim R/V_s$ ($R \sim$ 3 mm is a mean value of the radius of the cusped plasma column on the electrode axis, $V_s \sim 10^7$ cm/s is the sheath velocity in the convergent motion up to axial-column formation). So it seems that dissipation phenomena (e.g. bremsstrahlung) and stabilization effects are favored by the filamentary structure of the current sheath. The magnetic-field shear due to the filaments (on a filament axis the field becomes parallel to this axis[8]) can prevent some of the instabilities affecting an ordinary pinch with a simple azimuthal magnetic field B_θ. Time resolved photographs of the axial plasma column indicate an electron density distribution (by luminosity and laser light diffusion) typical of a hollow column.[8,9] Consistently with all available data we consider the current carrier (electrons[8]) as concentrated on the hollow column wall which is so formed by a texture of current channels (filaments) with a diameter of < 0.5 mm. The axial-column wall can so resist deformations, including localized pinching to smaller diameters for a relatively long time. If the field-energy increase via line-stretching in wall deformations can oppose instabilities due to the B_θ- field outside the column then the filamentary texture of the wall has to store a substantial amount of energy (this would also imply a very high energy density on the wall because the volume of the column wall with sharply-defined boundaries is much smaller than the volume with a larger B_θ outside the column). The energy of the filamentary texture can be estimated in terms of self-consistent $\underline{B}$ and electric field $\underline{E}$ by using a plasma theory we have developed in previous work[10,5] with source terms in the equation for electron phase-space density f_- and positive ion density f_+. Derivations and solutions of some of the field equations are first recalled for two typical steady-state conditions of the current sheath (i.e. current sheath motion between the electrodes and axial column) first, then a mechanism is analyzed by which peak values of B are built inside localized plasma regions with linear dimensions $\ell \sim 0.1$ -0.5 mm during the stage of decay of the filamentary current sheath. This process involves the formation of current loops, with a toroidal-vortex structure each from a fragment of the column wall[11] Transitions between different states of the current sheath span only a fraction of the duration time of these states so that a time-dependent treatment becomes necessary only for a transition stage. Since plasma filaments in a focus experiment can persist even[12] when the total electrode current vanishes on the electrodes they can be considered as plasmoids with a rather stable structure built by persistent local currents and fields. During the focus stage an anomalous increase of resistivity inside a filament or its sudden disruption can cause the increase of current flow in other filaments either by a rearrangement of the local structure or - when many filaments are involved - via dB_θ/dt inductive fields. This as well as the self-crowbaring of the plasma column between neighbouring cusps on the column (see paragraph 4) can illustrate the concept of current automodulation we intend to consider. Flow vorticity seems an essential element of the filament structure. Cause of vorticity and filament formation are examined in the next paragraph.

2. VORTICITY AND FILAMENT FORMATION

Mass-flow vorticity and filament formation are a result of the interaction between the magnetic field B_θ and the plasma streaming against B_θ which is produced by the amperian current I on the electrodes. The filaments are formed at an early stage of the discharge[13] when the current sheath is still at the breech of the coaxial accelerator. We recall that: (I) During the motion of the current sheath-a nonplanar shock-between the electrodes large filaments (diameter $2r \sim 0.5$ mm) along the electrode radius show a constant azimuthal position and a substantially-unchanged configuration. (II) Each filament produces a depression in B_θ[14] that can be observed as a groove in the shock front (i.e. the foremost luminous face Σ, say, of the current sheath) which separates the region of large B_θ (behind the shock)

from the region of vanishing B_θ ahead of the shock. This means that a concentration of amperian current occurs in a filament[14] By (I), (II) a filament has to be considered as a pinch rather than an instability of the current sheath. By the local nonplanarity of Σ (each filament is in a groove of Σ) and by imposing usual conservation laws[15] (shock condition across Σ as a discontinuity surface) a value of the vorticity $\underline{\omega} = \frac{1}{2}\underline{\nabla} \times \underline{u}$ can be derived in terms of the principal curvatures $1/r_a$, $1/r_b$ of Σ and of two shock-strength parameters

$\zeta = \rho_1/\rho_0 = u_{on}/u_{1n}$, $\alpha = B_1/B_0$,

$\underline{\omega}_1 = \underline{\tau}_a \zeta G/u_{on} r_b + \underline{\tau}_b \zeta F/u_{on} r_a$; $u_{on} = \underline{u}_0 \cdot \underline{n}$,

ρ=mass density, $\underline{u}_0 (\simeq \underline{u}_{0+})$ is the velocity of the incoming plasma (in the frame of Σ); $\underline{\tau}_a, \underline{\tau}_b$ are the unit vectors tangential to the curvature lines of Σ for a convenient system of orthogonal curvilinear coordinates γ_a, γ_b on Σ ; $\underline{n}$ is the unit vector orthogonal to Σ ; suffix 1,0 is used for quantities behind, ahead of Σ respectively; F, G are regular functions of α, ζ and of quantities on one side of Σ , $\rho_0, \underline{u}_0, B_0$. This expression for $\underline{\omega}_1$ follows by simply imposing the conditions of mass, momentum, energy conservation[15] across Σ, by the nonplanarity of Σ, and by using the infinite-conductivity approximation $\underline{\nabla} \times (\underline{u} \times \underline{B}) = 0$ as it is suggested by the observed current-sheath resistance $\lesssim 10^{-3} \Omega$ (during 1-2 μs of motion between the electrodes and rolling-off stage of the current sheath)[13] In the motion between the electrodes it is generally observed[14] $1/r_a \ll 1/r_b$ where the principal curvature $1/r_b$ of Σ on the plane orthogonal to τ_a (τ_a is taken along the bulk of the electric current on the current sheath, i.e. about parallel to the electrode radius) is determined by the depression of B due to a radial filament (during rolling off stage and axial-column formation it is observed in some location of Σ also $1/r_a \gg 1/r_b$ with a formation of a luminous azimuthal filament along the groove of Σ corresponding to this large $1/r_a$ value; these luminous filaments parallel to the background B_θ-due to the electrode current[3,14] do not seem to have the same nature of the radial filaments; see paragraph 6). Since $\underline{\omega}_1$ contains r_b in the combination τ_a/r_b ($F \sim G$, so that only one of the two $\underline{\omega}_1$-components, with $1/r_a$ or $1/r_b$, is present in one location of Σ) the vorticity is always parallel to a filament axis and has a maximum value where F/r_b (or G/r_a) is maximum i.e. in a filament region (both signs of F, G and so of ω are possible inside a filament; both signs for the magnetic-field-component along a filament axis have also been found by measurements inside different filaments[4,5,8]). One is tempted to conclude that the current pinching on the current sheath by causing nonplanarity of Σ is the cause of vorticity. However there are indications[16] that when a current sheath is periodically perturbed in shape then a transverse flow of charge in the sheath produce charge concentrations at alternating points of inflexion (the later process was proposed by Kyhl[17] to explain the formation of vortices in the auroral arcs; experiments with a thin electron beam of sinusoidal cross section have proved[16] that vortices form at the inflexion points as predicted by the Kyhl model). So ripples (with vortex filaments) in the current sheath can occur because of the cooperative effect of current concentration due to pinch effect and the Kyhl mechanism which can produce current concentration from ripples. Generation of mass-flow vorticity follows also by MFD equations[18] for a variety of conditions as: (i) non-constant tensor conductivity or strong Hall effect (i.e. $\omega_c \tau_c \gg 1$ in terms of electron cyclotron frequency ω_c and electron collision time τ_c; if the electrons are the current carriers $\sigma B u/en_- u_A \gg 1$ with scalar conductivity σ, Alfvén velocity $u_A \equiv B/(2\pi\rho)^{\frac{1}{2}}$, mass density $\rho \equiv m_+(\bar{n} + n_+)$, $e = 4.8 \times 10^{-10}$ e.s.u, $\bar{n}$ is the density of the neutral atoms with the same mass m_+ of the positive ions, $n_+ = \int f_+ dv$); (ii) uniform fractional ionization (i.e. ρ/n_- = const.); (iii) negligible ion slip(i.e. no motion of the ions relative to the neutrals); (iv) infinite conductivity or, alternatively, infinite mobility σ/en_-; (v) scalar pressure; (vi) quasineutrality ($n_+ \sim n_-$). Under those conditions the resulting vorticity of a conducting fluid by steady state and Newton equation satisfies[18]:

$(\rho\ \underline{\nabla} \times \underline{u}/n\ +\ e\underline{B}/c_o)\cdot d\underline{S} \simeq 0$

for an arbitrary element of surface $d\underline{S}$ (we use Gaussian units; $c_o = 3\times10^{10}$cm/s). Approximations made in deriving[18] this expression are not fully justified in all cases of interest. As an alternative also to gain a better insight it seems convenient to obtain an expression for the vorticity directly from the Lagrange invariant

(1) $\oint_s p_s ds = C_1$ = const ;

p_s is the component tangential to a closed curve s, of the single-particle canonical momentum $\underline{p}_\pm = m_\pm \underline{v} \pm e\underline{A}/c_o$; the integration is round any circuit s (with element ds; a continuous distribution of trajectories is considered) moving along with the particles[19]; by using Hamilton's equations it is easily proved that $d(p_s ds)/dt$ is an exact differential ($=d^s H$, H = Hamiltonian).[19] In terms of curl $\underline{p}$ (by Stokes theorem)and of any infinitesimal area $d\underline{S} = dS\underline{n}$ bounded by a curve s (in general the unit vector $\underline{n}$ will not concide with the direction of motion) equation (1) can be written in the differential form

(2) $(\underline{\nabla} \times \underline{p})\cdot d\underline{S}$ = const

Equations (1) and (2) can be generalized to a large class of phase-space densities $f(\underline{p},\underline{z})$; e.g. for $f = f(H)$ it is clear that also

$\oint_s f(H)p_s ds$ = const   and so

(3) $\{\underline{\nabla} \times (f\underline{p})\}\cdot d\underline{S} = C$ = const

An advantage of using this approach is that constraints on the flow can be imposed by using specific forms of the distribution of particles f, i.e. by specifying the "occupancy" of the continuous distribution of trajectories. We then apply eq. (3) to the filamentary current sheath with no ion slip and a $\rho/n_+ \simeq 1$. Ahead of the current sheath (where $\underline{B}=\underline{A}=0$) the particles stream toward the current sheath all with nearly the same velocity u_o ($\simeq 5\times10^6$ cm/s); in this case the vector field $\underline{p}$ and the distribution f must be such that the constants on the right-hand sides of eq.s (1),(2),(3) vanish; by integrating eq. (3) on all possible values of the single particle velocity $\underline{v}$ we obtain for the ions

(4) $\{\underline{\nabla}\times(m_+n_+\underline{u}_+ +\ n_+e\underline{A}/c_o)\}\cdot d\underline{S}=0$ i.e.

(5) $(\underline{\nabla}\times\underline{u}_+ + e\underline{B}/m_+c_o)\cdot d\underline{S}$ =

$\{(\underline{u}_+ + e\underline{A}/c_o m_+) \times \underline{\nabla}n_+/n_+\}\cdot d\underline{S}$;

variations along the filament axis can be neglected so ∇n_+ is orthogonal to a filament axis which is taken to coincide with the z axis ($\partial/\partial z= 0$). By considering axial symmetry for the filament structure then each $\underline{u}_+ d\underline{S}/n_+$ (and $\underline{A} d\underline{S}/n_+$)

will give a resultant vector in the direction of the filament axis after integration over a small surface S spanning the whole cross section of a filament so that

(6) $\int (\underline{\nabla} \times \underline{u}_+ + e\underline{B}/m_+c_0) \cdot d\underline{S} = 0.$

Equations (4),(5),(6) represent a generalized conservation law for the angular momentum and the flux of $\underline{B}$ together (angular momentum alone is of course not conserved). In deriving (3) - (6) each positive ion is considered as interacting with other ions only via electric and magnetic macroscopic fields $\underline{E}$, $\underline{B}$; short-range collisions with few neutral atoms ($\sim 1\%$) still existing near the current sheath (± 1-2 mm from Σ) makes the neutrals to move coherently with the ions. The motion of the electrons instead can be substantially affected by short range collisions near and on Σ so that it is doubtful the validity of an equation similar to (4) also for the electrons; in case of validity the summation of the two equations ($n_+ \sim n_-$) would yield

$$\{\underline{\nabla} \times (m_+n_+\underline{u}_+ + m_-n_-\underline{u}_-)\} \cdot d\underline{S} = 0.$$

The numerical value of this scalar product could then be taken in some case as an estimate of the electron collision effects inside the volume between two positions of dS as it is moving with the ions. The expression of the torque $\underline{G}$ in terms of magnetic stress[20] can help to identify the causes of the onset of vorticity on Σ. By taking $|\underline{E}| \sim u_0B/c_0$ ($\simeq 5 \times 10^2$ Volts/cm in our case with $B \sim B_\theta \sim 10^4$G; i.e. $\sim 10^2$ volt across the current sheath - in agreement with laboratory observations - during its motion between the electrodes) and an axially symmetric region (with the same axis as the filament) we have $|\underline{G}| \simeq G_z \simeq \int rB_\phi \underline{B} \cdot d\underline{S}$ by using

cylindrical coordinates (r,ϕ,z); the integration is on the surface of the plasma region for which we estimate the torque, B_ϕ is around the filament axis. The field structure in a filament seems to be substantially axially-symmetric so that this estimate would give $\underline{G} \sim 0$. A different value of $\underline{G}$ can be obtained e.g. if the voltage $\sim 10^2$ V across the current sheath corresponds to a value $E \gtrsim 10u_0B/c_0$ due to charge separation on a very thin layer(thickness $\simeq r_L$ electron Larmor radius) of the current sheath. Let us consider for simplicity plasma flowing toward each element of Σ along the corresponding normal direction and

and B-field lines laying on Σ; charge separation is caused by the sudden change of motion of the incoming electrons upon arrival on Σ where the electrons become tied to the magnetic-field lines (ions penetrate Σ by inertial crossing of field lines). Excitation of currents occur by this electron acceleration $-\dot{\underline{v}}$, with a maximum value for the component j_x say, of the current density in the direction orthogonal to Σ, at least in the plasma layer closest to Σ, where $\underline{v}$ is, in the bulk,orthogonal to Σ. The effective electric field is $\underline{E}' = \underline{E} - m_-\dot{\underline{u}}_-/e$; where $\underline{u}_- \equiv \int \underline{v} f d\underline{v}$; $\underline{B}$ on Σ diverts the kinetic energy ε of the electron random motion into the motion on Σ, by nearly suppressing the averaged value of $-\underline{v} \cdot \underline{n}$ for the electrons in a layer $\sim r_L$ thick contiguous to Σ (Σ-layer, say). The $\nabla|\underline{B}|$-drift is most effective in rearranging the current on this layer (consistently also with the previous shock theory we can here define Σ as the regular surface best fitting a two dimensional locus on the current sheath where $\underline{\nabla}|\underline{B}|$ is maximum). As an order-of-magnitude estimate the value $\Delta E \equiv |\underline{E}' - \underline{E}| \sim m_-\Delta v/e\Delta t \sim 10^4$ Volt/cm is obtained on Σ by $\Delta v \sim v$ where $v \simeq (2\varepsilon/m_-)^{\frac{1}{2}}$ is the dominant value of the electron velocity if $\varepsilon \simeq$ ionization energy ($\simeq$13eV for deuterium plasma)and by $\Delta t \sim r_L/v$; $r_L < 10^{-2}$cm for a $B \sim B_\theta/10 \sim 10^3$G in the Σ-layer during the current-sheath motion between the electrodes (Max $B_\theta \to 10^6 - 10^8$G, $\varepsilon \to 0.01 - 1$ MeV during the axial compression). For a point-by-point determination of the effective electric field $\underline{E}'$ on Σ we set $\dot{\underline{u}}_- = -\dot{\underline{u}}_+m_+/m_-$, enter $\underline{E} = \underline{E}' - 2\dot{\underline{u}}_+m_+/e$ in Maxwell's equations and we get

(7) $\underline{\nabla} \times \underline{E}' = -\partial\underline{B}/\partial t c_0 + m_+\underline{\nabla} \times \dot{\underline{u}}_+/e$.

We can decompose the motion of $\underline{u}_+$ as a translation and a rotation
($\underline{u}_+ = \underline{u} + \underline{\omega} \times \underline{r}$,
$\dot{\underline{u}}_+ = \dot{\underline{u}} + \underline{\omega} \times \underline{u} + \underline{\omega} \times (\underline{\omega} \times \underline{r}) + \dot{\underline{\omega}} \times \underline{r}$,
time variation following the motion of the ions i.e. the mass motion) in a sufficiently-small region of space (e.g. a sublayer within the Σ layer) so that distortions in the motion of the still-essentially-cold ions are negligible compared with the variations of a single-electron velocity; within this region both $\underline{u}$ and $\underline{\omega}$ can be considered as independent of position $\underline{r}$. Then $\underline{\omega} \times (\underline{\omega} \times \underline{r}) = -\frac{1}{2}\underline{\nabla}(\underline{\omega}x\underline{r})^2$, $\underline{\nabla}(\dot{\underline{\omega}} \times \underline{r}) = 2\dot{\underline{\omega}}$ and (7) becomes

(8) $\underline{\nabla} \times \underline{E}' = -\partial \underline{B}'/\partial t c_o$ with an

effective magnetic field $\underline{B}' = \underline{B} - 2\underline{\omega}m_+ c_o/e$

as in the well-known case of current excitation by non-uniform rotation of a rigid conductor[22]. As a summary the onset of vorticity in the Σ-layer is controlled by the interaction of different elements as: (i) $j_x = \underline{j} \cdot \underline{n}$ (due to a different penetration of electrons and ions in the Σ-layer); (ii) the random occurrence of local pinchings in the density component j_z (i.e. of electric current flowing within the current sheath from one electrode to the other); (iii) the ion back-flow (with sign $\underline{u} \cdot \underline{n}$ reversal on the downstream side-opposite to Σ-of the current sheath; this ion back-flow is a consequence of the observed snow-plow effect of the current sheath) contributing to j_x.

The net result of combining different j_x, j_y, j_z (locally $B_y \equiv B_\theta$ if we consider center electrode radius $R \gg r_o$ = typical-filament radius, consistently with $\partial/\partial z = 0$) is a helical current pattern on the current sheath, with a random sign of the self-consistent (by Eq. 6,8) B_-, $\underline{\omega}$ inside a filament, both along the axis of the helical flow in the direction of the electrode radius[5]. Formation of vortices is possible (not necessary) with $\omega \neq 0$; the typical linear dimensions of eddies forming on the current sheath could span from the observed[1-6] vortex-filament diameter $2r_o \sim 0.1 - 0.5$ mm down to the eddy-wavelength spectrum that is considered by turbulence theories[23]. The equation for $\underline{\omega}$ in an incompressible non-conducting fluid is identical in form[23] with that for $\underline{B}$ in a conducting, quasineutral liquid with $\omega_c \tau_c \gg 1$

if the kinetic viscosity ν is replaced by $c_o^2/4\pi\sigma$. The analogy between $\underline{B}$ and $\underline{\omega}$ can be used to determine how the magnetic energy distributes among eddies by interplay of hydrodynamic and electromagnetic effects;[23] when the nonuniformity of $\underline{u}$ is such as to increase the length of the lines of magnetic force, $|\underline{B}|$ increases, and vice versa (the magnitude of σ measures the firmness by which the medium grips these lines; energy dissipation occurs by particle collision among other causes as radiative and recombination processes). Energy can so be exchanged between $\rho\underline{u}$ and $\underline{B}$ fields and spontaneous appearance of magnetic fields can occur as

well, by the growth of small perturbations. A criterion for growth of electromagnetic energy in a turbulent medium (e.g. $4\pi\sigma\nu/c_o^2 > 1$) is known from the literature[23] where the plausible (even though not proven rigourously) point is made that equilibrium between the $\underline{B}$ and $\underline{u}$ fields is acheived when the large wave-number components (by writing $\underline{B}$, $\underline{u}$ as Fourier integrals) contain comparable amounts of kinetic and magnetic energy. By this analysis the kinetic energy ($\sim\rho u^2$) resides mainly in the large-size eddies and the magnetic energy resides mainly in the eddies of small size (sometimes called magnetic eddies) because the small-scale components of the turbulence by containing most of the vorticity are more effective in stretching field lines than large scale components. Equilibrium can be achieved, however, because the transfer of energy to smaller scales is limited by the non-existence of the stretching process for very small scales[23]. In turbulent flow $\underline{B}$ has a random variation with time so that the time-averaged value of $\underline{B}$ is zero (a non-zero mean field is incompatible with turbulence).[22] This implies that an external field with high intensity ($B^2 \gtrsim \rho u^2$) by penetrating a fluid in turbulent motion can supress the turbulence, i.e. under these conditions the fluid behaves as a superconductor. These arguments seem relevant for understanding the structure of the current sheath in a plasma-focus experiment. In fact it is observed[1-6] that filamentation disappears (as suddenly as it formed) behind the current sheath where the field $\underline{B}$ due to the electrode current reaches its maximum value $MaxB_\theta$ and the condition $B^2 \gtrsim m_+ n_+ u^2$ is satisfied (e.g. $MaxB_\theta \sim 2 \times 10^4$ G, $n_+ \sim 10^{17} - 10^{18}$ cm^{-3})[24]. The dominant size of vortex filaments is important also for deciding whether the observed filaments on the current sheath are clusters of smaller units or not. A method for describing different possible structures of the filaments is presented in:

3.   SELF-CONSISTENT FIELD

The sources
$\underline{j} = e(n_+\underline{u}_+ - n_-\underline{u}_-) = \underline{j}_+ + \underline{j}_-$, $\rho_c = e(n_+ - n_-)$

of $\underline{B} = \underline{\nabla} \times \underline{A}$, $\underline{E} = -\nabla\Phi - \partial\underline{A}/\partial t c_o$ in the

plasma are derived from the distributions f_+, $f_-(\underline{v},\underline{r},t)$ which satisfy
(9) $df_\pm/dt = S_\pm$

where

$$d/dt = \partial/\partial t + \underline{v} \cdot \underline{\nabla}_r \pm e(\underline{E} + \underline{v} \times \underline{B}/c_0) \cdot \underline{\nabla}_v$$

is the Liouville-Vlasov operator.
$S_\pm = d(f_{0\pm}F_\pm)/dt$ is a source (or a sink,

when $S_\pm < 0$) of particles in phase-space that can account for collisions, ionization, recombination, radiative processes and for any cause affecting the trajectory of a particle (e.g. microscopic fluctuations of the electric and magnetic field) other than the macroscopic fields $\underline{E}$, $\underline{B}$; $f_{0\pm}$ is any solution of

(10) $df_{0\pm}/dt = 0$

with some specific, predetermined fields $\underline{E}$, B (in particular the fields $\underline{E}$, $\underline{B}$ which satisfy (9), Maxwell's equations and the boundary conditions of our complete problem). The choice of $f_{0\pm}$ is a matter of convenience and in general we choose $f_{0\pm}$ among distributions which are suitable as a first approximation to a realistic distribution. We consider quasi-steady-state conditions in the frame of Σ ($\partial/\partial t = \partial/\partial z = 0$) so $p_{z\pm} = m_\pm v_z \pm eA_z/c_0$

and the single-particle energy ε(now ε stands for kinetic plus potential energy) are two constraints of the motion and any function of $f_0(\varepsilon, p_z)$ is a solution of Eq. 10 (indices $\pm$ are sometimes dropped). It is not possible to obtain realistic flow fields $\underline{u}$ by simply using a distribution $f_0(\varepsilon, p_z)$, even by assuming as negligible all particle interactions but the forces due to $\underline{E}$, $\underline{B}$ (so that $S = 0$). Other constants of the motion of a single particle are not readily available without imposing some other unjustified symmetry to the system. On the other hand the difficulty of solving the equations for self-consistent Φ, $\underline{A}$ are further increased if we take distributions $f(\varepsilon, p_z, C_1, C, \ldots)$

with a dependence on quantities which are determined by the simultaneous state of a group of particles as the invariants C_1, C of paragraph 2. We turn to S. S can be further specified with the objectives:

(I)   To get distributions f(and so n,$\underline{u}$) which satisfy at least some of the constraints on n, $\underline{u}$ (and/or on fields $\underline{E},\underline{B}$) which are known a priori, e.g. from the experiments.

(II)   Since $f_\pm = f_{0\pm}F_\pm$ is obviously a

solution of Eq. 9 we can specify $F_\pm$(i.e.S) simply with the objective of transforming a gross approximation f_0 into a more realistic f. This is useful also in the case that all interactions can be accounted for by macroscopic $\underline{E}$, $\underline{B}$, : In this case the transfer of particles by S from some phase-space cells to other cells is not the result of an actual physical process (e.g. bremsstrahlung or ionization) but a convenient procedure to transform f_0 to a new f_0F fitting e.g. a prescribed flow that is not obtainable by the tentative f_0.

An advantage of considering equations in phase-space (where problems are usually underdetermined) rather than macroscopic equations is that we can use the extra freedom to simplify the method of solution. We take $f_{0\pm} = \text{const} \cdot \exp(\alpha_\pm \mu_\pm)$ where

$$\mu_\pm(\varepsilon, p_z) = \{v_x^2 + v_y^2 + (v_z + c_\pm)^2\}m_\pm/2 + \Gamma_\pm,$$

c,α are constants and $\Gamma_\pm \equiv \pm e(\phi + c_\pm A_z/c_0)$;

$$F_\pm = 1 + \Sigma_{\zeta i} \int d\mu_i L_\pm(\mu_i) \, C(\psi_{\zeta i\pm}) \, C(D_{\zeta i\pm}),$$

$C(X)$ is the Heaviside unit function (=0 for $X < 0$), μ_i is any specific numerical value of μ, ζ specifies the branch of S (a singular function) that we take in a region of space where the two orthogonal (curvilinear or cartesian) coordinates in the plane orthogonal to z can (for convenience) be different from the orthogonal system in a neighbouring region;
$$\psi_{xi} = v_x - D_{xi}^{\frac{1}{2}} \, , \quad D_{xi} = (\mu_i - \Gamma)2/m - v_y^2 - (v_z + c)^2,$$

and ψ_ζ for curvilinear coordinates have been taken also in refer.10,11. The spectral function $L(\mu_i)$ calibrates the strength of the source which emits or removes particles on a trajectory that we will label by μ_i as in a Vlasov plasma; $\Sigma_{\zeta i} \int d\mu_i$ means integration on all intervals of continuous variation of μ_i and summation on all discrete values of μ_i and on all branches ζ (different branches could be taken simultaneously in the same region of space to account e.g. for ionization process and radiative emission). By taking e.g. only one term with $\psi = \psi_{xi}$ and a single value μ_i (which is sufficient for ion production by ionization ahead of Σ; the role of neutrals can be disregarded[10]) we have

$$S = L(\mu_i)f_0\{C(\psi_{xi})\delta(D_{xi})2v_xF_x/m_+ \quad +$$

$C(D)dC(\psi_{xi})/dt\}$; δ = Dirac's delta ,

$$d(\psi_{xi})/dt = \delta(v_x - D^{\frac{1}{2}})(D^{\frac{1}{2}}F_x + v_x F_y + v_z F_z$$

$$-e\underline{v}\cdot\underline{E})/m_+ D^{\frac{1}{2}} = 0$$

if $D \neq 0$ since $\underline{v}\cdot(\underline{F} - e\underline{E}) = 0$.
This means that with our choice of $F_\pm$
the source within a cell of phase space
is proportional to the power absorbed
by the plasma. The role of $C(\psi_{xi})$ is to
insure that there is a net stream of
particles $(u_x \neq 0)$ also in the x-direct-
ion; a flow $u_x \neq 0$ could not be derived
by an even function of v_x as $f_o(\varepsilon, p_z)$.
The given expression for F_x is in fact
equivalent to introducing a large number
of constants of the motion, specifically
sign v_x; one sign of the velocity component v_x dominates over the other for
many particles if $\psi = \psi_{xi}$). An excess
of particles can be ejected in one
direction of a trajectory or can be
removed from it without any of the
restrictions which are needed for turning
points. The source locus (where $S \neq 0$)
$D_{xi} = 0$ (a cylinder along the v_x axis)

can sweep the whole phase-space simply by
changing μ_i; a particle (after being
created in a cell of phase space at
the intersection of the two surfaces
$D_{xi} = 0$, $\psi_{xi} = 0$) will continue to move

on the sphere $\psi_{xi} = 0$ (i.e. $\mu = \mu_i$) until

it is eventually removed by S. S must be
consistent with the fact that the number
of particles which is produced or absorbed
at a specific time at each point $\underline{r}$ has to
be finite. This must be true not because
divergent contributions with opposite
algebraic signs are canceling each other:
$\int|S|d\underline{v}$ must be bounded if we want S to
have a simple physical meaning ($|S|$ is
the absolute value of S).This integrab-
ility property of S and all its components
(e.g. for a specific μ_i) permits the
exchange of the order of integration
(e.g. $\int d\mu_i$ with $\int d\underline{v}$) in multiple integrals
and so simplifies substantially the cal-
culations. The integral $\int d\mu_i \int d\underline{v}$ is usually more convenient than the reversed
order[10]. Spectral functions

$$L(\mu_i) = (\alpha_o + \alpha_1\mu_i + \alpha_2\mu_i^2 + \dots)e^{\beta\mu_i}$$
with different α, β for different branches

of S are sufficient to yield $n_+\underline{u}_+$
with current loops, specific
orientation or stagnation points on the
side of the filament facing the incoming
flow etc[10].
The resulting j_+ and densities n_+ are
proportional to $P_{1\pm}(\Gamma)\exp(\alpha_\pm\Gamma_\pm)^\pm$or $P_{2\pm}$

$\exp\{-(\alpha_\pm + \beta_\pm)\Gamma_\pm\}$, where $P_{1,2}$ are polinomials

of $\Gamma, |\beta| \leq \alpha$. Since all non trivial
solutions of the elliptic equations for
our potentials have singularities we take
solutions ϕ, A_z with singularities only
at the infinity $|\underline{x}| \to \infty$. To have $\underline{u}$, n
which are always bounded then we must
have $\Gamma_+ > 0$. This is obtained by taking
a different algebraic sign of the constant (first approximation) velocities
$c_+(=\int f_o v_z d\underline{v}/\int f_o d\underline{v})$ of the ions and $c_-$of

the electrons along the z(or filament)
axis. The signs of c_+, c_- are of course
determined by the polarities of the
electrodes. By combining the equations
for ϕ and A_z the potential equations can
be reduced to the form of the well known
non linear equation[10]

$$(11) \qquad \partial^2\Gamma/\partial\eta\partial\bar{\eta} = e^{\lambda\Gamma}$$

in the complex variables $\eta = x + iy$,
$\bar{\eta} = x - iy$, (λ = const) with a well known
solution. The form

$$(12) \quad c_o A_z = e(1/\alpha_+ + 1/\alpha_-)(c_- - c_+)^{-1}$$

$$\ln\{|\partial g/\partial\eta|^2(1 + |g|^2)^{-2}\}$$

where g is an arbitrary function of η,
can be derived from general integrals
for A_z and ϕ which contain two arbitrary
functions g_1, g_2 (see ref. 10) simply by
setting $g_1=g_2=g$. This last condition

implies also that $\phi = A_z(c_-\alpha_+ - c_+\alpha_-)/(\alpha_++\alpha_-)$
$\qquad\qquad$ + const .
By specifying the only arbitrary function
g left we can fit the observable
structural details on the current sheath.
The magnetic field $\bar{B}_o$ due to the current
in the center electrode can be accounted
for by a contribution $A_x = \bar{B}_o z$ to the
vector potential $\underline{A}$ (our condition $\partial/\partial z=0$
must have at least one exception to acc-
ount for the discontinuity of j_x at the
interface plasma-center electrode; here
the current sheath is orthogonal to the
surface of the center electrode); a
function $\bar{B}_o = \bar{B}_o(x)$ can account for the

nonuniform value of the electrode current
in the segment of the center electrode
where the current sheath is attached.
We will consider $\bar{B}_0$ = const ($\neq$MaxB$_\theta$) for
simplicity. The line of force of the
(B_x,B_y)-field are given by A_z = const

(A_z is the stream function of the $\underline{B}$-
field in the x,y plane) in Fig.1a

for g = a + be$^{m\eta}$ and in Fig. 1b for

g = a + be$^{m\eta}$ + ce$^{n\eta}$.

The condition $B_x(x \to \pm\infty,y) \to 0$ is satisfied with these g's and by a more general

g = $\Sigma_j a_j e^{n_j \eta}$; with the first choice for

g, the period $2\pi/m$ of the filamentary
structure along the y axis is determined
by the necessary conditions
$\lim B_y(x \to +\infty,y) = \lim(-\partial A_z/\partial x)+\bar{B}_0=B_0\equiv MaxB_\theta$

and $\lim B_y = 0$ for $x \to -\infty$, i.e. ahead of
the current sheath:
$2\pi/m = \pi c_0 16(\alpha_+ + \alpha_-)/\alpha_+\alpha_-(c_- - c_+)eB_0$.[10]

With the second choice for g we can have
e.g.: (i) pairing of loops,i.e. two loops
are closer to each other than to other
neighbouring loops;(ii) a multiplicity of
superposed periodic structures;(iii)closed
loops of different dimensions;(iv) a
multiplicity of mutually non-intersecting
loops, each with its center external to
other similar loops, but all inside a
single larger loop (see Fig.1a. left).
In each case we have to take different
values for the constants a,b,c,n,m;
restrictions on those constants for case
(iii) are given in ref.25 to determine
the distance between the points P_1,P_2 of
Fig. 1b. By each case the restrictions
on the constants a_i,n_i become milder and
the possible configurations with loops of
increasingly smaller dimensions (compared
e.g. with the current-sheath thickness
d $\sim$ MaxB$_\theta$/MaxdB$_y$/dx) can increase indefinitely by increasing the number of terms
in g with the given form. After reaching
the end of the electrodes the current
sheath rolls off the electrode interspace,
converges on the electrode axis by radial
collapse and forms a hollow column along
the electrode axis(or z axis ; all filaments in the pinched current sheath are
oriented in this direction but,eventual-
ly, for localized distortions). The magnetic field lines A_z = const on the (x,y)
plane can be obtained from the solutions

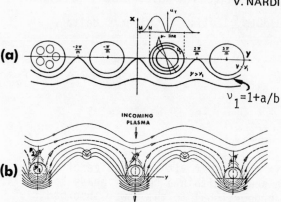

Fig. 1. $B_x B_y$ field lines (A_z = const).

(a): g = a + be$^{m\eta}$; in this case the field
lines are given by [cosh mx + (a/b) cos
my] =const=ν; u_τ($\|$ $\underline{B}$ - $\underline{B}_z$, collinear
flow) is plotted in the upper diagram for
a $\psi_{\tau i}$ in terms of orthogonal curvilinear
coordinates τ ($\|$ $\underline{B}$ - $\underline{B}_z$), ν ($\perp$ $\underline{B}$ - $\underline{B}_z$), z,
and a specific spectrum L(μ_i) from ref.
10 (see also Appendix I); small loops inside large loops at left can be obtained
only by adding one or more terms to g as
in:

(b) g = a + be$^{m\eta}$ + ce$^{n\eta}$. Different values
of a,b,c,m,n can give various structures.

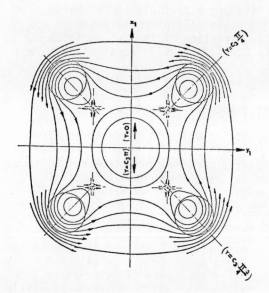

Fig. 2. $B_x B_y$ field lines (A_z = const)
after formation of the plasma column on
the electrode axis. This structure is
obtained by transforming the plasma current
sheath with g = a + be$^{m\eta}$ and $c_3 m$ = 4;
$c_3 m \sim 10^2$ or a g with more terms seem to
fit better laboratory observations.

MEGAGAUSS FIELDS BY AUTOMODULATING CURRENTS

A_z before roll-off and radial collapse, simply by a variable transformation η, $\bar{\eta} \to \eta_1$, $\bar{\eta}_1$ where

$$\eta = c_3 \ln \eta_1 = c_3 \ln r_1 + i c_3 \theta_1 \, ,$$

$$\eta_1 = r_1 e^{i\theta_1} \, , \quad r_1 = (x_1^2 + y_1^2)^{\frac{1}{2}}$$

In the case of Fig.1a this conformal transformation is mapping the plane (x,y) in which there is an infinitely-long row of filaments (with a constant spacing $2\pi/m$ between two filaments along the y axis) in a plane (x_1, y_1) in which there

are only k filaments on a circle of radius $r_1 = 1$ if $c_3 m = \pm k$ = integer (the equations

of the form of eq. 11 which yield the potentials A_z and ϕ are invariant with respect to these transformations of variables[26,10]). The result of the transformation with k = 4 is presented in Fig. 2. The fine structure of the current sheath remains substantially the same before and after (up to $\sim$ 50 ns) the axial column formation. During the motion toward the electrode axis the current sheath is deformed, it shrinks along the y axis (many filaments have to merge with others, or annihilate each other if the fields along the axis have the proper orientation, or fade away by other mechanisms with a relatively-low rate of current dissipation) and it is stretched along the z axis with an increase of the energy stored in the field along the filament axis (//z). The conformal mapping can give the correct structure in the (x,y) plane after pinch from the previous structure but cannot describe e.g. the increase of the filamentary axial field unless we specify the value of the transformation constants c_3 and k from physical conditions during rolling-off and axial-pinching. A possible procedure is to impose during this process the condition of minimum variation of the flux of: (i) the local magnetic field of the filaments (internal field) and simultaneously of (ii) the azimuthal field B_θ in the large volume behind the current sheath and outside the axial column during its pinched stage. An analysis based on this minimum principle has yet to be completed. It is clear that further information is needed from the laboratory to make a choice among possible structures from the available solutions. The underdetermination of the current-sheath problem in phase-space can be reduced,e.g.,by:

(i) a restriction to distributions $f_\pm$ by which invariants as in eq. (1) - (3) can be derived, (ii) the condition that at the electrode-plasma interface

$$\{j_z(x,y)_{plasma}\} = \{j_x(x,y)_{electrode}\}$$

holds.

4. BUILDING-UP OF $\underline{B}$

The cusped column which is formed by the imploding current sheath along the electrode axis may persist without major changes in some discharges over a time interval $\sim$ 50 ns[27,5] Apparent outflow of plasma from the cusps is observed by different diagnostic methods[28,8] The cusped profile ($90°$ view from the electrode axis) of the current sheath is observed also during its convergent motion toward the axis when the velocity is increasing to $\sim$ 2-4u_o, u_o = velocity between the electrodes. The loci of cusps on current sheath profile correspond by a $0°$ view to circular grooves (parallel-to-B_θ luminous rings on image-convertor photographs) which intersect without breaking the radial filaments[12] Energy dissipation increases in these points where field lines change rapidly their orientation. The trail due to the radial motion of these cusps can be observed on time-integrated-x-ray photographs ($90°$ view by pinhole cameras;[29] these trails on both sides of the pinched axial column seem to us the cause of the so called "herringbone" pattern observed[30] on x-ray photography of the plasma focus). The formation of cusps on the current sheaths is considered by a part of the literature-which disregards the current-sheath fine structure-as a long-wavelength Rayleigh-Taylor instability (see, e.g. ref. 27). In our view any estimate of the growth-rate of current-sheath instabilities has no bearing on understanding the sheath dynamics unless the fine structure is explicitly considered. The texture of current sheath with an internal axial field within each filament ($\perp B_\theta$) provides at least enough shear field to prevent an early disruption of the current sheath. This can account for the observed (otherwise unexplained) long life-time of the plasma column. Fig. 3 shows a cusped structure of the axial-column surface which is consistent with laboratory observations (by visible light[8] x-ray pinhole photographs[1,2,8] density profiles[27,28]). Some filaments can be broken at a cusp and can penetrate B_θ for a few mm's outside the current sheath as a result of the previous motion of the current sheath which leaves them behind (time-resolved image-convertor photographs show clearly that most of the radial filaments during the radial motion of the current sheath do not break inside the circular azimuthal groove which forms a cusp-apex locus;[12] this stability is observed also for the thin filaments on current sheaths in Ar or He with a relatively-high resistivity; filaments in He and Ar have smaller diameters than in deuterium). The quasi-steady state of the axial column is brought to an end by a secondary pinching, which reduces the column diameter to $\gtrsim 0.5$ mm. This process may take place on a 10-15 ns time interval and on different segments of the column which shows then typical neckings.[1,27,28] The beginning of this process and its progress as well are characterized by the formation of circular loops between neighbouring cusps:[1,8] This self-crowbaring of the axial-column compensates - at least in part - the increase of inductance due to the local shrinking of the column diameter. Current loops and $\underline{B}$ lines of force with a toroidal-vortex configuration[11] as in Fig.4 are produced by this mechanism.[8] A sequence of these structures can exist along the electrode axis with an inter-connecting current flow.[1,8] The current density has a peak value on the central channel of the torus (//z) formed by a segment of the column wall; these current-loops can eventually move at some distance from the place of production when segmentation of the axial column occurs. Cusps and toroidal structures can be described by considering $\partial/\partial z \neq 0$. A theory based on the method of paragraph 3 and without azimuthal symmetry is considered in ref. 11. The derivation of the toroidal structure field can be simplified by taking azimuthal symmetry so that $\varepsilon = $ const, $p_\theta = $ const, $f_o = f_o(\varepsilon, p_\theta)$; by cylindrical

coordinates r, θ, z, **a** toroidal field $\underline{B}_t$ can be added to a poloidal component $\underline{B}_p = \underline{\theta} \times (\nabla U)/r$ with a stream function $U^p = U(r,z)$, $\underline{\theta} = $ unit vector. The electric current flows mainly on the surface of the torus and in the axial channel; the region inside the torus (with a maximum B_θ) has low current and particle densities as in one of the "forbidden regions" of Stormer theory for charged particles in a magnetic dipole.[31] Several most intense bursts of hard x-rays (<5-20 ns duration each, 0.1-1

MeV) and of neutrons occur over a period of $\sim$ 40-60 ns at the end of the secondary pinching. Microwave pulses (3-10 cm wave-length,1-2 ns duration) starts at the same time[32] and may continue over a longer period of time.[33,34] A theory of the secondary pinching and of the successive intense emission activity requires a time-dependent analysis. By our picture (an axial column with filamentary walls) the secondary pinching is associated with the merging of neighbouring filaments. This process since it implies annihilation of the "internal" magnetic field of the filaments is equivalent to the onset of an energy-dissipation mechanism as ohmic resistive dissipation. The bremsstrahlung emission by 1-2 keV x-rays gives an image (time-integrated) of the cusped profile of the axial-column which coincides with the time-resolved image by visible light[1] (5 ns exposure). This implies that the onset of the dissipative resistive process is a relatively-fast process; the typical photon energy is also increasing with time so that x-ray differential filtering can give images of successive stages of the secondary pinching[1,2,33] It is suggested that two are the possible causes of this dissipative process (i) the increasing density of metallic atoms from the anode[35] in the column wall and possibly (ii) the electron cyclotron drift instability[36] with a relatively small growth rate when single electron velocities >> ion drift velocity.

Spectroscopic observations (Zeeman splitting of a CV line)[37] indicate that already during the axial column formation (i.e. before secondary pinching) the magnetic-field intensity reaches values of at least 0.8-1 MG. At a later time when ionization of the carbon atom is complete the determinations of the magnetic field is done by measurements of the emission-intensity anisotropy on bremsstrahlung x-rays.[1,2] From this anisotropy, the local value of the electron current $i \sim 3 \times 10^6$A($>> I \sim 7 \times 10^5$A total electrode current) is determined inside the localized x-ray sources ($\gtrsim$ 50-100 μm diameter) each of which form the axial channel of a toroidal vortex. Magnetic field intensities $B \sim 2 \times 10^8$ G are obtained by[1,2] by this method. The maximum of field intensity is reached during the rapid ($\gtrsim 10$ ns) decay of the toroidal vortices, which can have a life-time of $\gtrsim$ 10-60 ns. We describe this decay mechanism as a radiative instability due

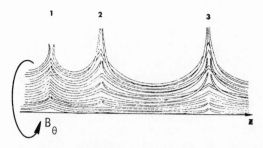

Fig. 3. Schematic profile of a segment of the axial column. Fine filamentary structure is indicated. Some cusps may span only an azimuthal angle $\theta < 360°$. During secondary pinching self crowbarring occurs by connection of some filaments from 1 with filaments from 2 and/or 2 with 3. Cusps are not equally spaced (ref. 1,2,8).

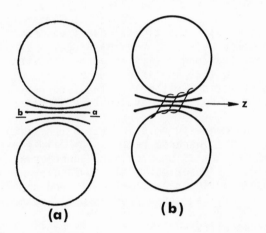

Fig. 4. Schematic profile of a toroidal vortex:

(a) each line represents a filament (or the line of force of the field <B> which is obtained by an average on a filament cross-section. The torus wall has a fine structure $_{Max} B \simeq \pi \bar{r}^2 n_u_e/\bar{r} c_o, \bar{r} = $ ax.channel rad.

(b) B-lines of force forming the torus as a single vortex without filamentary structure. Generally, x-ray pinhole photo-traphs fit (a).

to the rapid increase of the plasma transparency in the following manner. The electrons are tied to the field lines ($u\_$ // $\underline{B}$) with $r_L \lesssim 1$ μm (due to random

transversal velocity $v_\perp$, $m\_v_\perp^2 \lesssim 10$ keV,

$B \gtrsim 10^6$); by entering the axial channel in (a)(the region between points $\underline{a},\underline{b}$ in Fig. 4a)the velocity $v_{//}$ along the field lines is decreased by the mirror effect due to increasing field, with a corresponding increase of kinetic energy in the transversal motion[38] $v_{//}$ can recover its value at the exit ($\underline{b}$) by the reversed process as long as cyclotron emission is balanced by reabsorption and collisions are negligible. Fast electrons can proceed through the channel ($\underline{ab}$) but slow electrons are reflected before reaching the middle point of the channel. An accumulation of electrons (with charge separation) can so take place at the entrance ($\underline{a}$) of the channel and will progress (even by neglecting non linear effects as in the formation of a local virtual cathode) with the progressive shrinkage of the axial channel. This shrinkage may not continue at a uniform rate during the life-time of the toroidal vortex because of the onset of a faster charge-separation process. This process (the electronic ram effect that was presented years ago in an outstanding paper[39] by W. R. Raudorf) affects beams of electrons streaming along lines of force of an externally-produced magnetic field to a region of rapidly-increasing field intensity. The magnetic energy of the beam with an initially-high amperian current i_0 is transferred to the energy of the electric field which is generated by charge accumulation (and high voltage) in the region of large gradient of the external magnetic field. The electric-potential barrier arises at the front of the beam when the beam is suddenly decelerated on entering the zone of rapidly increasing B. A few electrons on the front layer of the beam (a virtual cathode) can then be accelerated in the forward direction by the high (MV) voltage with an energy gain of a factor $\gtrsim 10^2$ over the initial energy eV_0 of these electrons. The beam behaves as a series of inductive-capacitive-inductive elements in which the system energy is successively transferred. Effective values for inductance L_e, capacitance C_e and transfer time $\Delta t \propto \sqrt{L_e C_e}$ for the bunching stage of a cylindrical electron beam ($B_0 \sim 10^3$ G, $V_0 \sim 10^4$ volts, $i_0 \lesssim 50$ amps) was estimated by Raudorf[39] with the conditions of current (electron flow, displacement) and energy (Poynting's theorem) conservation. A similar estimate of L_e seems possible for $\underline{B}$ and toroidal geometry of the vortex described in ref. 11 (other constraints are further needed to determine the vortex structure). The focussed plasma has values B, i larger by a factor 10^5 than in Raudorf's system, and a $\underline{B}$ - near the axial channel of a toroidal vortex - which is entirely determined by the current i in the plasma. The increasing magnetic field in the axial channel reaches a critical value B_c when the electron-cyclotron frequency $\omega_c \equiv eB/c_o m_-$ reaches the local value of

the angular frequency $\omega_p \equiv (4\pi e^2 n_-/m_-)^{\frac{1}{2}}$

of the plasma (usually $\omega_p \gg \omega_c$).

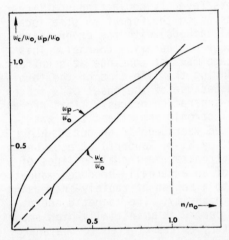

Fig. 5. ω_c ($\propto n_-$), ω_p ($\propto n_-^{-1/2}$) vs. n_- by assuming $\underline{u}_-$ as independent from n_-; here n_o is the maximum value that n_- can reach in the axial channel of a toroidal structure; $\omega_o \equiv \omega_p$ ($n_- = n_o$). The dependence of $\underline{u}_-$ on n_- cannot be ignored for large variations of n_- .

The absorption coefficient $\alpha(\omega, \omega_p)$ of the plasma is drastically decreased[40] when $\omega \sim \omega_p$. Specifically a critical frequency ω^* is defined from the equation $\ell\alpha(\omega = \omega^*, \omega_p) \sim 1$ in a plasma layer of thickness ℓ, such that[41]: (i) for $\omega < \omega^*$ the magnetic radiation is completely trapped i.e. $\ell\alpha(\omega, \omega_p) \gg 1$; (ii) for $\omega > \omega^*$ the plasma layer is transparent i.e. $\ell\alpha \ll 1$. The radiation with the fundamental frequency ω_c is the most intense, no matter how considerable is the fraction of the magnetic radiation which is concentrated in high-harmonic frequencies;[40] ω^* decreases by decreasing

ω_p for a given ℓ and the same electron-velocity distribution.[41] The value of ω_c reaches ω_p as in Fig. 5 for an increasing n_- because of the strong dependence of B^- ($\propto e n_- u_-$) on n_- in the axial channel of a torus. The sudden onset of magnetic-radiation losses when B reaches B_c has a cummulative effect with virtual-cathode formation in interrupting the electron flow in the vortex axial channel and in forming the MV accelerating potential over a distance < 1 mm. The kinetic energy of a nonrelativistic (or mildly relativistic) electron falls to $1/e$ of its initial value by free-space cyclotron emission in a time (see ref. 40 p.181) $\tau(B) = 2.58 \times 10^8/B^2$. We can take consistently $\Delta t = \tau(B_c)$ as an upper limit for the duration Δt of the magnetic field collapse in a toroidal vortex after the peak intensity field B_c has been reached. For $B_c \sim 2 \times 10^8$G we have $\tau(B_c) \lesssim 10$ ns;

this is also the typical duration of emission pulses of very hard ($\gtrsim 1$ MeV) bremsstrahlung x-rays in agreement with the idea that the life-time of the accelerating voltage $\simeq$ time of voltage building-up. The wavelength $\lambda_c \sim 2\pi c_o/\omega_c$

of the magnetic emission for $B_c \sim 2\times 10^8$G is in the infrared region ($\lambda_c \lesssim 0.6\mu$m). Infrared emission of clearly-nonthermal origin is observed in plasma focus experiments during the emission of hard x-rays (see ref. 34, these proceedings). Since B_c depends on the local value of n_- that can be different ($\sim 10^{19} - 10^{20}$ cm^{-3}) within different vortices, the pulses of infrared emission can be correspondently peaked at different frequency values (usually the observed infrared pulses -see bibliog. in ref. 34 -have a time duration $\sim$ 40-50 ns $\gg \Delta t$; emission from different sources can be the cause of it). A radiation temperature T_r can[40] be defined far from equilibrium conditions in terms of absorption coefficient α_ω and electron distribution f; $T_r < 0$ is frequently obtained with $f_- = f_0 F_-$ for a

variety of spectral functions $L(\mu_i)$ in F by which $\partial f/\partial \epsilon > 0$ and/or $\partial f/\partial v_i > 0$ for one (or more) component v_i of $\underline{v}$.

5.   DISCUSSIONS

Disruptions of high currents by current limitation, arc chopping, arc starvation[42,43]

is a common process under suitable
conditions. The formation of space
charge regions with high voltages have
been observed in a low-pressure arc
plasma (1 mtorr mercury, $\sim$ 10Amp); the
voltage is concentrated on a space-charge
sheath with a sharp visible boundary
across the discharge column (the formation
of this space-charge sheath is associated
with rapid current fluctuations)[43] It
has been suggested that[44] the particle-
acceleration mechanism in the plasma focus
might be the same as it occurs in the
space-charge sheath by these discharges.
Localization in space of the high voltage
and related features of the current
fluctuations give also a basis for the
similarity with a plasma focus operating
with current and density higher by a factor 10^5. It is observed that the space-
charge sheath is formed in an arc[43] also
when the current density is increased by
magnetic contraction (this is induced by
an externally-applied axial field
100-200 G)[43] To legitimate extrapolations
to the plasma focus it seems important to
understand whether electron-energy losses
by different processes (e.g. ionization,
and radiative combination in the low
pressure arc, bremsstrahlung and cyclotron
emission in the focus) can have similar
roles in accelerating the building-up
of space-charge and high-voltage. The
general phenomenon of current filamentation which affects cosmic[45] and
laboratory plasmas, becomes particularly
intense when a plasma streams against
B-field lines in a region with a large
gradient of B. Electrons are the current
carriers in the filamentary current sheath
of a coaxial accelerator[8] and laboratory
observations suggest the validity of
conditions as: (i) electrons tied to B-
lines of force;[5] (ii) lines of force of
ion/mass flow u_+ and B fields (not necessarily parallel to each other) form a layer
distribution of parallel helices (filaments) with a pitch increasing from the
periphery to the axis of each filament[5,8]
Consistently,well known properties of
force-free fields[46] can also be used to
account for the formation of helical u_+,
B patterns with $u_+//B$ or with a small
angle between u_+ and B: (i) a maximum of
magnetic energy density for a given mean-
square current density (exclusive of
surface currents) can exist in a stationary state only if the magnetic field is
force-free ($\nabla \times \underline{B} = \bar{\bar{\alpha}}\underline{B}$) and with a constant $\bar{\bar{\alpha}}$; (ii) force free fields characterize
plasma states of minimum dissipation of

magnetic energy by Joule heating (for a
fixed ohmic resistance of the plasma);
(i) can so describe the tendency of the
plasma to reach conditions of equilibrium
(with equipartition of kinetic energy and
magnetic energy among different degrees
of freedom) in a region where the
"external" field B_θ is rapidly increasing;
by (ii) the plasma can satisfy,locally,
the thermodynamic requirement of minimum
dissipation (see bibliog. in ref. 46).
By vortex filament formation the plasma
has, locally, still a directed motion;
from a purely thermodynamic-mechanical
point of view the flow can be considered
as isoentropic across Σ (Crocco's relation
connecting vorticity and entropy still
being valid)[47] if we take e.g. a "fine-
grained" density S_0 for the plasma entropy;
an entropy increase in this case can be
described by a coarse-grained density
$<S_0>$, with an average over a vortex-filament volume. Filaments form during the
motion of the current sheath and continue
to perform as effective current channels
also in the stage of maximum axial
compression of the plasma column (current
sheath velocity $\sim$ 0). By "hollowness"
of the axial column we refer consistent-
ly to the value of $|\underline{j}|$ (not necessarily
to n_+,n_-) which is peaked on the skin
layer (i.e. the filamentary wall) of the
column. Paramagnetic behavior of the
plasma on the current sheath and near
the torus axial channel can be recognized
by different criteria (including
$\partial^2 p_{//}/\partial B^2 > 0$ if $p//$ is the parallel-to-$\underline{B}$

pressure)[48] or in the sense which is
described in Tonks work.[49]

6.   CONCLUSIONS

(i) The building-up of a field intensity
$B \gtrsim 10^8$ G within localized regions of a
coaxial-accelerator discharge can be
described in terms of a relatively-slow
shrinking of the conducting wall of the
discharge axial column. This process
can occur by single filament decay
and/or by coalescence of neighbouring
filaments (resistive/radiative losses)
on the column wall where the current
is localized.
(ii) Formation of a multiplicity of
toroidal structures with close-loop
current circulation is the ultimate
result before complete collapse of the
magnetic field. The production of ion

and electron beams in the plasma focus follows the rapid decrease of the magnetic field either by virtual-cathode/charge-separation (ram) effect or by betatron[11] effect from the azimuthal field which is trapped inside the torus wall; a separation of these two accelerating processes is not necessarily meaningful in the plasma focus and probably a distinction can be made only for a group of particles, i.e. those particles that had enough energy to complete many turns of the current loop before complete collapse of the structure.

(iii) The filamentary structure of the pinched wall can well account for the relative stability of the axial column.

(iv) A nonstandard theory of strongly interacting plasmas is needed for an adequate description of the plasma focus. Such a theory is presented in paragraph 3.

(v) Large filaments which are formed by a multiplicity of smaller units are a consistent result of the theory and in some cases fit laboratory observations. For these composite filaments (or magnetic ropes) the observed dominant sign of B_z within a filament is then the net value of the average performed by the magnetic probe (dominance of one sign of B_z means greater stability for the component units). Further experimental data are necessary to decide on the typical dimensions of the component elements(if any) of large filaments.

(vi) No clear upper limit can be found for the current concentration in a filament; consistently,the intensity of B can have local maximum values which are order-of-magnitude larger than the conservative$\sim 10^8$G from observed intensity anisotropy of bremsstrahlung x-rays .

H.Alfven gave us reference 43, F.Gratton reference 19.

Work supported in part by C.N.R., Rome, NATO Scientific Affairs Division, Bruxelles, U. S. AFOSR, Arlington.

APPENDIX I

The structure of a single source term $S(\mu_i, \psi_{xi})$ with only one value μ_i is presented in Fig. 6. To calculate multiple integrals which contain $F(=f/f_0)$ the exchange of the order of integration on the variables v_x, v_y, v_z, μ_i {or $\gamma = (\mu_i - \Gamma)/2m$} simplifies the calculations. It is useful to recall the necessary changes in typical limits of integration (for $A > B > C$,

$$w \equiv v_y^2 + (v_z - c)^2, \quad v_x \equiv v):$$

$$\int_0^C d\gamma \int_0^B dw \int_0^A dv C(\gamma - w)C(v - \sqrt{\gamma - w}) =$$

$$\int_0^C d\gamma \{\int_0^{\sqrt{\gamma}} dv \int_{\gamma - v^2}^{\gamma} dw + \int_{\sqrt{\gamma}}^A dv \int_0^{\gamma} dw\} =$$

$$\int_0^C dw \{\int_0^{\sqrt{C - w}} dv \int_w^{v^2 + w} d\gamma + \int_{\sqrt{C-w}}^A dv \int_w^C d\gamma\}$$

$$= C^2(A/2 - 4C^{\frac{1}{2}}/15).$$

For computations it is sometimes convenient to replace e.g. $C(\psi_{xi})$ with

$\bar{C}(\psi_{xi}) \equiv C(\psi_{xi}) - \frac{1}{2}$, which means to add a negative source(sink) on the negative-v_x side and a positive source on the positive-v_x side of the source locus .

With $f_+ = f_0\{1 + \int d\mu_i L_+(\mu_i)\bar{C}(\psi_{xi+})C(D_{xi+})\}$

$$f_0 = L_0 \exp(-\alpha\Gamma), \quad L_0 = \text{const}$$

we have the typical integrals:

$$n = n_0 + n_s = n_0 + L_0 \int d\mu_i L(\mu_i)$$

$$(\frac{2\pi}{\alpha m})^{3/2} e^{-\alpha\mu_i}\{2(\frac{\alpha}{\pi})^{1/2}(\mu_i - \Gamma) -$$

$$e^{\alpha(\mu_i - \Gamma)} \text{erf}[\alpha^{\frac{1}{2}}(\mu_i - \Gamma)^{\frac{1}{2}}]\}C(\mu_i - \Gamma) ,$$

$$n_s u_{xs} = L_0 \int d\mu_i L(\mu_i)\frac{2\pi}{\alpha m^2} e^{-\alpha\mu_i}(\mu_i - \Gamma)C(\mu_i - \Gamma) ,$$

$$u_x = n_s u_{xs}/(n_0 + n_s) ,$$

$$n u_z = n_0 u_{zo} + n_s u_{zs} = n_0 c + n_s u_{zs}$$

By taking $L(\mu_i) = \Omega_0 e^{-\beta\mu}i$; $\Omega_0, \beta = const$

$\beta > 0$, we have:

$$n = n_0 + L_0 \left[\frac{2\pi}{m(\alpha+\beta)}\right]^{\frac{3}{2}} \frac{\Omega_0}{2\beta} e^{-(\alpha+\beta)\Gamma}$$

$$n_s u_{xs} = L_0 \frac{2\pi}{\alpha m^2} \frac{\Omega_0}{(\alpha+\beta)^2} e^{-(\alpha+\beta)\Gamma}$$

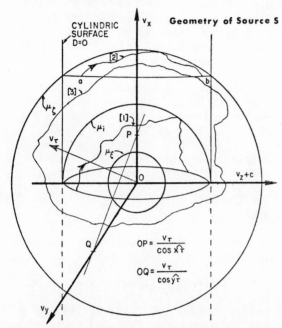

Fig. 6.Projection on v-space of the source locus $D_{xi} = (\mu_i - \Gamma) \, 2/m - v_y^2 - (v_z+c)^2 = 0$. f has a jump-discontinuity on the cylinder $D_{xi} = 0$ and on the characteristic sphere $\mu_i - \Gamma - [v_x^2 + v_y^2 + (v_z + c)^2] \, m/2 = 0$ for $v_x \geq 0$. A charged particle is produced by ionization in a and recombines in b (trajectory [2]). Sign of v_x is a constant of the motion on [1], [2]. μ_ζ - sphere $\equiv$ surface $\psi (v_x, \mu_\zeta) = 0$. By displacing $\psi_{\tau i}$ along the coordinate τ (see Fig. 1) the axis of $D_{\tau i} = 0$ rotates around the $\underline{z}$ axis.

REFERENCES

1 . W.H.Bostick, V.Nardi, W.Prior: Ann. New York Acad. Sci; Vol 251, N.Y.1975,p.2.

2 . V.Nardi, W.H.Bostick, W.Prior: Proc. Int. Conf. Physics in High Magnetic Fields, Grenoble, 18-21 Sept.1974,CNRS edit.

3 . I.F.Kvartskhava, K.N.Kervalidze, Y.S. Gvaladze & G.G.Zukakishvili. Nuclear Fusion 5, 181 (1965).

4 . W.H.Bostick, W.Prior, L.Grunberger, G. Emmert. Phys. Fluids, 9, 2078 (1966).

5 . W.H.Bostick, V.Nardi, L.Grunberger, W. Prior. Proc.I.A.U.Symp.43,Solar Magnetic Fields, Paris(ed.R.Howard),p.512.Reidel. 1970.

6 . W.H.Bostick, L.Grunberger, W.Prior, V. Nardi. Proc. 4th European Conf. Plasma Phys. and Nuclear Fusion, Rome, p.108. C.N.E.N. 1970.

7 . V.S.Imshennik, N.V.Filippov, T.I. Filippova. Nuclear Fusion, 13, 929 (1973).

8 . W.H.Bostick, V.Nardi, W.Prior. J. Plasma Phys. 8, 7 (1972). W.H.Bostick, L. Grunberger, V.Nardi, W.Prior. Proc. 4th Int. Conf. Thermophys.Properties, Newton, Mass.:495. American Association of Mechanical Engineering, New York, N.Y. 1970.

9 . V.A.Gribkov, O.N.Krokhin, G.V.Sklizkov, N.V.Fillipov, T.I.Fillipova. ZHETF Ps 18, 11 (1973).

10. V.Nardi. Phys. Rev. Lett. 25, 718 (1970).

11. V.Nardi. Proc. II Topic. Conf. Pulsed High-β Plasmas. 1972. Garching. W.Lotz edit.:163. 1972.

12. W.H.Bostick, V.Nardi, L.Grunberger, W. Prior. Proc. 10th Int. Conf. Ionized Gases, Oxford, p. 237. Parsons. 1971.

13. J.W.Mather in Methods of Plasma Phys, 9, 187 Academic Press, 1971.

14. W.H.Bostick, L.Grunberger, V.Nardi, W. Prior. Proc. 9th Int. Conf. Ionized Gases, Bucharest (ed. Musa et al), p. 66. 1969.

15. F.deHoffman, E.Teller. Phys. Rev. 80 , 692 (1950).

16. H.F.Webster, T.J.Hallinan. Radio Sci. 8, 475 (1973).

17. R.L.Khyl, H.F.Webster. I.R.E. Trans. Electron Devices, Ed.-3, 172 (1956).

18. E.Witalis. Plasma Phys. 10, 109 (1968). 10, 747, (1969).

19. D.Gabor. Proc. I.R.E. p.792, Nov. 1945.

20. J.W.Dungey. Nuclear Fusion, $\underline{1}$, 312 (1961). Cosmic Electrodynamics, Cambridge Un. Pr., 1958.

21. J.W.M.Paul et al. Nature, $\underline{208}$, 133 (1965); $\underline{216}$, 363 (1967).

22. L.D.Landau & E.Lifshitz. Electrodynamics of Continuous Media (Pergamon Press) Oxford 1960, p.210.

23. G.K.Batchelor. Proc. Roy. Soc. A $\underline{201}$, 405 (1950).

24. These are correct values during the motion of the current sheath between the electrodes. (see ref. 12).

25. W.H.Bostick, V.Nardi, W.Prior. Proc. I.U.T.A.M. Symp. on Dynamics of Ionized gases, Tokyo, 1971. Proc. Int. Conf. on Cosmic Plasma Phys.,Frascati, Plenum, p.175.

26. N.N.Komarov. Nuclear Fusion, $\underline{3}$,174,(1963). V.M.Fadeev, I.F.Kvartskhava, N.N. Komarov. Nuclear Fusion, $\underline{5}$,202 (1965).

27. N.J.Peacock, M.G.Hobby, P.D.Morgan. Proc. 4th Conf. Plasma Phys. and Controlled Nuclear Fusion Res.,Madison, 1971.

28. V.A.Gribkov, Korzhavin, O.N.Krokhin, G.V.Sklizkov, N.V.Fillipov, T.I. Fillipova. Zh.ETF, 15, 329 (1972).

29. M.J.Bernstein, D.A.Meskan, H.L.L.van Paassen. Phys. Fluids, $\underline{12}$, 2193 (1969). See e.g. Fig. 10 of this paper.

30. J.W.Mather, P.J.Bottoms, J.P.Carpenter, A.H.Williams, J.D.Ware. The Stability of Dense Plasma Focus. Los Alamos Sci. Lab.,University of California, Rep.LA-4088, TID-4500, 1969.

31. V.C.A.Ferraro, C.Plumpton. An Introduction to Magneto-Fluid Mechanics, Oxford 1966.

32. W.H.Bostick, V.Nardi, W.Prior. Proc. 4th Conf. Plasma Phys. and Controlled Nuclear Fusion Res.,Tokyo Nov. 1974, I.A.E.A. edit.,Vienna.

33. W.H.Bostick, V.Nardi, W.Prior. Proc. 6th Europ. Conf. on Controlled Fusion and Plasma Phys. Moscow, USSR. J.I.N.R.edit. Vol.2:395, 1973.

34. W.H.Bostick, V.Nardi, W.Prior, F. Rodriguez-Trelles, C.Cortese, W.Gekelman. These Proceed.

35. J.W.Mather, P.J.Bottoms, J.P.Carpenter, K.D.Ware, A.H.Williams. Proc. Conf. Plasma Phys. Controlled Nuclear Fusion Research. Vol.I. p.561, I.A.E.A.,Vienna 1971.

36. D.W.Forslund, R.L.Morse, C.W.Nielson. Proc. Conf. Plasma Phys. Controlled Nuclear Fusion Research. Vol.II. p.277, I.A.E.A. Vienna 1971. Phys. Rev. Letts. $\underline{26}$, 694 (1971).

37. M.J.Forrest, B.A.Norton, N.J.Peacock. Proc. 6th Europ. Conf. on Controlled Fusion and Plasma Phys. Moscow, 1973. J.I.N.R. edit. Vol. I. p.363.

38. L.Spitzer,Jr. Physics of Fully Ionized Gases, Interscience N.Y. 1962, p. 12.

39. W.R.Raudorf. Wireless Engineer, July 1951, p.215.

40. G.Bekefi. Radiation Processes in Plasmas, J. Wiley & S., Ch.6,7. N.Y. 1966.

41. B.A.Trubnikov, V.S.Kudzyavtsev. Proc. 2nd Int. Conf. Peaceful Uses Atomic Energy, Geneva, Vol. $\underline{31}$, 93 (1958). B.A.Trubnikov. The Phys. of Fluids, $\underline{4}$, 195 (1961).

42. H.Alfven, P.Carlquist. Solar Phys. $\underline{1}$, 220 (1967).

43. M.Babic, S.Torven. Current Limiting Space Charge Sheaths in a Low Pressure Arc Plasma, Rep. TRITA-EPP-74-02, Jan. 23. 1974, Dept. Electron Physics, Royal Inst. Tech.,Stockholm.

44. H.Alfven. Private communication (July1974).

45. H.Alfven. Proc. Royal Soc. A $\underline{233}$,296(1955).

46. S.Chandrasekhar, L.Woltjer. Proc. Nat. Acad. Sci., $\underline{44}$, 285 (1958).

47. L.Crocco. Z.Angew . Math. Mech. $\underline{17}$, 1 (1937). See also:G.K.Batchelor; An Introduction to Fluid Dynamics. Cambridge 1967, p.159-161. C.Truesdell, Velocity and the thermodynamic state in a gas flow, Mem. des Sci. Math., Vol. 119, Gauthier-Villars,Paris 1952.

48. H.Grad. Trans. New York Acad. Sci. $\underline{33}$, 163 (1971).

49. L.Tonks.Phys. Rev. $\underline{56}$, 360 (1939).

ON THE BURSTING OF FILAMENTS IN THE PLASMA FOCUS

Fausto T. L. Gratton

Instituto Nacional de Investigaciones Físicas para la Producción de Energía, SECYT
and Facultad de Ciencias Exactas y Naturales, Universidad de Buenos Aires
Laboratorio de Física del Plasma, Ciudad Universitaria, Buenos Aires, Argentina

ABSTRACT

Photographs of the current sheath of (low energy) plasma focus show a disruption of the filaments. We interpret this phenomenon as a vortex breakdown. Physical parameters which support this hypothesis are obtained from measurements, from the theoretical thickness of the current sheath given by Nardi and from some models of the plasma flow. The widening of a vortex due to axial velocity increase is analyzed by means of magnetohydrodynamic collinear models. The main results are: i) the existence of a limit separating supercritical from subcritical regimes (their character changes with the ratio between kinetic and magnetic energy); ii) the existence of flow regimes where the vortex radius remains approximately constant for moderate increments of the external velocity; iii) the structure of the vortex may change substantially for a sufficiently large increment of the external velocity, even in subcritical states; iv) the possibility that a burst of the vortex may occur when the external velocity suffers a slowdown.

INTRODUCTION

The sudden broadening of the filaments of the current sheath (CS) of the Plasma Focus (PF) along their axes is studied in this paper. In many articles, W. Bostick, V. Nardi and their collaborators[1-4] have demonstrated that the filaments possess a helical structure of magnetic field and velocity flow, with a maximum of vorticity on their axes. Thus they can be described as vortex filaments of plasma. The expansion of filaments has been interpreted by the present author in preliminary studies[5-6] as an effect similar to the breakdown[7] of vortices in ordinary fluids.

The fast decay of the magnetic fields of the filaments during the focus causes the acceleration of the ions up to fusion energies, according to the most plausible conception[1,4,5] The CS structure evolves until it produces during the collapse a number of 'hot spots' or configurations presumably of toroidal type[8,10] (which emit neutrons and X rays), bearing a high energy concentration. The knowledge of the fine structure of CS and filaments should allow us to understand the elements which produce the high energy centers, and hence, the efficiency of the PF as a fusion device.

In the present article we give estimates of the physical parameters of the filaments applying Nardi's[2,4] CS model (based upon his 'nascent plasma' kinetic theory, in which source terms for the particle distribution functions are introduced). Next we examine the mass flow over the CS surface, and show that this is the basic process that regulates the filaments structure. Finally we discuss the conditions which determine the widening of a single vortex by means of collinear MHD models (cgs units are used throughout).

The experimental data used here come from two low-energy PFs operating under the optimum conditions for the optical observation of filaments: Stevens Institute of Technology, N.J., USA,[SIT][1-4] (central electrode R_c = 1.7 cm, capacity c= 45.0 μF, period 8 μsec, $V_0 \approx$ 11 KV); Laboratory of the Universidad de Buenos Aires [BA] (R_c = 0.95 cm, c= 6 μF, period 4.2 μsec, $V_0 \approx$ 15 KV).

CURRENT SHEATH PROPERTIES

CURRENT SHEATH THICKNESS

Theoretical Model. Nardi's[2] CS theory assumes steady conditions and ignores changes along the filaments (idealized flat CS). Since the CS is actually curved (Fig.1), the theory is expected to be valid locally. The distribution function is the sum of two terms: one is a solution of the Vlasov equation (representing fast collisionless particles), the other satisfies a kinetic equation with particle sources which accounts for ionization, recombination, etc.

189

The theory gives the distance between filaments[4], or period T:

$$T = \frac{16 \, \pi\kappa \, (T_+ + T_-) \, c_o}{e \, B_\theta \, |c_- - c_+|} \qquad (1)$$

(here, κ: Boltzmann constant; c_o: speed of light; e: electron charge; B_θ: general magnetic field of PF; $T_\pm$, $c_\pm$: temperature and mean velocity of the high energy components of ions and electrons). When operating with a positive central electrode (which is the usual case) $c_+ > 0$ and $c_- < 0$. In order to derive a formula for T it is essential to measure the quotient $|B_\theta / B_\zeta| = s$ between the magnetic field of the PF and the maximum value of the axial magnetic field of the filaments. The empirical value $s \cong 3$, which leads to Eq.(1), was measured using an hexagonal central electrode that allows to separate the filaments in order to use magnetic probes[1]. The same value of s is assumed to hold also for circular electrodes.

The theory also renders the value of the thickness D of the CS (a length in which B_θ changes from zero to the maximum value behind the CS), which results proportional to T : $D = \gamma T$ (since the theory gives an exponential behaviour for the asymptotic trend of B_θ towards its maximum value, the definition of D depends upon the degree of accuracy desired).

Measurements of D and T. D has been measured with magnetic probes and T by 5 n sec image converter photographs during the coaxial stage. For deuterium, γ results approximately 1 (in agreement with the theory), see Table 1.(*) Equation (1) can be written as follows,

$$\frac{|c_- - c_+|}{U_o} = \gamma \frac{8\pi \, \kappa(T_+ + T_-)}{\frac{1}{2}M_+ \, U_o^2} \frac{R_L^+}{D} \qquad (2)$$

where U_o denotes the velocity of CS, and $R_L^+ \equiv U_o M_+ c_o / e \, B_\theta$ (M_+: mass of the ions).

The Snow-plow Hypothesis. A relationship between U_o and B_θ can be established with a fairly good approximation from the hypothesis (snow-plow neglecting CS accelerations)

$$\frac{B_\theta^2}{8\pi} = \rho_o V_{sp}^2 \qquad (3)$$

(*) For Ar (0.8 mm Hg, 11 KV) γ = 2.4, and for He (8 mm Hg, 11 KV) γ = 0.5,[SIT][3].

(where ρ_o is the initial density of the neutral gas) during the coaxial stage. The differences between U_o and V_{sp} do not exceed 20% and are usually smaller.(See Table 1.) Equation (3) is further corroborated by the fact that, at constant V_o, U_o turns out to be the same, within 10%, for different gases (Ar, He, H_2, D_2) modifying the initial pressure in order to keep ρ_o constant.

CS Profile. The snow-plow hypothesis allows also for a good description of the shape of the CS (idealized as a geometrical surface of null thickness; positive central electrode) in the coaxial stage and during part of the 'roll-off' at the end of the gun. The profile of the CS is shown in Fig.1 using a system of cylindrical coordinates z,r (z: PF axis), where the equation of CS r = f (z,t), (axial symmetry) is obtained by solving the equation:

$$ff_t = \frac{I(t)}{c_o\sqrt{2\pi\rho_o}} \, (1 + f_z^2)^{\frac{1}{2}} \quad , \quad (4)$$

which is derived applying Eq.(3) for the normal speed of displacement of CS, $f_t / (1 + f_z^2)^{1/2}$ (I: electric current). This model and its results are presented in detail in another paper (F. Gratton and M. Vargas, forthcoming). Figure 1 presents CS at different moments during the coaxial stage and the 'roll-off'. The profile propagates during the coaxial stage without changes of shape. The profile fits well with photographs of the CS taken sideways

Table 1. Measured and computed parameters of the current sheath.

Deuterium	SIT [3]	BA [11]
V_o KV	11	15
p_o (mm Hg)	8	2.34
I_M (MA)	0.388	0.125
B_θ (10^4 Gauss)[a]	4.6	2.6
D cm	0.30	0.7
T cm	0.28	0.6
V_{sp} cm	6.9	7.4
U_o (cm/μsec)	5.8	7.0
R_L^+ (cm)	0.026	0.055
$|c_- - c_+|/U_o$	0.36	0.33

[a] B_θ is computed from $2 \, I_M / c_o R_c$

and also with measurements of CS taken with magnetic probes (1 mm diameter, having a support allowing for movements in r,z) in Ar and air [BA][11]. The latter measures also show that D is approximately constant along CS, as r increases (at fixed t; the measures start at $\approx$ 0.5 cm from central electrode).

The Partition of Energy. The CS mass does not increase during its motion because the swept particles flow (towards the outer electrode) over the surface(*). Half of the work performed by the magnetic piston appears as kinetic energy of the CS and the rest goes into other kinds of internal energy of CS. Thus it can be stated that $E_o = 1/2\rho_o u_o^2 = E_{th} + E_i + E_m + E_k$ (being E_{th}, E_i, E_m, E_k, respectively, densities of thermal, ionization, magnetic and roto-translatory kinetic energy pertaining to the inner structure of CS). The current density can be estimated as $en|c_- - c_+| \approx I/2\pi R_c D = c_o B_\theta / 4\pi D$ and using Eq.(2) we have

$$\frac{B_\theta^2}{8\pi} = 8 n \kappa (T_+ + T_-) \qquad (5)$$

Taking $3/2 n \kappa (T_+ + T_-)$ as an estimate for E_{th} (together with Eq.(3), $V_{sp} \approx U_o$), we get $E_{th} = 3/8 E_o$. Measurements of the electron temperature in the CS (by means of double probes) give[1] $T_- \sim 7$ ev (being T_+ presumably smaller); the electron density n in the filaments (measured by Stark broadening of Hβ) is greater than n_o by a factor ≈ 3[1] ($\rho/\rho_o \equiv q \approx 3/2$ for D_2). Indirect estimates of temperature based upon the Joule effect and the equivalent resistance of CS,[BA][11], give ~ 7.5 ev for Deuterium. The last formula for E_{th} gives an average of 6.4 ev per particle (with q = 3/2). These estimations can be completed with $E_m \approx 2/9 E_o$ (assuming $B_\zeta \approx B_\theta / 3$), $E_i \sim 5/19 E_o$ (for $E_o \approx 7.5 \times 10^{18}$ ev/cm^3, Deuterium, BA), and $E_k \sim 7/50 E_o$ as a difference.

Axial Velocity. Equations (2), (3) and (5) yield

$$\frac{|c_- - c_+|}{U_o} = 4\pi \frac{\gamma}{q} \frac{R_L^+}{D} \qquad , \qquad (6)$$

(see numerical values on Table 1). Since $c_+ < |c_- - c_+|$ the axial velocity in the filaments is a fraction of U_o in the state of equilibrium given by Nardi's model. On the other hand, Eq.(6) can be written using Eq.(3) as
$$|c_- - c_+| / U_o = 2\pi\gamma\delta^+ / qD,$$
where $\delta^+ = (c_o / \omega_p)(M_+ /m)^{1/2} = \delta^- (M_+ /m)^{1/2}$ (for D_2, and $\omega_p = (4\pi e^2 n_o /m)^{1/2}$). Hence if c_+ increases with r on CS (at fixed t), D must decrease accordingly.

BROADENING OF FILAMENTS

Increase of Axial Velocity. The existence of a mass flow along CS has been verified experimentally[1]. The luminosity of the filaments remain approximately constant during the coaxial stage, which would indicate that $\rho \approx$ constant, and we have already mentioned that the measurements of D show that it is constant along the CS. Under these circumstances and owing to the cylindrical geometry of the PF, we have to admit that v_ζ, the mass flow parallel to the filament axes, must increase along CS (in order to maintain a steady state). An elementary balance of mass flow yields $D \rho d (r v_\zeta) / dr = U_o \rho_o r \cos \theta$. ($\theta$ is the angle between the normal to CS and the z axis) Near $r = R_c$, $\cos \theta \approx 1$ (Fig.1), and since $v_\zeta = 0$ at $r = R_c$, we have
$$v_\zeta = R_c U_o (r/R_c - R_c /r) / Dq .$$

At small distances from R_c, v_ζ becomes an important fraction of U_o and hence D should decrease below the observed values, unless another phenomenon arises to modify the predicted CS structure. In fact, photographs of CS at the coaxial stage show a sudden broadening of all the filaments a few millimeters away from the central electrode, with a remarkable change in the visible structure of CS.

Breakdown of Vortices. This process has

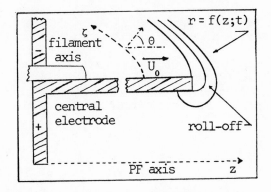

Fig. 1. The current sheath profile.

(*) This allows to assume steady conditions in the CS theory.

been interpreted as a breakdown of vortices[5,6]. If such is the case then the broadening is accompanied by a decrease of velocity (which is just what is needed to reestablish the self-consistent equilibrium of CS). During the 'roll-off', the breakdown point begins to drift towards the outer electrode, while the filaments acquire an obviously regular structure (at least for D_2 in the cases observed at low energies). (*) According to Benjamin[7] the breakdown phenomenon is a transition from a supercritical state of the vortex (i.e. that in which infinitesimal standing waves cannot occur), towards a subcritical one, which occurs spontaneously when the flow velocity exceeds the velocity of the waves of the system.

Breakdown Criterion. Following this line of thought in the case of a plasma filament, the criterion for the breakdown must be $v_\zeta \gtrsim V_A$. Here V_A is the Alfven velocity and it represents the velocity of energy propagation of hydromagnetic disturbances. When $v_\zeta > V_A$, the Alfven waves are swept down by the flow and thus the breakdown occurs with energy dissipation. Equation (3) allows to estimate
$$V_A = B_\zeta / (4\pi\rho)^{1/2} \approx (2/q)^{1/2} U_o /s,$$
and hence

$$\frac{v_\zeta}{V_A} \approx \frac{s}{\sqrt{2q}} \frac{R_c}{D} \left(\frac{r}{R_c} - \frac{R_c}{r}\right) \qquad (7)$$

From the photographs we can measure the breakdown position r' and compute v_ζ / V_A from Eq.(7). We obtain for deuterium $r'/R_c = 1.12$, $v_\zeta/V_A \approx 2.2$ [SIT] and $r'/R_c = 1.37$, $v_\zeta/V_A \approx 1.5$ [BA].These results support the breakdown interpretation.

BREAKDOWN OF A COLLINEAR MHD VORTEX

COLLINEAR FLOWS

A complete theory for the breakdown of a single plasma vortex (like the Benjamin[7,12] theory for fluids) is still missing. On this account, the attempts to study

(*) In heavier gases, e.g.:Ar (or at higher current values) the breakdown, or reorganization of the flow is also visible near the middle of CS before the collapse. Some authors tend to classify this effect hastily as a MHD instability of the pinch. But the fact that we are dealing with a complex structure of fields and flows prevents us from drawing a simple analogy with the theory of the pinch.

the even more complex behaviour of the filaments are necessarily incomplete. Here some results are given which can be translated from ordinary fluid theory to a particular case of MHD flows. The incompressible MHD equations of stationary motions ($\partial/\partial t = 0$, div $\underline{v} = 0$) become of the Euler type when collinear flows, $\underline{B} = \lambda \underline{v}$, are considered:

$$\tilde{\rho} (\underline{v}.\nabla)\underline{v} = -\nabla \tilde{p}, \quad \underline{v}.\nabla \lambda = 0; \qquad (8)$$

here $\tilde{\rho} = \rho(1 - \lambda^2/4\pi\rho)$, $\tilde{p} = p + \lambda^2|\underline{v}|^2/8\pi$.

Among the collinear flows with constant λ there is a subset with the property of being extremals of the total energy for closed systems (Woltjer[13])(*). These flows are also force-free: $\underline{j} \times \underline{B} = 0$. Collinear force-free flows attain their minimum energy where $\lambda = \sqrt{4\pi\rho}$. These are called equipartition solutions of the MHD equations ($E_k = E_m$), and are globally stable according to Woltjer criterion (i.e. against finite perturbations). There are other equipartition solutions which are not force-free. Chandrasekhar[15] has shown that, in general, they are stable against infinitesimal perturbations.

VARIATIONS IN THE STRUCTURE OF A VORTEX WITH AXIAL VELOCITY

Effect of Changes in the External Velocity.
Given cylindrical co-ordinates r, ϕ, ζ with ζ along the vortex axis, let $\underline{v} = (v_r , v_\phi , v_\zeta)$ and $\underline{B} = (B_r , B_\phi , B_\zeta)$.The MHD flows which are independent of ϕ and ζ ($\partial/\partial\phi = \partial/\partial\zeta = 0$) are called cylindrical. In this case the equation for the freezing of the B lines in the plasma reduces to d [$r(v_\phi \overline{B}_\zeta - v_\zeta B_\phi)$] /dr=0 (for steady states). As a consequence, a cylindrical vortex is necessarily collinear, although λ may be a function of r. For axisymmetric motions ($\partial/\partial\phi = 0$) a Stokes function ψ may be used with $r v_r = -\psi_\zeta$, $r v_\zeta = \psi_r$, to satisfy identically div $\underline{v} = 0$. Assuming also that $\underline{B} = \lambda \underline{v}$, with constant λ we have from Eq.(8)

$$\underline{\nabla} \tilde{H} = -\underline{\omega} \times \underline{v} \quad , \qquad (9)$$

where $\tilde{H} = (p/\rho + |\underline{v}|^2/2)/\alpha = \tilde{H}(\psi)$ (Bernouilli's theorem) and $\alpha = \tilde{\rho}/\rho$. Following standard procedures[16] Eq.(9) reduces to

$$\psi_{\zeta\zeta} + \psi_{rr} - \frac{1}{r} \psi_r = r^2 \frac{d\tilde{H}}{d\psi} - c \frac{dc}{d\psi}, \quad (10)$$

(*) Wells and Norwood[14] essayed to extend those results to more general boundary conditions.

here $v_\phi r = c(\psi)$ represents the conservation of angular momentum. For a cylindrical flow it turns out that $\tilde{H}(\psi) = v_\zeta^2/2 + \int cdc/r^2$.

Therefore we obtain the same equation as in the case of fluids and α does not intervene explicitly. The Batchelor model[16] can be used without further alterations to study the effect of changes in the external velocity on a cylindrical vortex. The following velocity profile is assumed (for $\zeta < 0$);

$$v_r = 0 \ , \ v_\phi = \Omega r \ , \ v_\zeta = U_1 \ \text{for} \ r \leqslant a,$$
$$\tag{11}$$
$$v_r = 0 \ , \ v_\phi = \frac{\Omega a}{r} \ , \ v_\zeta = U_1 \ \text{for} \ r > a,$$

i.e., a core rotating like a rigid body and an irrotational flow in the outer region, with an axial current superposed.

After a finite length (for $\zeta > 0$) the hypothesis is done that the velocity of the external flow has changed to a value $v_\zeta = U_2$ (for $r > a$), and the flow remains irrotational there, having the same circulation as before. The vortex goes through a transition (described by Eq.(10)) and it is supposed that it becomes cylindrical again with another radius b, and a different velocity profile in the core. Due to the conservation of mass flow in the core, we have a boundary condition at $r = b$, $\psi = \frac{1}{2} U_1 a^2$, that leads to the equation

$$A = \frac{U_2 - U_1}{U_1} = [(\frac{ka}{kb})^2 - 1] \ \frac{kb}{2} \ \frac{J_0(kb)}{J_1(kb)} \ , \tag{12}$$

which determines the new radius b, given the velocity increment and the original radius (here $k = 2\Omega/U_1$ and J_0, J_1 are Bessel functions).

The axial velocity v_ζ undergoes in some cases an inversion in the core and the model is no longer valid since some current lines would enter at the system from $\zeta = +\infty$, where there is no reason to assume that the same ψ distribution should hold. However the appearance of an inversion is probably heralding a fundamental change in the structure of the vortex; it is observed in fluids[17,18] that the breakdown is followed by reversals of v_ζ along with the development of a globular structure in the vortex core. Hence, the possible reversal of v_ζ should be controlled by means of

$$\frac{v_\zeta}{U_2} = \frac{1}{1 + A} (1 + A \frac{J_0(kr)}{J_0(kb)}). \tag{13}$$

Increment of External Velocity. In Fig.2, A is represented as a function of kb for several values of ka. The roots of Eq.(12) can be read from the graph. We comment first upon the case A>0 which might be of interest for the filaments of CS. Starting with A=0, where b=a, let us consider a continuous increment of U_2. The radius of the vortex decreases for ka<2.4$^+$ (the first zero of J_0), while v_ζ in the core increases. When 2.4< ka<3.8$^+$, b increases but the changes become almost imperceptible at ka $\approx$ 3.8$^+$. The change in b is larger when

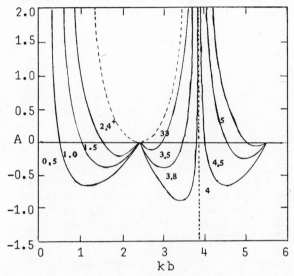

Fig. 2. Roots of Equation (12).

ka is close to 2.4$^+$. This point marks a critical state of the vortex; in its neighbourhood even small values of A can give flow reversal (see Eq.(13)). Values of ka in the interval 3.8$^+$ - 5.5$^+$ give decreasing kb. Values of ka in the proximity of 3.8$^+$ define regimes of permanence of the radius, since kb tends to this number from both sides. Nevertheless values of A must be moderate, otherwise inversion occurs at r=0 (e.g., for ka $\approx$ 3.8, A $\leqslant$ 0.4). Similar considerations can be repeated for the other zeros of J_0 and J_1.

Spontaneous Breakdown. The breakdown of the vortex can also occur spontaneously, (i.e., without external variations). We lack an extension to plasmas of the Benjamin theory, but there are reasons to believe that it can be applied to collinear flows when $\lambda < \sqrt{4\pi\rho}$. According to Benjamin[7] the flow given by Eq.(11) is supercritical for ka<2.4 and will suffer a breakdown. When $\lambda < \sqrt{4\pi\rho}$ the situation varies and the theory cannot be applied as it has been formulated for fluids. On one hand, the flow of linear momentum in the vortex (the 'flow-force' of Benjamin) changes its sign when $\lambda > \sqrt{4\pi\rho}$ (it bears the factor α), and on the other hand, the theory of infinitesimal disturbances applied to the profile of Eq.(11) shows instability when $E_k < E_m$[15]. The approximate model which is presented under the next heading supports these conclusions and shows that the breakdown occurs for v_ϕ greater than v_ζ, when $\lambda > \sqrt{4\pi\rho}$. Summing up, for $\lambda < \sqrt{4\pi\rho}$ and $ka > 2.4$ the vortex does not suffer spontaneous breakdown. But, as we have already shown, if the external velocity grows enough, a reversal of the flow appears. Conversely, for $ka < 2.4$, b decreases as U_2 augments, but the vortex is liable to decay towards subcritical states with radius enlargement and energy dissipation. For $\lambda > \sqrt{4\pi\rho}$ the breakdown condition is reversed and $ka < 2.4$ defines the states which do not decay spontaneously. These states are also stable against infinitesimal axisymmetric disturbances[15]. The approximate model shows energy dissipation together with radius decrease when $v_\phi > v_\zeta$. However it is doubtful that this circumstance may occur in practice since the vortex is in this case unstable under infinitesimal axisymmetric perturbations.

If one supposes that this analysis has some relevance for CS during the coaxial stage, the following cases may occur: i) $v_\phi/v_\zeta \geqslant 1$ at $r=Rc$ (where $v_\zeta \approx 0$); then if $E_k > E_m$, the filaments may stabilize around a configuration of small changes in radius until the increase of velocity over CS produces slowing-down and structural changes; conversely, if $E_k < E_m$ the filaments should be unstable (which does not seem to be the case because the filaments suffer only a change of structure); ii) the other alternative is that the filaments form near $r=Rc$ (where v_ζ has already grown) with $v_\phi < v_\zeta$; then if $E_k > E_m$ the spontaneous breakdown should occur, while if $E_k < E_m$ the decay should not take place (although there exist unstable modes for infinitesimal non axisymmetric disturbances). During the roll-off, B_θ and U_o increase and a stretching of the filaments (due to the curvature of CS) takes place. They show a more stable aspect and the splitting point drifts towards the outer electrode.
The data at our disposal are not sufficient to clarify these points. From Eq.(5) we have obtained E_k somewhat less than E_m, but this is only an approximate estimation and furthermore, the value $B_\theta/B_\zeta \approx 3$ has

not been directly measured for circular electrodes. A value somewhat greater that 3 would be sufficient to give $E_k > E_m$. (*)

Slow-down of External Velocity. Figure 2 shows that for $A < 0$, kb increases for ka < 2.4 (and v_ζ decreases at $r=0$). When ka is in the interval (2.4; 3.8), kb decreases (and v_ζ increases at $r=0$). In both cases a new fact arises: for sufficiently negative A there are no roots of kb, which means that there exists no solution for the steady state problem. In that case (which for $2.4 < ka < 3.8$ occurs before the reversal of the axial flow sets in) the vortical flow should undergo a drastic change.
Eventually, the case $A < 0$ would not apply to the CS during its displacement to the focus. But at the moment of collapse the snow plow action suddenly ceases ($U_o = 0$), and the source of flow along the CS disappears. As a consequence, it may be speculated that, if a strong slow-down of velocity occurs, the filaments should burst and a recomposition of the plasma flow should take place instantaneously. This would give rise to high rate changes of the magnetic fields trapped in the filaments and to the acceleration of ions.

We do not know about the possible configurations of plasma vortices after the burst. But it may be worthwhile to keep in mind that, in the breakdown of fluids, globular structures appear which contain a vortex ring inside. This has been revealed in beautiful photographs by Sarpkaya[18]. It is an open question whether this process is related to the appearance of the toroidal 'hot spots'[18-10] in the plasma column formed at the focus.

APPROXIMATE MODEL OF BREAKDOWN

Conservation Laws. Some results on the energy variations during spontaneous breakdown can be obtained by means of an approximate method. It consists in applying the integral theorems for the conservation of mass, linear momentum and angular momentum to the transition, in order to obtain data about the final state, given the initial vortex conditions[19]. The shortcoming of this method is that the profile after the breakdown must be postulated 'a priori'. Here we include the results for a vortex with the following velocity distribution:

(*) There exist some preliminary measurements that point towards this conclusion[20].

$v_\zeta = V = const., v_\phi = \Omega r, \text{ for } r < a ;$
$$\text{(14)}$$
$v_\zeta = 0 , v_\phi = \Gamma/2\pi r \; (\; \Gamma = 2\pi R^2 \Omega) \text{ for } r > a.$

The magnetic field must be parallel to $\underline{v}$ with constant λ. Equation (9) yields

$$H = p/\rho + v_\phi^2/2 = p_\infty/\rho \quad \text{for } r > R$$
and $\quad 2\Omega r = d\tilde{H}/dr \qquad \text{for } r < R.$

By integrating this last equation and applying the boundary condition

$(p + B_\zeta^2/8\pi)_{int} = p_{ext} \quad \text{at } r = R, \text{ we have}$

$p + \lambda^2 V^2/8\pi = p_\infty - \rho\Omega^2 R^2 (1 - \lambda^2/4\pi\rho) +$

$(1 - \lambda^2/2\pi\rho) \rho \Omega^2 r^2/2 \text{ for } r < R.$

The same velocity profile is assumed after the transition, although with different values of the parameters. The following conservation equations for the steady state with the control surface shown in Fig.3, which encloses the transition region are applied

$$\oint \rho \; \underline{v} . \underline{n} \; ds = 0 ,$$

$$\oint [\rho \; \underline{v} \; (\underline{v} . \underline{n}) - \frac{B^2}{4\pi} \; (\underline{B} . \underline{n}) + (p + B^2/8\pi) \; \underline{n} \;] ds = 0 ,$$

$$\oint [(p + \frac{B^2}{8\pi})(\underline{r} \times \underline{n}) + \rho(\underline{r} \times \underline{v})(\underline{v} . \underline{n})$$
$$- \frac{1}{4\pi} (\underline{r} \times \underline{B})(\underline{B} . \underline{n})] ds = 0 .$$

Besides the contributions of sections 1 and 2 of the vortex (see Fig.3), these formulae have a lateral surface contribution, where it can be assumed that $\underline{B} . \underline{n} = \underline{v} . \underline{n} = 0$ (the flow is supposed to remain axisymmetric). Hence, the only lateral contribution derives from the linear momentum equation, and its value is

$$\int_{lat.} (p + \frac{B_\phi^2}{8\pi}) \; ds = p_\infty \pi (R_2^2 - R_1^2) - \tilde\rho \frac{\Gamma^2}{4\pi} \ln\frac{R_2}{R_1}. \text{(15)}$$

The subindexes $_1, _2,$ refer to the initial and final state, respectively. From the mass and angular momentum conservation equations

$$V_1 R_1^2 = V_2 R_2^2 , V_1 \Omega_1 R_1^4 = V_2 \Omega_2 R_2^4 , \quad (16)$$
we obtain

$$\Omega_1 \quad R_1^2 = \Omega_2 \quad R_2^2 \quad (17)$$

(this implies conservation of the circulation Γ in the outer region, and therefore the conservation of the collinearity constant λ). Using Eqs.(15) to (17), the momentum balance equation reads:

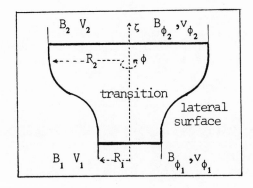

Fig. 3. Control Surface.

$$R_2^2 \; V_2^2 - R_1^2 \; V_1^2 = -\frac{\Gamma^2}{4\pi^2} \ln \frac{R_2}{R_1} \quad . \text{ (18)}$$

Radius Variations. Writing $x = R_2/R_1$, then $V_2/V_1 = \Omega_2/\Omega_1 = 1/x^2$, and x satisfies the equation

$$1 - \frac{1}{x^2} = \epsilon^2 \ln x , \qquad (19)$$

where $\epsilon = \Omega_1 R_1/V_1 = (v_\phi/v_\zeta)_1$. This equation has the following properties (roots can be obtained graphically from

$$x^{\epsilon^2} = e \; . \; \exp(-1/x^2)):$$

i) For every ϵ^2 there exists a root $x = 1$ (no breakdown); ii) a positive root $x \geqslant 1$ exists in the interval $0 \leqslant \epsilon^2 \leqslant 2$; iii) a positive root $x < 1$ exists for $\epsilon > 2$. Some of the values for the roots (ϵ, x) are: (2.0; 1.0), (1.5; 1.35), (1.0; 2.22), (0.707; 3.70), (0.15; 7.13) and for $\epsilon^2 \ll 1$ we can take $x \approx \exp(1/\epsilon^2) \gg 1$.

Changes in the Energy. The variations of radius must be compatible with the changes of energy during the transition. The flow of energy through the control surface is

$$\oint [(p + \rho | \underline{v} |^2/2)\underline{v} . \underline{n} + \underline{Py} . \underline{n}] \; dS,$$

where $\underline{Py} = (| \underline{B} |^2 \underline{v} - (\underline{B} . \underline{v}) \underline{B})/4\pi$ is the Poynting's vector (null in this case because $\underline{B} = \lambda \underline{v}$). The difference in the flow of energy between sections 2 and 1 (the lateral surface does not contribute) reads:

$$\dot{E} = -\pi V_1 R_1^2 \rho \frac{V_1^2}{2}(1 - \frac{\lambda^2}{4\pi\rho})(1 - \frac{1}{x^2})(1 + \frac{1}{x^2} - \epsilon^2)$$

The physically acceptable roots of Eq.(19) occur with losses of energy during the transition, i.e., $\dot{E} < 0$ (*). This could happen by processes (not included in the ideal MHD balance) of turbulent dissipation, radiation, selective acceleration of a small fraction of the plasma particles, etc. When $\lambda^2 < 4\pi\rho$ (i.e. $E_k > E_m$), the breakdown with vortex broadening and energy dissipation ($\dot{E} < 0$) occurs for $\varepsilon^2 < 2$ (i.e.: $\Omega_1 R_1 / V_1 < 1.4^+$ are supercritical states). We must disregard the solutions with decreasing radius for $\varepsilon^2 > 2$ since then $\dot{E} > 0$ (i.e.: there is no transition). This situation is reversed when $\lambda^2 > 4\pi\rho$ ($E_k < E_m$), then there is no breakdown for $\varepsilon^2 < 2$, and the transition with dissipation ($\dot{E} < 0$) may occur instead for $\varepsilon^2 > 2$ with radius decrease. The equipartition case $\lambda^2 = 4\pi\rho$ gives $\dot{E} = 0$.

AKNOWLEDGEMENTS

Thanks are due to Dr. W. Bostick and his collaborators for generously offering to me photographs and data of their experiments, and to Dr. V. Nardi for clarifying aspects of his theory to me. I am also grateful to H. Bruzzone, H. Kelly and J. Pouzo for their experimental data and to M. Vargas for his numerical computations.

This work has been supported in part by a grant of the Argentine 'Comisión Nacional de Estudios Geo-Heliofísicos' and by the 'Istituto Avogadro di Tecnologia', c.p.10757, Roma.

REFERENCES

1 W.H. Bostick, L. Grunberger, V. Nardi and W. Prior, Proc. V Symp. on Thermophys. Prop., Am. Soc. Mech. Eng., New York, 1970, p. 495.

2 V. Nardi, Phys. Rev. Lett. 25, 718 (1970).

3 W.H. Bostick, V. Nardi, L. Grunberger and W. Prior, Proc. X Int. Conf. on Phenomena in Ionized Gases, Oxford, 1971, p.237. Parsons.

4 W.H. Bostick, V. Nardi and W. Prior, Proc. Int. Conf. on Cosmic Plasma Phys., Frascati, 1971, p.175. Plenum.

5 W.H. Bostick, V. Nardi and W. Prior, Proc. Int. Symp. on Dynamics of Ionized Gases, IUTAM, Tkyo, 1971, p.375.

6 F. Gratton, Proc. II Topical Conf. on Pulsed High-β Plasmas, Garching, 1972, Max Planck Inst. f. Plasmaphysik.

7 T.B. Benjamin, J. Fluid Mech., 14, 593, (1962).

8 W.H. Bostick, V. Nardi and W. Prior, J. Plasma Phys., 8, 7 (1972).

9 W.H. Bostick, V. Nardi, W. Prior and F. Rodriguez Trelles, Proc. II Topical Conf. on Pulsed High-β Plasmas, Garching, 1972, p.155.

10 V. Nardi, Proc. II Topical Conf. on High-β Plasmas, Garching, 1972.

11 J. Pouzo, 'Análisis Experimental de la Descarga en un Acelerador Coaxial de Plasma'. Trabajo de Seminario, FCEyN, UBA, Buenos Aires, 1974.

12 T.B. Benjamin, J. Fluid Mech., 28, 65 (1967).

13 L. Woltjer, Proc. Natn. Acad.Sci. USA, 45, 769 (1959 a).

14 D.R. Wells and J. Norwood Jr., J. Plasma Phys., 3, 21 (1969).

15 S. Chandrasekhar, Hydrodynamic and Hydromagnetic Stability, Oxford, 1961, Clarendon.

16 G.K. Batchelor, Fluid Dynamics, 1970, pp.543, 550. Cambridge Univ. Press.

17 J.K. Harvey, J. Fluid Mech., 14, 585 (1962).

18 T. Sarpkaya, J. Fluid Mech., 45, 545 (1971).

19 A. Barcilon, J. Fluid Mech., 27, 155 (1967).

20 H. Bruzzone, R. Gratton and H. Kelly, 'On the Current Sheath in a Coaxial Accelerator', Internal Report, DFCE 2/72, UBA, Buenos Aires, 1972.

(*) Although this criterion is largely acceptable, one could conceive some special cases in which $\dot{E} > 0$, e.g. by ionization of fast neutral particles and recombination of slow ions preserving the mass flow and giving an increase of kinetic energy.

ADIABATIC COMPRESSION OF PLASMA VORTEX STRUCTURES

D.R. Wells, E. Nolting, F. Cooke, Jr., J. Tunstall, P. Jindra, and J. Hirschberg
Department of Physics, University of Miami
Coral Gables, Florida, USA

ABSTRACT

Plasma vortex structures generated by conical theta pinch guns have been success-fully compressed and amplified by secondary adiabatic compression. A pair of vortex rings meet at the center of a primary magnetic mirror. A secondary mirror starts to compress them as soon as the collision has occurred. Peak ion temperatures of 170 eV have been obtained at an n_τ of 10^{12} sec/cm^3 utilizing a capacitor bank that stores 250 kilojoules. As an aid to the understanding of the stability of vortex structures, an outline of the theory of quasi-force-free collinear equilibria is given in the appendix.

Over the past fourteen years, an extensive series of experiments concerned with naturally occurring stable plasma states has been performed.[1-6] We have intensively studied, both theoretically and experimentally, a whole new field of plasma physics concerned with the problem of production, heating and application of these states.

A theory has been developed which predicts a spectrum of naturally occurring states.[5,7,8] An outline of this theory is given in Appendix A of this paper. The particular stable state studied at Miami consists of a closed plasma structure produced by conical theta pinch guns. This state is the force-free collinear vortex ring or spheroid first described by Wells in 1962.[1,2]

It has been shown elsewhere that, by variational methods, one can predict the current and mass flow patterns of the stable states of lowest free energy.[5,7,8] These structures, or plasma cells, have minimal free energy available for driving instabilities and thus can persist for long periods of time, even when they are violently perturbed by the surrounding plasma and electromagnetic fields. An important part of this work is concerned with the inclusion of the mass flow terms in the equation of motion of the plasma.[5,8] The most stable plasma states are structures with a large part of their total energy in the form of mass flow energy. Some of these properties of the cells have been extensively investigated by others, and the results obtained verify in detail the results of Wells and his co-workers.[9-11,15]

In Appendix A, group theoretical and symmetry formalisms have been used to derive and discuss the Chandrasekhar-Woltjer-Wentzel variational approach to finding equilibrium solutions of the conservation equations of a MHD fluid plasma model. The theory has been derived from first principles utilizing the multiple integral techniques of Caratheodory and Weyl. The principle of least constraint is shown to emerge naturally and rigorously from these formalisms. This principle combined with the fundamental variational formula for multiple independent integrals, allows calculation of the global stability of closed plasma structures. The reference Nolting, Jindra and Wells presents experimental verification of the most important predictions of this theory.

Difficulties have been encountered in the past in attempting to couple the primary currents in the conical theta pinch guns to the secondary currents in the plasma rings. The rings move away from the guns so fast that the coupling coefficient decreases rapidly and the rings move away before much energy has been transferred to them. This problem has been overcome by abandoning the attempt to couple a very big capacitor bank directly to the rings.[4] Instead, a set of rings originating at each end of a primary magnetic mirror are produced in and guided by steady state mirror field (D.C. mirror coils shown in Fig. 1) to the center of the machine. The ring moving parallel to the mirror field has its velocity and magnetic fields antiparallel (contra-rotational). The ring moving antiparallel to the primary mirror field has its velocity and magnetic fields parallel (co-rotational). Both rings are force free (e.g., $\vec{j} \times \vec{B} = 0$ and $(\nabla \times \vec{v}) \times \vec{v} = 0$). Thus, the contra-rotational ring is right handed in $(B_\theta + B_p)$ and the co-rotational ring is left handed in $(B_\theta + B_p)$, where B_θ is the toroidal component of

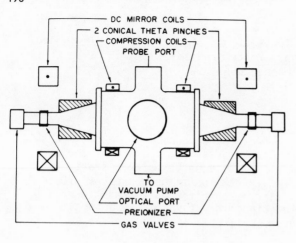

FIG. 1. Schematic diagram of machine used
for secondary compression of plasma vortex
structures. Primary-mirror field strength
was 900 G.

the trapped magnetic induction field in
the rings and B_p is the corresponding
trapped poloidal field.[3,4,13] They then
collide and are amplified and compressed
by a secondary mirror system located at
the center of the primary mirror system.
The current flow in the secondary compression coils is in the same direction
as the currents flowing in the primary
mirror coils. Thus (by Lentz's Law) they
both compress the vortex rings and amplify their currents. The diamagnetic currents in the rings increase as the current in the secondary compression coils
increases. The toroidal currents in the
two rings are in the same direction. The
poloidal currents components are in opposite directions. The ring currents are
left handed and right handed helices.
This geometry is illustrated in Fig. 1.
The cylindrical vacuum chamber has an 8
inch outside diameter. The center line
distance between the compression coils is
12 inches.

The base pressure in the vacuum sys-
tem is 3×10^{-6} torr. Deuterium gas is
admitted by the pulsed gas valves. The
operation of these valves is described in
detail in Ref. 13. The preionizers consist of a set of conical arc guns which
strike an arc in the gas when it drifts
into the preionizer region of the chamber.
The effective gas pressure in the preionizers is 40 microns.

The conical theta pinches are powered by a single General Electric "clam-
shell" capacitor rated at 1 µf, 50 kV. The
quarter cycle rise time on the conical

theta pinches is .5 µsec. They ring out
in three microseconds. The peak magnetic
fields in the throats of the conical theta
pinch guns were 20 kG at 18 kV. The secondary compression coils are powered by a
quarter megajoule capacitor bank consisting of a series parallel combination of 15
µf capacitors rated at 20 kV. This bank
and its operation are described elsewhere.[13]
The current rises to peak value in 19.6µsec.
A crowbar is activated at peak current.
The decay time for the circuit is 30 microseconds. The peak magnetic field produced
by the secondary compression coils is
35,000 G at 20 kV. The bank was not fired
at voltages over 15 kV because of crowbar
switching difficulties at higher voltages.

Since the rings, after collision, are
stationary in the laboratory frame, the
coupling problem is no longer critical and
very large currents and mass motions can be
induced in the double ring system. The
macroscopic toroidal conduction currents
in the rings produce a long range attrac-
tion force which draws them together, the
vortex forces (mass flow forces) and poloidal conduction currents (which are
strongest near the surface of the rings
and produce a short range force), force
them apart.[4,6,12,13] Thus, they oscillate
axially at the center of the primary mirror when the secondary mirror is applied.
This oscillation has been observed in earlier experiments without secondary compression. We observe the same characteristic streak pictures of these phenomena
when the secondary compression is applied.

There is no apparent limit to the
size of the currents that can be induced
or the amount of compression that can be
obtained without producing any instability
in the rings. (See Fig. 2) The compression mirror acts as its own magnetic containment bottle and no special auxiliary
fields are required. Currents as high as
120 kA have been induced in rings compressed to 2 cm major diameter. In earlier experiments[4,6] the induced ring currents were measured directly with
Rogowski coils. In the very high field
experiments described here, the ring currents are calculated by the use of the
coupling coefficients between the compression coils and the current rings which
were calculated from earlier direct measurements.[13] There is no change in these
coefficients at the higher field levels
since the ring separation and oscillation
as observed by streak pictures do not
change at higher field levels. These
phenomena depend directly on the coupling
coefficients.

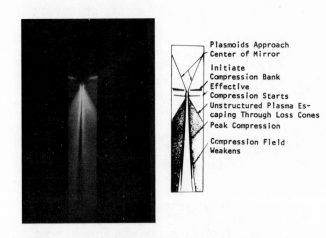

Plasmoids Approach
Center of Mirror

Initiate
Compression Bank
Effective
Compression Starts
Unstructured Plasma Es-
caping Through Loss Cones
Peak Compression

Compression Field
Weakens

FIG. 2. Streak picture and schematic diagram illustrating collision of vortex structures at center of secondary mirror system. Time goes from top to bottom of the picture and diagram. Streak time is 50 μsec. The horizontal slit was placed along the centerline of the machine, between the secondary compression coils.

Thermalization of the macroscopic currents and mass motions can be accurately controlled. Thermalization begins when the compression field becomes high enough to overcome the short range repulsive forces that hold the rings apart. The macroscopic flow energy and the energy trapped in the rings by conduction currents is then turned into random thermal energy as the rings slowly decay. The ring decay rate is a function of mirror ratio and the time that the compression field remains at values high enough to contain the high temperature plasma. The ion temperatures reported below were obtained at peak secondary compression before the rings thermalized and formed a normal mirror plasma. Complete flow breakdown and thermalization has not occurred in the compression shown in Fig. 2. This streak picture was taken for a compression at relatively low voltage for the purpose of a clear picture of the event.

Classical diffusion of plasma held in the mirror but not in the rings can be observed but the plasma trapped in the rings remains in the system for times that are orders of magnitude greater than those predicted for classical mirror diffusion. These decay rates can be easily computed from streak pictures of the type illustrated in Fig. 2. Spectroscopic examination of the line widths for neutral emission indicates that there is no appreciable line broadening. This indicates that there is minimal charge exchange between energetic ions and cold neutrals during heating and confinement.

The rings reduce the magnetic field at the center of the mirror system by 80% and constitute a "plugged" open ended system which is stable and has small end losses.

Measurements of the ion temperature by observation of Doppler broadening in a deuterium plasma for various voltages on the compression coil capacitor bank have been made using a multichannel Fabry-Perot interferometer[12] especially developed for the purpose. Since the bank is switched by air-gap switches, one would expect that switch efficiency would increase rapidly with increasing bank charging voltages. The voltage drop across the gap is approximately 3.0 kV at 9 kV bank voltage. It decreases rapidly to a fraction of a kV at 14 kV. The switch loss is the product of the voltage and switch current. At low bank voltages, this is an appreciable fraction of total bank voltage. At high voltages it is not. Thus, the maximum magnetic field is a nonlinear function of bank voltage. The peak temperatures should increase more rapidly than the square of the voltage. Figure 3 shows the results of these measurements. Ion temperatures of 170 eV have been observed at 14.3 kV on the 20 kV capacitor bank. It should be emphasized that this is a slow bank with a quarter cycle rise time of 19.6 μsec. Appropriate corrections have been made for instrument broadening, macroscopic motion of the plasma, turbulence effects and pressure broadening.[12,14]

Ion density was measured by scanning the D_β line of 4861 A. Peak density at 14.3 kV on the compression bank was 10^{16} ions/cm^3. Observed oscillations in density at the center of the machine are in phase with the axial oscillation of the rings observed in the streak pictures (Fig. 2). Electron temperature at 8 kV on the compression bank was measured by taking line intensity ratios of Helium II at 4686 A and He I at 5875 A. At peak compression, the ion temperature is 30 eV and the electron temperature is approximately 10 eV. There is preferential heating of the ions during secondary compression.

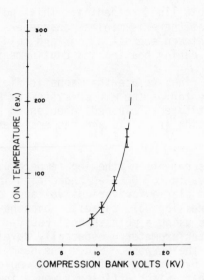

FIG. 3. Ion temperature of deuterium versus voltage applied to capacitor bank that powers the secondary-mirror coils. The temperatures were measured with a Fabry-Perot interferometer.

Peak temperatures and densities persist for times of the order of 20 μsec. No attempt has yet been made to optimize the magnetic mirror ratio of the compression mirror. The data were taken with a secondary mirror ratio of 1.4. There are strong theoretical reasons to believe that this ratio must be increased to at least 3.5 for optimum containment of the rings during compression.

We note that for operation at a mirror ratio of 1.4, $n\tau \cong 10^{12}$, $T_i = 170$ eV, at 14.3 kV on the compression bank.

In two previous papers[4],[5] hereafter designated I and II respectively, a new method of calculating the structure of naturally occurring stable plasma cells was outlined. The method invokes a variational principle in which the total energy of a closed (bounded) plasma configuration (cell) is varied subject to a set of constraint integrals on the flow. The resulting Euler-Lagrange equations describe the magnetic and flow fields in the plasma "bunch" or plasmoid. These differential equations can be solved subject to appropriate boundary conditions to give a quantitative description of the structure of globally stable plasmoids; i.e., the stability is calculated for the whole plasmoid (plasma cell) in its actual geometric configuration and ambient plasma surroundings.

The low-lying cell energy states are analogous to the low-lying stable states of an atom in which the lowest lying and most stable atomic configurations correspond to minimum energy and the higher energy states correspond to structures that are more easily perturbed and broken down when subjected to external disturbances. The type of stable cell-state changes as one changes the type and number of constraint integrals used in the variational calculation.

The total energy is used in making the calculations because nearly all plasmoids (cells) are interacting and exchanging energy with their surroundings. This means that the structures are nonconservative systems, and the usual methods of stability calculation utilizing effective potentials have no meaning. The method is interesting because the calculations include a consideration of all nonlinear states, they are global and not local, they make no assumption about the strength of coupling between various modes in the plasma and include the dynamic nonlinear terms in the equations of motion or equivalently in the equations describing conservation of formal currents and charges in the plasma.

The principle of least constraint (discussed in II) is invoked in order to find the various lowest lying cell energy states. The fewer the number of constraints applied to the system in performing the variational calculation, the more stable the corresponding plasmoid. It was also shown in II that the lowest lying states correspond, under certain conditions, to superposable flows and fields. All other states are nonsuperposable and nonlinear. Two of these nonlinear states will interact to form new states, rather than simply superpose to form a composite structure.

ACKNOWLEDGEMENTS

The authors gratefully acknowledge useful discussions with Drs. Joseph Mather, James Phillips and Joseph Di Marco of the Los Alamos Scientific Laboratories and Mr. Marvin Young of the Naval Research Laboratories.

This work was supported by AFOSR Grant 72-2295 and National Science Foundation Grant GP-43733.

REFERENCES

[1] D.R. Wells, Phys. Fluids 5, 1016 (1962).
[2] D.R. Wells, Phys. Fluids 7, 826 (1964).
[3] D.R. Wells, Phys. Fluids 9, 1010 (1966).
[4] D.R. Wells and J. Norwood, Jr., Phys. Fluids 3, 21 (1969).
[5] D.R. Wells, J. Plasma Phys. 4, 645(1970).
[6] E. Nolting, P. Jindra, and D.R. Wells, J. Plasma Phys. 9, 1 (1973).
[7] D.R. Wells and J. Norwood, Jr., Phys. Fluids 11, 1582 (1968).
[8] D.R. Wells, University of Miami Report No. MIAPH-PP-70.13, 1971 (see Appendix A).
[9] W.B. Jones and R.D Miller, Phys. Fluids 11, 1550 (1968).
[10] R. Turner, Phys. Fluids 13, 2398 (1970).
[11] R.L. Small and W.H. Bostick, Gruman Aircraft Engineering Corporation, Research Department, Report No. RE-284, 1967.
[12] F.N. Cooke, Jr., thesis, University of Miami, 1974.
[13] E. Nolting, U.S. Air Force Office of Scientific Research Scientific Report No. AFOSR-TR-71-2905.
[14] J. Hirschberg, private communication.
[15] W.H. Bostick, private communication.

APPENDIX A

THE CURRENT ALGEBRA OF GLOBAL MHD STABILITY

Daniel R. Wells

In order to aid in the understanding of the material discussed above, we outline some important aspects of the theory of naturally occurring stable plasma states. A theory of nonlinear global magnetohydrodynamic stability is described. The formalism is an entirely new approach to the problem. The concepts of space-time and generalized gauge symmetries of the flow fields are invoked to find constants of the motion. The constants correspond to charge operators in a theory of the current algebra of the fields. The charges, in turn, are defined by integrals that are determined by the symmetries of the fields. The strengths of the individual components of the currents and charges determine the amount of symmetry breaking in each physical situation. The constants of the motion corresponding to the charge operator are used in conjunction with the principle of least constraint to generate the Euler-Lagrange equations corresponding to stable plasma motion. For every symmetry there is a corresponding conserved integral or charge (Noether's Theorem). The principle of least constraint states that if the total energy of the flow field of a bounded plasma cell is varied, subject to a set of constraint integrals, then the fewer the number of constraint integrals applied, the more stable the resulting flow. The constraint integrals which generate a linear (superposable) field yield a set of equations which describe force-free collinear flow. If linearity is sacrificed, then fewer constraints can be used and many other types of flow structures are possible.

The symmetries and corresponding flow structures are classified by the Lie algebra of the currents and charges. The formalism leads to uniqueness theorems necessary to calculate the type of structure present for a given set of boundary conditions. It is demonstrated that a multiple integral variation problem can be related to the principle of least constraint. The fundamental variational formula for the appropriate tensor field is developed. The group generators for a particular space time and gauge transformation are then used to demonstrate that the variational approach of Woltjer and Wenzel is a special case of the problem of Lagrange with expanded integral constraints.

An understanding of the experimental results reported in this paper is strongly dependent on an understanding of this theory. The stability of the rings during compression and the reasons for their complex interaction when they meet at the center of the primary mirror are easily interpreted in terms of the theory of collinear force-free plasma tori. The theory predicts that the rings coming from each end of the primary mirror are all either contra or co-rotational. The composite rings formed by superposition of all rings generated at each end are thus co and contra-rotational and the theory predicts that they should interact and not superpose. Thus the stable septum is formed as observed in the experiment.

Proof that the theory actually describes the lowest lying and most stable energy states of the flow structures is given elsewhere (Nolting, Jindra & Wells, 1973) in the form of experimental data which is a measure of the actual magnetic fields trapped in the linear structures. Agreement with theory is excellent.

FORMAL CHARGES AND CURRENTS

In I and II (Wells & Norwood, 1969, and Wells, 1970) the lowest lying stable plasma cell states are derived. The constants of the motion corresponding to space-time and gauge symmetries are derived using the Clebsch potentials and the concept of a "generalized gauge transformation". In this section we consider all of the constants of the motion corresponding to space-time and gauge symmetries. We discuss the expansions of the corresponding functionals and utilize them in a fundamental variational formula to show that minimum constraint corresponds to minimum total energy. We then show that minimum total energy corresponds to minimum "free energy" available to drive instabilities in the case of the lowest lying linear states.

In II we discussed the fundamental importance of Noether's theorem in relating the symmetry properties of the Lagrange density of the MHD fields to the constants of the motion for closed plasma structures. It is necessary to give a precise formulation of the basic problem of the calculus of variations if further progress is to be made in understanding the global stability problem for bounded plasma cells. We merely outline the necessary theorems. A rigorous treatment of the problem is given by Rund (Rund, 1966).

Given n real variables x^i, together with m independent real variables t^α, (Latin and Greek indices run from 1 to n and 1 to m respectively) consider the space R_{n+m} of n+m dimensions of x^i, t^α. Denote a subspace C_m in R_{n+m}. A set of n equations of the type:

$$x^i = x^i(t^\alpha) \qquad (1)$$

defines this subspace. It is assumed that we can form the derivatives $\dfrac{\partial x^i}{\partial t^\alpha}$ on C_m. We denote these derivatives as:

$$\dot{x}^i_\alpha (t^\beta) = \frac{\partial x^i(t^\beta)}{\partial t^\alpha} .$$

Assume that G_t denotes a fixed simply-connected domain in the m-dimensional space of the t^α, bounded by a hypersurface ∂G_t. Each point of this surface corresponds to a set of values of the t^α. Consider a second set of equations of the type $\bar{x}^i = \bar{x}^i(t^\alpha)$ representing another subspace $\bar{C}_m$ of R_{n+m}. We require that this second subspace coincide with C_m for those values of t^α

which define the boundary ∂G_t of G_t. Then:

$$\bar{x}^i(t^\alpha) = x^i(t^\alpha) = f^i(t^\alpha) \quad \text{for } t^\alpha \epsilon \partial G_t ,$$

where functions $f^i(t^\alpha)$ are separately prescribed.

Consider a suitably differentiable function $L = L(t^\alpha, x^j, \dot{x}^j_\alpha)$ with m+n+mn arguments. This function is defined as a function of t^α over each subspace of the type of Eq. (1).

We now form the integral:

$$I(C_m) = \int_{G_t} L(t^\alpha, x^j, \dot{x}^j_\alpha) \, d(t), \qquad (2)$$

where

$$d(t) \overset{\text{def}}{=} dt^1 \cdots dt^m .$$

The value of Eq. (2) depends on C_m, i.e., the choice of the functions defined by Eq. (1) together with their derivatives. The fundamental problem is to find the necessary and sufficient conditions the $x^i(t^\alpha)$ must satisfy in order to yield an extreme value of the integral (2).

In order to proceed with the proof, another set of n equations of the type:

$$x^i = x^i(t^\alpha, u)$$

is considered. These represent a 1-parameter family of m-dimensional subspaces $C_m(n)$ of R_{n+m}. Consider two neighboring subspaces $C_m(u_0)$ and $C_m(u)$, where $(u-u_0)$ is considered to be small, and quantities of order $|u-u_0|^2$ are neglected. Let P and P' be points on $C_m(u_0)$ and $C_m(u)$ respectively, corresponding to the same t^α-values. The components of the displacement PP' in R_{n+m} are given by $(0, \cdots, 0; \delta x', \cdots, \delta x^m)$, where:

$$\delta^* x^i = (u-u_0) \left[\frac{\partial x^i}{\partial u} \right]_{u=u_0} + \cdots . \qquad (3)$$

It can then easily be shown (Rund, 1966, p. 213) that:

$$\delta x^i = (\dot{x}^i_\alpha)_{u=u_0} \delta t^\alpha + \delta^* x^i \qquad (4)$$

and

$$\delta^* \dot{x}^i_\alpha = \frac{\partial}{\partial t^\alpha} (\delta^* x^i) .$$

If one defines the first variation of I (as given by Eq. (2)) as:

$$\delta I = I(u) - I(u_o) = (u-u_o) \left(\frac{dI}{du}\right)_{u=u_o} \, ,$$

then it can be shown that:

$$\delta I = \int_{G_t} \left(\frac{\partial L}{\partial x^i} \delta^* x^i + \frac{\partial L}{\partial \dot{x}^i_\alpha} \delta^* \dot{x}^i_\alpha\right) d(t)$$

$$+ \int_{G_t} \frac{d}{dt^\alpha} (L \delta t^\alpha) \, d(t) \, .$$

This can be written as

$$\delta I = \int_{G_t} \left(\frac{\partial L}{\partial x^i} - \frac{d}{dt^\alpha}\left(\frac{\partial L}{\partial \dot{x}^i_\alpha}\right)\right) \delta^* x^i \, d(t)$$

$$+ \int_{G_t} \frac{d}{dt^\alpha} \left(L \delta t^\alpha + \frac{\partial L}{\partial \dot{x}^i_\alpha} \delta^* x^i\right) d(t) \, , \quad (5)$$

using Eq. (4). The first variation takes the final form,

$$\delta I = \int_{G_t} \left(\frac{\partial L}{\partial x^i} - \frac{d}{dt^\alpha}\left(\frac{\partial L}{\partial \dot{x}^i_\alpha}\right)\right) \delta^* x^i d(t)$$

$$+ \int_{G_t} \frac{d}{dt^\alpha} \left[(L \delta^\alpha_\beta - \frac{\partial L}{\partial \dot{x}^i_\alpha} \dot{x}^i_\beta) \delta t^\beta \right.$$

$$\left. + \frac{\partial L}{\partial \dot{x}^i_\alpha} \delta x^i\right] d(t) \, . \quad (6)$$

Equations (5) and (6) are the fundamental variational formulas for multiple integrals. The integrand of the second integral is formally a divergence. It is now easily shown from Eq. (6) that an extremum for the integral defined by Eq. (2) is uniquely determined by the solutions of:

$$\frac{d}{dt^\alpha}\left(\frac{\partial L}{\partial \dot{x}^k_\alpha}\right) - \frac{\partial L}{\partial x^k} = 0 \, ,$$

where $k = 1, \cdots, n.$

This set of n equations reduces to the Euler-Lagrange equations for multiple integrals. Written out in full they take the form:

$$\frac{\partial^2}{\partial \dot{x}^k_\alpha \partial \dot{x}^i_\beta} \frac{\partial^2 x^i}{\partial t^\alpha \partial t^\beta} + \frac{\partial^2 L}{\partial x^i \partial \dot{x}^k_\alpha} \frac{\partial x^i}{\partial t^\alpha} + \frac{\partial^2 L}{\partial t^\alpha \partial \dot{x}^k_\alpha}$$

$$- \frac{\partial L}{\partial x^k} = 0 \, . \quad (7)$$

It can be shown (Rund, 1966) how certain invariance properties of multiple integrals imply that various quantities are constant along an extremal. These results can be formulated in terms of Noether's Theorem. The so-called conservation laws then easily follow. Application of Noether's Theorem to MHD global stability is discussed by Wells in II (Wells, 1970). Here we utilize the formalism of the fundamental symmetry transformations to directly tie together Noether's Theorem and the principle of least constraint.

Consider an r-parameter Lie group operating on the variables (t^α, x^i). A typical element of the group is the transformation:

$$\bar{t}^\beta = \bar{t}^\beta(t^\gamma, x^j, \alpha_s), \quad \bar{x}^i = \bar{x}^i(t^\gamma, x^j, \alpha_s) \quad (8)$$

where $\alpha_s(s,t=1,\cdots,r)$ represents the r parameters of the group. We assume that the identity transformation of the group is given for $\alpha_s = 0$. The infinitesimal transformations corresponding to the finite transformations given by Eq. (8) are:

$$\delta t^\beta = \xi^\beta_{(s)} \alpha_s \, , \quad \delta x^i = \zeta^i_{(s)} \alpha_s$$

(summation over s), (9)

where

$$\xi^\beta_{(s)} = \left[\frac{\partial \bar{t}^\beta(t^\gamma, x^j, \alpha_t)}{\partial \alpha_s}\right]_{\alpha_t=0} \, ,$$

$$\zeta^i_{(s)} = \left[\frac{\partial \bar{x}(t^\gamma, x^j, \alpha_t)}{\partial \alpha_s}\right]_{\alpha_t = 0} \, .$$

Corresponding to the infinitesimal increments given by Eq. (9) we have, from Eq. (4),

$$\delta^* x^i = \delta x^i - \dot{x}^i_\beta \delta t^\beta = \eta^i_{(s)} \alpha_s$$

(summation over s) ,

where

$$\eta^i_{(s)} = \zeta^i_{(s)} - \dot{x}^i_\beta \xi^\beta_{(s)} \quad \text{(summation over s)}.$$

The variations (9) induce a variation δI of the fundamental integral (2) which we evaluate according to (5). This gives:

$$\delta I = \int_{G_t} \left[\frac{\partial L}{\partial x^i} - \frac{d}{dt^\alpha} \left(\frac{\partial L}{\partial \dot{x}^i_\alpha} \right) \right] \eta^i_{(s)} \alpha_s d(t)$$

$$+ \int_{G_t} \frac{d}{dt^\beta} \left[L\xi^\beta_{(s)} + \frac{\partial L}{\partial \dot{x}^i_\beta} \eta^i_{(s)} \right] \alpha_s d(t)$$

$$\text{(summation over s)} . \quad (10)$$

We now require that this fundamental integral be invariant under Eq. (9) up to an "independent integral". This means that L transforms under Eq. (9) according to:

$$\int_{G_t} L(\overline{t}^\beta, \overline{x}^j, \dot{\overline{x}}^j_\beta) \, d(t) =$$

$$\int_{G_t} \left[L(t^\beta, x^j, \dot{x}^j_\beta) + \Phi \right] d(t) \quad ,$$

where $\Phi = \Phi(t^\beta, x^j, \dot{x}^j_\beta, \alpha_s)$ is the integrand of an independent integral. We must now examine carefully the meaning of Φ because much of what follows depends on this concept.

A thorough discussion of the concept of an independent integral is given by Rund (Rund, 1966). We begin our discussion by stating that the integral of a divergence depends solely on the values of its argument functions on the boundary ∂G_t of the domain G_t over which we perform the integration. Divergences are not the only integrands which have this property. There are a large class of integrands giving rise to this type of independence. These are referred to as "independent integrals". Integrands which are divergences provide a satisfactory and relatively simple theory. This approach was extensively developed by Weyl (Weyl, 1935). There is another special sub-class, furnished by certain determinants, which yield a theory for multiple integral problems. This work is commonly associated with Carathéodory (Carathéodory, 1935). There is a more general theory discussed by Rund (Rund, 1966, p. 250) that includes the theories of Weyl and Carathéodory as special cases. This theory is interesting because it classifies the independent integrals by expressing their integrands as homogeneous polynomials. This is extremely useful in treating MHD stability problems for the case of broken symmetries since, as shown below, it is necessary to expand the integrands of the constraint integrals in order to treat these problems. It will also enable us to prove the uniqueness of the constraint integrals corresponding to the generalized gauge symmetries.

In order to greatly simplify the discussion of the general treatment of symmetry transformations of the type given in Eq. (9), we assume now that the functions Φ are divergences, i.e., there exist m functions $\Phi^\beta_{(s)}(t^\varepsilon, x^j, \dot{x}^j_\beta)$ such that

$$\Phi = \Phi_{(s)} \alpha_s = \frac{d\Phi^\beta_{(s)}}{dt^\beta} \alpha_s \quad , \quad (11)$$

where

$$\frac{d\Phi^\beta_{(s)}}{dt^\beta} \overset{\text{def}}{=} \frac{\partial \Phi^\beta_{(s)}}{\partial t^\beta} + \frac{\partial \Phi^\beta_{(s)}}{\partial x^i} \dot{x}^i_\beta + \dots.$$

and

$$\dot{x}^i_\beta \overset{\text{def}}{=} \frac{dx^i}{dt^\beta} \quad .$$

If Φ depends on t^α only, $\frac{d\Phi}{\partial t^\alpha}$ and $\frac{\partial \Phi}{\partial t^\alpha}$ have the same meaning.

We also assume that Φ is linear in the parameters α_s ;

$$\Phi = \Phi_{(s)}(t^\beta, x^j, \dot{x}^j_\beta) \, \alpha_s \quad . \quad \text{(summation over s)}$$

Then we obtain:

$$\delta I = \int_{G_t} \Phi_{(s)} \alpha_s \, d(t) \quad . \quad (12)$$

Comparison with Eq. (10) yields:

$$\int_{G_t} \left[\frac{d}{dt^\alpha} \left(\frac{\partial L}{\partial \dot{x}^i_\alpha} \right) - \frac{\partial L}{\partial x^i} \right] \eta^i_{(s)} \alpha_s \, d(t) =$$

$$\int_{G_t} \left\{ \frac{d}{dt^\beta} \left[L\xi^\beta_{(s)} + \frac{\partial L}{\partial \dot{x}^i_\beta} \eta^i_{(s)} \right] - \Phi_{(s)} \right\} \alpha_s d(t) \ .$$

(summation over s)

By hypothesis, this equation is valid for any region G_t. The α_s are the r parameters of an r-parameter Lie group and therefore must be independent. We have:

$$\left[\frac{d}{dt^\alpha} \left(\frac{\partial L}{\partial \dot{x}^i_\alpha} \right) - \frac{\partial L}{\partial x^i} \right] \eta^i_{(s)} =$$

$$\frac{d}{dt^\beta} \left[L\xi^\beta_{(s)} + \frac{\partial L}{\partial \dot{x}^i_\beta} \eta^i_{(s)} \right] - \Phi_{(s)} \ , (s=1,\cdots,r).$$

Combining this with Eq. (11), we have, finally:

$$\left[\frac{d}{dt^\alpha} \left(\frac{\partial L}{\partial \dot{x}^i_\alpha} \right) - \frac{\partial L}{\partial x^i} \right] \eta^i_{(s)} =$$

$$\frac{d}{dt^\beta} \left[L\xi^\beta_{(s)} + \frac{\partial L}{\partial \dot{x}^i_\beta} \eta^i_{(s)} - \Phi^\beta_{(s)} \right] \ ,$$

(s = 1,$\cdots$,r). (summation over s) (13)

In II we have discussed the concept of generalized gauge symmetries as developed by Calkin. We have related the corresponding constants of the motion to the linear superposable states of closed plasma structures. We will now show that Eq. (13) ties together the concepts of Noether's theorem, least constraint, and symmetry breaking (i.e., expansions of $\Phi^\beta_{(s)}$ in complete sets of functions of the spatial coordinates). We will also show that application of the generalized theory of independent integrals of Rund plays an important role in selecting the proper complete set of constraint integrals.

From Eq. (11) we note that there exists a system of m properly behaved functions $\Phi^\alpha = \Phi^\alpha(t^\beta,x^j)$ obeying the divergence equation. In order to simplify the discussion and make the formalism directly comparable to that developed in II, we now introduce the following change of variables:

$$t^\alpha \rightarrow x^j(x_1,x_2,x_3,t) \ ,$$

where t is the time. We assume that there are N dependent functions ψ^a of these independent variables, and these, in turn, replace the functions $x^i(t^\alpha)$ used previously:

$$x^i(t^\alpha) \rightarrow \psi^a(x^j), \ (a,b,\cdots = 1,\cdots,N) \ .$$

A detailed discussion of Noether's theorem and related problems in this notation is given by Hill (Hill, 1951). The reader is referred to this paper for a rigorous treatment of the conservation equations which we now develop.

Considering transformations that arise continuously from the identity transformation, it is sufficient to consider the infinitesimal transformations

$$x'^k = x^k + \delta x^k$$

and   (14)

$$\psi'^\alpha(x') = \psi^\alpha(x) + \delta\psi^\alpha(x).$$

The finite transformations are found by iteration. It is then easily shown (Hill, 1951, p. 258) that one can associate a differential conservation equation with each infinitesimal symmetry transformation in the form:

$$\frac{d}{dx^k} \left\{ \left[L\delta^k_\ell - \frac{\partial L}{\partial \frac{\partial \psi^\alpha}{\partial x^k}} \frac{\partial \psi^\alpha}{\partial x^\ell} \right] \delta x^\ell + \frac{\partial L}{\partial \frac{\partial \psi^\alpha}{\partial x^k}} \delta\psi^\alpha \right.$$

$$\left. + \delta\Omega^k \right\} = 0 \ .$$
                                                                                      (15)

Then we can write

$$\frac{\partial \sigma}{\partial t} + \nabla \cdot \vec{S} = 0 \ ,$$ (16)

where σ, the "formal charge density" associated with our symmetry transformation (14) is:

$$\sigma = \left[L - \frac{\partial L}{\partial \nabla\psi^\alpha} \frac{\partial \psi^\alpha}{\partial t} \right] \delta t - \frac{\partial L}{\partial \frac{\partial \psi^\alpha}{\partial t}} \delta\underline{x} \cdot \nabla\psi^\alpha$$

$$+ \frac{\partial L}{\partial \frac{\partial \psi^\alpha}{\partial t}} \delta\psi^\alpha + \delta\Omega_t.$$
                                                                                      (17)

and the "current density" $\underline{S}$ is:

$$\underline{S} = -\frac{\partial L}{\partial \nabla \psi^\alpha} \frac{\partial \psi^\alpha}{\partial t} \delta t + \left(L \delta \underline{x} - \frac{\partial L}{\partial \nabla \psi^\alpha} \delta \underline{x} \cdot \nabla \psi^\alpha\right)$$

$$+ \frac{\partial L}{\partial \nabla \psi^\alpha} \delta \psi^\alpha + \delta \underline{\Omega} \quad . \tag{18}$$

Then σ and the three components of $\underline{S}$ can be associated with the ms functions $\phi^\beta(s)$ of Eq. (11). Integrating over a closed volume τ with surface $\sum$ and using Gauss's theorem, we have:

$$\frac{\partial}{\partial t} \int_\tau \sigma d(x) = - \oint_\sum \underline{S} \cdot d\underline{\sum} \quad .$$

Calkin has shown (Calkin, 1963; Wells, 1970) that one can write a Lagrange density for an MHD fluid in the form:

$$L = \frac{1}{2} \kappa_0 |\vec{E}|^2 - \frac{1}{2} \mu_0^{-1} |\vec{B}|^2 + \vec{P} \cdot \vec{E}$$

$$+ \rho \left(\frac{\partial \chi}{\partial t} - \eta \frac{\partial \zeta}{\partial t} - \frac{1}{2} |\vec{v}|^2 - \int \rho^2 p d\rho\right) \quad ,$$

where $\vec{E}$ is the electric field intensity, μ_0 and κ_0 are the permeability and permittivity of the fluid, $\vec{v}$ is the velocity of the center of mass of a fluid element, ρ is the mass density, p is the scalar pressure, χ, η, ζ are the Clebsch potentials defined by:

$$\vec{v} + \rho^{-1} \vec{B} \times \vec{P} = -\nabla \chi + \eta \nabla \zeta$$

and $\vec{P}$ is a "polarization" vector. The vector $\vec{P}$ is in reality a type of "vector potential" that defines the current density according to:

$$\vec{j} = \frac{\partial \vec{P}}{\partial t} + \nabla \times (\vec{P} \times \vec{v}) + (\nabla \cdot \vec{P})\vec{v}, \quad \varepsilon = \nabla \cdot \vec{P} \quad ,$$

and ε is the electric charge density.

Then $\vec{P}$ is defined only up to a generalized gauge transformation of the form:

$$\vec{P}' \rightarrow \vec{P} + \vec{\Lambda}, \quad \text{where} \quad \frac{\partial \vec{\Lambda}}{\partial t} + \nabla \times (\Lambda \times v) = 0$$

$$\text{and} \quad \nabla \cdot \vec{\Lambda} = 0 \quad .$$

The last two equations imply that $\vec{\Lambda}$ is frozen into the fluid. In II it is shown that the three vector fields that move with the fluid are $\vec{B}$, $\vec{z}$, and $\vec{v}$ where

$$\vec{z} \stackrel{\text{def}}{=} \nabla \times (\vec{v} + \rho^{-1} \vec{B} \times \vec{P}) \quad .$$

Then one can make the following infinitesimal transformations:

(I) $\delta \vec{\Lambda} = 0$ The resulting finite transformations are just the canonical transformations of the fields.

(II) $\delta \vec{\Lambda} = \delta \alpha \vec{B}$
$\vec{P} \rightarrow \vec{P}' = \vec{P} + \delta \alpha \vec{B}$
This transformation leads to conservation of the integral

$$\int \vec{A} \cdot \vec{B} d\tau \quad .$$

(III) $\delta \vec{\Lambda} = \delta \beta \vec{z}$
$\vec{P} \rightarrow \vec{P}' = \vec{P} + \delta \beta \vec{z}$
This leads to conservation of the integral

$$\int \vec{B} \cdot \vec{v} d\tau \quad .$$

(IV) $\delta \vec{\Lambda} = \delta \gamma \vec{v}$
This leads to conservation of the integral

$$\int \vec{A} \cdot \vec{v} d\tau \quad .$$

For given symmetry transformations arising continuously from the identity, we can write the conservation law (16). The corresponding density σ and current $\underline{S}$ are found from (17) and (18) respectively. The appropriate conserved integrals (constants of the motion) are then found by the application of vector identities. In this way the conserved integrals in (II), (III), and (IV) above have been found.

In II there is a discussion of the completeness of this set of constraints. It is pointed out that there is a constraint integral (constant of the motion) corresponding to each generalized gauge symmetry. There is a gauge symmetry corresponding to each field that is convected with the flow. Since the set $\vec{B}$, $\vec{A}$, and $\vec{v}$ are the lowest order independent fields convected with the flow, it is assumed that the set is unique and complete. It is assumed that the velocity $\vec{v}$ is appropriately expressed in terms of the Clebsch potentials, that the electric field $\vec{E}$ has been eliminated from the Lagrange density by setting κ_0 equal to zero, and that the displacement current is negligible. The equations which determine convection with the flow require that the "frozen in" field be solenoidal. (e.g., $\nabla \cdot \vec{A} = 0$). This condition is satisfied by $\vec{B}$ and $\vec{z}$. The velocity field is solenoidal for an incompressible or isocloric flow ($\nabla \rho \cdot \vec{v} = 0$). One can now ask why higher order fields cannot generate symmetries that, in turn, will generate additional independent constraint integrals. While $\nabla \times \vec{B}$ is ruled out by the assumption of negligible displacement current (i.e., $\vec{j}$ is not solenoidal to this approximation), higher order derivatives of $\vec{A}$ and $\vec{v}$ would seem to satisfy all requirements and add to the allowed set of constants of the motion. In order to show that these higher order fields are not allowed as generators, one must examine more closely the concept of the independent integral and its interpretation in the theories of Weyl, Carathéodory and Rund. (Rund, 1966, p. 250) The fundamental differences between the theories of Weyl and Carathéodory is in the method of defining the independent integrals. Rund generalizes these two theories by examining the Euler-Lagrange equation itself. He gives a rigorous development of his theory (Rund, 1966). Space limitations do not permit a rigorous discussion of these concepts here but the resulting formalism allows a rigorous proof that these higher order derivatives of $\vec{A}$ and $\vec{v}$ do not satisfy all the requirements and do not add to the allowed sets of constants of the motion.

The complete analysis shows that one is dealing with a simple divergence transformation. This concept is discussed by Hill (Hill, 1951, p. 256). It is demonstrated that for such a transformation, the equation

$$\frac{\partial \sigma}{\partial t} + \nabla \cdot \vec{S} = 0$$

always holds and thus if the surface terms involving $\vec{S}$ are zero, the corresponding formal charge density σ is a constant of motion. Then.

$$\frac{\partial}{\partial t} (\vec{A} \cdot \vec{B}) + \nabla \cdot (\vec{E} \times \vec{A} + \phi \vec{B}) = 0.$$

If $\hat{n}$ is the unit normal to a boundary surface $\sum$, then

$$\frac{\partial}{\partial t} \int_\tau \{\delta\alpha(\vec{A} \cdot \vec{B})\} d\tau + \int_\tau \delta\alpha[\nabla \cdot (\vec{E} \times \vec{A} + \phi \vec{B})] d\tau$$

$$= \frac{\partial}{\partial t} \int_\tau \{\delta\alpha(\vec{A} \cdot \vec{B})\} d\tau + \int_{\sum} \delta\alpha(\vec{E} \times \vec{A} \cdot \hat{n}$$

$$+ \phi \vec{B} \cdot \hat{n}) \ d\sum \quad , \tag{19}$$

where, for the transformation (II) above, we see that

$$\sigma = \delta\alpha(\vec{A} \cdot \vec{B}) \quad \text{and} \quad \underline{S} = \delta\alpha\{\vec{E} \times \vec{A} + \phi\vec{B}\} .$$

We see that σ plays the role of a "formal charge density" and $\underline{S}$ plays the role of a "formal current density". Thus one can discuss the current algebra of global stability if one can relate these formal charges and currents to the equilibrium and stability of the plasma (magnetofluid) inside the boundary surface Σ of Eq. (19). For nested surfaces in an infinitely conducting fluid, the surface integral goes to zero, since $\vec{v}$, $\vec{B}$, $\vec{j}$ and $\vec{A}$ are all assumed to lie in the surfaces (see discussion of boundary conditions in II) and $\int \vec{A} \cdot \vec{B} d\tau$ is a constant of the motion (see I, page 27), e.g., for the transformation II one has,

$$\Psi = \frac{\delta\alpha \partial (\vec{A} \cdot \vec{B})}{\partial t} + \delta\alpha\{\nabla \cdot [(\vec{E} \times \vec{A}) + \phi\vec{B}]\} .$$

Since the divergence terms drop out in any application to a bounded plasma of the type under discussion, one has:

$$\frac{\partial}{\partial t} \int \Psi d\tau = \frac{\partial}{\partial t} \int (\vec{A} \cdot \vec{B}) \ d\tau = 0$$

then

$$\Psi = \frac{\delta\alpha \partial}{\partial t} (\vec{A} \cdot \vec{B}) = \delta\alpha[\frac{\partial \vec{A}}{\partial t} \cdot \vec{B} + \vec{A} \cdot \frac{\partial \vec{B}}{\partial t}] .$$

The expansions of the charge density σ and the current $\vec{S}$ by multiplication by the entirely arbitrary function $f(x^k(t))$, where $x^k = x,y,z$ for $K = 1,2,3$ can be used to introduce the concept of symmetry breaking into the theory.

If both sides of Eq. (19) are multiplied by an arbitrary function of the space coordinates (see I, page 32) arranged in a convergent series, then $\sigma \to \tilde{\sigma}$ where

$$\tilde{\sigma} \overset{\text{def}}{=} \frac{\delta\alpha}{2} \{f(x^k(t))\} \, \sigma.$$

Then

$$\frac{\partial}{\partial t} \int \tilde{\sigma} d\tau = \frac{\partial}{\partial t} \int \frac{\delta\alpha}{2} f(x^k(t))(\vec{A}\cdot\vec{B}) \, d\tau$$

$$= -\frac{\delta\alpha}{2} \int f(x^k(t))\{\nabla\cdot[(\vec{E}\times\vec{A}) + \phi\vec{B}]\}d\tau$$

$$= +\frac{\delta\alpha}{2} \int \nabla\cdot f(x^k(t))[(\vec{E}\times\vec{A}) + \phi\vec{B}] \, d\tau$$

$$\qquad -\frac{\delta\alpha}{2} \int \nabla f(x^k(t))\cdot[(\vec{E}\times\vec{A}) - \phi\vec{B}] \, d\tau$$

$$= +\frac{\delta\alpha}{2} \int f(x^k(t))[(\vec{E}\times\vec{A}) + \phi\vec{B}]\cdot \hat{n}d\Sigma$$

$$\qquad -\frac{\delta\alpha}{2} \int \nabla f(x^k(t))\cdot[(\vec{E}\times\vec{A}) + \phi\vec{B})] \, d\tau. \quad (20)$$

We assume that $\vec{A}$, $\vec{B}$, and $\vec{v}$ lie on nested surfaces of constant ρ, p, and ϕ. Thus $\nabla\rho$, ∇p, and $\nabla\phi$ are assumed normal to those surfaces. Since

$$\vec{E} = -\vec{v}\times\vec{B} = -\nabla\phi - \frac{\partial\vec{A}}{\partial t},$$

$\frac{\partial\vec{A}}{\partial t}$ must be collinear with $\nabla\phi$ if surfaces of constant ϕ are to be coincident with surfaces of constant ρ, p, $\vec{A}$, $\vec{B}$ and $\vec{v}$ and if the gauge symmetry is exact (not broken). After equilibrium is achieved, $\frac{\partial\vec{A}}{\partial t}$ goes to zero. The requirement that $\frac{\partial\vec{A}}{\partial t}$ be normal to the surfaces of constant $\vec{A}$ is relaxed in the case of broken symmetries since then $\vec{v}\times\vec{B}$ is no longer normal to the nested surfaces (there is leakage of magnetic field and mass through the surface of the plasma structure). We now make the additional assumption that $\nabla f(x^k(t))$ is normal to the nested surfaces. This is easily accomplished by using a complete set of functions of the electric scalar potential, ϕ. Then $f(x^k(t))$ becomes $\phi(x^k(t))$ and the term in $\nabla f(x^k(t))$ drops out. In the P, T, U, V representation for azimuthally symmetric fields, one can use a complete set of functions of $(\tilde{\omega}^2 P)$ for the expansion (Wells, 1970). This function is constant on a flux tube and its gradient is normal to the nested surfaces.

Returning now to Eq. (13) and Noether's theorem, we consider a special group of transformations defined by Eq. (9). If we make just the gauge transformations corresponding to $\delta\vec{A} = \delta\alpha\vec{B}$, we have from Eqs. (19) and (20):

$$\frac{\partial}{\partial t} \int \tilde{\alpha}d\tau = +\frac{\delta\alpha}{2} \int f(x^k(t))[(\vec{E}\times\vec{A}) + \phi\vec{B}]$$

$$\qquad\qquad \times \hat{n}d\Sigma \quad . \qquad (21)$$

If $f(x^k(t))$ is a polynomial with constant term equal to unity, then one term of the integrand in Eq. (21) is $(\vec{A}\cdot\vec{B})$. If only this single term in the integrand is retained, we will say that the gauge symmetry corresponding to $\delta\vec{A} = \delta\alpha\vec{B}$ is exact. Since $(\vec{E}\times\vec{A})$ and $\vec{B}$ both lie in the nested surfaces, the right hand terms in Eq. (21) are zero and $\tilde{\sigma}$ is a constant of the motion. If other terms in $f(x^k(t))$ are included in the expansion of the integrand, we will say that the symmetry is broken. As we allow more terms in the expansion of $f(x^k(t))$, we put more constraints on the system. The more terms other than the constant term that are included, the more badly broken the symmetry. The right hand terms in Eq. (21) are no longer zero but determine the coefficients in the expansion in the following way. Let

$$I = \int f(x^k(t))\{\vec{A}\cdot\vec{B}\} \, d\tau$$

then

$$\int \frac{\partial}{\partial t} \{f(x^k(t))(\vec{A}\cdot\vec{B})\}d\tau = -\oint f(x^k(t))$$

$$\{(\vec{E}\times\vec{A}) \cdot n + \phi\vec{B}\cdot\hat{n}\} \, d\Sigma$$

and

$$\int \frac{\partial}{\partial t} \frac{f(x^k(t))}{\{[1 + \text{---} A_n(\phi)\}\{\vec{A}\cdot\vec{B}\}d\tau = \eta(t),}$$

where $f(x^k(t))$ is a polynomial in $x^k(t)$ and $\eta(t)$ corresponds to the initial value of the surface integral.

Now

$$\int \frac{\partial F}{\partial t} \, d\tau = \frac{d}{dt} \int F d\tau - \oint F\vec{v}\cdot\hat{n}d\Sigma$$

for convective flow and any space-time function F (Wells and Norwood, 1969, p. 27). Therefore

$$\int \frac{\partial}{\partial t} [\{1 + --- A_n(\phi)\}\{\vec{A}\cdot\vec{B}\}] \, d\tau$$

$$= \frac{d}{dt} \int [\{1 + --- A_n(\phi)\}\{\vec{A}\cdot\vec{B}\}] \, d\tau$$

$$- \oint [\{1 + --- A_n(\phi)\}\{\vec{A}\cdot\vec{B}\}]\vec{v}\cdot\hat{n}d\Sigma \ .$$

Therefore

$$\frac{d}{dt} \int [\{1 + --- A_n(\phi)\}\{\vec{A}\cdot\vec{B}\}]d\tau = \eta(t)$$

$$+ \oint [\{1 + --- A_n(\phi)\}\{\vec{A}\cdot\vec{B}\} \ ]\vec{v}\cdot\hat{n}d\Sigma \ ,$$

or

$$\frac{d}{dt} \int [(M_n(\phi))(\vec{A}\cdot\vec{B})]d\tau = \eta(t) + \oint [M_n(\phi)$$

$$\times (\vec{A}\cdot\vec{B})]\vec{v}\cdot\hat{n}d\Sigma \ .$$

This gives

$$\Delta \int M_n(\phi)(\vec{A}\cdot\vec{B})d\tau = + \int M_n(\phi) \ \{(\vec{E}\times\vec{A}) - \phi B$$

$$- (\vec{A}\cdot\vec{B})v\}\cdot\hat{n}d\Sigma\Delta t. \quad (22)$$

We can solve Eq. (22) for $\Delta \int M_n(\phi) (\vec{A}\cdot\vec{B})d\tau$. The right hand side depends on initial and boundary conditions.

Knowing $\Delta \int M_n(\phi)(\vec{A}\cdot\vec{B})d\tau$ for a given Δt, we can calculate a corresponding $\Delta\Psi(\phi)$ (Wells and Norwood, 1969, p. 44) or $\{\Delta\eta (\phi), \Delta\beta(\phi), \Delta\omega(\phi), \Delta\Omega(\phi)\}$ and can then insert them into Eqs. A(30) through A(33) (of Wells and Norwood) to find $\Delta\vec{B}$, $\Delta\vec{j}$, $\Delta\vec{v}$, Δp, and $\Delta\rho$ for a given Δt. This determines $\frac{\Delta\vec{B}}{\Delta t}$, etc., for a given set of initial and boundary conditions and hence gives the explicit growth-rates toward stability of the fields for given initial conditions, boundary conditions, and degree of symmetry-breaking. The method is independent of the conventional linearization restrictions and seems quite tractable for a numerical study.

It is convenient to define $M(\phi)$ in some applications as a polynomial with constant term unity. In other applications $M(\phi)$ can be expanded in some convenient complete set of functions of the electric scalar potential (see II, p. 648).

If one wants to calculate directly the final cell configurations that will result for a given degree of symmetry breaking, then a modified constant of the motion must be found. In this case

$$\tilde{\sigma} \overset{def}{=} M_n(\phi)\sigma \ ,$$

where $M_n(\phi)$ is an appropriate complete set of functions of the electric potential, ϕ, or some other convenient set of functions whose gradient is normal to the nested surfaces: i.e., $(\tilde{\omega}^2 P)$. Then we obtain

$$\frac{\partial}{\partial t} \int \tilde{\sigma}d\tau = \frac{\partial}{\partial t} \int \frac{\delta\alpha}{2} M_n(\phi)(\vec{A}\cdot\vec{B})d\tau$$

$$= - \frac{\delta\alpha}{2} \int M_n(\phi)\{\nabla\cdot[(\vec{E}\times\vec{A})+\phi\vec{B}]\}d\tau \ .$$

This gives

$$\frac{\partial}{\partial t} \int \tilde{\sigma}d\tau = +\frac{\delta\alpha}{2} \int \nabla M_n(\phi)\cdot[(\vec{E}\times\vec{A}) + \phi\vec{B}] \ d\tau$$

$$- \frac{\delta\alpha}{2} \int \nabla\cdot M_n(\phi)[(\vec{E}\times\vec{A}) + \phi\vec{B}] \ d\tau$$

$$\frac{\partial}{\partial t} \int \tilde{\sigma}d\tau = + \frac{\delta\alpha}{2} \int \nabla M_n(\phi)\cdot[(\vec{E}\times\vec{A}) + \phi\vec{B}] \ d\tau$$

$$- \frac{\delta\alpha}{2} \int \{M_n(\phi)[(\vec{E}\times\vec{A})+\phi\vec{B}]\}\cdot\hat{n}d\Sigma. \quad (23)$$

One must now assume that during the time that the cell decays to its final lowest energy state, the time average fields $(\vec{E}\times\vec{A})$ and $\phi\vec{B}$ remain on the nested surfaces. (Good only to first order in $\frac{B_T}{B_N}$ and $\frac{(\vec{E}\times\vec{A})_T}{(\vec{E}\times\vec{A})_N}$ i.e., it is good only for almost exact symmetries).

Then the first term on the right hand side drops out because $\nabla M_n(\phi)$ is normal to the nested surfaces. The second term on the right is zero because $(\vec{E}\times\vec{A})$ and $\phi\vec{B}$ lie on the surfaces. This procedure will always yield a set of Euler-Lagrange equations corresponding to an equilibrium. The interpretation of the corresponding physical situation must be carefully considered because the proof that $\int \tilde{\sigma}d\tau$ is a constant of the motion is only as good as the assumption that $\nabla M_n(\phi)$ remains normal to the nested surfaces during the formation of the final structure.

It is interesting to note that a straightforward discussion of the fact that the formal charge density can contain the independent and dependent variables but not the derivatives of the dependent variables is given by Hill (Hill, 1951, p. 256) using a different and straightforward approach. The close similarity between Hill's proof and that given by Wells in (Wells, 1970) is self-evident.

Now consider a different group of transformations of the type defined by Eq. (9) which include not only the gauge transformation corresponding to, say, $\delta\vec{A} = \delta\alpha\vec{B}$ but also a simultaneous transformation of the time. Then by Eq. (9)

$$\delta t = \xi^t_\nu \, \nu \quad \text{and} \quad \delta x^{\vec{P}} = \zeta^{\vec{P}}_{\vec{\Pi}} \, \vec{\Pi} \tag{24}$$

$$\text{(summation over } \nu \text{ and } \vec{\Pi})$$

where

$$\bar{t} = t + \nu \quad \text{and} \quad \overline{x^{\vec{P}}} = x^{\vec{P}} + \vec{\Pi} = \vec{P} + \delta\alpha\vec{B}.$$

One must examine the meaning of this transformation. Consider the change in a function $F(x)$ under an infinitesimal transformation of the form

$$x'_i = f_i(x_1 \cdots x_n; \, a_1 \cdots a_r),$$

$$i = 1, \cdots, n \quad \text{(Hammermesh, p. 296)}.$$

$$dF = \sum_{i=1}^n \frac{\partial F}{\partial x_i} \, dx_i = \sum_{i=1}^n \frac{\partial F}{\partial x_i}$$

$$\sum_{\ell=1}^r u_{i\ell} (x)\delta a_\ell$$

where

$$u_{i\ell}(x) = \sum_{k=1}^r \left(\frac{\partial f_i}{\partial a_k}\right)_{a=0} .$$

Then

$$dF = \sum_{\ell=1}^r \delta a_\ell (\sum_{i=1}^n u_{i\ell}(x) \frac{\partial}{\partial x_i}) F$$

$$= \sum_{\ell=1}^r \delta a_\ell \, X_\ell \, F .$$

The operators

$$X_\rho = \sum_{i=1}^n u_{i\rho} (x) \frac{\partial}{\partial x_i}$$

are the group generators.

In the particular case of the group described by the divergence transformation, one has

$$x_1 = t, \quad \delta a_1 = \delta\nu, \quad u_{11} = 1 \quad \text{and}$$

$$d\vec{P} = \frac{\partial \vec{P}}{\partial t} \, \delta\nu .$$

We have, therefore,

$$\frac{\partial \vec{P}}{\partial t} \, \delta\nu = \begin{cases} \vec{B}\delta\alpha \\ \vec{z}\delta\beta \\ \vec{v}\delta\gamma . \end{cases}$$

One can now easily anticipate from this relationship, derived from the properties of the group generators, the form of the Euler-Lagrange equations of the corresponding flow.

For a nonmagnetized medium moving with a velocity $\vec{v}$ which is small compared with the velocity of light

$$\nabla\times\vec{B} = \frac{\partial P}{\partial t} + \nabla\times(\vec{P}\times\vec{v}) + \vec{v}(\nabla\cdot\vec{P}) .$$

Then

$$\frac{\partial\vec{P}}{\partial t} = \nabla\times\vec{B} - [\nabla\times(\vec{P}\times\vec{v}) + \vec{v}(\nabla\cdot\vec{P})] .$$

The requirement on the group generators is that $\frac{\partial\vec{P}}{\partial t}$ be parallel to $\vec{B}$, $\vec{z}$, and $\vec{v}$. Then

$$\frac{\partial\vec{P}}{\partial t} = \alpha\vec{B} = \{\nabla\times\vec{B} - [\nabla\times(\vec{P}\times\vec{v}) + \vec{v}(\nabla\cdot P)]\} ,$$

$$\frac{\partial\vec{P}}{\partial t} = \beta\vec{z} = \beta\{\nabla\times(\vec{v}+\frac{1}{\rho}\vec{B}\times\vec{P})\} = [\nabla\times\vec{B} - \nabla\times(\vec{P}\times\vec{v})$$

$$+ \vec{v}(\nabla\cdot\vec{P})] ,$$

$$\frac{\partial\vec{P}}{\partial t} = \gamma\vec{v} = [\nabla\times\vec{B} - \nabla\times(\vec{P}\times\vec{v}) + \vec{v}(\nabla\cdot\vec{P})]$$

where α, β, and γ are scalar constants.

These equations are trivially satisfied for a force-free collinear flow on closed nested surfaces. This is the flow described by the Euler-Lagrange equations resulting from a variation of the total energy of a closed plasma volume subject to the constants of the motions corresponding to these three symmetries.

The coefficients of the group generators then become

$$\xi^t_\nu = \left(\frac{\partial t}{\partial \nu}\right)_{\nu=0} = 1 , \quad \text{and}$$

$$\zeta^{\vec{P}}_{\vec{\Pi}} = \left(\frac{\partial \vec{x}}{\partial \vec{\Pi}}\right)_{\vec{\Pi}=0} = 1|$$

We now write Eq. (13) in the form

$$\left[\frac{d}{dt^\alpha}\left(\frac{\partial L}{\partial \dot{x}^i_\alpha}\right) - \frac{\partial L}{\partial x^i}\right]\eta^i_{(s)} \equiv -\frac{d}{dt^\beta}\left(\theta^\beta_{(s)} + \phi^\beta_{(s)}\right)$$

$$\text{(summation over (s)),} \qquad (25)$$

where

$$\theta^\beta_{(s)} = -L\xi^\beta_{(s)} - \frac{\partial L}{\partial \dot{x}^i_\beta}\eta^i_{(s)} .$$

From the definition of $\eta^i_{(s)}$, we have

$$\theta^\beta_{(s)} = -\left[L\delta^\beta_\alpha - \frac{\partial L}{\partial \dot{x}^i_\beta}\dot{x}^i_\alpha\right]\xi^\alpha_{(s)}$$

$$-\frac{\partial L}{\partial \dot{x}^i_\beta}\zeta^i_{(s)} \text{ (summation over s).}$$

It is now convenient to define a Hamiltonian tensor (Rund, 1966, p. 240) in the form

$$H^\beta_\alpha(t^\varepsilon, x^j, p^\varepsilon_j) = -\delta^\beta_\alpha L(t^\gamma, x^j, \phi^j_\gamma(t^\varepsilon, x^h, p^\varepsilon_h))$$

$$+ p^\beta_i \phi^i_\alpha(t^\varepsilon, x^h, p^\varepsilon_h) .$$

This implies

$$\theta^\beta_{(s)} = H^\beta_\alpha \xi^\alpha_{(s)} - \frac{\partial L}{\partial \dot{x}^i_\beta}\zeta^i_{(s)} .$$

From the transformations Eq.(24) and the definition of $\eta^i_{(s)}$ given above

$$\eta^{\vec{P}}_{(\delta\alpha\vec{B})} = \zeta^{\vec{P}}_{(\delta\alpha\vec{B})} - \dot{\vec{P}}_t \xi^t_{(\nu)}$$

or

$$\eta^{\vec{P}}_{\vec{\Pi}} = \zeta^{\vec{P}}_{\vec{\Pi}} - \dot{\vec{P}}_t \xi^t_\nu$$

and

$$\eta = 1| - \dot{P}_t .$$

We then obtain

$$\theta^\beta_{(s)} = H^\beta_t \xi^t_{(s)} - \frac{\partial L}{\partial \dot{P}_\beta}\zeta^{\vec{P}}_{(s)}$$

(summation over s)

which reduces to

$$\theta^\beta_{(s)} = H^\beta_t - 1|\frac{\partial L}{\partial \dot{P}_\beta} = H^\beta_t$$

since L does not contain terms in $\dot{P}_\beta$.
We have assumed that $\frac{\partial \vec{P}}{\partial t}\delta\nu = \vec{B}\delta\alpha$.

The left hand side of Eq. (25) is zero on any extremal subspace (i.e., the Euler-Lagrange equations are obeyed). Thus, it takes the form

$$\frac{d}{dt^\beta}\{H^\beta_t + \phi^\beta_{\vec{\Pi}}\} = 0 .$$

Utilizing the Lagrange density given above, the Hamiltonian density, H, is proportional to the <u>total energy per unit volume</u> (Rund, 1966, p. 300) of the fluid; i.e., we have

$$H^4_4 = -\varepsilon \qquad \text{where } t^4 = t .$$

We may write

$$\frac{d}{dt}\{-\varepsilon + \phi^\beta_{\vec{\Pi}}\} = 0$$

where

$$\varepsilon \overset{def}{=} \text{ energy density of the plasmoid}$$

$$-\varepsilon + \phi^\beta_{\vec{\Pi}} = f(t^1, t^2, t^3) = f(x, y, z)$$

$$\int\{-\varepsilon + \phi^\beta_{\vec{\Pi}}\}d\tau = \text{constant}$$

$$- E + \int_{\vec{\Pi}} \phi^{\beta}_{(s)} d\tau = \text{constant (summation over s)}$$

$$E \overset{\text{def}}{=} \text{the total energy of the closed plasma structure}$$

$$\int_{(\vec{\Pi})} \tilde{\sigma} \, d\tau = E + \text{constant (summation over s)}$$

$$\int_{(\vec{\Pi})} \sigma \, f(\phi) d\tau = E + \text{constant} .$$

From the discussion above, it follows that the smaller the number of terms included in $f(\phi)$, the polynomial expansion of the constraint integrand, the lower the total energy of the closed plasma structure under consideration. If the symmetries are all exact, then the larger the number of symmetries included in the summation over (s) that are included, (the larger the number of constraint integrals) the higher the energy.

The arguments presented above apply specifically to the generalized gauge symmetries. For the space-time symmetries, the Lagrange density is invariant. Thus the independent integrals $\phi^{\beta}_{(s)}$ are identically zero for these transformations. The constants of the motion corresponding to the latter symmetries are also independent integrals, however, and can be written as divergences in the sense of Eq. (15). If we perform the space-time transformations, Eq. (25) can be shown (Rund, 1966, p. 296 and 297) to take the form

$$\frac{dH^{\beta}_{\epsilon}}{dt^{\beta}} = 0.$$

Any other independent integral of the form

$$\Phi_{(s)} = \frac{d\phi^{\beta}_{(s)}}{dt^{\beta}} = 0$$

can be added to

$$\frac{dH^{\beta}_{\epsilon}}{dt^{\beta}} = 0.$$

Since the constants of the motion corresponding to the space-time symmetries are of this form, the relationship be-

tween the total energy and the independent integrals is of the same type for these symmetries and the generalized gauge symmetries discussed above. We have here an alternate proof of the principle of least constraint (see II, page 663). If the total energy of the structure can be directly related to the free energy available to drive various types of instabilities, then this principle of least constraint can be in turn, related to the stability of the plasma structure. This relationship must be established independently for each type of plasma structure formed for the various possible combinations of constraint integrals (see I, page 45). In I and II we have described the force-free, collinear structure as a very low energy and very stable structure. This structure corresponds to exact gauge symmetry and totally broken space-time symmetry, i.e., the constants of the motions are

$$\int \vec{A} \cdot \vec{v} d\tau ,$$

$$\int \vec{B} \cdot \vec{v} d\tau ,$$

and

$$\int \vec{A} \cdot \vec{B} d\tau .$$

Linear momentum, angular momentum and total energy are not conserved. The resulting Euler-Lagrange equations are found to be

$$\nabla \times \vec{B} = \kappa \vec{B}$$

and

$$\vec{v} = \pm \beta \vec{B} ,$$

where κ and β are scalar constants.

If we write Euler's equation for a steady state, incompressible, fluid in the form

$$0 = -\nabla \left(p + \frac{1}{2} \rho v^2 \right) + \vec{j} \times \vec{B} - \rho (\vec{\zeta} \times \vec{v}) ,$$

the force-free, collinear equilibrium equation becomes

$$0 = - \nabla \left(p + \frac{1}{2} \rho v^2 \right) . \qquad (26)$$

It can be shown (see Shercliff, 1965, p. 187) that the complete thermodynamic energy equation can be written

$$\iiint \{ \rho \, \frac{D}{Dt} (U + \frac{v^2}{2}) - \vec{E} \cdot \vec{j} - div \, Q$$

$$- \, div \, \rho \vec{v} + viscous \, terms \} \, d\tau = 0 \, ,$$

where $\iiint \vec{E} \cdot \vec{j} d\tau$ is the rate at which electrodynamic energy is supplied to any instabilities growing in the structure, $\int div Q d\tau$ is the heat flux available for driving instabilities, $\int div \, \rho \vec{v} \, d\tau$ is the flow work available for driving instabilities and the viscous term is the energy available for resistive mode instabilities.

From Eq. (26) and the relation

$$U = \frac{3}{2} \, p,$$

where U is the internal energy of the fluid, we have

$$\nabla (\frac{2}{3} U + \frac{1}{2} \rho v^2) = 0 \qquad (27)$$

for the force-free collinear structure. But the rate of change of the energy available to drive all the instabilities (the free energy for our plasma model) is

$$\rho \, \frac{D}{Dt} (U + \frac{v^2}{2}) = \rho \, \frac{\partial}{\partial t} (U + \frac{v^2}{2})$$

$$+ \, \rho \vec{v} \cdot \nabla (U + \frac{1}{2} v^2).$$

For a steady state force-free collinear structure,

$$\frac{\partial}{\partial t} (\frac{2}{3} U + \rho \, \frac{v^2}{2}) = 0.$$

This implies

$$\frac{D}{Dt} (\frac{2}{3} U + \rho \frac{v^2}{2}) = \vec{v} \cdot \nabla (\frac{2}{3} U + \frac{1}{2} \rho v^2) = 0 \, ,$$

where we have used Eq. (27). We see that for this model, the lowest lying linear plasma state has no free energy available to drive instabilities.

Rewriting Eq. (24) in the form

$$\frac{d}{dt^{\beta}} \{ \theta^{\beta}(s) - \Phi^{\beta}(s) \} = 0$$

for an extremal subspace, we see that the exact form that the integrands will take depends not only on the amount of symmetry breaking present in a given physical situation, but also on the particular representation of the independent vector fields that is utilized in writing the Lagrange density and making the Lie group of transformations represented by Eqs. (9). The number of terms retained in the expansion $f(\phi)$ is determined by how badly the assumption underlying the symmetry is violated. For example, if there is some ohmic resistance present in the fluid, then the lines of magnetic induction will slip slowly through the fluid and the symmetry represented by the transformation $\delta \vec{A} = \delta \alpha \vec{B}$ will be weakly broken. This amount of line slip must be represented by the terms retained in

$$\int f(\phi) \vec{A} \cdot \vec{B} \, d\tau \, .$$

If there is no slip, i.e., an ideal fluid, the symmetry is exact and $f(\phi) = 1$, etc. If one uses a special representation of the fields, for example, the representation for azimuthally symmetric fields described in II, (Wells, 1970, p. 653), then the group parameters α_s in Eqs. (9) will be different and we have an entirely different representation of the same Lie group. This will, in turn, mean that the integrands in the constraints will change and there will not be a one to one correspondence in the integrands corresponding to the same degree of symmetry breaking. This is why we have written in II, page 655, Eq. (32),

$$I_{21} = \int T(\tilde{\omega}^2 P)' d\tau \, \sim \, \int \vec{A} \cdot \vec{B} \, d\tau$$

i.e., $\int \vec{A} \cdot \vec{B} d\tau$ replaces $\int T(\tilde{\omega}^2 P)' d\tau$ but the mapping is not one to one. The resulting Euler-Lagrange equations are the same, however, (compare Eq. (38) page 656 in II with Eq. (14), page 29 of I). This must be so since the two group representations have the same group algebra. It is interesting to note that the expansion utilized by Woltjer for azimuthally symmetric systems, i.e.

$$\sum_{n=0}^{\infty} a_n (\tilde{\omega}P)^n,$$

meets our requirement that

$$\nabla f(\phi) = \nabla \{ \sum_{n=0}^{\infty} a_n (\tilde{\omega}P)^n \}$$

is normal to the nested surfaces.

We have shown that Eq. (25) can be written in the form

$$\int \{ - \mathcal{E} + \Phi^{\beta}_{(s)} \} d\tau = \text{constant} \qquad (28)$$

$$(\text{summation over s})$$

where $\mathcal{E}$ is the total energy density and $\Phi^{\beta}_{(s)}$ is the integrand of an independent integral representing the sum of the constraints. These integrands have been assumed to be linear in the parameters, α_s. Indeed, for our special problem, we found that the integrands $\Phi^{\beta}_{(s)}$ took the special form

$$\frac{d}{dt} \left[\frac{\delta\alpha}{2} (\vec{A} \cdot \vec{B}) \right], \quad \frac{d}{dt} \left[\frac{\delta\beta}{2} (\vec{B} \cdot \vec{v}) \right] \quad \text{and}$$

$$\frac{d}{dt} \left[\frac{\delta\gamma}{2} (\vec{A} \cdot \vec{v}) \right]$$

where $\delta\alpha$, $\delta\beta$ and $\delta\gamma$ are proportional to the group parameters $\delta\alpha\vec{B}$, $\delta\alpha\vec{z}$ and $\delta\gamma\vec{v}$.

If Eq. (25) is considered to represent a relationship between the total energy of a closed plasma configuration and the appropriate constraints on the flow, then one can vary the integrands and replace the coefficients $\delta\alpha/2$ etc., with a new set of parameters, the Lagrange multipliers. These will now be proportional to the original group parameters. The old parameters α_s, and the first Lie group of transformations transform the arbitrary vector fields x^i, or equivalently ψ^α, into a subspace which represents equilibrium flow, i.e., the Euler-Lagrange equations are satisfied and the left hand side of Eq. (25) is identically zero. If we write the equation for this second variation, Eq. (28) takes the form,

$$-\delta \int \mathcal{E} d\tau - \delta \int \tilde{\alpha}_{(s)} \Phi^{\beta}_{(s)} d\tau = 0 \qquad (29)$$

$$(\text{summation over s}) .$$

For the case of the generalized gauge symmetries, this equation is in the form of Eq. (21) of II, i.e.

$$\delta E + \sum_{i=1}^{4} \sum_{n=0}^{\infty} \alpha_{in} \, \delta I_{in} = 0 . \qquad (30)$$

For the case of space time symmetries, $\Phi^{\beta}_{(s)}$ is zero, i.e., the Lagrange density is invariant with respect to the space time transformations (Calkin, 1961). Thus, if constants of the motion corresponding to space time symmetries are to be incorporated into the variations, they must be added separately to Eq. (30). As we observed above, however, this does not affect any of the present arguments.

Equation (29) is a special case of "the problem of Lagrange" (Rund, 1966, p. 323). In this problem, the curves which afford extreme values to the fundamental integral are required to satisfy certain subsidiary conditions. These may be in terms of first order differential equations or, in our case, constraint integrals.

The Lagrange multipliers $\tilde{\alpha}_{(s)}$ are a new set of parameters corresponding to a second Lie group of transformations that take the equilibrium vector fields of the first variation and transform them into the special subset of fields describing the equilibria corresponding to the constrained fields. These may or may not be relatively stable, depending on the number of constraints employed and the depth and slope of the corresponding total energy well. The variational calculation of Woltjer and Chandrasekhar (see I and II) is simply a method of incorporating the Lie group of transformations on $\vec{P}$ and the time into another second group of transformations of the equilibrium fields into constrained equilibrium fields.

It is interesting to note that the reason each integrand in the constraints must be expanded in an analytic series if one is to recover all possible equilibria is now quite clear. We have already determined all possible equilibria when we obtained the fundamental variational formulas, i.e., Eq. (6) and (7). When we seek constants of the motion by performing the Lie group of transformations given by Eq. (9), we are projecting out a subgroup of constrained equilibria. When we perform the second group of transformations given by Eqs. (29) and (30), we make a second projection to a manifold corresponding to a further specialization of constraints. When we again demand all solutions of Eqs. (6) and (7), we must reverse the whole projection process by complete expansions of our constraint integrals. This reverse process is of practical interest only for fields close to minimum energy when one is studying the rate of change of the stable structure for weakly broken symmetries.

We should note here that if one performs the second variation to solve our problem of Lagrange using many terms in an expansion of one or more of the constants of motion (independent integrals) corresponding to the gauge symmetries, combined with a constant of the motion corresponding to say one exact space-time symmetry, the resulting energy may be higher than that resulting from another situation in which all symmetries of whatever type are nearly exact. In this sense, the energy is not a monotonically decreasing function of the number of classes of constraints employed.

The details of these transformations for the case of the exact gauge symmetries is most easily developed in terms of the "formal charge density" defined by Eq. (17) and the corresponding formal charge, Q.

In a field-theoretical formalism, a formal charge Q is defined as the space integral of the zeroth component of a local four-vector current,

$$Q(x_0) = \int dV_3 j_0(x) \ . \tag{31}$$

These quantities appear in discussions of symmetries and broken symmetries in quantum field theory and are used in the "current-algebraic" approach to particle theories.

In quantum field theory, relations of the type given by Eq. (31) have rather troublesome convergence properties. There are other major difficulties in quantum theory which do not arise here as long as the Lagrangian and the fields remain unquantitized.

ACKNOWLEDGMENTS

The author wishes to acknowledge helpful discussions with Dr. Joseph Wojtaszek and Dr. Joseph Norwood, Jr., of the Department of Physics, University of Miami, and Dr. Behram Kursunoglu and Dr. Geoffrey I. Iverson of the Center for Theoretical Studies, University of Miami, Coral Gables, Florida.

This work was supported by the Air Force Office of Scientific Research Propulsion Division under grant number AF-AFOSR-72-2295, the United States Air Force Cambridge Research Laboratories, under contract numbers USAF F 19628-69-C-0072-P001, and by the National Science Foundation under grant number GP-43733.

REFERENCES

Calkin, M.G., Thesis, University of British Vancouver, British Columbia, 1961.

Calkin, M.G., Canadian Jour. of Phys. 41, 2241 (1963).

Caratheodory, C., Acta Szeged Sect. Scien. Mathern 4, 193 (1929).

Hammermesh, M., Group Theory, Addison Wesley, 1964.

Hill, E.L., Revs. of Modern Phys. 23, 253 (1951).

Kursunoglu, B., Modern Quantum Theory, W.H. Freeman & Co., 1962.

Nolting, E., P. Jindra, and D.R. Wells, Jour. Plas. Phys. 9, 1 (1973).

Orzalesi, C., Rev. of Mod. Phys. 42, 381 (1970).

Rund, H., The Hamilton-Jacobi Theory in the Calculus of Variations, D. Van Nostrand Co., Ltd., London, 1966.

Shercliff, J.A., A Textbook of Magnetohydrodynamics, Pergamon Press, N.Y., 1965.

Wells, D.R. and J. Norwood, Jr., J. Plasma Phys. 3, 21 (1969).

Wells, D.R., J. Plasma Phys. 4, 645 (1970).

Wells, D.R., Phys. Fluids 9, 1010 (1966).

Wells, D.R., Cooke, F.N., Tunstall, J. Nolting, E., Jindra, P., and Hirschberg, J., Phys. Rev. Letters 33, 1203 (1974).

Weyl, H. Ann. Math. (2) 36, 607 (1935).

THE PLASMA FOCUS EXPERIMENT. SURVEY ON THE PRESENT STATE OF THE RESEARCHES AND POTENTIAL FUSION APPLICATIONS

Ch. Maisonnier, F. Pecorella, and J. P. Rager

Laboratori Gas Ionizzati

(Ass. CNEN - EURATOM)

Frascati, Italy

ABSTRACT

This paper gives a survey of the Plasma Focus activities, summarizes the present results and knowledge, and considers potential scaling up to thermonuclear applications.

The initial development phase (1960-1970) has mainly dealt with the building of experimental facilities in the 5-150 kJ range of stored energy and the study of the plasma focus (PF) as a powerful source of D-D neutrons and X and gamma-rays. This phase has given the scaling laws of these devices in which the neutron yield varies as the square of the value of the stored energy.

The second phase (after 1970), is mainly concerned with the study of the physical properties of the plasma and neutron emission mechanisms, resorting to more sophisticated diagnostic techniques. The main achievement has been the recognition that the neutron production was not associated with the existence of the dense phase, but with that of a turbulent one, at the beginning of which intense electron beams are observed.

In the Filippov type devices it is thought that this turbulent mechanism actually heats up a diffuse plasma to temperature in the 5-10 keV range, and that neutrons are mainly due to thermonuclear D-D reactions. In Mather type devices, the situation is not as clear; the existence of several different regimes has been put in evidence; in some of these, intense ion beams are responsible for the neutron production.

Prospective for thermonuclear scaling can now reasonably be considered, on the basis of thermonuclear mechanisms. On both experiments, it is shown that the efficiency of energy transfer from the energy stored in the condenser to the electron beams is of the order of 10-20 per cent.

The scaling law for optimized devices is shown to vary approximately as the square of the value of the stored energy divided by the square of the final plasma radius. The only efficient ways to achieve greater thermonuclear efficiency are to go to higher energy levels or to decrease the hot plasma radius.

Assuming that the scaling of an optimal experiment is performed, the break-even condition is reached at a 300 MJ stored energy level, which places the reactor in the 1 GJ range.

TWO METHODS OF SPACE-TIME ENERGY DENSIFICATION[*]

Harry L. Sahlin

Lawrence Livermore Laboratory, University of California

Livermore, California 94550 U.S.A.

ABSTRACT

With a view to the goal of net energy production from a DT microexplosion, we study two ideas (methods) through which (separately or in combination) energy may be "concentrated" into a small volume and short period of time--the so-called space-time energy densification or compression. We first discuss the advantages and disadvantages of lasers and relativistic electron-beam (E-beam) machines as the sources of such energy and identify the amplification of laser pulses as a key factor in energy compression. The pulse length of present relativistic E-beam machines is the most serious limitation of this pulsed-power source. The first energy-compression idea we discuss is the reasonably efficient production of short-duration, high-current relativistic electron pulses by the self interruption and restrike of a current in a plasma pinch due to the rapid onset of strong turbulence. A 1-MJ plasma focus based on this method is nearing completion at this Laboratory. The second energy-compression idea is based on laser-pulse production through the parametric amplification of a self-similar or solitary wave pulse, for which analogs can be found in other wave processes. Specifically, the second energy-compression idea is a proposal for parametric amplification of a solitary, transverse magnetic pulse in a coaxial cavity with a Bennett dielectric rod as an inner coax. Amplifiers of this type can be driven by the pulsed power from a relativistic E-beam machine. If the end of the inner dielectric coax is made of LiDT or another fusionable material, the amplified pulse can directly drive a fusion reaction--there would be no need to switch the pulse out of the system toward a remote target.

INTRODUCTION

DT MICROEXPLOSIONS

It is useful to consider the conditions required to attain net energy release from a DT-fusion microexplosion, because these define particularly demanding space-time energy compression goals. For purposes of discussion we will assume as the primary energy source, a 1-MJ capacitor bank with a 1- to 10-μsec quarter-cycle time when driving an appropriate load.

The theory of high-density fusion has been expertly reviewed by Linhart.[1] To attain the $\rho r = 1$ g/cm^2 burn condition for uncompressed solid-density fuel requires input energy of about 10^9 J in time of 2×10^{-9} sec to a pellet with an approximately 1-cm radius, i.e., a power of 5×10^{17} W and power flux of 5×10^{16} W/cm^2. An initial value $\rho r < 1$ can be increased by convergent compression to $\rho r = 1$ because ρ increases as the compression ratio η, while for spherical compression r varies as $1/\eta^{1/3}$, and thus ρr is proportional to $\eta^{2/3}$. For $\eta = 1000$, the uncompressed ρr of the solid-density target can be $\rho r = 0.01$. Consequently, the total energy requirement can be reduced by a factor of 10^6 to about 1000 J, the energy delivery time decreases to 2×10^{-11} sec and the power requirement to 2×10^{14} W, and the power-flux remains unchanged.

If compression is achieved by the Nuckolls ablation-driven rocket,[2] then the efficiency of energy coupling to the compressed-pellet core is only about 5%, making the energy and power requirements given above too low by a factor of 20. More efficient compression is possible as noted by Hora,[3] and in Section 5 we will discuss the possibility of 100% efficient ablation-free compression.

The $\rho r = 1$ criteria arises from two independent factors: the requirement that inertial confinement times be equal to the burn time, and the demand that the α particles from the DT burn are deposited in the fuel. It is possible to independently satisfy these two criteria, and thus obtain exothermic burn conditions for ρr of the fuel <1. The inertial confinement time can be increased by means of a tamper, or by transient containment of the hot plasma in a strong field such that the microinstability time of the contained plasma is equal to its burn time. On the other hand, a magnetic field with Br = 1 MG-cm may couple burn α's to the system

---

[*]Work performed under the auspices of the U. S. Energy Research and Development Administration.

even though $\rho r \ll 1$.

In a tamped system with $\rho r < 1$, a field is required to reduce electron conduction losses to the tamper. It may also be possible to catch the low-density ablated fuel used to compress the pellet core in a strong field for a Bohm diffusion time and achieve yield from this ablation product. In fact it may be easier to obtain net yield from the contained corona than to burn the dense core produced by the ablation process. The idea that it is easier to reach exothermic burns of the rocket fuel than to achieve breakeven energy release from the rocket is suggested by study of the plasma focus.

It may also be possible to decrease the input energy to obtain a net DT yield by relatively unexplored direct conversion of fusion α to field energy in a field-contained explosive with $\rho r < 1$. For example, a pellet that is relatively transparent to its own α particles and surrounded by a strong magnetic field will permit the energetic α's to diffuse across the field more easily than the electrons. As a result a net charge will develop on this quasi-contained pellet to the point where the α loss time is increased and the crossfield electron diffusion time is decreased so that a quasi-neutral plasma wind becomes possible. This results in the radius of positive charge becoming slightly larger than that of the negative charge, and as a result causes electrostatic implosion of the pellet. In addition, the DT-burn-driven anomalous diffusion of the target electron across the magnetic field might appear as a negative resistance and amplify the confining field.

At least one of the above means, or an as yet undiscovered equivalent trick, is mandatory to reduce the energy and power requirements well below the 10^9 J, 5×10^{17} W values for an inertially confined, uncompressed solid pellet.

ENERGY SOURCES

The pulsed power sources that most nearly approach the energy and power requirements for driving a microexplosion with a potential yield $Y \geq 1$ MJ are the laser and the relativistic E-beam machine. Existing E-beam machines have greater pulse energy and higher efficiency than high-power pulsed lasers, but also have a lower power level due to their 30- to 200-

pulse length and smaller power flux density--there is greater difficulty in focusing large relativistic currents, compared to the relative focusing ease of the laser. It is often argued that an advantage of the laser relative to the relativistic electron-beam pulsed power source is that the range for deposition of the laser light is shorter than that for relativistic electrons.

It is presently possible to convert 5% of a relativistic E-beam energy to collectively accelerated ions and 15% of the energy to few-cm microwaves.[4]

The efficiencies of these processes are likely to increase soon, and efficient conversion to microwaves can probably be extended to wavelengths shorter than 1 cm. Thus microwave pulses and energetic ion bursts must be considered as pulsed power sources that are competitive with relativistic E-beam and pulsed lasers for fusion-microexplosion applications.

Short-wavelength microwaves have only recently been explored seriously[5] for laser applications because of the widely accepted but erroneous belief that far shorter wavelengths are necessary, and that a submicron laser is ideal for microexplosion applications.[6] We will suggest that the "brand x" laser should actually be in the 0.01- to 1-cm wavelength range, and thus modern extension of well established microwave technology could be a more effective source of the coherent radiation in this wavelength range than a long-wavelength laser (maser). Combined use, by preexcitation of a mode of a microwave generator cavity with a maser followed by amplification of this field energy of the mode with a relativistic electron burst, might provide an effective hybrid pulsed power source.

The present low efficiency of the pulsed laser relative to that for the production of relativistic electron bursts is a transitory advantage of electron beams. The efficiencies of lasers will increase, and no doubt approach the values of present E-beam machines. On the other hand only small improvements in E-beam efficiency are possible, and efforts to reduce the multinanosecond pulses of these machines into the subnanosecond range will certainly result in some reduction in source to E-beam energy conversion efficiency. The ability to focus laser light is also not a very fundamental advantage, since present lasers can be focused to spot sizes substantially smaller than the pellet diameter, while experimental efforts to focus relativistic electron bursts[7] have achieved results approaching the focusings required to reach the diameter of a microexplosive having a potential yield at 20% burn efficiency of 1 MJ. The electron bursts obtainable from a 1 MJ plasma focus will have a diameter in the range 0.1 cm > d ≥ 100 μm, and thus may even exceed the required E-beam focusing.

The argument that deposition of relativistic E-beam energy is a problem relative to the range of laser light in a target cannot be maintained. The reason for this statement is the fact that E-beam, pulsed power sources that approach breakeven requirements will need a ν/γ of ≥ 100, and thus the energy is primarily in the field rather than in the kinetic energy of the electrons. When a plasma is subjected to an electromagnetic field the E field is screened out in a Debye length $\lambda_d = v/\omega_p$, where ω_p is the electron plasma frequency, while the B field is screened in a collisionless skin depth $\Delta = c/\omega_p$, a distance larger than λ_d by a factor c/v. Thus, from the point of fiew of the plasma, the electromagnetic field is an oscillatory B field, i.e., in effect one is dealing with an ac pinch.

Coupling of the field energy is not governed by a corona with a density corresponding $\omega = \omega_p$, where ω is the field frequency, but by the corona density that can be held by the radiation pressure of the field at a temperature where the plasma becomes collisionless. The resistivity is then governed by strong turbulence. We have estimated this critical field intensity to be B = $2\pi \times 10^7$,[8] sufficient to contain a corona with density n = 5×10^{22} at a temperature of 1 kV; at this temperature and density, the plasma becomes sufficiently collisionless to become strongly turbulent.

The greatly increased resistivity of the skin layer in the turbulent state results in rapid heating of the skin layer, and the skin's driven Bohm-like diffusion across the field is an ablation process that multiplies the field's radiation pressure by a factor of order v_α/v, where v_α is the Alfvén velocity for inward transport of field energy in the low-density plasma outside the corona, and v is the ion thermal velocity of the

material ablated from the dense corona surface. Since the Alfvén velocity v_α approaches c, the speed of light, the ablation amplification of the field pressure becomes a factor of order 100, identical to the results for laser driven ablation.

The anomalous resistivity leading to the ablation is due to Langmuir turbulence, i.e., to the current driven amplification of the skin-layer thermal electron plasma oscillation, and thus the frequency associated with energy deposition is not the electromagnetic driving field frequency ω, but the plasma frequency ω_p of the driven corona. ω_p is proportional to $\sqrt{n_e}$, where n_e is the corona density and is determined by the radiation pressure of the driving field, not by the frequencies of this field.

Thus, one realizes that large-energy-density fields result in skin currents that can drive strong turbulence in the corona whose density depends on the field's radiation pressure, and the energy transport to the pellet is due to field diffusion governed by the turbulent resistivity. These skin currents are in effect dc phenomena on the time scale of ω_p of the corona, and the Langmuir oscillation can become large in the time required for a few plasma oscillations. Thus for $\omega < \omega_p$, in first approximation, effective coupling of field energy to the pellet is governed by the driving field's intensity, not by its frequency. This frequency-independent equivalence of fields of the same energy density is valid provided $\omega < \omega_p$, and the field power flux density approaches 10^{15} to 10^{16} W/cm² , the value required to drive a fusion microexplosion.

It is an accidental coincidence that the critical density for 1-μm Nd radiation of n = 10^{21} is high enough that radiation pressure effects at current power flux densities make a difference of only about a factor of 2 in corona density, and thus energy coupling occurs at approximately the place where $\omega = \omega_p$. It is apparently this accidental coincidence that has given rise to the widespread belief that short-wavelength radiation is required for effective energy coupling without superthermal electrons because energy is deposited where $\omega = \omega_p$.

This accidental coincidence is not scalable to other field frequencies, since a decrease in frequency at constant Poynting vector power flux density will still result in a dense corona (where the radiation pressure now plays a dominant rather than an insignificant role), and since the Langmuir turbulent-field diffusion rate (which governs the rate of energy transport from the field into skin layer heating) is governed by ω_p of the corona; effective energy deposition at fixed field intensity will continue as the driving field frequency is decreased.

The most extreme form of this argument is achieved by reducing the field frequency to the point where the burn time is a quarter cycle time of the field, and the "ac pinch" becomes an ordinary dc pinch. For dc pinch it is well known that the field plasma interface occurs at the point of pressure balance $B^2/8\pi = 2nkT$, provided the resistivity is not strongly anomalous, and thus the Alfvén velocity (snowplow velocity) exceeds the field diffusion velocity.

If we consider as an example a 10-kV plasma, then a field of $2\pi \times 10^7$ G will be in pressure balance with the plasma for $n = 5 \times 10^{21}$, for which the burn time is 2×10^{-8} sec, corresponding to a quarter-cycle time of a field with frequency $\omega = 2.5\pi \times 10^7$. If we assume energy deposition occurs at $\omega = \omega_p$, the corresponding electron density for the plasma target would be $n = 2 \times 10^6$, a factor of 2×10^{15} less than the actual density of the skin layer, the location of the real resistive heating. This most extreme case clearly indicates that in general the idea that energy is deposited in a corona at or near $\omega = \omega_p$ is fallacious.

If the use of very-short-wavelength radiation is not fundamental, then we are led to ask what, if anything, does determine the optimum field frequency? If one goes beyond the question of energy deposition and considers stability, then it is possible to arrive at an optimum frequency range. Ensley[5] has shown that the interface between the field and the pellet is stabilized if the wavelength of the field is equal to or greater than the pellet radius. Evidence for instabilities for $\lambda \ll R$ has been provided both by actual laboratory experiment and numerical simulation computer experiments.[9] On the other hand, the dynamic stabilization of MHD instabilities[10] such as the $m = 0$ instability demand a field frequency ν such that $1/\nu < R/v$, where R

is the pellet radius and v is the ion thermal velocity, which for 10-kV burn condition becomes $v \geq 10^8$ cm/sec. Thus we can conclude that the wavelength γ of the driving field should be in the range $R \leq \gamma \leq 100\ R$. We conclude that γ should be in the range: 250 µm to 1 cm.

Thus the pulse power problem becomes that of producing pulses in this frequency range with total energy $E \leq 10^6$ J, at a power of $E/T > 10^{14}$ W, and flux $E/TA \geq 10^{16}$ W/cm^2, where A is the surface area of the uncompressed solid-density DT pellet.

OUR PROPOSED ENERGY SOURCE

As a result of the above discussion we are left with only the short pulse length as a unique advantage of the laser as a pulsed power source for driving fusion microexplosion. This advantage is fundamental and cannot be dismissed. The object of this paper is to present two methods that separately or in combination can approach the short pulse requirements with a nonlaser source. The production of small-diameter electron bursts at a pulse length less than that of existing E-beam machines by self-interruption and relativistic restrike of the current of a pinch due to the sudden onset of a strongly turbulent anomalous resistance is the basis for the first energy densification idea.

The short-pulse production in a laser amplifier is the particularly unique feature that results in the attractiveness of the laser for pulsed fusion application. In essence this process is the amplification of a self-similar or solitary wave in an active medium and results in an effective pulse width $\tau \ll L/C$, where L is the length of the energy storage device, in this case the active medium. The pulse width resulting from this parametric amplification is much shorter than the maximum passive delivery time τ of the energy stored in a reservoir of length L, which is $\tau \geq L/C$.

Parametric amplification of a solitary wave is not unique to the laser but can be formulated for a much broader class of wave processes. The most important idea contained in the present paper is a specific means of amplification of a transverse magnetic solitary wave in a coaxial cavity with a relativistic electron beam. If the idea presented here proves viable in practice then presently unique laser-pulse amplification can be extended to microwave generation technique in the optimum wavelength range 250 μm to 1 cm.

Section 1 of the paper discusses briefly the concepts of transit time amplification, the production of strong turbulence in a plasma, the resulting collapse of Langmuir solitons, and the idea of Alfvén and Lawson limiting currents. These concepts are central to the idea developed in the remainder of the paper. Section 2 presents the idea for short-duration intense relativistic-electron burst production in a plasma by self interruption of a pinch current. Section 3 discusses a specific target for the short electron burst produced by a 1-MJ plasma focus. In Section 4 analogs between fundamental atomic and molecular process and their macroscopic counterparts, high-Q cavities, are developed. The results of Sections 1 and 4 are then used to develop the second energy-compression idea, parametric amplification of a solitary wave pulse. The mathematical details of this parametric solitary-wave amplification are given in the Appendix, and Section 6 summarizes the paper.

1. SOME BASIC CONCEPTS

In this section we will discuss briefly several concepts that are fundamental to developments in later sections of this paper. Specifically we will discuss the idea of a transit-time oscillator, and, with the aid of the result, introduce the concept of strong turbulence and the Zakharov collapse of plasmons, and also introduce the various limiting currents for electron beams.

TRANSIT-TIME OSCILLATOR

The idea of a transit-time oscillator is an important cornerstone for the study of parametric amplification. In simplest form one can imagine a beam of randomly distributed electrons passing through small holes in a pair of capacitor plates so that the electron current path is perpendicular to those plates.[11] The capacitor is assumed tied to an inductor to form a microscopic harmonic oscillator in the form of an L-C resonant circuit with frequency $\omega = 1/\sqrt{LC}$, where L and C are respectively the inductance and capacitance of the system. An oscillating E field is present between the plates and directed perpendicular to them, parallel to the electron beam threading the plates through the small hole drilled in each plate.

Thus the separate electrons of the beam experience a force along their line of motion pointing either in the direction of motion or against the direction of motion, depending on the phase of the L-C circuit during the transit of the electrons between the plates. For simplicity we consider nonrelativistic electrons. Those electrons that pass through the plates when the E field is pointed in a direction opposed to that of their motion have work done on them, being accelerated at the expense of the energy of the oscillator. On the other hand, electrons that transit the active region between the plates when the E field points in their direction of motion do work on the field increasing the energy of the L-C circuit at the expense of their kinetic energy. Electrons that do work on the field are slowed down and have an increased transit time relative to that of the electrons when the oscillator is unexcited. However, an electron encountering the oscillator in the opposite phase will accelerate, thus decreasing the transit time relative to that for the case of an unexcited oscillator.

If the transit time of an electron is in the vicinity of an odd number of half cycles, then about half the electrons will undergo varying degrees of acceleration, while the other half undergo varying amounts of deceleration. Because of the larger transit timt of the decelerated electrons than the accelerated electrons, the net effect of the electrons in the beam, even for a completely random distribution, is to increase the excitation level of the L-C circuit.

A string of such L-C circuits with variable capacitor plate spacing to compensate for changing transit time provides a simplified model of a microwave generator, where, in such a generator, the analog of our L-C circuit becomes a mode of a resonant cavity. Nonlinear effects of such a string of oscillators can act to bunch the initially randomly distributed electrons in the beam. The bunching greatly increases the effectiveness of coupling between the beam and the oscillator. If the bunches encounter the oscillator with phase such that the bunched electrons decelerate, putting energy in the field, then the oscillator string behaves like a microwave generator, and if the beam becomes bunched with opposite phase then the bunches are accelerated at the expense of energy stored in the oscillators, and the system becomes a model of a linear accelerator.

In the case of the relativistic electron the spacing of the plates does not have to be changed because the transit time of relativistic electrons is not changed by small gains or losses of electron energy. In this case, if the beam consists of a set of bunches that do work on the oscillators and an alternate set of bunches that receive energy from the oscillators, one has a system analogous to that involved in autoacceleration of portions of a relativistic electron beam at the expense of energy supplied by other portions of the same beam.[12]

The oscillators in these examples can represent specific modes of macroscopic accelerator cavities, or one can think of them as the equivalent harmonic oscillator corresponding to transition between two energy levels of "microscopic cavities," i.e., atoms or molecules. If the L-C circuit in the above example is initially unexcited it will nonetheless possess a zero-point oscillation and may also have shot-noise-like thermal fluctuations. The small random field will become transit-time amplified by the electron beam, and thus it is possible to amplify the thermal or zero-point quantum noise to obtain macroscopic coherent excitation of the oscillators. Of course, the thermal or zero-point excitation will also be present when the oscillator is macroscopically excited.

ZAKHAROV COLLAPSE OF PLASMONS

If one considers as a model of a plasma a space-filling set of nonoverlapping Debye spheres, then in first approximation the ions in a sphere constitute a fixed "uniform" neutralizing background. Each thermally excited electron will oscillate with random phase and direction in the field of the ions and the remaining electrons. The motion, like that of a rock dropped in a hole drilled through the earth, will be harmonic. It will have an amplitude proportional to $\sqrt{T}$, where T is the temperature. The electric field due to the similar independent thermal motion of all the electrons in the Debye sphere will have an average value of zero and an rms value proportional to $\sqrt{NT}$ where N is the number of electrons in a Debye sphere.

Now if a string of randomly distributed electrons constituting a current with drift velocity equal to the thermal velocity of the Debye-sphere electrons passes through the Debye sphere, the electrons' transit time is $1/\omega_p$ and it can cause transit-time amplification of initially random thermal-plasma oscillation. Since the population of electrons in a Debye sphere is changed in a plasma period, the amplified oscillation will convect to a neighboring Debye sphere, where amplification can continue if the diameter of the electron current is greater than the Debye length. Nonetheless it is clear that the electron-ion collisions, and random thermal motion will act to damp coherent oscillation produced by transit-time amplification. Thus to obtain a significant nonthermal excitation of the plasma oscillation, the density of electrons in the driving current must not be too much smaller than the electron density of the medium.

Numerical simulations show that for a beam density greater than 1% of the background plasma density the plasma oscillation can be driven to a large amplitude in less than 100 plasma periods, provided the plasma temperature is high enough to make the electron-ion large-angle collision frequency less than the electron plasma-frequency. This transit-time amplification of plasma oscillation is in effect the Buneman[13] or two-stream instability. For large-amplitude oscillation the nonlinear effect can act to bunch the electrons of the driving current.

The amplified plasma oscillations correspond to a large coherent electric field in the Debye sphere, and this field can accelerate electrons out of the sphere, resulting in a space charge that will also cause ions to be expelled. If the plasma oscillations are strongly driven, the plasmon pressure that remains after damping of the wave by particle expulsion from the Debye sphere is large enough to more than compensate for the thermal pressure of the expelled particles. For this condition the system is unstable towards developing a local cavity of decreased density that is held open

against the particle pressure of the surrounding medium by radiation pressure of
the coherent Langmuir oscillation of the
cavity.

Virtually any low-entropy, intense
energy source can drive the plasma oscillation, and thus the collapse of Langmuir
oscillation first predicted by Zakharov[14]
is a relatively universal phenomenon in
strongly driven plasmas. Nonlinearities
in frequency space provide the means by
which energy is coupled across the density
and hence the plasma-frequency gradient
as a cavity develops.

Both laboratory experiment and computer simulation have now firmly established the reality of the collapse phenomenon,[15] and local density depressions
approaching a factor of 2 have been
observed experimentally. The cavity
development, which tends to have a diameter of about 10 λ_d, or roughly the order
of the collisionless skin depth, will stop
decreasing in density despite the decreased damping in the density-depressed
region because the nonlinearities may not
be strong enough to couple energy across
too great a change in local plasma frequency.

It is nonetheless instructive to consider the extreme state of collapse where
the cavity density goes to zero and the
longitudinal plasma excitation is replaced
by the transverse lowest-standing-wave-
mode excitation of the vacuum cavity in
the plasma, and the radiation pressure of
the excitation holds open the cavity
against the particle pressure of the surrounding medium. In this limiting case
we can speak of the local cavities or
"plasma-contained microwaves."

For a strongly driven medium the
number of initially randomly distributed
collapse regions will become large, and
this can lead to conglomeration of cavities into a larger collapse region. The
location of the bubbles that arise as the
separate cavities fall into each other
should be determined by the state of minimum dissipation, i.e., the condition of
highest cavity Q. If the cavity is in
the center of the surrounding experimental
cavity we have the plasma-contained
microwave state. If on the other hand
the resonator has a very high Q, e.g.,
because it is a superconductor, then it
may be energetically favorable for the
bubble to develop between the supercon-

ducting wall and the plasma, thus turning
the plasma-contained microwave state
inside out to produce a microwave-con-
tained plasma.

We suspect that the state of minimum
dissipation may correspond to confining
the plasma in a shell between a centered
and a surrounding electromagnetic field
standing-wave bubble. It is of considerable interest to study this system to see
if the state of minimum dissipation is in
fact the one postulated here because we
are dealing with a phenomenon that is
unstable towards forming in a strongly
driven plasma.

There is a problem associated with
this conceptual cavity limit where one
approaches vanishingly small particle
density. The longitudinal plasma oscillations are scalar waves and can be a
node-free excitation in a singly connected
region. On the other hand the electromagnetic-wave excitation of a particle-
free cavity satisfies a vector equation,
and must have an axial nodal line in a
singly convected region such as a sphere.
Thus as the collapse of the plasma
density goes to completion a nodal line
must develop, and plasma will tend to
accumulate in this nodal region, rather
than be expelled.

For this reason the natural limiting
case for complete collapse may not be a
spherical-Bessel-function spherical-
harmonic product that is an eigenstate of
a simple spherical cavity and has an axial
node. Instead the mode may be a spheri-
cal-Neumann-function spherical-harmonic
product that can exist only if there is
a dense-plasma axial stem along a diameter of the cavity, which in fact provides a collapse that is multiply connected, i.e., a torus or doughnut with a
small hole, where the doughnut is a
plasma-free region of strong standing-
wave excitation and the plasma fills the
space not occupied by the doughnut-shaped
standing wave.

In the case where the collapse is
driven by a relativistic electron beam,
the return current that forms along the
electron beam path when the relativistic
beam is injected into the plasma will be
expelled from the relativistic beam for
current densities high enough to drive
strong turbulence. The return-current
expulsion will occur in the form of nonrelativistic filaments, and this return-

current expulsion is known as the Weible instability.[16] If the path of the initial relativistic beam's return-current composite has a region of slightly lower density than the surrounding plasma, then the return-current expulsion will occur in the local region to form a pseudo-spherical region with a relativistic current along a diameter of the depressed density sphere, and the return-current filaments will flow in the spherical surface of the region of depressed density.

The return currents will drive the Langmuir turbulence, which will cascade into the depressed-density spherical region, and the resulting field will expel electrons and ions to further depress the density. In the zero-density limit one obtains a spherical cavity with nonrelativistic return currents in the form of filaments flowing on the vacuum-plasma interface, with the relativistic beam forming an axial current along the diameter of the sphere, and the sphere will contain a standing-wave excitation whose radiation pressure will hold open the spherical cavity and at the same time confine the axial relativistic beam to a tightly pinched small-diameter path. It is this type of spherical collapse that we postulate as the form the Langmuir collapse in three dimensions must approach. Such a cavity is similar to a toroidal vortex[17] with the added feature that the doughnut-shaped region of vanishingly small particle density is held open by a standing-wave electromagnetic excitation.

The confinement of the relativistic beam to a small-diameter thread passing through the center of the standing-wave-excited vacuum bead also facilitates the conceptual complete collapse. It may lead to an actual complete collapse, because this axial relativistic current can now directly drive the cavity excitation, rather than indirectly cause collapse by driving plasma oscillations in the surrounding dense plasma that must then cascade into the lower density collapsing region due to nonlinearity of the intense plasma oscillation. The indirectly driven collapse will not go to completion because the nonlinearities will be ineffective in driving plasma energy across too steep a density gradient. The cavity directly driven by a radiation-pressure-confined, axial relativistic electron beam does not depend on the nonlinearity to supply the

energy to drive the collapse.

At a high level of excitation it will become energetically favorable for random local collapse regions to become a regular array of collapse beads threaded by the axial relativistic-electron-beam string. Such a threaded string of collapse beads is a natural linear acceleration structure for relativistic electrons, and if formed in a plasma with an appropriate density gradient becomes a string of variable-diameter, relativistic-E-beam-threaded beads that could provide coherent linear acceleration of ions.

Collapse in the form of a string of cavities has been suggested recently by several theoretical studies and also has been found in a experimental study.[19] The new point contained in the present discussion is the argument that in the case of collapse driven by a relativistic beam, the beam will become an electromagnetically pinched axial current along the cavity diameter. Again, such structure is one of particular interest precisely because such a collapse phenomena appears to be mandatory for strongly driven plasmas.

LIMITING CURRENTS FOR ELECTRON BEAMS

In the remainder of this section we turn our attention to the topic of limiting currents for relativistic electron beams. A beam of charged particles possesses kinetic energy per unit length proportional to number of particles per unit length. When the beam exists in a conducting plasma, both the electric and magnetic fields are screened out of existence. On the other hand, such a beam in a plasma-free region will have electric and magnetic field energy proportional to the square of the number of particles per unit length. In the absence of space-charge neutralization, the current can be space-charge or Langmuir-Chiles limited, and in the case of space-charge neutralization the current can become magnetic-field or Alfvén-Lawson limited.

Relativistic electron beams are particularly interesting because the magnetic field is less than the electric field by a factor v/c and an ion density $n_i = (1/\gamma^2)n_e$ is all that is necessary in the lab frame to balance the electric and magnetic fields of the beam. In the above expression, γ is the relativistic contraction factor $\gamma = 1/\sqrt{1 - \beta^2}$, where $\beta = v/c$; v is the electron drift velocity, c is the speed of light, and n_e is the density of the electron beam as viewed in the lab frame.

One must consider the limiting currents in two different contexts. First the question arises as to whether a particular current configuration can exist without being cut off by its self-field, independent of the question of how the field and particle energy were initially established. A second question arises when one asks how the current was produced in the first place.

If the only source of field energy is the kinetic energy of particles emitted from a field-screened region, e.g., from the cathode of a diode on the relativistic E-beam machine, then the electric and magnetic fields of the resulting particles cannot exceed the initial kinetic energy of the particle that produced this field, and as a consequence the total field energy and the total particle energy in an impedance-matched diode are roughly equal, i.e., the system has a "global" ν/γ of 1 even though the local ν/γ may be much higher than 1. The factor ν/γ is the ratio of the field energy to the kinetic energy of a beam of charged particles.

Ordinary currents have very high values of ν/γ, e.g., $\nu/\gamma > 10^4$, while highly relativistic beams have $\nu/\gamma \leq 1$. The value of ν/γ is given by Nr_0/γ where r_0 is the classical electron radius, and $N = 2 \times 10^8$ I is the number of electrons per unit length, where I is the beam current in amperes. The factor γ should actually be written as $\gamma - 1$, which subtracts the electron rest energy from its "relativistic kinetic energy."

We will consider first the limiting currents in the case of space-charge neutralization so that one need consider only the magnetic field. For a solid beam of radius a with concentric return-current path of radius R the Alfvén or Lawson limiting current is $I = 1.7 \times 10^4$ $\beta\gamma$ amperes. Physically this is the current whose self-field becomes strong enough for the Larmor radius of a single electron moving in the self-consistent beam field to become equal to $\frac{1}{2}(a)$, so that the beam field is able to prevent further progress of the electron, i.e., the beam is stopped by "self magnetic resistance."

For the concept of this and related limiting currents to be applicable, it is necessary for the beam to be sufficiently collisionless that the electron-cyclotron frequency ω_c exceeds the electron-ion large-angle collision frequency ω_{ei}. For beams with $\omega_{ei} > \omega_c$ the system is collision dominated, the resistance is due to particle collision rather than to the beam's self field, and the concept of the Alfvén or Lawson limit does not apply.

If one considers a hollow cylindrical shell beam of radius R and thickness Δ surrounded by a concentric cylindrical return path where, e.g., Δ could be the collisionless skin depth $\Delta = c/\omega_p$, then the limiting current becomes $I = 1.7 \times 10^4$ $\beta\gamma$ R/Δ, a factor R/Δ larger than that of the solid beam. In this case the limiting current corresponds to the Larmor radius of a beam electron in the self-field of the beam becoming equal to $\Delta/2$, i.e., the beam is limited by its own "self magnetic resistance." A beam with both B_θ and B_z fields corresponding to a force-free configuration has essentially no Alfvén or Lawson limit.[19]

If we consider the emission of a relativistic beam from a cathode into a vacuum region, then the particle kinetic energy provides the source for the field energy, and the diode current can be both space-charge limited and magnetic-field limited. We will assume that a sufficient number of ions are evolved from the anode, the cathode, and the residual background gas (which typically has a pressure $p \leq 10^{-4}$ Torr) to provide space-charge neutralization so that we need consider only the magnetic field. If the cathode of such a relativistic-beam machine has a radius R, then the impedance matched anode-cathode spacing for the diode corresponds to a distance d such that the outermost electrons of the beam will be moving tangential to the anode. The diode current corresponding to this condition is $I = 8500$ β $\gamma R/d$ amperes, the impedance Z of the diode is $Z = 30$ (ν/γ), and the ratio $R/d = \nu/\gamma$.

A plasma exists at both the cathode and the anode, and the electrons are field emitted from the cathode plasma. When the beam enters the anode plasma the remaining E-field will be screened out in a Debye length and the magnetic

field will be screened out in a collisionless skin depth, i.e., a return
current is generated back along the
relativistic beam as it penetrates into
the anode plasma and this return current
flows radially outward in the skin layer
of the growing anode plasma. A self-
consistent anode plasma will result that
is contained by the magnetic field in the
diode, prevented from penetrating deeply
into the anode plasma because of the
presence of a return current. The
temperature of the anode plasma will
increase, eventually reaching a value of
the order of 1 kV so that it becomes
collisionless, and then the return
current is able to drive the Langmuir
oscillation resulting in an anomalous
resistance that diffusively rips the
return current out of the relativistic
current path, permitting the field to
penetrate the anode in a small-diameter
channel surrounding the relativistic
current.

This complex sequence of events
requires a careful and extensive two-
dimensional simulation to properly follow
the initial development of the return
current--the self-consistent evolution of
the anode plasma against the excluded
magnetic field, the onset of strong turbulence as the anode plasma reaches a
temperature such that it is collisionless,
and the resulting axial field penetration.
A complete self-consistent simulation of
this process has not been made to date.
We believe that the self-consistent
evolution of the anode plasma will result
in the relativistic beam pinching to the
density of the anode, with field penetration in a concentric channel along
the pinched beam, where the channel
diameter is of the order of a collisionless skin depth of the anode. Such a
channel will be unstable toward development of a set of Langmuir collapse beads
strung along the pinched relativistic
beam. One interpretation of recent
microscopic study of anode targets is
that such a string of cavities forms
along current filaments that have pinched
to solid or near-solid density.[20]

2. PRODUCTION OF RELATIVISTIC ELECTRON BURSTS VIA A PLASMA FOCUS

ENERGY STORAGE

To provide space-time concentration
of energy by passive switching of the
energy content of a reservoir, it is
necessary both to have a high energy
density, and a large effective sound speed
of the energy storage medium. The
highest sound speeds attainable approach
the speed of light and occur when the
energy storage medium is a vacuum
electromagnetic field in a cavity.

The desirability of high energy
density favors energy storage in magnetic
fields rather than electric fields, since
magnetic fields do not have magnetic pole
sources while the sources of the electric
fields are charged particles. Thus large
electric fields tend to be limited by
electrical breakdowns, while magnetic
fields experience this problem only in
second orders of v/c due to induced
electric fields resulting from any rate
of change of the magnetic field. Thus,
in efforts to produce short-duration,
high-energy-density, pulsed power sources,
one is naturally driven to consider
inductive storage, and the desirability
of a large effective sound speed dictates
that the magnetic field exist in either
a low-density plasma or ideally in vacuum.

Despite the limitation on energy
density, one advantage of energy storage
in capacitors is the fact that switching
energy out of a capacitor bank involves
the relatively straightforward problem
of closing a switch to establish current
flow to a load. Switching an inductive
storage device requires solution of the
more difficult problem of opening a
switch, i.e., interrupting the current
flow on one path, followed by restrike of
the current at higher voltage along a
different path that carries the stored
energy to the load. This asymmetry in
case of switching is a direct consequence
of the asymmetry in field source, i.e.,
the fact that electrons exist but magnetic
poles apparently do not exist.[21]

ION AND ELECTRON ACCELERATION

The earliest efforts on Project
Sherwood concentrated on Bennett Z-pinch
experiments,[22] and resulted in unexpect-
edly large neutron yields that proved to
be due to fast-ion acceleration, which was
often correlated with m=0 instabilities.
The fast ions were usually accompanied by
hard x rays due to bremsstrahlung from
accelerated electrons, and also by high-
frequency microwave bursts whose origin
is more difficult to explain. Considerable effort to associate the ion acceleration with IL voltages due to MHD

instabilities, such as the m = 0 sausage instability, have not been fully success- ful, and it is now clear that microinsta- bilities, strong turbulence, and the associated large anomalous resistance play a central role, and that m = 0 instabilities tend to be the natural region for nucleation of Langmuir and other turbulent phenomena.

Although a large variety of mecha- nisms for ion and electron acceleration in plasma, beam plasma, and beam neutral- gas systems are known,[23] two specific types of acceleration occur in plasma that are of primary importance in general, as well as to the specific objectives of this paper. Type I acceleration is associated with the turbulent-resistance interruption of an existing current, often in a pinch, followed by restrike of the current with more energetic electrons. Type II acceleration occurs in processes related to establishing a current in an initially field-free region by, for example, injection of a relativistic beam into a target plasma. In the Type II process the initially induced return cur- rent is later expelled from the relativi- stic-beam path by the turbulent resisti- vity associated with amplification of Langmuir oscillations.

Both Type I and Type II acceleration can be associated with the Langmuir col- lapse of plasma due to a very high level of driven plasma oscillations.[24] In both the Type I and Type II acceleration schemes, the ions can be given an energy 10 to 25 times the internally applied voltage. The distinguishing feature of Type I acceleration relative to Type II is that the Type I ion acceleration is in the direction of the externally applied field, while the Type II ion acceleration is in the direction of relativistic cur- rent flow and opposite to that of the internally applied potential. This lat- ter type of ion acceleration is often termed collective acceleration.

We wish to suggest that the Type I and Type II acceleration are actually basically the same process because in each case the nonrelativistic current is instrumental in driving Langmuir turbu- lence. The apparent differences arise because the nonrelativistic current in Type I acceleration is driven by the externally applied potential, while in the case of Type II the nonrelativistic current is driven by the induced electric field due to the changing magnetic field of the advancing primary relativistic electron burst, and flows in a direction opposite that of the relativistic beam, and thus against the electron flow caused by the externally applied potential.

PLASMA PINCH: PLASMA FOCUSSING

The most extreme form of the classic pinch is provided by the r-Z pinch of a plasma focus, which in a sense can be thought of as a giant m = 0 instability. For detailed discussion of the plasma focus, consult the review article by Mather.[25] The detailed structure of the energy of the focus and its release into heat at maximum pinching is intimately associated with the force-free vortex filaments studied by Bostick[26] and co- workers.

A fully correct, detailed theory of electron or ion burst production in a plasma focus must involve these filaments and their creation and annihilation in a fundamental way. In the present discus- sion we will emphasize the importance of strong turbulence in the fast-particle production in the focus. The sudden onset of strong turbulence near maximum pinching and a resultant 2-3 order-of- magnitude jump in resistivity is now firmly established experimentally.[27] The onset of the turbulent resistance must be intimately associated with the annihila- tion of Bostick filament pairs. In the present note our description will ignore this filamentary structure. We nonthe- less believe that the present description treats the particle acceleration pheno- mena, at least in gross detail, in a correct way. We hope to present a detailed theory that fully incorporates the filaments and associated filament- annihilation process in a fundamental way in a future publication.

The current-sheet snowplow of the plasma focus driven supersonically, relative to the low-density field gas, by JxB forces, results in half of the bank energy going into inductively stored energy behind the snowplow, while half of the energy goes into particles. This particle energy occurs half in the form of directed kinetic energy of the snow- plow, and half goes into direct heating of the snowplow ions and indirect heating of the electrons through electron-ion collision. The thermal energy is carried away both by hydrodynamic flow of the

snowplowed plasma parallel to the snow-
plow density step and perpendicular to
its directed motion, and also by radiation processes: bremsstrahlung plus
bound-bound and bound-free radiation of
high-Z partially ionized impurity ions in
the current sheet. The net result of
these processes is to produce at maximum
pinch a dense plasma column with density
of 10^{19} to 10^{20} with an ion temperature
of about a kilovolt and an electron
temperature that is less than the ion
temperature by a factor of 2 to 10. The
pinch current in a megajoule system
driven by a 40-kV capacitor bank will be
several megamps.

Two modes of operation, a high
density mode and a low density mode,
result from different initial fill-gas
pressures.[28] The basic distinction
between the two modes is due to the fact
that in the high density mode the diffusion of the skin currents into the back
side of the snowplow is slower than the
snowplow growth, and thus at maximum
pinch the high-density shock front leads
the magnetic field, and at maximum pinch
a relatively field-free, dense plasma
"wire" is formed on axis. By decreasing
the fill-gas pressure, the relative ratio
of field diffusion into the snowplow is
increased, and the magnetic-field front
and the shock front tend to coincide, so
that at maximum pinch one has a hollow
plasma shell with density of about 10^{19}
with a low-density center of ionized fill
gas, which typically will have a density
$n \le 7 \times 10^{17}$ corresponding to an initial
fill-gas pressure of $\le$ 1 Torr.

As maximum pinching is approached,
the energy demand exceeds the supply rate
from the capacitor bank and the system
begins to draw on the inductive energy
behind the snowplow. This is accomplished
by the addition to this applied bank
voltage of an IL voltage with the same
direction as the external field. The
energy supply is provided by $L\dot{I}$, where $\dot{I}$
is not localized but occurs throughout
the system.

The production of fast particles,
both electron and ions, in a plasma focus,
particularly when operated in the low-
density mode, is now well documented
experimentally.[29] A number of theoretical
explanations of this and similar phenomena
in vacuum sparks[30] have been published.
The complex behavior that follows the
development of maximum pinching has been
clarified due to the early work of
Bostick et al.,[31] and recently by double-
exposure holography in a number of laboratories. The work of Bernard et al.
deserves special mention.[32]

These measurements show that there
is sudden onset of Langmuir oscillations,
and simultaneously the dense plasma stem
at maximum pinch disappears, starting at
the anode face and proceeding away from
the anode. The result of this sequence
of events is a larger diameter plasma,
with a diameter a factor of 10 greater
than the diameter at maximum pinch, and
with density of 10^{18} to 5×10^{17}, or two
orders of magnitude smaller than that at
maximum pinching. The energy per particle
in this lower density plasma approaches
10 to 20 kV, and in the high-density mode
the plasma may not differ too greatly
from a Maxwellian. The primary contribution to the neutron yield of the focus
is a "quasi-thermonuclear" reaction in
this low-density hot plasma. In the low
density mode, substantial electron and ion
acceleration occurs, the accelerated ions
are concentrated along the axis, and the
neutron yield is substantially of a non-
thermonuclear beam-target nature. Experimentally the largest neutron yields are
produced in the high density mode and
the yield falls off as the fill-gas
pressure is decreased. It is normally
concluded that the high density mode
favors neutron production. The conclusion
must be carefully reexamined in view of
the fact that sufficiently energetic ions
$E \ge 1$ MeV are produced so that the yield
may fall off simply because there is not
enough target gas to stop the fast ions.

Another factor that reduces the yield
at low pressures is ion starvation, i.e.,
the supply of ions available to be accelerated becomes limited and the ion-starved
condition should favor electron acceleration. Low-density-mode operation of a
1-MJ system should result in about 10%
of the bank energy going into a small-
diameter, 10-nsec relativistic electron
burst, and 1% of the bank energy may go
into an oppositely directed axial MeV
ion burst. Careful efforts directed
toward deeper understanding and optimization of the electron and ion burst
production could result in increasing
the coupling efficiency to relativistic
electrons and to energetic ions by about
a factor of 3.

A further increase in efficiency requires the elimination of the snowplow assembly, which squanders half of the bank energy in the form of particle kinetic energy, and then only 50% of the bank energy exists as inductive energy storage behind the snowplow. It is the inductive energy that drives the pseudo-thermonuclear yield in the high-density mode, where up to 60% of the inductive energy is converted into relatively random 10-to-20-kV ions; in the low-density mode at least 20% of the inductive energy goes into fast-ion and electron production during a period of 10 nsec or less.

The rise times of the electron and ion bursts in the low density mode are very short and may be in the 100-psec range. The actual initial diameters of the electron and ion bursts are unknown. The diameter must be at least as small as the focus, i.e., a diameter of the order of 1 mm. Work by Bostick et al.[33] suggests that the diameter of these energetic particle beams could be as small as 100 μm.

In order to understand the focus phenomena we first consider the current sheet well before maximum pinching. As discussed in Part 1, the Alfvén or Lawson limiting current for a hollow electron beam is $I = 1.7 \times 10^4 \, \beta\gamma \, R/\Delta$, where R is the beam radius and Δ the collisionless skin depth c/ω_p: c is the speed of light and ω_p is the electron plasma frequency. The current that leads to self pinching in a vacuum diode of radius R and anode-cathode spacing d, in the case of space-change neutralization, is $I = 8500 \, \beta\gamma \, R/d$. These concepts apply for a drifting beam in the case where the electron-cyclotron frequency for electrons in the self-field of the beam is greater than the electron-ion large-angle collision frequency, and then the "resistance" is field rather than collision dominated.

In the case of the collisionless limiting current for the diode the electrons are first accelerated by the external field at the surface of the cathode plasma. These fast electrons then are decelerated to provide the field energy, and consequently the field energy of the diode must be of the same order as the initial total kinetic energy of electrons undergoing transit between the cathode and anode.

In either the case of self-field-dominated resistance or in the case of collision-dominated resistance, the presence of an axial E-field can drive the particles against the resistance, continuously resupplying the energy to the particles, or they give their energy to the field or to plasma heating, and in this way the field energy can greatly exceed the initial kinetic energy of the drifting electrons. The resulting high-ν/γ beam can continue to exist in the collisionless case in the absence of an applied voltage if the current field configuration is less than the appropriate Alfvén or Lawson limiting current, i.e., if the configuration is relatively force free so the self magnetic resistance of the system is small.

The current is collision dominated during most of the collapse phase of the plasma focus. As maximum pinch condition is approached the sum of the bank voltage and the IL contribution to the voltage accelerates current-carrying electrons, between large-angle collisions with ions, to approach a drift velocity comparable to the random thermal velocity of the electrons, and one reaches the condition for transit-time amplification of the longitudinal plasma oscillation. The amplified plasma oscillation rapidly reaches a significant amplitude, and causes turn-on of a turbulent resistivity in at most a few dozen plasma periods. The resistance increases due to the observed plasma oscillations by a factor of 100 to 1000, and this results in a rate of diffusion of the field into the snowplow plasma that approaches 5×10^8 cm/sec, or about 10 times the rate of snowplow motion. The sudden increase in resistivity greatly reduces the particle current, resulting in a large displacement current, and the field diffusion through the plasma produces rapid heating.

EXPLODING PLASMA

Thus the pinched plasma in the high density mode is in effect transformed into an exploding plasma wire, and in the low density mode into an exploding plasma shell, i.e., the rapid particle heating drives the particle pressure well above the local field pressure, and the plasma undergoes driven Bohm-like diffusion across the field, further increasing its temperature. The density of the plasma drops by a factor of 100, the radius increases by a factor of 10, and the energy per particle (i.e. the effective temperature) increases by a factor of 10 or more to a value of 10-20 kV. More than half of the inductive energy stored behind the snowplow is transformed into heat, and the hot, expanded low-density plasma achieves near pressure balance with the remaining field, where it can be contained for the order of a Bohm diffusion time, during which the primary yield of the focus is produced.

The explosion of the plasma wire or plasma shell must be accompanied by a momentum-conserving recoil, i.e., a convergent implosion of some of the plasma toward the axis. Thus one can look upon the explosion of the plasma wire as an ablation process that causes implosion. It is not clear if the creation of a dense plasma on axis by the implosion recoil to the outward-directed plasma has been observed experimentally, although it is possible that observation of repinching reported in some experimental studies is due to this process.

The ρr of the imploded portion of the current sheet in the plasma focus is far too small to produce significant yield. The surprising fact is that the exploding portion of the plasma is caught by the remaining magnetic field and the yield actually comes from the rocket fuel rather than from the convergently compressed rocket recoil with larger ρr. The high-density-mode yield from a 50-50 DT plasma in a 1-MJ focus will approach 0.5% of the bank energy, 1% of the inductively stored energy, or about 2% of the field energy that is destroyed to produce plasma heating. This leads to the conclusion that it is easier to produce net yield from the exploded rocket fuel than it is to obtain yield from the imploded rocket.

If this exploding-plasma-wire phenomenon can be applied to the same number of particles but at an initial density of about 10 times solid density, then the initial density is increased by 10^4, the burn rate is increased by 10^4, the disassembly time is decreased by a factor of 10^2, and the net yield from the ablated rocket fuel becomes 100 times larger, thus approaching breakeven relative to the 1-MJ bank energy. By directing the electron burst from the low density focus mode

over a small diameter ($d = 0.01$ cm), solid-density DT wire,[68] it may be possible to achieve scaling approaching that of the above example.

The recoil compression of the central portion of the wire even in this case will have a ρr too small to give significant yield in the absence of an exotic effect such as direct conversion to field energy of the α particles produced by fusion. However if the exploding-wire-surface, imploding-wire-core phenomenon is extended to the condition where the field energy initially possesses 10 to 100 MJ, the wire core can reach about 100-fold compression at $\rho r = 1$ burn conditions and one may obtain breakeven yield from both the imploded core and the exploded corona. In addition, the detonation of the core will drive a shock outward through the field-contained corona and can produce additional yield by "refrying" this corona.

The high-density-mode, exploding-plasma-wire phenomenon is analogous in a number of respects to ordinary or relativistic-E-beam-driven exploding-wire work[34] and the exploding plasma shell in the low density mode has similarity to exploding cylindrical foils.[35]

In the low density mode the field diffusion precedes the inward imploding half of the plasma shell and in the low density core the restrike of the hollow cylindrical current is initially impossible because the current would exceed the Alfvén limit $I = 1.7 \times 10^4$ $\beta\gamma R/\Delta$ until the induced axial E field (in other words, the displacement current) is able to increase the value of $\beta\gamma$. As the plasma-foil explosion proceeds outward from the anode, the unexploded portion of the plasma cylinder will act as a plasma cathode that will emit a relativistic electron beam that has $\nu/\gamma = 100$, and this beam will self pinch because of space-charge neutralization in a distance d about 1/100 of the plasma cathode radius, producing a virtual cathode. Because the electrons are temporarily prevented from propagating due to the beam's self-field, the normally more sluggish ions have a chance to respond to the axial field, and ions are accelerated until the ion-containing region between the anode and the unexploded end of the plasma shell are exhausted; the system becomes ion-starved and the space-charge neutralization is destroyed, and then the current is reestablished between the virtual cathode and the anode as a small-diameter relativisitic electron burst.

The triggering of the complex sequence of events that leads to the ion and electron burst is provided by the acceleration of electrons of the current shell to a drift velocity equal to the electron thermal velocity. Since an applied voltage will distribute itself proportional to the resistance, both the material and shape of the anode center can produce significant variation in the voltage drop across the skin layer of the metal anode, and this anode-skin-layer potential drop subtracts from the voltage applied across the pinch, and consequently can alter ("fine tune") the point during final pinching at which amplification of Langmuir oscillation begins, and causes the sudden increase in resistivity of the current sheet.

The different hydrogen-adsorption properties of different anode-center materials could alter the availability of ions, and thus change the ratio of energy that goes into the electron burst and the oppositely directed ion burst. A hollow anode should favor ion acceleration over electron-burst production in low-density-mode operation. The presence of impurity ions in the current sheet changes the radiation rate, and this affects the extent to which the electron temperature lags behind the ion temperature, and this in turn can change the time of onset of strong Langmuir turbulence. Turbulent enhancement of bound-bound and bound-free impurity radiation could also be important.

As pointed out by Gribkow,[36] one can enter the region of strong turbulence either by accelerating the current-carrying electrons to a drift velocity equal to the thermal velocity, or by reducing the thermal velocity to a value equal to or less than the drift velocity. The rapid cooling that will occur when the current sheet reaches the rod, if a small-diameter rod is placed on axis extending from the anode, may help to trigger the strong turbulence by quenching the electron thermal energy.

Careful experimental study of the variables could lead to increased efficiency of electron and ion burst production, and probably also permit increase of the size of the ion burst relative to that of the electron burst, and vice versa.

The rate of increase of turbulent resistivity could be very large, the change by a factor of 100 or more taking place in as few as 10 electron plasma periods, i.e., in a time $\tau = 100$ psec. However in most cases the strong turbulence will probably start at one point on the current sheet, and then spread through a full 2π angle. Thus the cutoff of the current will usually require several nanoseconds instead of only 100 psec. In occasional high-symmetry shots, the 100-psec cutoff time may be approached, and this could explain the occasional very large, energetic, very-short-duration electron burst that is observed experimentally. Effort to achieve high-symmetry with respect to angle of onset of strong turbulence and current cutoff could help in increase the intensity, and decrease the rise time of the electron burst. A rod extending from the anode center with a disk mounted perpendicular to the rod end, where the disk has a razorsharp edge, might achieve the desired result for highly symmetrical collapsing current sheets.

Since the current sheet actually consists of many small filaments, the unexploded plasma shell end, which acts as a cathode, should actually emit a large number of separate electron beams, one from each filament end. It is tempting to visualize as an orderly phenomenon the self pinching of these separate beams to form a virtual cathode, so that each current stream winds up in a coil equal to a Larmor radius, to produce a toroidal vortex-like structure with a circulating current greater than the net pinch current. The decay of toroidal vortices of that type have been observed by Bostick and co-workers.[37] The model for the process suggested here is presently under study.

3. A TARGET FOR THE PLASMA-FOCUS ELECTRON BURST

In this section we consider some key features of one type of target to be utilized for the focus-produced relativistic-electron burst. An extended treatment of this and other targets will be presented elsewhere. The purpose of the present discussion is to provide an answer to the question, "Why produce relativistic electrons?" The self interruption and relativistic restrike of the plasma-focus current discussed in the previous section is, in effect the transformation of a "super-high"-ν/γ ordinary nonrelativistic current into a high-ν/γ relativistic current with $\nu/\gamma = 100$. What is the fundamental use of such a ν/γ reduction?

The answer to this question is best provided by discussing the electron-burst target shown along with the plasma-focus anode in Fig. 1. The target is a hollow glass shell with an inside diameter of roughly 1/2 cm, with a dielectric axial rod of DT, LiDT, or perhaps CDT with diameter $d \leq 0.02$ cm. The space inside the sphere is filled with DT gas at a pressure of 1 atm or less. The collapsing current sheet of the focus will wrap around the sphere as shown, and produce a pinch of axis beyond the end of the axial target rod just outside the glass shell. For low-density-mode operation this will result in restrike of the current as a hollow, relativistic electron beam that will travel along the surface of the dielectric rod, by virtue of the phenomenon discovered by Bennett et al.[38] A nonrelativistic return current will be generated and it will flow back along the relativistic current path. Thus, the initial consequence of the transduction of the pinched nonrelativistic current into a relativistic beam is that the relativistic beam is able, by virtue of its greater electron inertia and energy, to penetrate into an initially current-free region, and because of the induced return current the new charged-particle path remains current free.

What in effect has been created is a zero-inductance current loop folded back upon itself. By appropriate choice of the gas pressure in the sphere, it is possible to achieve plasma conditions along the relativistic-nonrelativistic current path for the return current to drive longitudinal Langmuir oscillations, once the plasma becomes hot enough along the current path to prevent strong collisional damping of plasma oscillation, i.e., when the temperature reaches roughly 1 kV.

The onset of strong turbulence results in the Weible instability expelling the return current as a set of filaments from the return relativistic path until they flow along the inner surface of the glass shell. The shape of this target has been chosen to have the doughnut-shaped limit of spherical Zakharov-Langmuir collapse discussed in Section 1. The net effect of driving the return current filaments from the relativistic current path on the surface of the dielectric rod is to rapidly fill the sphere with 10^5 J of field energy that was initially stored outside the sphere.

<u>It is this switching of field energy, and only this switching of field energy, that is the real benefit of restrike of the super-high-ν/γ nonrelativistic pinch current as a high-ν/γ relativistic current.</u> The ν/γ of the field energy inside the sphere will be of order 100, and thus the effect this energy has on the central rod will be due to turbulent diffusion of the field into the rod, <u>not the deposition of single relativistic electrons</u>. We believe that this example contains the <u>real</u> <u>essence</u> of the usefulness of high-ν/γ relativistic-electron bursts to pulsed fusion.

The pressure of the field switched into the cavity will cause the dielectric rod to undergo an insulator-to-metal transition, become an exploding-imploding wire and, as indicated in the previous section, there is a chance at the 10^5 J level of obtaining nontrivial yield from

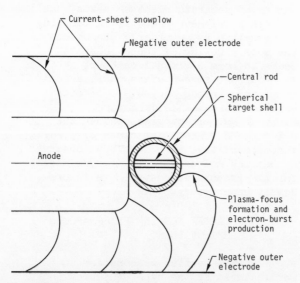

Fig. 1   Plasma-focus anode and
         electron-burst target.

the low-density-ablated corona even though the ρr of the compressed rod core may well be much less than the burn value ρr = 1. It is presently unclear if the Langmuir collapse will drive the fill-gas plasma density to zero, faster than the rod explodes. If this turns out to be the case, then the explosion-implosion of the rod will be driven by standing-wave electromagnetic energy of one or more modes of this multiply connected rod-sphere cavity.

It is amusing to note that the target discussed here is in effect a miniature version of the imploded-belt-pinch concept of Velekov,[39] except that we are interested in destroying the central rod and he wants the rod to remain intact long enough to burn the compressed belt pinch. By changing the objective of his implosion scheme to destruction of a central fusionable dielectric rod, a means of scaling the present target to the multi-megajoule driving energy range might result. Such an approach must be preceded by study of the Bennett rod phenomenon over periods of time of 10^{-3} sec, and also at high fields, to learn where the ability of the rod to carry high surface currents is destroyed and the rod becomes an exploding-imploding wire. For the scheme to work the dielectric rod phenomenon must persist to field pressure of at least 1 to 10 Mbar, where the insulator-to-metal transition will take place, in order that sufficient field energy is created by the slow implosion process in the Velikov scheme prior to onset of rod destruction, so that enough energy is available in the field for the explosion-implosion phase to reach burn conditions both in the imploded core and in the field-contained exploded corona.

4. ATOMS AND GIANT ATOMS

In this section we wish to present a picture that will facilitate our discussion in the following section of parametric amplification of solitary waves. We wish to emphasize the parallel between the fundamental processes in atoms and molecules and analogous processes in macroscopic analogs that are most naturally provided by superconducting cavities such as those developed for linear accelerations at Stanford.[40]

To work toward optimum energy compression concepts it is useful to begin by considering fundamental atomic processes and the results obtained by combining these processes in various ways. The discrete energy levels of an atomic system may be excited from the ground state by a photon of the appropriate energy, by an inelastic collision with an electron or other charged particle, or by a combination of radiation and charged particles acting together. It is even possible to excite energy levels with an uncharged particle such as a neutron. Level excitation and deexcitation are subject to selection rules that are a consequence of symmetry considerations, which determine the allowed processes connecting various energy levels. The excitation of atomic energy levels by inelastic scattering was first seen in the famous Frank-Hertz experiment.[41]

An excited state of an atom has a finite lifetime because of its coupling to zero-point field fluctuations and charge-density fluctuations of the vacuum that are the cause of spontaneous decay. An excited level can decay by photon emission, an Auger process in which energy is transferred to electron, by an electron capture by the nucleus, or in numerous other more complex, less probably ways. In the present discussion we will concentrate on level excitation and deexcitation by single photon emission or absorption, or by inelastic electron scattering.

It is nonetheless instructive to keep in mind the full range of excitation-deexcitation mechanisms, not only for bound-bound transitions, but also for bound-free and free-free transitions of atoms, as well as the even richer electronic and nuclear transitions of diatomic and polyatomic molecules and the optical and acoustical branch transitions of giant molecules, such as single crystals, because each of these fundamental processes can have a macroscopic analog in our giant-atom superconducting cavities, and the microscopic-atom, macroscopic-atom parallel is a useful tool for thinking on either the microscopic or the macroscopic level.

In addition to the spontaneous decay of an energy level, the level decay can be induced by a photon of appropriate energy corresponding to $\hbar$ times the difference frequency of the two levels ω_{12}. The stimulated emission is the inverse of the photon excitation. A particularly important aspect of the stimulated emission process is that the Bose nature of photons favors growth of the final state. This also applies to deexcitation by plasma emission, or photon or other excition-emission processes that are possible for an atom in a plasma, solid, or other dielectric medium.

Energy-level decay by the inverse of the inelastic Frank-Hertz excitation mechanism exists, and is oftern referred to as a collision of the second kind or a superelastic collision. In this case, a free electron is accelerated at the expense of the excitation energy of the atomic level. Processes involving electrons do not favor growth of the final state because electrons obey Fermi statistics, and it is this fundamental difference that makes lasers possible, while the analogous amplification phenomenon for single-electron wave functions is not presently known to exist.

It is possible to consider deexcitation of a level with a two-electron state, with the electrons in either a singlet or a triplet state. The wave function of these quasi-Bose particles can be amplified, and a near perfect realization of this process is provided by certain Josephson-junction phenomena in the gap between two superconductors. The growth of charged Bose-particle final states also requires appropriate charge neutralization to prevent the amplification process from being space-charge limited.

Near perfect analogs of the fundamental process are provided by considering superconducting cavities as giant atoms, whose longitudinal standing-wave excitations provide the analog of atomic energy levels. The analogy approaches perfection because the macroscopic quantum states of the quasi-Boson electron pair in a superconductor find their most appropriate description in terms of states, known as coherent states, that are eigen functions of the annihilation operator. The eigen functions in this case have the forms of single-particle wave functions, known as the complex order parameters, and apart from a self-consistent potential term obey a single-particle Schroedinger-like equation known as the Ginsberg-Landau equation. The details of this picture are provided in the Appendix, in conjunction with the mathematical formulation of the solitary wave amplification that is presented in Section 5.

The Q of our macroscopic-atom superconducting cavity is a natural analog of the spontaneous decay of an atom. The excitation of a standing-wave cavity mode by means of a traveling wave, produced, for example, by a microwave generator, provides the analog of photon excitation of an energy level, and the use of the stored energy in the cavity to feed a traveling wave provides an analog of stimulated emission. Electron acceleration and deceleration by the field of a transverse magnetic standing-wave mode, depending on the field phase during the electron transit time of the cavity, provides the parallel to Frank-Hertz level excitation and superelastic level excitation collision, or collision of the second kind.

With laser light it is possible to excite a coherent superposition of nearly degenerate atomic energy levels, and the decay of such a state has considerable structure due to the interference terms between different states of the coherent superposition. The mode-locking of nearly degenerate cavity levels provides a macroscopic analog of this process.

A string of giant-atom cavities may be excited by a random beam of electrons so that no particular phase relation exists between cavities, a process analogous to creating a population inversion in a lasing medium with an incoherent electron beam. The deexcitation could in both cases be by a stochastic electron acceleration process. An appropriately phased, bunched electron beam can coherently excite a cavity string, and the coherent deexcitation of the cavities with bunches of the opposite phase is a linear accelerator. An analog of bunching for a single electron is probably possible for the microscopic analog, provided the density of electrons occupying the electron wave function is less than two particles per phase-space volume $\hbar^3$, at which point Fermi statistics play a decisive role. The more natural analog of a bunched beam requires the use of quasi-Boson electron pairs. The cavity string can be coherently excited by means of a traveling wave, and a string of atomic resonators can be similarly excited by use of coherent laser light. The linear acceleration of an electron or of a relativistic ion by such a coherently excited atom string via a phased string of superelastic collisions has been considered in an unpublished Laboratory report by Hofsteadter, and some closely related work has appeared in the Russian literature.[42]

An initially randomly distributed electron beam passing through a string of coherently excited cavities can become bunched by nonlinear effects. An incoherent electron beam scattering from a crystal of atoms in the ground state will produce a diffraction pattern that is the Fourier transform or reciprocal lattice of the crystal. This is possible with incoherent electrons (the Davison-Germer experiment) or with incoherent x rays because the periodic array acts like a passive filter, selecting the accidentally coherent electrons or photons of the initial beam to produce the diffraction pattern. It is of considerable interest to study incoherent electron scattering from a periodic crystal array of coherently excited atoms, because in this case the crystal can act as an active filter and one should obtain an analog of bunching. An initially incoherent electron beam interacting with an active generalized crystal, produced by interference in an active medium of a reference wave and a wave scattered from some object, is of particular interest. The active generalized crystal in this case is a structured population inversion or active three-dimensional hologram whose reciprocal lattice is a three-dimensional image of the object.

One process is significantly absent in present practice from our analogy. The most important means of deexcitation of excited microscopic cavities, i.e., atoms, is provided by stimulated emission. The amplification of a traveling wave by the standing-wave energy stored in a string of linear acceleration cavities is a perfectly viable process that has not to our knowledge been considered previously. Such a macroscopic stimulated-emission process provides the ultimate in series/parallel switching of energy storage reservoirs, and forms the basis for the discussion in the next section.

5. PARAMETRIC AMPLIFICATION OF SOLITARY WAVES

In this section we develop the second and most fundamental of our energy-compression ideas. In section 1 we identified laser pulse amplification as the feature primarily responsible for the unique advantage of lasers for pulsed fusion applications.

PULSE AMPLIFICATION IN LASERS

The pulse amplification is obtained by first Q-switching and then mode-locking an oscillator, i.e., a laser with mirrors, to produce a string of very short temporal width pulses. One of these pulses is switched out of the oscillator and sent to an amplifier that is a pre-inverted volume of atoms of the same species as those of the oscillator. The population inversion is provided by means of flashlamp, electrons, etc., driven by a primary energy source, such as the 1-MJ capacitor bank discussed previously. The amplifier is a laser without mirrors. The initial pulse, if it is intense enough, can extract a significant fraction of the energy stored in the active medium, and in a single pass become amplified in a self-similar fashion to very high intensity. A self-similar nonlinear solution of a wave equation is known as a solitary wave[43] or soliton.

One type of pulse with self-similar behavior arises in the study of the concept of self-induced transparency.[44] If an intense pulse passes through an initially unexcited medium, energy can be extracted from the front of the pulse to excite the medium oscillators, i.e., atoms, molecules, etc., and the back half of the pulse can then stimulate emission back into the pulse from the coherently excited atoms. In this way a self-similar pulse shape evolves that can pass through the normally opaque medium in an almost lossless fashion known as self-induced transparency.

For the losses to be small the medium must be pure, and the transient energy storage in excited state of the medium oscillator must be short relative to the spontaneous-decay time of the excited energy levels. The spontaneous decay will result in gradual damping or decreases in amplitude of the solitary wave, assuming that other loss mechanisms are negligible. If a separate energy source is employed to re-excite levels that decay spontaneously, then compensation for the spontaneous-decay damping is provided and the pulse will in principle undergo lossless transmission through the medium.

If the separate energy source excites energy levels so that more energy is provided than is lost by the spontaneous decay or other loss mechanisms, then the self-similar pulses can experience inverse damping, i.e., become parametrically amplified. Thus the process of pulse amplification in the laser amplifier is in effect self-similar parametric amplification of a solitary wave by externally driving the medium in which the pulse propagates. The concept of parametric amplification of a solitary wave is a phenomenon more general than the specific example provided by the laser pulse amplification, and we now turn our attention to finding a concrete realization of the process for microwave generation.

UBITRON MICROWAVE GENERATOR

One of the most successful microwave generators for short-wavelength radiation was invented by Phillips and was known as the Ubitron.[45] The Ubitron is a coaxial system where an electron beam collimated by a B_z field forms the inner coax, and the outer coax is a smooth-walled cylinder fitted with concentric rings of periodically spaced magnets. The purpose of the periodic magnetic field is to dynamically modulate the electron-beam inner coax and thus avoid the process of careful machining of a periodic cavity structure on the outer coax, the procedure normally employed in microwave tubes and which becomes increasingly difficult at shorter and shorter wavelengths. The resulting slow-wave structure proved to be an efficient source of short-wavelength radiation. The B_z field in the system served only to prevent space-charge blowup of the electron-beam inner coax, and played no fundamental role in the microwave generation.

Recent work on the microwave generation with relativistic electron bursts in cavity structures similar to the Ubitron have successfully produced efficient conversion of energy to 3-cm microwaves, and prospects for high efficiency in the millimeter and sub-millimeter range appear promising.[46] The radiation produced in the relativistic analog of the Ubitron, in some cases, does show B_z field dependence, perhaps because the cyclotron mode that plays a central role in the Drummond-Sloan auto-resonant acceleration concept is involved.[47] A beam in the absence of a B_z field can have both longitudinal and traverse oscillations, while the traverse plasma modes tend to be suppressed by the stiffness provided by the B_z field.

OUR VERSION OF THE UBITRON

We wish to consider a version of the Ubitron in which the B_z field is replaced by a Bennett cathode.[38] Such a dielectric rod is known to develop a plasma layer and electrons embedded in the dielectric rod float the electron current off the surface of the rod. Such a "plasma silvered" rod can carry a large current without damaging the dielectric.

When the rod is employed as a cathode for a relativistic E-beam machine, the plasma surface rapidly drops to the cathode potential, and the current flow along the rod is nonrelativistic. The relativistic electrons are produced at the growing plasma tip, off the end of the rod. If no anode/cathode gap is provided at the rod end, then the rod would constitute a short circuit.

The large magnetic field of the small-diameter-rod surface current provides self magnetic insulation, inhibiting the formation of a radial emission or shank fire. For this reason a concentric return path can be present with a spacing of a few rod diameters without shorting problems, a fact established by unpublished work of Smirnov et al.[48] It is interesting to make the outer coax out of a dielectric also, so that it acts as a return current path only after it evolves a similar plasma layer. By providing this plasma-silvered dielectric coax with the periodic Ubitron magnets, one should obtain a tube that will work as well as the original Ubitron, the basic change being merely replacing the collimating B_z field with the collimating Bennett rod.

At sufficiently high excitation levels, the radiation pressure of the amplified slow wave begins to become significant, and will be able to help cause the periodic beam modulation, and for strong self-excitation it should be possible to remove the periodic magnetic field and still have a working system.[66] One might use a maser to produce the initial level of slow-wave excitation sufficient for the system to become self driving, and thus avoid the periodic magnetic structure altogether. For strong electromagnetic excitation it should also be possible to remove the dielectric rod. This would result in a system that is essentially identical to the pinched-relativistic-beam-threaded string of Langmuir collapse cavities discussed in the first section of this paper.

The slow-wave structure for a flared outer coax provides an increased propagation velocity with distance, and for transverse magnetic excitation one obtains a method of coherent ion acceleration with the E field of this transverse magnetic mode, rather than the cyclotron mode of the Drummond-Sloan autoresonant accelerator. The present structure is an ion accelerator that should be as viable as the Drummond-Sloan approach, and has the virtue that the ion acceleration from a single Langmuir collapse region has been demonstrated by computer simulation,[49] the results of which agree with those from collective-acceleration experiments.[50] Further it has a structure known from experiment and from computer simulation to be unstable towards forming.

Once this Ubitron-like tube becomes loaded, the potential difference will become distributed proportional to the load, and for a dielectric rod of appropriate length, the potential drop of the driving E-beam power source should occur along the rod. Then the rod can be extended to touch the anode without resulting in a short circuit, i.e., the loaded rod/cavity system itself becomes an impedance-matched load for the relativistic E-beam machine.

MATHEMATICAL TREATMENT

Now we wish to consider a cylindrical-plasma surfaced, coaxial dielectric cavity of inner radius a, outer radius R, length L, and with vacuum spacing between the coaxial electrodes of $\Delta = (R - a)$. We will initially assume that the cavity also has end plates so that it is a completely closed resonator.

The lowest cylindrically symmetric, transverse magnetic mode of the cavity has the form

$$B_0(r,z) = B_0 \sin (k_z z)$$

$$x \frac{\sin [k_r(r - a)]}{(r - a)} e^{-i\omega t}$$

where $k_z = \pi/L$, and $k_r = \pi/\Delta$. The substitution of this form into the wave equation yields the dispersion relation

$$\omega^2 = \omega_c^2 + (k_z c)^2,$$

where ω_c, the cutoff frequency, is given by $\omega_c \equiv \pi c/\Delta$. Since $\Delta \ll L$, it follows that $\omega_c^2 = \omega^2 \gg (k_z c)^2$, and thus the dispersion relation for the "nonrelativistic" photon (i.e., $v_g = \partial\omega/\partial k \ll c$, where v_g is the group velocity) is approximately

$$\omega = \omega_c + \frac{k_z^2 c^2}{2\omega_c} + \dots$$

If we multiply this expression by h, we obtain $E = E_0 + p_z^2/2m_0$, where $\hbar\omega = E$, $\hbar\omega_c = E_0$, and $m_0 = E_0/c^2$.

Now assume that near the coaxial guide center ($z_0 = L/2$), an (m = 0)-like local pinched region is slowly introduced so that the waveguide mode adapts adiabatically to the perturbation. When this local constriction increases beyond the point where $\int p_z dz$ over the constriction exceeds $\hbar$, then: (1) the original mode, which was a "bound state" of the entire guide, becomes a bound state of the constriction, (2) the electromagnetic energy originally distributed proportional to the unperturbed $(E_0^2 + B_0^2)/8\pi$ in the entire guide becomes localized near the constriction, and (3) the perturbed mode decays exponentially outside the constriction, becoming vanishingly small at the guide ends. The end plates of the coax, having served their conceptional purpose, can now be imagined away.

In reality we are interested in the case where this localized, solitary electromagnetic excitation is the cause, via its radiation pressure, of the constriction in which it is self-trapped. The radiation pressure causes a local depression in the electron density of the plasma silvering layer on the dielectric rod, and this depression becomes the well in which the electromagnetic field is localized.

A fully correct treatment of this self-trapping requires a self-consistent formulation that is provided in the Appendix. The Appendix also provides a coherent-state formulation of the waveguide problem for the case of superconducting guides. A remarkable similarity between the superconducting treatment and the self-consistent collisionless-plasma treatment is noted. This similarity arises because for both cases, Ohm's law is replaced by a London-like relation where the current $\bar{J}$ is proportional to the vector potential A. In this section we will employ a simplified treatment of the self-trapped wave which is shown in the Appendix to be the lowest order term in the more nearly correct self-consistent formulation.

The $(m = 0)$-like adiabatic local pinch in the inner coax causes Δ to become a function of z, i.e., $\Delta = \Delta(z)$, and $E_0 = E_0(z)$ now plays the role of a potential. For convenience we will place the origin of the coordinate system at the minimum of the constriction, and then Taylor expand $\omega_c(z) = \pi c/\Delta(z)$ about the minimum at $z = 0$:

$$\omega_c(z) = \omega_c(0) + \frac{\partial \omega_c}{\partial \Delta} \frac{\partial \Delta}{\partial z}\bigg|_{z=0} z$$

$$+ \frac{1}{2}\left\{\left[\frac{\partial^2 \omega_c}{\partial \Delta^2} (\frac{\partial \Delta}{\partial z})^2\right]\bigg|_{z=0}\right.$$

$$\left. + (\frac{\partial \omega_c}{\partial z} \frac{\partial^2 \Delta}{\partial z^2})\bigg|_{z=0}\right\} z^2 + \ldots$$

Since $\frac{\partial \Delta(0)}{\partial z} = 0$ we obtain

$$\omega_c(z) = \omega_c(0)$$

$$+ \frac{z^2}{2} \omega_c(0) \frac{1}{\Delta(0)} \left|\frac{\partial^2 \Delta(0)}{\partial z^2}\right| + \ldots ,$$

and thus $\varepsilon = p^2/2m_0 + 1/2\ m_0\omega'^2 z^2$, the Hamiltonian of a harmonic oscillator, where $\varepsilon = E - E_0(0)$, $E = \hbar\omega'$,* $E_0 = \hbar\omega_0(0)$, $m_0 = \hbar\omega_c(0)/c^2$, and

$$m_0\omega'^2 = \frac{E_0(0)}{\Delta(0)} \left|\frac{\partial^2 \Delta(0)}{\partial z^2}\right| ,$$

and we have neglected a higher order term proportional to $p^2 z^2$. Thus, in this

approximation the $\sin(k_z z)$ portion of the wave evolves into $\exp(-\alpha z^2/2)$, the ground state of a harmonic oscillator, and the localized excitation becomes

$$B_0(r,z) = B_0 \frac{\sin\left[\frac{\pi}{\Delta}(r - a)\right]}{(r - a)}$$

$$\exp\left[-\frac{1}{2}(\alpha z^2 + i\hbar\omega' t)\right],$$

where the eigenvalue ε is now equal to $\hbar\omega'/2$, and this in turn implies $\hbar^2\alpha/2m_0 = \hbar\omega'/2$, or $\alpha = \omega'\omega_c(0)/c^2$, and in addition

$$\frac{\hbar^2\alpha^2}{2m_0} = \frac{1}{2}\frac{\hbar\omega_c(0)}{\Delta(0)}\left|\frac{\partial^2\Delta}{\partial z^2}\right| ,$$

or

$$\alpha^2 = \frac{\omega_c^2(0)}{\Delta(0)}\left|\frac{\partial^2\Delta(0)}{\partial z^2}\right| .$$

Thus we obtain two expressions relating α to ω', Δ, and $\partial^2\Delta(0)/\partial z^2$.

SOLITARY-WAVE AMPLIFICATION

A current in the surface plasma can result in transit-time amplification of the localized field excitation because it has an axial electric field, and in this way we may obtain parametric amplification of the standing solitary-wave pulse. If the localized well and its cause, the self-trapped wave, are moving with velocity U << c in the lab system, then the pulse will be viewed as a propagating solitary wave that is undergoing transit-time amplification by the electron current in the plasma silvering layer.

The initial pulse to be amplified in this way could be produced by a microwave analog of the laser oscillator, which could be a closed cavity (like that of our initial discussion in this section) whose nearly degenerate modes are mode locked to produce a finite number of new states of equal spacing ΔE. The coherent superposition of such a set of states will result in a string of equally spaced

*Note: ω', the oscillator frequency, should not be confused with ω, the wave frequency.

narrow-width pulses. By rendering the coaxial-oscillator end wall temporarily transparent, one pulse in the string could be sent into the amplifier section of the coax. Alternatively, one might use a mode-locked Q-switched maser to provide the initial pulse.

The viability in practice of this parametric amplification must be determined by experiment. We have made an effort to present the idea in a form that we believe may actually be workable experimentally. The earliest form of this idea is contained in an unpublished note on "An Electromagnetic Brake."[51]

One potential application of such large-amplitude solitary-wave pulses would be for coherent ion acceleration, where one would employ a flared outer coax to permit gradual acceleration of the solitary wave and thus of the ions riding the wave. As the amplitude of the solitary becomes very large, the radiation pressure of the pulse will reach the point where it can begin to cause an insulator-metal transition and thus transform the dielectric rod into a metal rod, followed rapidly by it becoming an exploding -imploding wire similar to the focus plasma of section 2 or the electron-burst target rods of section 3.

Typically, insulator-metal transition occurs in the pressure range of 1 to 10 Mbar. Because of the short transit time of the solitary pulse, the transition pressure in this case may exceed the "static" pressure for such a transition by as much as a factor of 10. It is clear that an interesting means of employing the amplified solitary wave for pulsed fusion is to make the high-amplitude end of the dielectric inner coax from DT, LiDt, or perhaps CDT, or from a hollow glass tube containing DT gas at high pressure.

By appropriate choice of parameters it is possible to set the solitary-wave propagation velocity U at a desired value. This opens the possibility of obtaining highly efficient ablation-free compression of the rod. Compression that is subsonic with respect to the electron Fermi velocity, and supersonic relative to the ion thermal-vibrational velocities of the solid, presents an interesting and genuinely complex problem. The compression of the electrons of the solid in this case will be adiabatic. Space-charge effects

will force the ions to also be compressed, and at the same time will drive them to high vibrational levels which ultimately would melt the solid, but not necessarily destroy the electron degeneracy.

The ionic vibrations could relax by emitting radiation or, because large molecules tend to readily couple energy between electronic and nuclear motion, the highly excited nuclear vibrations might relax by Auger emission, resulting in heating of electrons at the Fermi surface, thus tending to destroy the degeneracy. However, polyatomic molecules tend not to have an electronic spectrum but to relax by nonadiabatic effects into excitation of nuclear vibration and rotation states.[52] Such a decay mechanism for excitation at the Fermi surface would result in cooling, tending to preserve the electron degeneracy. Thus nonradiative cooling of the Fermi surface is analogous to the role played by bremsstrahlung in cooling electrons in the plasma-focus snowplowed plasma, so that the electron temperature lags by as much as a factor of 10 behind the ion temperature.

The effect of the time-dependent radiation pressure in this case would be to drive coherent standing waves into the Fermi gas, and the ions would take on the same structure due to space-charge effects, so that the electron density would vary as $1/r^2$ and have a periodic modulation of the standing wave in the Fermi gas "resonator;" the electrostatic waves in the degenerate medium are the extension of the electromagnetic-wave excitation in the vacuum cavity region. A self-consistent interface between the field and the plasma should develop that could provide impedance matching, so that the time-dependent radiation pressure could feed energy at the speed of light from the vacuum continuum as it evolves at the radiation-plasma interface into the Fermi-gas continuum. The pressure amplification of the ablation-driven compression is replaced in this case by the presence of many more waves in the degenerate gas than would exist for the electromagnetic field in the cavity in the absence of the degenerate plasma. As a result the outer-wall radiation pressure is amplified by a factor equal to the number of standing waves as one passes radially inward.

Analogous possibilities exist for

microwave confinement of hot plasmas, and have been considered by Ensley[53] and by the present author. The idea of Hora[54] on increased compression efficiency are also apparently of a similar nature. From the point of view of the field-free plasma region electromagnetic (i.e., radiation pressure) confinement of a plasma is similar to the electrostatic confinement invented by Farnsworth and studied experimentally by Hirsch and others.[55] The reason for this fact is that the electrons directly follow the time-dependent radiation pressure while the ions indirectly follow the field due to space-charge effects.

The preceding discussion is a suggestion that a new type of highly efficient compression may be possible that is in essence an isentropic radiation-pressure-driven-electrostatic confinement of ions by standing charge-density waves in a degenerate Fermi gas. For this type of compression, burn would be initiated when the decreased ion spacing and increased vibrational amplitude reached the level at the rod center where barrier penetration becomes sufficiently probable. Since this penetration probability varies exponentially with nuclear separation, the onset of burn in the rod center could be very rapid.

6. SUMMARY

The pulsed power sources that most nearly approach fusion-microexplosion requirements (pulse energy $\geq 10^5$ J, power $\geq 10^{14}$ W, and power flux density $\geq 10^{16}$ W/cm^2) are relativistic E-beam machines and high-power pulsed lasers. The ability to convert 5% of existing E-beam energy to accelerated ions and 15% to short-wavelength microwaves makes collective ion bursts and microwave pulses competitive alternatives for microexplosion applications.

The greater efficiency of relativistic E-beam, the greater ability to focus laser light relative to electron bursts, and the apparent greater ease of energy deposition of laser light in a target than electron beam energy are not decisive differences. For fields in excess of

$2\pi \times 10^7$ G, and for power-density fluxes of $>10^{15}$ W/cm^2 the deposition of high-ν/γ energy is in first approximation a function of energy density, not of field frequency, because the radiation pressure determines the corona density and above 1 kV the skin currents drive Langmuir oscillations that result in large anomalous resistivity, and enhanced field diffusion, rapid energy deposition and thus causes ablation-driven compression.

When stability is considered, the optimum wavelength range for pulsed power sources intended for pulsed fusion application is $100 R \geq \lambda \geq R$, where R is the pellet radius. This singles out the wavelength range 1 cm $\geq \lambda \geq 100$ μm as particularly interesting. This is a wavelength range accessible to advanced microwave generation techniques utilizing both relativistic E-beam machines, or to long-wavelength laser and masers, or to a combination of these sources.

The short-pulse capability of the lasers is a decisive advantage. The laser-pulse amplification is a special case of a more widely applicable phenomenon of parametric amplification of a solitary wave.

One means under development at LLL for production of short-duration, small diameter, high-ν/γ ($\nu/\gamma = 100$) relativistic E-beam pulses is by the strong-turbulence-triggered self interruption of the current of the pinch produced by low-density-mode operation of a 1-MJ plasma focus. This current interruption is followed by restrike of the initial "super high"-ν/γ ordinary current as a lower ν/γ, relativistic electron burst. In this way, one can produce small-diameter, 10^5-J, 10-nsec multi-megamp electron bursts and oppositely directed 10^4-J-MeV ion bursts with diameter d of 1 mm and perhaps as small as 100 μm. The rise time of these pulses under the most ideal conditions can be as short as 100 psec.

Study of the neutron-yield-producing mechanisms of the focus suggests that a significant increase in the focus yield could be obtained by scaling the focus phenomena to the case of exploding-imploding DT wire at initial density 10 times solid density by employing the relativistic electron burst produced by the focus. The energy delivery to the fusionable rod occurs by the relativistic electron flow over the rod being initially cancelled out by a return current. This phase is followed by expulsion of the return current from the relativistic beam path by the Weible instability due to Langmuir turbulanece enhanced resistivity.

The return-current expulsion results in rapid transfer of 10^5 J of magnetic field energy to the region around the DT rod. The primary yield will be due to the turbulent, heated, lower density plasma ablated from the rod, and temporarily contained in the field that remains after a significant portion of the field energy has been transferred into particle energy, to produce a 10-kV plasma at 1/10 solid density. These conclusions are based on scaling of the experimentally relatively-well-studied plasma-focus deflagation burn mode, and are presently untested experimentally. Experiment as always must provide the acid test for these theoretical scaling arguments.

An alternative approach, utilizing the focus-produced electron burst or the longer duration bursts of a relativistic E-beam machine, is provided by the parametric amplification of a solitary transverse magnetic wave in a plasma-coated-dielectric coaxial cavity. This idea may permit the fundamentally important pulse amplification of laser light to be extended to microwave generation techniques. The inner coax of the proposed parametric amplifier is a Bennett dielectric rod.

The important applications of this solitary-wave parametric amplification are coherent ion acceleration and pulsed fusion. For ion acceleration the outer coax is gradually increased in diameter to permit acceleration of the solitary wave and thus of any ions riding the wave. The application to pulsed fusion is provided in the simplest way by making the high-energy end of the dielectric central coax of a fusionable dielectric material or of a glass tube containing high pressure DT gas. The amplified soloton radiation pressure results in an insulator-metal transition in the dielectric rod, and this will be followed by rapid explosion-implosion self destruct of the rod end, which for pulse fields of about $2\pi \times 10^7$ G could result in significant DT yield. This latter idea has no experimental antecedents, and its value in practice must await appropriate experimental tests. We believe the approach could be tested experimentally in the specific form discussed in this paper.

APPENDIX

NONLINEAR ELECTROMAGNETIC WAVES

In this appendix we develop the mathematical details of the treatment of nonlinear electromagnetic waves both for the case of interaction with a superconducting medium and for the case of a collisionless plasma, and apply the nearly identical results of the two cases to a more complete treatment of a solitary wave, to provide support for the oversimplified discussions in Section 5 of the main body of this paper.

We will simplify the treatment by assuming only nonrelativistic electrons; we also assume the ions merely serve as a neutralizing background, i.e., we use a quasi-neutral formulation. The central problem is to determine the fields from Maxwellian equations self-consistently from the current and charge-density sources whose distribution depends on the fields caused by these sources.

We begin with the Maxwell equation

$$\bar{\nabla} \times \bar{H} - \frac{1}{c} \frac{\partial \bar{E}}{\partial t} = \frac{4\pi}{c} \bar{J} \quad , \qquad (1)$$

where $\bar{H}$ is the magnetic field, $\bar{E}$ the electric field, c the speed of light, and $\bar{J}$ the current density. Now introduce the vector and scalar potential $\bar{A}$ and ϕ. In terms of these potentials, the field becomes

$$\bar{H} = \bar{\nabla} \times \bar{A} \ , \ \text{and} \ \bar{E} = -(\bar{\nabla}\phi + \frac{1}{c} \frac{\partial \bar{A}}{\partial t})$$

which, substituted into (1) with the use of the identity $\bar{\nabla} \times \bar{\nabla} \times \bar{A} = \bar{\nabla}(\bar{\nabla} \cdot \bar{A}) - \bar{\nabla}^2 \bar{A}$, gives

$$\bar{\nabla}(\bar{\nabla} \cdot \bar{A} + \frac{1}{c} \frac{\partial \phi}{\partial t}) = \nabla^2 \bar{A} - \frac{1}{c^2} \frac{\partial^2 \bar{A}}{\partial t^2} + \frac{4\pi}{c} \bar{J}. (2)$$

The left side of (2) vanishes by virtue of the Lorentz condition. Recall that the potentials $\bar{A}$ and ϕ may be changed to the potentials $\bar{A} + \bar{\nabla}\chi$ and $\phi - (1/c)(\partial\chi/\partial t)$, and become $\bar{\nabla} \times \bar{\nabla}\chi = 0$. This gauge transformation causes the new potentials to produce the same fields as the old potentials.

To solve Eq. (2) in specific cases we must obtain an expression for $\bar{J}$:

$$\bar{J} = \frac{e}{m} <m\bar{v}> = \frac{e}{m} <\bar{p} - \frac{e}{c} \bar{A}>$$

where $m\bar{v}$ is the particle momentum and $\bar{p}$ is the canonical momentum. The equation of motion for $\bar{p}$ is

$$\frac{d\bar{p}}{dt} = -\bar{\nabla}(\phi + \frac{e}{c}\,\bar{V}\cdot\bar{A})$$

and thus for force-free motion $d\bar{p}/dt = 0$, $\bar{p}$ is a constant, and $\bar{\nabla}\times\bar{p} = 0$. For $\bar{\nabla}\times p = 0$, $\bar{p}$ is the gradient of some scale λ, i.e., $\bar{p} = \nabla\times\lambda$, and $\bar{p} - (e/c)\,\bar{A}$ may be looked upon as a new vector potential $-(e/c)\,A'$, i.e., $\bar{p}$ provides a gauge transformation, and consequently Eq. (2) becomes

$$\nabla^2\,\bar{A} - \frac{1}{c^2}\frac{\partial^2\bar{A}}{\partial t^2} - \frac{\omega_p^2}{c^2}\,\bar{A}\;, \tag{3}$$

where $\omega_p^2 = 4\pi n e^2/m$; $n = \int f(\bar{x},\bar{p},t)\,d\bar{p}$ and the electron distribution function satisfies the self-consistent truncation of the many-particle Liouville equation, i.e., the Vlassov equation. We will obtain an identical equation for the superconducting case.

Anticipating the results we note that the similarity between the collisionless plasma and the superconductor arise because in both cases $(4\pi/c)\,\bar{J} = (\omega_p^2/c^2)\,\bar{A}$, a London-like relation[64] that replaces Ohms law. The difference between the problem for the superconductor treated as a coherent change Boson fluid, and the collisionless Vlassov plasma is contained in the definition of the density n. The parallel between the collisionless plasma and the quantum fluid can be made particularly close by introducing the Wigner distribution function $f_w(\bar{x},\bar{p},t)$ given by

$$f_w(\bar{x},\bar{p},t) = \int\psi^*(\bar{x} + \frac{\bar{y}}{2},t)\,e^{i\bar{p}\cdot\bar{y}}\psi(\bar{x} - \frac{\bar{y}}{2},t)\,d\bar{y}.$$

It is the most nearly analogous function for a quantum system of the classical phase-space distribution function, and in the limit $\hbar \to 0$, $f_w \to f$. The Wigner distribution function has the peculiarity in general that unlike a classical distribution function it can sometimes be negative. The expectation value of a quantum operation $\langle\hat{O}(\hat{x},\hat{p})\rangle$ expressed in terms of the Wigner distribution has the form $\langle\hat{O}(\hat{x},\hat{p})\rangle = \int f_w(\bar{x},\bar{p},t)\,O(\bar{x},\bar{p})\,d\bar{x}d\bar{p}$, which is the same form as the classical phase-space average in statistical mechanics. Formulation of a quantum problem in terms of the Wigner distribution function is particularly similar to the classical problem in the special case of a single harmonic oscillator in a coherent state.

We will summarize briefly the results for a single oscillator at the end of the appendix. The coherent-state picture we will employ for the second quantized field is equal to a superposition of countable infinity of oscillators in coherent states, because of the well known fact that a second quantized Bose field and an infinite set of noninteracting harmonic oscillators are isomorphic.

We assume that the charged Bose fluid is a model of the quasi-Boson electron pairs of a superconductor. The second quantized field for such a system has annihilation and creation operators, $\hat{\psi}(\bar{x},t)$ and $\hat{\psi}^+(\bar{x},t)$, that satisfy the equal-time Bose commutation relations

$$[\hat{\psi}(\bar{x}',t),\hat{\psi}^+(\bar{x},t)] = \delta(\bar{x}' - \bar{x})\;.$$

The Heisenberg equation of motion of the annihilation operation is

$$i\hbar\,\frac{\partial\hat{\psi}}{\partial t} + [\hat{H},\hat{\psi}] = 0\quad,$$

where $\hat{H} = \hat{H}_0 + \hat{H}_1$ and

$$\hat{H}_0 = \frac{1}{2m}\int\hat{\psi}^+(\hat{p} - \frac{e}{c}\,\bar{A})^2\,\hat{\psi}d\bar{x}$$
$$+ e\int\hat{\psi}^+\phi(x)\,\hat{\psi}d\bar{x}$$

and

$$\hat{H}_1 = \frac{1}{2}\iint\hat{\psi}^+(\bar{x})\,\hat{\psi}^+(\bar{n})\,V(\bar{x} - \bar{y})$$
$$\hat{\psi}(\bar{y})\,\hat{\psi}(\bar{x})\,d\bar{x}d\bar{y}$$

where V is a pair potential and $e\phi$ is an externally applied potential. Substitutions of these forms of $\hat{H}$ into the equation of motion for $\hat{\psi}$ yields a Hartree-like operator equation

$$i\hbar\,\frac{\partial\hat{\psi}}{\partial t}(\bar{x},t) = \frac{(\bar{p} - \frac{e}{c}\,\bar{A})^2}{2m}\,\hat{\psi}(\bar{x},t) + U\hat{\psi}(\bar{x},t)$$
$$+ \frac{1}{2}\int\hat{\psi}^+(\bar{y})\,V(\bar{x}-\bar{y})\,\hat{\psi}(\bar{y})\,d\bar{y}\,\hat{\psi}(\bar{x},t)\quad.$$

Now we make the basic assumption that superfluidity is the condition corresponding to all the Bosons occupying the same wave function, and thus the system is in a coherent state or Glauber state.[57,65] Such a state $|\psi\rangle$ is an eigenstate of the annihilation operation $\hat{\psi}$, i.e., $\hat{\psi}\,|\,\psi\rangle = \psi(\bar{x},t)\,|\,\psi\rangle$; the wave-function-like quantity ψ is the eigenvalue, and is known in the jargon of superfluidity and superconductivity as the complex order parameter. Projecting the equation for $\partial\hat{\psi}/\partial t$ onto the coherent state $|\psi\rangle$, we obtain the

Hartree-like equation for the complex order parameter known as the Ginsberg-Landau equation:

$$i\hbar \frac{\partial \psi}{\partial t} = \frac{(\hat{p} - \frac{e}{c}\bar{A})^2}{2m}\psi + e\phi\psi$$
$$+ \frac{1}{2}\int |\psi|^2 V(\bar{x} - \bar{y})\, dy\, \psi(y). \quad (4)$$

In obtaining Eq. (4), higher terms that arise because the overcomplete set of states $|\psi(\bar{x},t)\rangle$ are not orthogonal.

We should also treat the electromagnetic field as a second quantized field and introduce the operator $\hat{E}(\bar{x},t)$, which is a superposition of positive frequency field components and annihilates a photon at a point. Then for a coherent (superfluid-like) photon field we would introduce the coherent state $|E(\bar{x},t)\rangle$, where

$$\hat{E}(\bar{x},t)|E(\bar{x},t)\rangle = E(\bar{x},t)|E(\bar{x},t)\rangle;$$

the eigenvalue order parameter $E(\bar{x},t)$ looks like and satisfies the same equation as a classical electric field.

We omit the details of the procedure outlined above and simply state that we arrive at the same mathematical form as if we had used an unquantized electromagnetic (EM) field in the first place with the slightly subtle distinction that $E(\bar{x},t)$ is an order parameter for the positive frequency part of the field, not a classical field. The only apparent consequence of this fact is that complex exponential solutions for waves are used as a mathematical device in the classical case, and only the real part of the answer has meaning, whereas $\bar{E}(\bar{x},t)$, the order parameter, must have complex solutions corresponding to directed energy and momentum. The distinction is usually conveniently overlooked in classical EM books when they cover the topic of an evanescent wave, e.g., one at a throat below cutoff between two cavities. The book's author, having carefully pointed out in some earlier part of the book that complex fields are a mathematical device, will then go on to deduce the wave "inside the quantum barrier" ignoring this fact.

Returning to the complex order parameter, Eq. (4), we note that $V(\bar{x}-\bar{y})$ will be a screened Coulomb potential and by oversimplifying this potential into a delta function of strength α we may obtain the nonlinear Schrödinger equation.

$$i\hbar \frac{\partial \psi}{\partial t} = \frac{(\bar{p} - \frac{e}{c}\bar{A})^2}{2m}\psi + e\phi\psi + \alpha|\psi|^2\psi = 0 .$$

We will obtain an equation of precisely this form for the nonlinear wave equation Eq. (3) presently.

The complex order parameter, or macroscopically occupied wave function, has the property that $|\psi|^2$ and

$$\bar{J} = \frac{1}{2m}\{\psi[(\bar{p} - \frac{e}{c}\bar{A})\psi]^* - \psi^*(\bar{p} - \frac{e}{c}\bar{A})\psi\}$$

are "honest to God" particle and current densities rather than probability densities as is the case for the mathematically identical Hartree equation for a single particle. For this reason the Hartree-like equation for the order parameter is "exact" given the validity of the assumption of a coherent state as an exact description of superfluidity.

If we take ψ of the form $\psi = \sqrt{n}\exp(-(i/\hbar)S)$, then $|\psi|^2 = n$, and $\bar{J} = n/m$ $(\bar{\nabla}S-(e/c)\bar{A})$, and from the real and imaginary parts of the complex order parameter equation for this form of ψ we obtain

$$\frac{1}{2m}(\bar{\nabla}S - \frac{e}{c}\bar{A})^2 + e\phi + \alpha n^2 + \frac{\partial S}{\partial t} = \frac{\hbar^2}{2m}\frac{\nabla^2 n}{\sqrt{n}} ,$$

which is the usual Hamilton-Jacoby equation with quantum potential $(\nabla^2\sqrt{n})/\sqrt{n}$. As noted by Feyman[56] the requirement of space-charge neutralization makes the quantum term small almost everywhere. We also obtain the continuity equation

$$\frac{\partial n}{\partial t} + \frac{\bar{\nabla}}{m}\cdot(\bar{\nabla}S - \frac{e}{c}\bar{A}) = 0 \qquad .$$

If we interpret the role of $\bar{\nabla}S$ and $\partial S/\partial t$ as providing a gauge transformation, then these two equations become

$$\frac{e^2}{2mc^2}A^2 + e\phi + \alpha n^2 = \frac{\hbar^2}{2m}\frac{\nabla^2\sqrt{n}}{\sqrt{n}} ,$$

and

$$\partial n/\partial t = \bar{\nabla}\cdot[n(e/mc)\bar{A}] ,$$

and the quantum-mechanical order-parameter current reduces to $(4\pi/c)\bar{J} = (\omega_p^2/c^2)\bar{A}$, which, as promised, leads to Eq. (3). Thus for either the collisionless plasma or the superconductor the problem is to solve

$$\nabla^2\bar{A} - \frac{1}{c^2}\frac{\partial^2\bar{A}}{\partial t^2} - \frac{\omega_p^2}{c^2}\bar{A} = 0 .$$

We will assume that the problem of interest will involve only a single component of $\bar{A}$ and a single field frequency ω so that we may take A in the form $A = a(\bar{r}) \exp(i\omega t)$, for which the wave equation becomes

$$\left[\nabla^2 + \left(\frac{\omega}{c}\right)^2 - \left(\frac{\omega_p}{c}\right)^2 \right] a(\bar{r}) = 0 \quad .$$

We must now find an expression for n the density in ω_p in order to have a completely defined problem. In the case of the Vlassov plasma we assume that $f(\bar{x}, \bar{p}, t)$ is an equilibrium distribution function

$$f = \exp\left[-\beta \left(\frac{\bar{p} - \frac{e}{c}\bar{A}}{2m} \right)^2 + e\phi \right]$$

and then interpret $\bar{p}$ as a gauge

$$\left(\bar{p} - \frac{e}{c}\bar{A}\right)^2 \rightarrow \frac{e^2}{2mc^2} A'^2 \equiv \frac{e^2}{2mc^2} A^2 \quad .$$

The quasi-neutrality condition demands that $e\phi = -\frac{1}{2}(e^2/2mc^2) A^2$, and approximating A^2 by its time average a^2 we obtain $\omega_p^2 = (\omega_p^0)^2 \exp(-\beta e^2 a^2/4mc^2)$, where $(\omega_p^0)^2$ is $4\pi n_0 e^2/m$, n_0 being the density in the asymptotic region where the fields are small. Thus we obtain the equation

$$\nabla^2 a + \left[\left(\frac{\omega}{c}\right)^2 - \left(\frac{\omega_p^0}{c}\right)^2 \exp\left(-\frac{\beta e^2}{4mc^2} a^2\right) \right] a = 0.$$

If we further specialize this to a one dimensional problem then we obtain

$$\frac{\partial^2 a}{\partial z^2} + \left[\left(\frac{\omega}{c}\right)^2 - \left(\frac{\omega_p^0}{c}\right)^2 \exp\left(-\frac{\beta e^2}{4mc^2} a^2\right) \right] a = 0.$$

If this equation is multiplied by $\partial a/\partial z$, it becomes

$$\frac{\partial}{\partial z}\left(\frac{\partial a}{\partial z}\right)^2 + \left[\left(\frac{\omega}{c}\right)^2 - \left(\frac{\omega_p^0}{c}\right)^2 \exp\left(-\frac{\beta e^2}{4mc^2} a^2\right) \right]$$

$$\times \left(\frac{\partial}{\partial z}(a^2) \right)$$

or

$$\frac{\partial}{\partial z}\left[\left(\frac{\partial a}{\partial z}\right)^2 + \left(\frac{\omega}{c}\right)^2 + \frac{2mc^2}{\beta e^2}\left(\frac{\omega_p^0}{c}\right)^2 \exp\left(-\frac{\beta e^2 a^2}{4mc^2}\right) \right]$$

$$= 0.$$

The integral of this equation is

$$\left(\frac{\partial a}{\partial z}\right)^2 + \left(\frac{\omega}{c}\right)^2 + \frac{4mc^2}{\beta e^2}\left(\frac{\omega_p^0}{c}\right)^2 \exp\left(-\frac{\beta e^2 a^2}{4mc^2}\right)$$

$$= E,$$

where E is the constant of integration. The solution of this equation for z is given by

$$Z = Z_0 + \int_{a(z_0)}^{a(z)} \frac{da}{\sqrt{E - \left(\frac{\omega}{c}\right)^2 - \frac{2mc^2}{\beta e^2}\left(\frac{\omega_p^0}{c}\right)^2 \exp\left(-\frac{\beta e^2 a^2}{4mc^2}\right)}}$$

It is not possible to obtain an analytic solution to this integral. However, in the approximation where we may linearize the exponential, the integral becomes

$$Z = Z_0 + \int_{a(z_0)}^{a(z)} \frac{\partial a}{\sqrt{E - \left(\frac{\omega}{c}\right)^2 - \frac{4mc^2}{\beta e^2}\left(\frac{\omega_p^0}{c}\right)^2 + \frac{4mc^2}{\beta e^2}\left(\frac{\omega_p^0}{c}\right)^2 a^2}}$$

If we now consider the problem corresponding to this equation as that of a one-dimensional Langmuir collapse, where the particle-free field region is at the origin and the field-free particle region is at $\pm z$ for large z, i.e., $z \rightarrow \pm\infty$, $n \rightarrow n_0$, then the first integration can be taken as $\int_0^z$ and then E is the integrand at $z = 0$, where $\partial a/\partial z = 0$ and $\omega_p = 0$, and consequently $E = (\omega/c)^2$; and for this value of E the second integral becomes

$$z = \alpha \int_{a(0)}^{a(z)} \frac{a}{\sqrt{a^2 - 1}}, \quad \text{where } \alpha = \frac{1}{4\sqrt{4nkT}} \cdot 1$$

$$z = \alpha[\cosh^{-1} a(z) - \cosh^{-1} a(0)]$$

or

$$a(z) = a(0) + \cosh(z/\alpha).$$

We note that this is the solution of the linearized equation

$$\frac{\partial^2 a}{\partial z^2} + \left[\left(\frac{\omega}{c}\right)^2 - \left(\frac{\omega_p^0}{c}\right)^2 \left(1 - \frac{\beta e^2}{4mc^2} a^2\right) \right] a$$

an equation of exactly the same form as the nonlinear Schrödinger equation

$$i\hbar \frac{\partial \psi}{\partial t} = \frac{\left(\bar{p} - \frac{e}{c}\bar{A}\right)^2}{2m} \psi + \alpha|\psi|^2 \psi + e\phi\psi$$

if $\psi = a e^{-i\omega t}$.

The nonlinear equation for a is also of the form employed by Shiff[58] as the basis of a model of an elementary particle, and the treatment in this section is also similar in a number of respects to the work of de Broglie.[59] The linearization is actually not valid near z = 0 in the case treated above, which corresponds to the case where the electromagnetic field via its radiation pressure holds open its own wave guide.

In the case where the vacuum well exists even in absence of the radiation pressure, such as for a superconducting cavity or the plasma-coated dielectric-wall cavity for the parametric amplifier in Section 5 we may introduce an effective potential u that simulates the field-free guide, i.e.,

$$\frac{\omega_p^2}{c^2} = \frac{\omega_p^{0^2}}{c^2} \exp(-\beta u) \exp\left(-\frac{\beta e^2}{4mc^2} a^2\right) .$$

and for this case the major burden of producing the vacuum region is due to the "known" potential u, the radiation pressure provides a perturbation.

In this case the linearization is valid since $(\omega_p/c)^2$ vanishes at small r values because of u, i.e., in the region where n vanishes in the absence of the linearization would be invalid. For this case one may write ω_p^2/c as

$$\left(\frac{\omega_p}{c}\right)^2 = \frac{\omega_p^{0^2}}{c^2} \left[\exp(-\beta u)\left(1 - \frac{e^2}{4mc^2} a^2\right)\right]$$

where u can be taken as a strong function of r, so that ω_p drops rapidly to zero at the guide surface in a physically realistic manner.

Note that if we assume a is of the form $\exp(-\lambda x^2/2)$, e.g., as a trial function for a variational approach to the nonlinear Schrödinger-like equation for (a), and in addition if we approximate the nonlinear term as $(\omega_p^0/c)^2(1 - \lambda x^2/2)$, then the equation for a becomes

$$\frac{\partial^2 a}{\partial z^2} + \left[\left(\frac{\omega}{c}\right)^2 - \left(\frac{\omega_p^0}{c}\right)^2\left(1 - \frac{\lambda x^2}{2}\right)\right] a ;$$

this equation has the form of a harmonic oscillator equation from which λ can be determined in terms of ω and ω_p^0. Apart from the present problem being a one-dimensional linear formulation instead of a one-dimensional radial problem, the problem and the result are identical to the simplified treatment of a standing, solitary transverse magnetic wave and this establishes the fact that the simplified treatment is the lowest-order self-consistent solution.

Finally we turn to the determination of the form of n for the superconduction problem. The equations relating $\bar{A}$ and n in this case are

$$\frac{e^2}{4mc^2} A^2 + \alpha n = \frac{\hbar^2}{2m} \frac{\frac{\partial^2}{\partial z^2}\sqrt{n}}{\sqrt{n}} \quad , \text{ or}$$

$$\frac{e^2}{4mc^2} A^2 + \alpha n + \frac{\hbar^2}{8m}\left[\left(\frac{1}{n}\frac{\partial n}{\partial z}\right)^2 - \frac{2}{n}\frac{\partial^2 n}{\partial z^2}\right] ,$$

$$\frac{e^2}{4mc^2} A^2 + \alpha n = \frac{\hbar^2}{4m}\left[\frac{1}{n}\frac{\partial^2 n}{\partial z^2} - \frac{1}{2}\left(\frac{1}{n}\frac{\partial n}{\partial z}\right)^2\right] .$$

If we neglect the Hartree term αn and use the fact that for quasi-neutrality $e\phi = -(e^2/4mc^2) A^2$, then we obtain

$$\left(\frac{e}{\hbar c}\right)^2 A^2 = \left[\frac{1}{n}\frac{\partial^2 n}{\partial z^2} - \frac{1}{2}\left(\frac{1}{n}\frac{\partial n}{\partial z}\right)^2\right] .$$

Now let $n = n_0 e^{-f}$ to obtain

$$\left(\frac{e}{\hbar c}\right)^2 A^2 = \left[\left(\frac{\partial f}{\partial z}\right)^2\frac{1}{2} - \frac{\partial^2 f}{\partial z^2}\right] .$$

To obtain a first approximation we assume $\partial^2 f/\partial z^2$ is negligible, and thus

$$f = \frac{\sqrt{2}\,e}{\hbar c}\int_{z_0}^{z} A dz' \quad \text{which implies}$$

$$n = n_0 \exp - \frac{\sqrt{2}\,e}{\hbar c}\int_{z_0}^{z} A dz' .$$

OSCILLATOR COHERENT STATES

Finally we outline the coherent-state picture in the case of a single harmonic oscillator. The Hamiltonian H of a one-dimensional oscillator in dimensionless form is $H = \frac{1}{2}(\hat{p}^2 + \hat{x}^2)$, or, in terms of creation and annihilation, operator $H = \hat{n} + \frac{1}{2}$, where $\hat{n} = \hat{a}^+\hat{a}$, $[\hat{a},\hat{a}^+] = 1$, and $|n>$ are the number eigen states; $\hat{n}\,|n> = n\,|n>$, $\hat{a}^+\,|n> = \sqrt{n+1}\,|n+1>$, $\hat{a}\,|n> = \sqrt{n}\,|n-1>$, and the Schrödinger wave functions are the position representatives of $|n>$, i.e., $<x|n> = \phi_n(x)\alpha \exp(-x^2/2) H_n(x)$, and the time-dependent wave functions are

$$\psi_n(x,t) = A \exp\left[-(n + \frac{1}{2})\,t\right] \exp\left(-\frac{x^2}{2}\right)H_n(x)$$

where A is a normalization constant and H_n a Hermite polynomial.

The coherent state of the oscillator $|\alpha\rangle$ is an eigenstate of annihilation operator $\hat{a}$, $\hat{a}|\alpha\rangle = \alpha|\alpha\rangle$. The position representative of $|\alpha(t)\rangle$ is given by

$$\langle x|\alpha(t)\rangle = \exp\left[-\frac{1}{2}(x - x(t))^2\right]$$

$$\exp\left[ixp(t)\right] \exp\left[-\frac{1}{4}x(t)\,p(t)\right]$$

$$\exp\left(-\frac{it}{2}\right) \quad ,$$

where $x(t)$ and $p(t)$ are the classical oscillator position and momentum coordinates; $x(t) = x_0\cos(\phi - t)$ and $p(t) = x_0\sin(\phi - t)$. Thus we see that the coherent-state wave function $\psi(\alpha(t),x) = \langle x|\alpha(t)\rangle$ is a gaussian, nonspreading wave packet that travels on the classical path with classical momentum, and, if expressed in dimensional rather than dimensionless form, shows that as $\langle n\rangle$ increases the gaussian becomes more sharply peaked and the packet approaches a delta-function-like probability density appropriate to a classical particle undergoing a harmonic oscillation. By Fourier transform of $\psi(x,\alpha(t))$, we obtain the momentum representation $\phi(p,\alpha(t))$, where

$$\phi(p,\alpha(t)) = B\exp\left[-\frac{1}{2}(p - p(t)]^2\right.$$

$$\exp\left[-ipx(t)\right] \exp\left[\frac{i}{4}p(t)\,x(t)\right]$$

$$\exp\left(+\frac{it}{2}\right) \quad .$$

The Wigner distribution function for this state is $f_W(x,p,t) = |\phi(x,\alpha(t))|^2 |\phi(p,\alpha(t)|^2$, a genuine probability that satisfies Liouville's equation: $f(x,p,t) = C\exp\left[-(x-x(t)]^2 + (p - p(t))^2\right]$, where C is a normalization constant. We may also write $f(x,p,t)$ in the form $f(x,p,t) = \chi^*(x,p,t)\,\chi(x,p,t)$, where $\chi = C\exp\left[-\frac{1}{2}(x-x(t))^2 + \frac{1}{2}(p-p(t))^2\right]\exp\left[i(x(p(t) - px(t))\right]\,e^{-it/2}$, a "phase-space wave function" that satisfies Liouville's equation and represents a wavepacket traveling with directed motion on the classical phase-space path with the classical energy and momentum. These packets and their excited states are of the type studied by Della Riccia and Wiener[61] in Wiener's last published work. Excited coherent states or Kennard packets also exist and are of the form

$$\psi_n(x(t)) = C_n\exp\left[-\frac{1}{2}(x - x(t))^2\right]$$

$$\exp\left[ixp(t)\right] \exp\left[-\frac{i}{4}x(t)\,p(t)\right]$$

$$\times H_n(x - x(t))\exp\left[-i(n + \frac{1}{2})\,t\right]$$

with Fourier transform

$$\phi_n(x(t)) = C_n\exp\left[-\frac{1}{2}(p - p(t))^2\right]$$

$$\exp\left[-ipx(t)\right] \exp\left[\frac{i}{4}x(t)\,p(t)\right]$$

$$H_n(p - p(t)) \quad ,$$

and Wigner distribution function

$$f_n(x,p,t) = x_n^*(x,p,t)\,x_n(x,p,t) = |\psi_n|^2|\psi_n|^2,$$

where

$$x_n(x,p,t) = \exp\left[-\frac{1}{2}(p - p(t))^2\right.$$

$$+ \frac{1}{2}(x - x(t))^2\right]\exp\left[i(xp(t) - px(t)\right]$$

$$\times H_n\left[(x - x(t)) + i(p - p(t))\right]$$

$$\exp\left(-i(n + \frac{1}{2})\,t\right) \quad ;$$

the f_n and x_n satisfy Liouville's equation.

The Boson annihilation operation $\hat{\psi}(x,t)$ in the discussion of the coherent-state picture of a superconductor is a superposition of harmonic-oscillator creation operators $\hat{a}_n$, i.e.,

$$\psi(x,t) = \sum_n \hat{a}_n\exp(i\omega_n t)\,\phi_n(x) \quad .$$

The $\phi_n(x)$ are a complete set of states, and the coherent state $|\psi(x,t)\rangle$ is a product $\prod_n|\alpha_n\rangle$ of the separate oscillator states. Thus, the position representation of $|\psi\rangle$ is a product of the separate oscillator position representations. The Fourier transform of the product states is a product of the simple-oscillator Fourier transforms, and the Wigner distribution function is a product of the separate-oscillator Wigner distribution functions. The product of coherent states, and the sum or coherent superposition of such states are equivalent — a particularly unique property of the coherent state.[62]

The form of the eigenvalue $\psi(x,t)$ of annihilation operator $\hat{\psi}$ is the complex order parameter

$$\psi(x,t) = \sum_n \alpha_n(t)\,\phi_n(x) \quad ,$$

where $\alpha_n(t)$ is the eigenfunction of the separate oscillator state; $\hat{a}_n|\alpha_n(t)\rangle = \alpha_n(t)|\alpha_n(t)\rangle$; and $\phi_n(x)$ is any complete set of states.

ACKNOWLEDGMENTS

I would like to thank John Luce, Oved Zucker, John Wyatt, Richard Gullickson, Winston Bostick, Dick White, and Victorio Nardi for helpful discussions and Robert Wright for careful reading and correction of the manuscript. The numerous discussions of microwave confinement of plasmas with Donald Ensley, extending over a period of several years, were particularly stimulating. A two-day discussion group on microwave generation and collective acceleration of ions with relativistic electron beams sponsored by AFSOR on June 5 and 6, 1975, provided timely information during the final stages of the preparation of this paper.

REFERENCES

1. J. G. Linhart, (this conference).

2. J. Nuckolls, L. Wood, A. Thiessen, and G. Zimmerman, Nature 239, 139 (1971).

3. H. Hora, Atomkernenergie 24, 187 (1974).

4. J. S. Luce, H. L. Sahlin, and T. R. Crites, Proc. IEEE NS-20, 336 (1973); J. S. Luce, Conf. Proc. Annals N.Y. Academy of Sci. 251, 217 (1975); H. L. Sahlin, Conf. Proc. Annals N.Y. Academy of Sci. 251, 238 (1975); H. F. Kovalev, M. I. Petelin, M. D. Raiser, A. V. Smorgonskii and L. E. Tsopp, Sov. Phys. JETP Lett. 18, 138 (1973).

5. D. L. Ensley, Conf. Proc. Annals N.Y. Academy of Sci. 251, 30 (1975).

6. J. Nuckolls, J. Emmett, and L. Wood, Phys. Today 26, 46 (1973); Scientific American 230, 24 (1974).

7. D. L. Marrow, J. D. Phillips, R. M. Stringfield, Jr., W. O. Doggett, and W. H. Bennett, Appl. Phys. Lett. 19, 441 (1971); A. E. Blangrund, G. Cooperstein, Phys. Rev. Letters 34, 461 (1975); J. Chang. M. M.Widner, G. W. Kuswa, and G. Yonas, Phys. Rev. Letters 34, 1266 (1975); V. P. Ignatenho, Sov. Phys.-Usp. 4, 96 (1967); G. Yonas, J. W. Poukey, K. R. Prestowich, J. R. Freeman, A. J. Toepfer, and M. J. Glauser, Nuc. Fusion 14, 731 (1974); B. L. Freeman, (Ph.D. Thesis) published as

Lawrence Livermore Laboratory Rept. UCRL-51608 (1974); W. C. Condit, D. O. Trimble, G. A. Metzger, D. G. Pellinen, S. Heurlin, and P. Creely, Phys. Rev. Letters 22, 123 (1973); J. G. Clark, J. R. Kerns, and T. E. McCann, Conf. Proc. Annals N.Y. Academy of Sci. 251, 273 (1975); P. S. P. Wei, W. M. Leavens, and J. L. Adamski, J. Appl. Phys. 45, 2163 (1974).

8. H. L. Sahlin, paper presented at IEEE Int. Conf. on Plasma Science, May 14-16, 1975 (UCRL-76584-Abstract).

9. F. W. Perkins, and E. J. Valeo, Phys. Rev. Letters 32, 1234 (1974); C. E. Max, J. Arons, and A. B. Langdon, Phys. Rev. Letters 33, 209 (1974); J. J. Thomson, C. E. Max, and K. Estabrook, Lawrence Livermore Laboratory, Rept. UCRL-76690; K. Estabrook, M. Widner, I. Alexeff, and W. D. Jones, Phys. Fluids 14, 2355 (1971); A. S. Kingsep, L. I. Rudakov, and R. N. Sudan, Phys. Rev. Letters 31, 1482 (1973); J. F. Drake, P. K. Kaw, Y. C. Lea, G. Schmidt, C. S. Liu, and M. N. Rosenbluth, Phys. Fluids 17, 778 (1974); E. J. Valeo and W. L. Kruer, Phys. Rev. Letters 33, 750 (1974); A. A. Galeev, R. Z. Sagdeev, Nucl. Fusion 13, 603 (1973); K. O. Abdulloev, I. L. Bogolyubskij, and V. G. Makhankov, Nucl. Fusion 13, 21 (1975); E. J. Valeo and K. G. Estabrook, Phys. Rev. Lett. 34, 1008 (1974); S. Jonas, Phys. Fluids 17, 745 (1963); S. E. Bodner, Phys. Rev. Letters 33, 761 (1974); L. V. Dubovoi et al., Soc. Phys.-Tech. Phys. 13, 1477 (1975); W. Kruer et al., Fifth Int. Conf. on Plasma Physics and Controlled Fusion, Tokyo (1974); B. H. Ripon et al., Phys. Rev. Letters 34, 1313 (1975).

10. S. M. Osovets, Sov. Phys. Uspekhi 17, 239 (1974).

11. D. Marcuse, Engineering Quantum Electrodynamics, (Harcourt, Brace and World, New York, 1970), Chap. 4.

12. M. Friedman, Phys. Rev. Letters 31, 1107 (1973); W. R. Raudorf, (this conference); L. I. Gudgenko and E. M. Moroz, Sov. Phys.-JETP Letters 8, 265 (1968); N. E. Belov, A. V. Kisletsov, and A. N. Lebedev, Sov. Phys.-Atomic Energy 36, 251 (1974); L. N. Kazanskii, A. V. Kisletsov,

and A. N. Lebedev, Sov. Phys.-Atomic Energy 30, 30 (1971); L. Smith, Lawrence Livermore Laboratory, Rept. UCID-16199 (1973); I. I. Koba, S. G. Kononenko, L. D. Lobzov, N. I. Mocheshnikov, and N. N. Naugolnyi, Sov. Phys.-JETP Letters 20, 278 (1974); I. A. Grishaev and A. M. Shenderovich, Sov. Phys.-Tech. Phys. 17, 1871 (1973); R. J. Briggs, T. J. Fessenden, and V. K. Neil, Proc. Fourth Int. Conf. on High Energy Accelerators, IEEE Proc. on Nuc. Sci. NS-22, p. 278 (1974); A. P. Ishkov, Sov. Phys.-Tech. Phys. 16, 471 (1971); I. A. Grishaev, A. N. Dedik, V. V. Zakutin, I. I. Magda, Yu. V. Tkach, and A. M. Shenderovich, Sov. Phys.-Tech. Phys. 19, 1087 (1975); J. M. J. Madey, H. A. Schwettmen, and W. M. Fairbank, IEEE Trans. Nuc. Sov. NS-20 980 (1973); M. Friedman, Conf. Proc. Annals N.Y. Academy of Sci. 251, 294 (1975).

13. O. Buneman, Phys. Rev., 115, 503 (1959); Yu. B. Fainberg, Sov. Atomic Energy 11, 958 (1961); M. V. Nezlin, Sov. Phys.-Usp. 13, 608 (1971).

14. V. E. Zakharov, Sov. Phys.-JETP 35, 908 (1972); see also Zakharov et al., Sov. Phys.-JETP Letters 21, 4 (1975); 20, 3, 164 (1974); 19, 156 (1975); 18, 243 (1973); Sov. Phys.-JETP 39, 285 (1974); 38, 494 (1974); 34, 621 (1972).

15. H. C. Kim, R. L. Stenzel, and A. Y. Wong, Phys. Rev. Letters 33, 886 (1974); A. Y. Wong, and R. L. Stenzel, Phys. Rev. Letters 34, 727 (1975); R. L. Stenzel, A. Y. Wong, and H. C. Kim, Phys. Rev. Letters 32, 654 (1974); A. Y. Wong, R. L. Stenzel, H. C. Kim, and F. F. Chen, IAEA-CN-33/H preprint from Fifth Int. Conf. on Plasma Physics and Controlled Fusion, Tokyo (1974); G. J. Morales and Y. C. Lee, preprint of talk at Jan. 30, 1975, APS Meeting, Anaheim, Calif., p. 211; J. D. Lindl and W. C. Mead, Phys. Rev. Letters 34, 1273 (1975); E. J. Valeo and W. L. Kruer, Phys. Rev. Letters 33, 750 (1974); K. Nishihawa, Y. C. Lee, and C. S. Liu, UCLA, preprint (1973); A. Ishida and K. Nishihawa, "On Transit Time Damping of Envelope Solutions," preprint (1975).

16. R. Lee and M. Lampe, Phys. Rev. Letters 31, 1390 (1973); S. J. Gitomer, Phys. Fluids 14, 1591 (1971); R. Lee and M. Lampe, Proc. Sixth European Conf. on Plasma Physics and Controlled Fusion 2, 426 (1973); E. S. Weibel, Phys. Rev. Letters 2, 83 (1959); G. Bendford, Phys. Rev. Letters 28, 1242 (1972); R. L. Morse and C. W. Nielson, Phys. Fluids, 14, 830 (1971); also see P. Kaw, G. Schmidt, and T. Wilcox, Phys. Fluids 16, 1522 (1973); R. White and P. Kaw, preprint Trieste Conf., Matt-1103 (1975); A. B. Langdon and B. F. Lasinski, Phys. Rev. Letters 34, 934 (1975); E. Ott, W. M. Manheimer and H. H. Klein, Phys. Fluids 17, 1757 (1974); G. Schmidt, Phys. Rev. Letters 34, 724 (1975); Y. C. Lee, C. S. Liu, H. H. Chen, and K. Nishikawa, Fifth Int. Conf. on Plasma Physics and Controlled Fusion, Tokyo (1974).

17. V. Nardi, Second Topical Conf. on Pulsed High-Beta Plasmas, p. 163 (1972); Phys. Rev. Letters 25, 718 (1970); V. S. Komelkov et al., Proc. Fifth Int. Conf. on Ionized Gases, Munich (1961); Proc. of Second Int. Conf. on Peaceful Uses of Atomic Energy 31, 2504 (1959); Sov. Phys. Doklady 7, 779 (1963).

18. A. Y. Wong and B. H. Quon, Phys. Rev. Letters 34, 1499 (1975); M. S. Sodha, S. Prasad, and V. K. Tripathi, Appl. Phys. 6, 119 (1975), J. Z. Wilcox and T. J. Wilcox, Phys. Rev. Letters 34, 1160 (1975); C. E. Max, Lawrence Livermore Laboratory, Rept. UCRL-76843 PREPRINT (1975); E. J. Valeo and K. G. Estabrook, Phys. Rev. Letters 34, 1008 (1975); J. Denavit, N. R. Pereira, and R. N. Sudan, Phys. Rev. Letters 33, 1435 (1974).

19. O. S. F. Zucker and W. H. Bostick, (this conference); J. W. Poukey and J. R. Freeman, Phys. Fluids 17, 1917 (1974); K. Ikuta and A. Mohri, Nucl. Fusion 14, 444 (1974).

20. B. I. Dmitrenko. L. V. Leskov, G. P. Miksimov, S. L. Nedoseev, V. V. Savichev, V. P. Smirnov, and A. M. Spektor, Sov. Phys. 19, 1224 (1975).

21. E. Teller, in Physics of High Energy Density, R. Caldirola and H. Knoepfel, Eds., (Academic Press, New York, 1971), p. 10.

22. H. L. Sahlin, Conf. Proc. N.Y. Academy Sci. 231, 238 (1975).

23. A. A. Kolomensky, Particle Accelerators 5, 73 (1973); G. Yonas, Particle Accelerators 5, 81 (1973); C. L. Olson Phys. Fluids 18, 585, 598 (1975); B. Echer, S. Putnam, and D. Drickey, IEEE Trans. Nuc. Sci. NS-20, 301 (1973); G. W. Kuswa, L. P. Bradley, and G. Yonas, IEEE Trans. Nucl. Sci. NS-20, 305 (1973); C. L. Olson, IEEE Trans. Nucl. Sci. NS-22, 962 (1975); G. T. Zorn, H. Kim, and C. N. Boyer, IEEE Trans. Nucl. Sci. NS-22, 1006 (1975); R. B. Miller and D. C. Straw, IEEE Trans. Nucl. Sci. NS-22, 1022 (1975); A. A. Kolomensky, V. M. Likhachyov, and B. N. Yablokov, IEEE Trans. Nucl. Sci. NS-22, 1983 (1975); A. K. Berezin, Lawrence Livermore Laboratory translation UCRL-Trans-1487 (Jan. 1972); J. M. Peterson, Lawrence Berkeley Laboratory, Rept. LBL-704 Preprint (1972); J. D. Lawson, Particle Accelerators 3, 21 (1972); V. I. Veksler et al., Sov. Phys.-Atomic Energy 24, 395 (1967); A. V. Gurevich, Sov. Phys.-JETP 11, 1150 (1960); F. C. Young and M. Friedman, J. Appl. Phys. 46, 2001 (1975); A. A. Plyutto et al., Sov. Phys.-Tech. Phys. 18, 1026 (1974); D. C. Straw and R. B. Miller, Appl. Phys. Letters 25, 379 (1974); B. A. Tskhadaya, A. A. Plyutto et al., Sov. Phys.-Tech. Phys. 19, 1108 (1975); V. Y. Davydovskii and A. S. Ukolov, 19, 1303 (1975); V. Y. Davydovskii and E. M. Yakushev, Sov. Phys.-JETP 25, 709 (1967); C. B. Wheeler, J. Phys. D: Appl. Phys. 8, 12 (1975); V. B. Krasovitskii, Sov. Phys.-JETP 32, 98 (1971); S. P. Gray and H. W. Bloomberg, Appl. Phys. Letters 23, 112 (1973); L. E. Gurevich, A. A. Runyantsev, Sov. Phys. JETP 20, 1233 (1965); Masahiko Sumi, Jap. J. of Appl. Phys. 12, 1772 (1973); V. V. Pustovalov and V. P. Silin, Sov. Phys.-JETP Letters 121, 299 (1971); S. P. Gray, J. Geophys. Res. Space Phys. 73, 7524 (1968); D. W. Swift, J. Geophys. Res. Space Phys. 75, 6324 (1970); N. G. Kovalskii, B. I. Khripunov, and S. A. Chuvatin, Sov. Phys.-Tech. Phys. 16, 232 (1971); C. J. Chen, Astrophys. Letters 15, 1135 (1973); B. A. Zhigailo and N. S. Zinchenko, Sov. Phys.-Tech. Phys. 11, 1594 (1967); N. M. Gavriclov and A. V. Nesterovich, Sov. Phys.-Tech. Phys. 18, 785 (1973); A. N. Karkhov, Sov. Phys.-Atomic Energy 26, 560 (1969); V. I. Karpman, J. N. Istomin, and D. R. Shklyar, Phys. Rev. Letters 53A, 101 (1975).

24. B. B. Godfrey and L. E. Thode, Conf. Proc. Annals N. Y. Academy Sci. 251, 582 (1975).

25. J. W. Mather, Methods of Experimental Physics, R. H. Lovberg and H. R. Griem, Eds., (Academic Press, New York, 1971), Vol. 9b, p. 197.

26. W. H. Bostick, W. Prior, L. Grunberger, and G. Emmett, Phys. Fluids 9, 2078 (1966); J. W. Mather and A. H. Williams, Phys. Fluids 9, 2080 (1966); W. H. Bostick, W. Prior, and E. Farber, Phys. Fluids 8, 745 (1965); D. R. Wells, (this conference); W. H. Bostick Proc. of Fifth Int. Conf., Vol. 2, 1562 (1961); W. H. Bostick, V. Nardi, L. Grunberger, and W. Prior, in Solar Magnetic Fields, R. Howard, Ed. (D. Reidel Publishing Company, Doedrecht-Holland, 1971), p. 512; W. H. Bostick, V. Nardi and W. Prior, in Cosmic Plasma Physics, Karl Schindler Ed. (Plenum, New York, 1970), p. 175; W. H. Bostick, L. Grunberger, V. Nardi and W. Prior, Proc. of the Fifth Symposium on Thermophysical Prop, p. 495 (1970); (1) W. Prior, W. H. Bostick, L. Grunberger, P. Palmdesso, and J. Zorskie; and (2) F. Grotton; Proc. Second Topical Conf. on Pulsed High Beta Plasmas, pp. 155, 159 (1972); K. Braun, H. Fischer, and L. Michel, Proc. Second Topical Conf. on Pulsed High Beta Plasmas, p. 183 (1972).

27. W. L. Kruer and J. M. Dawson, Phys. Fluids 15, 446 (1972); A. T. Lin, J. M. Dawson and B. D. Fried, paper presented at Fifth Annual Anomalous Absorption Conference, UCLA, April 22-24 (1975); W. L. Kruer, J. Katz, J. Byers and J. DeGroot, Phys. Fluids 15, 1613 (1972); V. A. Skoryupin, Sov. Phys.-JETP 26, 711 (1968); L. V. Dubovoi, V. P. Fedyakov and V. P. Fedyakova, Sov. Phys-JETP, 32, 805 (1971); T. R. Soldatenkov, Nucl. Fusion 10, 69 (1973); K. Minami, K. P. Singh, M. Masuda and K. Ishii; Phys. Rev. Letters 33, 740 (1974); K. F. Sergeirchev and V. E. Trofimov, Sov. Phys.-JETP Letters 13, 166 (1971); B. A. Demidov, N. I. Elagin, and S. D. Fanchenko, Sov. Phys.-Dokl. 12, 467 (1967).

28. A. Bernard, A. Coudeville, J. Durantet
     A. Jolas, J. Launspach, J. de
     Mascureau, and J. P. Watteau, Proc.
     Second Topical Conf. on Pulsed High-
     Beta Plasmas, p. 147 (1972).

29. R. L. Gullickson, Lawrence Livermore
     Laboratory, Rept. UCRL-76831 PREPRINT
     (1975); paper presented at IEEE Conf.
     on Plasma Sciences, May 14-16, 1975;
     R. L. Gullickson and R. H. Barlett,
     Lawrence Livermore Laboratory Rept.
     UCRL-75910 PREPRINT (1974); to appear
     in Proc. Conf. on Appl. of X-Ray
     Analysis, Denver, Colorado (1974);
     D. J. Johnson, J. Appl. Phys. 45,
     1147 (1974); J. H. Lee, D. S. Loebbaka
     and C. E. Roos, Plasma Phys. 13, 347
     (1971); H. Conrads, P. Cloth,
     M. Demmeler, and R. Hecker, Phys.
     Fluids 15, 209 (1972); H. Conrads,
     and P. Cloth, Fifth European Conf.
     on Plasma Phys. and Controlled Fusion
     2, 67 (1972); P. Cloth, H. Conrads,
     M. Dammeler and R. Hecker, Sixth Symp
     on Fusion Technology, p. 525, (1970);
     V. A. Gribkov, O. N. Krokhin,
     G. V. Sklizkov, N. V. Filippov, and
     T. I. Filippova, Sixth European Conf.
     Plasma Phys. and Controlled Fusion,
     375 (1974); V. A. Gribkov, O. N.
     Krokhin, G. V. Sklizhov, N. V.
     Filippov, and T. I. Filippova, Sov.
     Phys.-JETP Letters 18, 5 (1973);
     JETP Letters 18, 319 (1973); JETP
     Letters 15, 232 (1972); S. V.
     Bazdenkov, K. G. Gureev, N. V.
     Filippov and T. I. Filippova, JETP
     Letters 18, 118 (1973); I. F. Belyaeva
     and N. V. Filippov, Second Topical
     Conf. on Pulsed High Beta Plasma,
     p. 191 (1972); I. F. Belyaeva and
     N. V. Filippov, Nucl. Fusion 13, 881
     (1973); G. E. Notkin, N. V. Filippov
     and D. A. Shcheglov, Fifth European
     Conf. on Plasma Physics and Controlled
     Fusion 2, 69; V. A. Gribkov, V. M.
     Korzhavin, O. N. Krokhin, V. Ya.
     Nikulin, G. V. Sklizhov, N. V.
     Filippov and T. I. Filippova, Fifth
     European Conf. on Plasma Physics and
     Controlled Fusion 2, 64 (1972);
     N. J. Peacock, M. J. Forrest, M. G.
     Hobby, and D. D. Morgan, Fifth Euro-
     pean Conf. on Plasma Physics and
     Controlled Fusion 2, 66 (1972);
     V. I. Agafonov, G. V. Golub, L. G.
     Golubchikov, V. F. Dyachenko, V. D.
     Ivanov, V. S. Imshennik, Yu A.
     Kolesnikov, E. B. Svirsky, N. V.
     Filippov, and T. I. Filippova,
     Special Suppl. to Nucl. Fusion,

     p. 121 (1969); M. J. Bernstein,
     C. M. Lee, and F. Hai, Phys. Rev.
     Letters 27, 844 (1971); R. C. Elton
     and T. N. Lee, Space Sci. Rev. 13,
     747 (1972); Ch. Maisonnier et al.,
     Proc. Fifth Int. Conf. on Plasma
     Physics and Controlled Fusion, Tokyo
     (1974); N. V. Filippov et al., Fifth
     Int. Conf. on Plasma Physics and
     Controlled Fusion, Tokyo (1974);
     N. Lee, Conf. Proc. Annals, N.Y.
     Academy of Sci. 251, 112 (1975).

30. J. Fukai, E. J. Clothiaut, Phys.
     Rev. Letters 34, 863 (1975); M. D.
     Raizer and V. N. Tsitovich, Plasma
     Phys. 7, 203 (1965); H. M. Epstein,
     W. J. Gallagher, P. J. Mallozzi, and
     T. F. Stratton, Phys. Rev. A2, 146
     (1970); S. P. Gray, Phys. Fluids 17,
     2135 (1974); S. P. Gray and F. Hohl,
     Phys. Fluids, 16, 997 (1973); C. E.
     Newman, V. Petrosian, Phys. Fluids
     18, 547 (1975); M. J. Bernstein,
     Phys. Fluids 13, 2858 (1970); M. J.
     Bernstein, G. G. Comisar, Phys.
     Fluids 15, 700 (1972); G. A. Askariyan,
     Sov. Phys. Atomic Energy 6, 487
     (1959); S. Nagoo, Third Int. Conf.
     on Ionization Phenomena on Gases,
     p. 736 (1957).

31. W. H. Bostick, V. Nardi, and W. Prior
     Conf. Proc. Annals N.Y. Academy of
     Sci 251, 2 (1975).

32. A. Bernard, A. Coudeville, J. P.
     Garconnet, P. Genta, A. Jolas,
     Y. Landure, J. De Mascoreau, C. Nazet,
     and R. Vezin, Lawrence Livermore Lab-
     oratory translation UCRL-Trans-10801
     of IAEA-CN-33/E6-1 (Dec. 1974); A.
     Bernard, A. Coudeville, A. Jolas,
     J. Launspach, and J. de Mascureau,
     Phys. Fluids 18, 180 (1975).

33. W. H. Bostick, V. Nardi, W. J. Prior,
     and F. Rodriquez-Trelles, Proc.
     Second Topical Conf. on Pulsed High
     Beta Plasmas, p. 155 (1972); W. H.
     Bostick, V. Nardi, W. Prior, Fifth
     Int. Conf. on Plasma Physics and
     Controlled Fusion, Tokyo (1974).

34. S. K. Handel, B. Stenerhag, I. Holm-
     strom, Nature 209, 1227 (1966);
     U. Fischer, H. Jager, and W. Lochte-
     Holtgreven, Proc. Sixth European
     Conf. on Plasma Physics and Con-
     trolled Fusion, p. 442 (1974);
     D. Mosher, L. S. Levine, S. J. Step-
     hanakis, I. M. Vitkovitsky, and

F. Young, Proc. Sixth European Conf. on Plasma Physics, and Controlled Fusion, p. 419 (1974); W. G. Chance and M. A. Levine, J. Appl. Phys. 31, 1298 (1970); B. Stenerhag, S. K. Handel, and I. Holmstrom, Z. Physik 198, 172 (1967); B. Stenerhag, S. K. Handel, and B. Gohle, Rev. Sci. Instr. 40, 563 (1969); I. Holmstrom, S. K. Handel, and B. Stenerhag, J. Appl. Phys. 39, 2998 (1967); S. O. Nyberg. S. K. Handel, and B. Stenerhag, J. Appl. Phys. 45, 1746 (1974); A. E. Vlastos, J. Appl. Phys. 44, 106 (1973).

35.   D. Y. Cheng, Nucl. Fusion 13, 129 (1973); J. W. Shearer, C. W. Hartman, R. H. Munger, R. L. Gullickson, D. O. Trimble, and D. Y. Cheng, Lawrence Livermore Laboratory Rept. UCID-16738, (1975); E. C. Cnare, J. Appl. Phys. 32, 1275 (1961); D. Y. Cheng, W. J. Loubsky, and V. E. Fausekis, Phys. Fluids 14, 2328 (1971); V. S. Komelkov and V. I. Modzolevskii, Sov. Phys.-JETP Letters 15, 210 (1972); J. Lipiec, W. Soszka and Taraszkiewiez-Plasma Phys. 13, 537 (1971); A. M. Timonin and V. A. Churaev, Sov. Phys.-Tech. Phys. 17, 788 (1972); M. G. Haines, Proc. Phys. Soc. (London) 74 576 (1959); I. F. Kvartskhava, Yu V. Matveyev, E. Y. Khautiyev, AEC-tr-7305, Sept. (1970).

36.   V. A. Gribkov, (this conference).

37.   W. H. Bostick, V. Nardi, W. Prior, J. Plasma Phys. 8, 7 (1972); W. H. Bostick, Proc. Fifth European Conf. on Plasma Physics and Controlled Fusion 2, 239 (1972); W. H. Bostick, V. Nardi, W. J. Prior, and F. Rodriquez-Trelles, Proc. Second Topical Conf. on Pulsed High Beta Plasmas, p. 155 (1972); W. H. Bostick, V. Nardi, and W. Prior, Sixth European Conf. on Plasma Physics and Controlled Fusion 2, 395 (1973).

38.   W. R. Bennett, summary of talk, Conf. Proc. Annals N.Y. Academy of Science 251, 213 (1975).

39.   E. P. Velikhov, A. A. Vedenov, A. D. Bogdanets, V. S. Golubev, E. G. Kasharskii, A. A. Kiselev, F. G. Rutberg, and V. V. Chernukha, Sov. Phys.-Tech. Phys. 18, 274 (1973); E. P. Velikhov, V. S. Goloubev, and V. V. Chernovkha, Paper delivered at Culham, Sept. (1974), preprint.

40.   H. A. Schwettman, IEEE Trans. Nuc. Sci. NS-22, 1118 (1975); P. H. Ceperley, I. Ben-Zvi, H. F. Glavish, and S. S. Hanna, IEEE Trans. Nuc. Sci. NS-22, 1153 (1975); P. Kneisel, C. Lyneis, and J. P. Turneaure, IEEE Trans. Nuc. Sci. NS-22, 1197 (1975).

41.   J. Franck and G. Hertz, Verhandl. Deut. Physik Ges. 16, 457 and 512 (1914).

42.   R. Hofstadter, "The Atomic Accelerator," Stanford Univ., HEPL Rept. 560 (1968); Y. V. Petrov, Sov. Phys.-JETP 36, 395 (1973); E. K. Zavoiskii and V. I. Kurilko, Sov. Phys.-Dokl. 14, 790 (1970); V. B. Krasovitskii, Sov. Phys.-JETP 30, 951 (1970); A. P. Kazantsev, JETP Letters 17, 150 (1973); V. P. Gavrilov and A. A. Kolomenskii, JETP Letters 14, 431 (1971); P. L. Csonka, Particle Accelerators 2, 39 (1971).

43.   A. C. Scott, F. Y. F. Chu, and D. W. McLaughlin, Proc. IEEE 61, 1443 (1973).

44.   S. L. McCall and E. L. Hahn, Phys. Rev. Letters 18, 908 (1967); Phys. Rev. 183, 457 (1969).

45.   R. M. Phillips, IRE Trans. on Electron Devices ED-7, 231 (1960); C. E. Enderby and R. M. Phillips, Proc. IEEE 53, 1648 (1965).

46.   N. F. Kovalev, M. I. Petelin, M. D. Raizer, A. V. Smorgonskii, and L. E. Tsopp, Sov. Phys-JETP Letters 18, 138 (1973); V. L. Granatstein, M. Herndon, P. Sprangle, Y. Carmel, and J. A. Nation, Plasma Phys. 17, 23 (1975); Y. Carmel, J. Ivers, and R. E. Kribel, J. A. Nation, Phys. Rev. Letters 33, 1278 (1974); N. I. Zavtsev, T. B. Pankratova, M. I. Petelin, and V. A. Flyagin, Sov. Phys.-RadioPhysics 19, 103 (1975); V. L. Granatstein, M. Herndon, R. K. Parker, and S. P. Schlesinger, Int. Conf. on Submillimeter Waves and Their Applications, p. 21 (1974).

47.   M. L. Sloan and W. E. Drummond, Phys. Rev. Letters 31, 1234 (1973); also see V. P. Indykul, I. P. Panchenko, V. D. Shapiro, and V. I. Shevchenko, Sov. Phys.-JETP Letters 20, 65 (1974).

48. V. P. Smirnov, private communication, August 1974.

49. See Ref. 24.

50. See Ref. 4.

51. H. L. Sahlin, Lawrence Livermore Laboratory, unpublished note, "The Electromagnetic Brake," Oct. 9, 1971.

52. H. Sahlin and E. Teller, Chap. 1 and 2 in Vol. 5 of Physical Chemistry, An Advanced Treatise, H. Eyring, D. Henderson, and W. Jost, Eds., (Academic Press, New York, 1970).

53. D. L. Ensley, Lawrence Livermore Laboratory, to be published.

54. H. Hora, see Ref. 3 above.

55. R. L. Hirsch, J. Appl. Phys. $\underline{38}$, 4522 (1967); Phys. Fluids $\underline{11}$, 2486 (1969); also see papers on Part II, Electrostatic Confinement of Plasmas, Conf. Proc. Annals of N.Y. Academy Sci. $\underline{251}$, 126-190 (1975).

56. R. P. Feynman, in The Feynman Lectures on Physics, Feynman, Leighton, and Sands, Eds., Addison-Wesley, Reading, Mass. (1965) Vol. III, Chap. 21.

57. P. Carruthers and M. Nieto, Rev. Mod. Phys. $\underline{40}$, 411 (1968); Am. J. Phys. $\underline{33}$, 537 (1965).

58. H. Schiff, Proc. Roy. Soc. (London) $\underline{269}$, 277 (1962).

59. Louis de Broglie, Foundations of Phys. $\underline{1}$, 5 (1970).

60. G. K. Batchelor, Proc. Roy. Soc. (London) $\underline{A201}$, 405 (1950).

61. G. Della Riccia and N. Wiener, J. Math Phys. $\underline{7}$, 1372 (1966).

62. D. Falkoff, in Quantum Optics (Academic Press, New York, 1969), p. 147; Y. Aharonov, D. Falkoff, E. Lerner, and H. Pendleton, Ann. Phys. $\underline{39}$, 498-512 (1966).

63. B. Freeman, J. Luce, and H. Sahlin, Proc. Fifth European Conf. on Plasma Physics and Controlled Fusion $\underline{2}$, 68 (1972).

64. F. London, Macroscopic Theory of Superconductivity, (Dover, New York, 1960); F. London, Superfluids, (John Wiley & Sons, New York, 1954).

65. R. J. Glauber, Ed., Quantum Optics (Academic Press, 1967); Selected papers on Coherence and Fluctuations of Light, Vols. 1 and 2, L. Mandel and E. Wolf, Eds. (Dover, New York, 1970); J. S. Langar, Phys. Rev. $\underline{167}$, 183 (1968); G. A. Casher and M. M. Revzen, Am. J. Phys. $\underline{35}$, 1154 (1967); M. M. Nieto, Phys. Rev. $\underline{167}$, 416 (1968); J. R. Klauder and E. C. G. Sudershan, Fundamentals of Quantum Optics (Benjamin, New York, 1968); A. C. Biswas, C. S. Warke, Phys. Rev. $\underline{A}$, 2568 (1970); P. W. Anderson, Rev. Mod. Phys. $\underline{38}$, 298 (1966).

66. This strongly nonlinear operating region may have been attained in the Resonatron tube, a highly efficient microwave tube employing a grid-screened cathode driven by a phased feedback signal. (L. C. Marshall, Sea Ranch California, private communication, 1975.) The Resonatron tube was discovered by L. C. Marshall, D. Sloan, and H. Salisberry.

EXPERIMENTAL RESULTS OF A LOW ENERGY PLASMA FOCUS

H. Bruzzone, R. Gratton, H. Kelly, M. Milanese and J. Pouzo

Laboratorio de Física del Plasma

Facultad de Ciencias Exactas y Naturales - UBA - Departamento de Física

Pabellón I - Ciudad Universitaria - Buenos Aires - Argentina

ABSTRACT

We study the structure, motion and collapse of the current sheath (CS) in a small low pressure Plasma Focus ($\approx$ 1 KJ, 0.4 to 3 Torr with D_2, 0.03 to 1.4 Torr with A). The observation of a radial component of the magnetic field and the image converter pictures show a complex structure of the CS. However, a snow-plow model is adequate to describe the CS motion as a whole, which in turn, determines the CS thickness regardless of the gas employed. In the focus stage we observe bursts of non thermal microwaves, hard X-rays (0.5 - 1.0 MeV), and neutrons.

INTRODUCTION

We report some results obtained with a small Plasma Focus device. A fast capacitor bank (V $\leq$ 16 KV, C= 6 μF, T= 4.2 μsec) is discharged between two coaxial electrodes (r_i= 0.9 cm, r_e= 2.7 cm, effective length l= 7 cm). We used D_2, A, He and air. Pressure ranges are given in Table 1. We study: 1) the propagation of the CS and its structure by means of miniaturized magnetic probes and an image converter camera (time of exposure $\geq$5 nsec); 2) the sheath collapse and the focus phase using the image converter camera, X-ray, microwave and neutron detectors. In addition, we measure with a Rogowsky coil the derivative of the total discharge current. The CS structure appears to be fairly complex: a pattern of filaments is clearly present at least in photographs of deuterium discharges, a radial component of the magnetic field has been detected and the thickness of the magnetic field transition through the sheath is about 1 cm. Nevertheless, the motion of the CS as a whole is well explained in terms of a simple snow-plow model, just assuming that in its advance, the sheath sweeps a constant fraction of the gas. In a preliminary study of the focus phase we detected bursts of non thermal microwave emission ($\lambda \approx$ 3 cm), besides the usual X-rays and (with D_2) neutron pulses.

MOTION OF THE SHEATH

The time interval t(z) between the onset of the discharge current and the onset of the signal given by a magnetic probe located at a distance z from the end of the electrodes (oriented in order to select

Table 1. Pressure ranges and η values.

Gas	Pressure range (Torr)	η
Argon	0.03 - 1.40	0.32
Air	0.15 - 0.80	0.45
Helium	0.20 - 2.80	0.55
Deuterium	0.40 - 3.00	0.64

the azimuthal component B_ϕ of the field) has been determined within the pressure ranges of Table 1. Some typical data are shown in Fig.1. The results agree with a snow-plow model provided the filling density ρ is multiplied by a factor η (η < 1) constant for each gas; η may be thought of as giving the efficiency of the sweeping process, and appears to be larger for

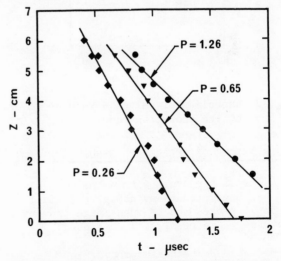

Fig. 1. Position of the CS vs t. Gas A, P in Torr.

lighter gases. The values of η are given in Table 1.

A sheath velocity ν' may be defined in terms of the slope of the t(z) vs. z curves, which are roughly linear. On the other hand, the snow-plow model gives a velocity ν (which turns out to be nearly time independent after a short initial transient):

$$\nu \approx \frac{VC}{T}\left[\frac{\mu_0 \ln(r_e/r_i)}{\eta\rho (r_e^2-r_i^2)}\right]^{1/2} \approx 7.5 \times 10^{-4} \frac{V}{\sqrt{\eta\rho}} \text{ (MKS)}$$

The agreement with the experimental values is very good, as shown in Fig.2, where ν' is plotted vs ν for different gases for a wide range of pressures. The dependence of ν on V and ρ agrees with the results obtained by several authors[2,3] However, Mather[4] and Dattner and Eninger[5] found different relations. These discrepancies can be explained in terms of currents flowing on the insulator wall[6], which are likely to vary from one device to another. In our case these currents are negligible, as we checked by comparing the magnetic field intensity actually measured and that computed from the total discharge current.

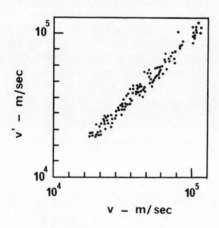

Fig. 2. Experimental values ν'vs.ν as given by the snow-plow model.

OBSERVATIONS OF THE SHEATH STRUCTURE

The CS structure within the coaxial gun has been studied by using very small magnetic probes and through image converter pictures (time resolution of 5 nsec).

1. Magnetic Probe Measurements.
a) The rise time Δt of B_ϕ when the sheath impinges on the probe, times ν' gives the thickness D of the CS. In Fig.3 D is given as a function of t(z). The position of the

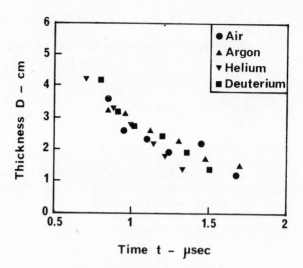

Fig. 3. Thickness of the CS vs. t.

probe was kept fixed (r = 14 mm, z =20 mm) in the measurements reported, so that different values of t correspond to different pressures for each gas. Our results show that for each fixed t, D has roughly the same value for all the gases employed.This fact can be stated as follows: if the filling pressures of different gases are adjusted in such a way that the global CS dynamics is the same for all (i.e. they have equal ν'), then also the CS thicknesses will have the same value.

One is then led to conclude that D is determined by the global dynamic processes of the CS and is not directly related to the microscopic parameters of the gas (such as cross-sections, molecular masses, etc.).

b) The position of the leading and trailing edges of the CS as a function of r have been determined from $B_\phi(t)$ measured at a fixed z near the muzzle of the coaxial gun, where ν is constant, with the assumption that D does not change in the time intervals involved. In Fig.4 a typical result is shown: we note that D is almost independent of r.

c) A radial component B_r of the magnetic field has been detected and measured by using magnetic probes with external dimensions less than 1 mm; however we failed to detect this component when larger probes (≈2.5 mm) were employed. For a given z, B_r is different from zero only during a time interval $\Delta t_r < \Delta t$, hence the radial component is limited to a thin inner region of the CS.

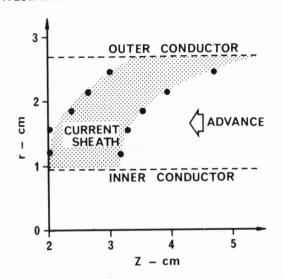

Fig. 4. Leading and trailing edges of the CS. Argon, 0.26 Torr.

We observed strong fluctuations and sign inversions of B_r from shot to shot, but $|B_r/B_\phi|$ was always less than 0.2. These results suggest that B_r is related to small scale structures within the CS.

2. Image Converter Pictures. Frontal pictures of the CS showed a well defined pattern of filaments when deuterium was employed (see Fig.5). There is a neat, sudden transition between an inner region where the filaments are very thin and difficult to be resolved and an intermediate region where they are thick and well resolved; finally, near the external electrode we did not observe filaments but a ring-like structure of bright and dark bands which might be due to reflections.

This situation changes when the filling pressure is varied: in the 2-3 Torr range, the intermediate region is larger and practically all the space between electrodes is bridged by the filaments; when

the filling pressure is lowered, the intermediate region becomes gradually smaller and practically disappears at 0.5 Torr.

In the intermediate region the apparent diameter of the filaments is ≈1 mm, their number about 30 and they show some tendency to appear grouped in pairs. No filaments have been observed with Argon and air as filling gases in all the range of pressures studied (1000-18mTorr), however the sheath appeared clearly not uniform. It should be noted that in our experimental conditions, filaments spaced 1 mm or less could not be resolved.

OBSERVATIONS ON THE FOCUS PHASE

The focus phase is characterized by a sudden variation of the total discharge current, which produces a well defined dip in the dI/dt curve. Almost at the same time or with slight delays, microwaves, X-rays and (if D_2 is used), neutrons are emitted as revealed by the appropriate detectors. Frontal and sideview sequences of image-converter pictures show the collapse of the CS towards the z-axis.

The sheath collapses over a wide range of pressures. The dI/dt dip has been observed in A at pressures as low as 30 mTorr, while the upper pressure limit (0.5 Torr) is given by the condition that the sheath must arrive at the end of the gun well within the first half cycle of the discharge (≈ 2 μsec).

1. Microwave Detection. An X-band detector was employed to record emission in the direction of the gun axis (time resolution ≈ 10 nsec). Measurements were performed with A and D_2 as filling gases and yielded similar results. The emission takes place in several non completely resolved bursts (typically 2-3 bursts per shot) ≈ 10 nsec wide, distributed over a time of 50-100 nsec, the first burst being simultaneous with the vertex of the dI/dt dip. The peak power radiated per sr reached values of 0.1 - 1 mW, much in excess with respect to the thermal emission level.

2. X-Ray Detection. A plastic scintillator was employed; discrimination with respect to neutrons in the case of D_2 was accomplished by time of flight measurements. Time resolution was about 20 nsec. Two peaks were usually observed. The first one (which sometimes was not detected) is simultaneous with the vertex of the dI/dt dip, its duration is 20 nsec and corresponds to soft X-rays in the KeV range.

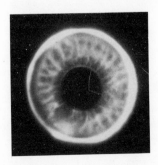

Fig. 5. Frontal view of the CS. Deuterium, 2.4 Torr.

The second peak comes 40 nsec later, its duration being 20-30 nsec. By measuring the attenuation produced by lead absorbers we estimated that a considerable fraction of the second peak was in the 0.5-1.0 MeV range.

3. Neutrons. To discriminate against hard X-ray radiation, it was necessary to place the detector far from the focus. The neutron emission was almost simultaneous with that of the hard X-rays as it was checked by varying the position of the detector.

Total neutron production was about 10^6 per pulse. Since this was near the lower limit of detection we could not perform detailed measurements.

FINAL REMARKS

Our work deals with a régime of energy and pressures considerably different with respect to most of the current experimental work on Plasma Focus. However, the main features of the phenomenon are essentially the same. In particular it is worth mentioning several facts, such as observation of filamentary patterns, detection of a radial component of the magnetic field, detection of microwave bursts and hard X-rays of clearly non thermal origin, etc.

These facts confirm the presence of fine structures of plasma and magnetic field within the CS, of the kind already observed by other workers.[7-8]

We believe that our results may contribute to the development of models which relate the focus phenomena to the CS structure[9].

AKNOWLEDGEMENTS

We are deeply grateful to Drs. W. Bostick and V. Nardi (Stevens Institute of Technology, Hoboken N.J., USA), for their constant encouragement, and, together with Mr. W. Prior, for their material support in the construction of the apparatus.

This work was supported by the Comisión Nacional de Estudios Geo-Heliofisicos (Argentina) and by the Fondo Especial para la Investigacion, Universidad Nacional de Buenos Aires. One of us (J.P.) was benefited by a fellowship from the Servicio Naval de Investigación y Desarrollo, A.R.A.

REFERENCES

1 C. Patou, A. Simonnet, J.P. Watteau. Le Journal de Physique, 29, 973 (1968).

2 J. Keck. The Physics of Fluids, 7, S 16 (1964).

3 T. Butler, I. Hennis, F. Jahoda, J. Marshall, R. Morse. Phys. Fluids, 12, 1904 (1969).

4 J. Mather. Phys. Fluids, 8, 366 (1965).

5 A. Dattner, J. Eninger. Phys. Fluids, 7, S 41 (1964).

6 J. Jelen, Plasma Phys., 14, 1117 (1972)

7 W. Bostick, V. Nardi, W. Prior. Journal of Plasma Physics, 8, 7 (1972).

8 W. Bostick, V. Nardi, W. Prior, "Decay of the Magnetic Structures of Dense Plasma and X-Ray and Microwave Emission". Preprint (1973).

9 W. Bostick, L. Grunberger, V. Nardi, W. Prior. Proceedings 4th. European Conference on Plasma Physics and Nuclear Fusion. Rome, p.108 (1970).

TIME RESOLVED X-RAY FINE STRUCTURES, NEUTRON EMISSION, AND ION VELOCITIES IN A FAST 1 kJ PLASMA FOCUS

K. H. Schoenbach, L. Michel, H. Krompholz,
and H. Fischer
Applied Physics, Technische Hochschule
Darmstadt, Germany

ABSTRACT

One to four (mostly 2) X-ray spot structures are present in hydrogen and deuterium along the z-axis during the dense pinch phase; approximately 4% argon is required for soft radiation pin hole photography. Spots have an observed 5 to 10 ns lifetime and develop in succession within the first 30 ns of the X-ray emission. The hard > 80 keV radiation shows the spot time structure which controls the neutron pulse. The X-ray spots represent $m = 0$ plasma instabilities.

The neutron energy spectrum emitted during the dense pinch phase (2×10^8 n per shot) was determined from 30 m flight histograms. The neutron producing runway ions (plasma temperature ~ 1 keV) have an average 100-120 keV energy within a 15-25 degree beam angle to the z-axis when the target beam model is applied in the theoretical analysis.

Experimental magnetic analysis of the ion beam results in an ion energy spectrum between 50 - 150 keV. The neutrons are produced at the far end of the focus and beyond within about 4 cm distance from the front face of the center electrode. This is shown directly by neutron cut-off.

NONUNIFORM ENERGY CONCENTRATION IN FOCUSED PLASMAS*

W. H. Bostick, V. Nardi, W. Prior, F. Rodriguez-Trelles**
Stevens Institute of Technology, Hoboken, N. J.

C. Cortese, W. Gekelman†
I.E.N.G.F., Torino, Italy

ABSTRACT

The dynamics of the dense plasma column ($\sim 10^{20}$ particles/cm^3) produced in a deuterium plasma focus experiment is studied by the emission spectrum from hard x-rays to infrared and microwaves. A maximum of space and time resolution shows energy lumping within time and space intervals $\Delta\tau \sim 1 - 5$ ns $\Delta\ell \sim 0.1 - 0.5$ mm. This energy granulation involves a fraction $\sim 10\%$ of the energy of the whole system and occurs by localized conversion of magnetic field energy into particle energy. The non-thermal distribution of ions and electrons as it is derived from radiation anisotropy and polarization, fits the observed beaming of electrons and ions that is observed by other methods (including anisotropy measurements of the neutron emission).

1. INTRODUCTION

Observations of the EM emission spectrum from focused plasmas indicate that the radiation source in the plasma is localized either in space[1] or in time[2] or both. A typical example of space localization are the x-ray sources in plasma regions with linear dimensions ℓ ($\sim 0.01 - 1$ mm) which depend on the minimum value ε_{min} of the photon energy we use for observation (e.g. to form an image on a film): in general the larger we choose $\varepsilon_{min} > 1$ keV the smaller becomes ℓ. Microwave radiation ($\lambda = 3 - 10$ cm) from focused plasma shows an extreme localization in time, with emission peaks $1 - 3$ ns FWHM[2]. A peak of infrared emission (IR; $\lambda \sim 1 - 4$ μm) is also observed over a time interval ~ 50 ns in a coincidence with the emission of very hard x-rays ($\varepsilon \gtrsim 1$ MeV). Our data on infrared emission are reported in paragraph 3. Observations without time resolution[3] or at somewhat different wavelengths ($\lambda \gtrsim 5$ μm)[4] by other laboratories have already indicated the nonthermal origin of this IR emission. The filamentary structure of the current sheath,[1,5] the observed polarization of microwaves pulses[2] and the emission anisotropy of bremsstrahlung x-rays[1,6] and D-D neutrons [7,8] are also typical of a nonthermal plasma. We want to check whether or not a correlation exists among bursts of different radiation. The aim is to understand better particle-acceleration mechanisms which cause the emission pulses (and the underlying energy concentration in the plasma). Our data specify to what extent a correlation exists and lead to an estimate of the correlation parameters. We have used two coaxial accelerators with the same geometry and the same linear dimensions one at IEN and one at SIT (a Marshall gun with a Mather-type pyrex insulator).[1] The two electrodes are coaxial cylinders with a circular cross-section; the hollow center electrode-anode- is a brass pipe with 2 mm thick wall, diameter 2 R_c = 3.4 cm; external electrode dia. 2 R_e^c = 10 cm; by our operation conditions the current sheath reaches the electrode end at the time of peak value I_M of the electrode current I for an electrode length ~ 9 cm. Nearly the same energy $\gtrsim 5$ kJ was used to operate the focus device at IEN (75 μF capacitor bank at 12 kV) and at SIT (45 μF at 15 kV). Under those conditions I_M was about 10% higher at IEN than $I_M \sim 0.6$ MA at SIT. Usually the neutron yield sharply increases with I_M[9] and somewhat decreases by increasing the D_2 filling pressure P above optimum conditions[9,10]. With P = 8 Torr at IEN the neutron yield n was decreased to the same mean value $\bar{n} = 1.0 - 1.1 \times 10^8$ neutrons per shot (average over ~ 100 discharges) we had at SIT with P = 5.6 Torr. The following data have all been obtained specifically under this variety of operation conditions, with large fluctuations of n about the same n ($10^{-1} \times \bar{n} \lesssim n \lesssim 3.5 \times \bar{n}$). The observed IR emission fits $\omega_{el.cycl.}$ for the magnetic field intensity B $\sim 10^8$G we estimate by x-ray bremsstrahlung anisotropy[5,6].

2.   X-RAY AND NEUTRON EMISSION

Our observations are made during the time interval ($\sim$ 500 ns) of most intense emission activity in the plasma when: (A) the current sheath pinches on the electrode-axial region forming a cusped column (mean diameter 3-4 mm) with a relatively-long life time ($\sim$ 50-60 ns); (B) a typical necking (dia. << 1 mm) can be detected in one or in several points of the plasma column; (C) the column is disassembled. The details in this stage (A,B,C) of the discharge have been observed best by image converter (IC) photographs (5 ns exposure by visible light; essentially low frequency bremsstrahlung[11,1,5]) by shadowgraph (with ruby-laser light[12,13]), and by x-ray pinhole-camera photographs, with quite consistent results. Pinhole-camera photographs give a maximum of space resolution with a limit ($\sim$ 10 μm) due only to silver-grain size. A variety of structural details of a localized x-ray source can be recorded on films by a convenient choice of the filter thickness, i.e. of ε_{min} photons reaching the film (beryllium filters are used to avoid k-edge effects). Time integration on films does not prevent a time resolution of the recorded source activity because different regions of plasma usually radiate on different time intervals with a peak intensity by different photon-energies ε.[5]

The x-ray emission in a discharge is simultaneously recorded by pinhole photographs from different directions ($\varepsilon \stackrel{>}{\sim} 2$ keV) an by a x-ray (NE-102) scintillation detector ($\varepsilon \stackrel{>}{\sim} 30$ keV; 3 mm Aℓ + 8 mm pyrex filter). A set of three pinhole cameras is introduced inside the discharge chamber in all experiments and the films are exposed only to a single discharge, after firing three or four discharges for electrode conditioning (a neutron yield n $\stackrel{\sim}{\sim}$ n was obtained in at least one of these preliminary discharges). NE-102 detector and photomultiplier (RCA-8275, rise time $\sim$ 2 ns) are inside a box with lead walls 2.5 cm thick on all sides to screen out diffused x-rays. The lead has a 1-mm-dia. window facing the NE-102 crystal (0.1 mm thick, less than a D-D neutron mean-free-path in the plastic detector) sealed behind 3 mm Aℓ. X-ray signal and photographs from the same discharge are presented in Fig. 1 (discharge with a low neutron yield) and in Fig. 2 (high yield discharge).

A multiplicity of localized sources on a film and of peaks in the corresponding scintillator signal are observed by each discharge. By photographs with differential-filters we know[1] that the image on film is formed by $\varepsilon \stackrel{>}{\sim} 2$ - 10 keV photons with somewhat harder photons ($\varepsilon \stackrel{>}{\sim} 5$ - 10 keV) from the central part of a localized source (the source core, say; $\ell \sim 10$ - 100 μm). This core is embedded in a more diffuse source of softer photons ($\varepsilon \sim 2$ keV) and lower intensity. The time-resolved NE-102 signal indicates that the very hard photons are produced by two or three initial emission peaks on a time interval $\sim$ 50 ns (photons with $\varepsilon \stackrel{>}{\sim} 1$ MeV have been observed by D_2 or H_2 filling also in previous experiments [14] with lead walls 6 cm thick - no windows - on all sides of the detector system). The number N_F of localized sources on each film is usually matched by an equal number N_s of peaks in the scintillator signal. For a quantitative analysis each peak in a NE-102 scintillator signal is classified by its height h[h= h_w, 50 mV < $h_w \stackrel{\sim}{\sim} 100$ mV weak; 100 mV < $h_m \stackrel{\sim}{\sim} 250$ mV medium; h_s > 250 mV strong; see Fig. 1,2; N(w), N_s(m), N_s(s) are, respectively, number of weak, medium, strong peaks in a single discharge; N_s (s+m+w) $\equiv N_s$(s) + N_s(m) + N_s(w), etc.] . A similar analysis is performed on N_F by micro-densitometer reading of film density d; the exposure E erg/$cm^2 \sim$ d (mean value on a film area 50 μm x 50 μm centered on the image of a localized-source core) is used instead of h (same delimitations for E_s, E_m, E_w as for h_s, h_m, h_w).

Number of shots N vs. $\Delta = N_s - N_f$ is reported in Fig.3 histogram for a succession of 42 shots. Each group of weak, medium, strong signal-peaks/localized sources is separately considered. Even though different spectral regions determine N_s (ε > 30 keV) and N_F ($\varepsilon \sim 2 - 10$ keV) most of these shots have $\Delta = 0$ except for the weak group, in which data accumulation occurs for N_s(w) usually exceeding N_f(w) (a value $\Delta_w \equiv N_s$(w) - N_f(w) > 0 can also be accounted for by the difficulty - greater for films than for traces - of resolving a cluster of low-intensity x-ray sources[15]). A regression analysis[15] for these 42 shots gives a linear correlation coefficient $r = [N \Sigma N_f N_s - \Sigma N_f \Sigma N_s]/$ $[N \Sigma N_s^2 - (\Sigma N_s)^2]^{\frac{1}{2}} [N \Sigma N_f^2 - (\Sigma N_f)^2]^{\frac{1}{2}}$ with values $r_1 = 0.67$ for N_s(s), N_f(s);

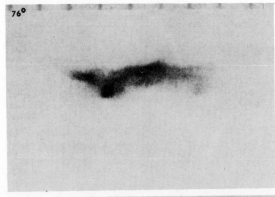

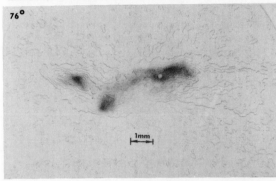

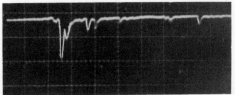

Fig. 1. X-ray pinhole-camera photographs
(50 μm Be) from -4°, 42°, 76° with respect
to discharge axis and x-ray detector
signal (Tektronix 7904 display time increasing to the right, 100 ns/cm) by the
same discharge. Distance pinhole-source
∿ 8 cm, pinhole-film 4 cm. Circular pinhole diameter 50 μm at -4°, 76°; pinhole
area at 42° is about half the value for
other directions. Arcs in 42° photograph
are due to x-rays from anode-edge bombardment (ε ∿ 10 keV) leaking through low-Z
pinhole material; x-ray leakage is not
visible by platinum pinholes at -4°, 76°.
Small divisions in the printed scale
(various magnifications) are spaced 2 mm.
Equidensity contour plots are reported
below photograph at -4° (contour interval
0.12) and 76° (contour interval 0.15).
Smoothing effect of scanning aperture
(100 μm at 76°, 50 μm at 50°) can be
assessed by film flaw (white dot dia.

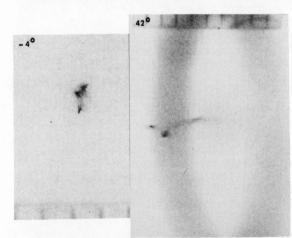

30 μm) on 76° photograph.(K-No screen film).
Low neutron yield n ≅ 0.2 n̄. $N_s(s) = 2$;
$N_s(m) = 3$, $N_s(w) = 4$ (ε ≳ 30 keV); $N_f(s) =$
$N_f(m) = 2$, $N_f(w) = 3$ i.e. same localized
x-ray sources can hardly be identified on
film (ε ∿ 2 - 10 keV). Ar doping (0.5%
by pressure) was used in all shots to reen-
force x-ray emission. IEN data.

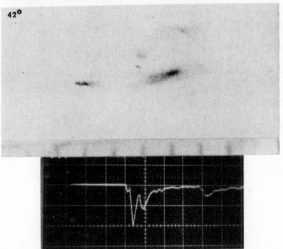

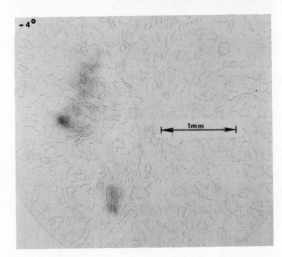

Fig. 2. Pinhole dia. 50 μm (50 μm Be),42°
view. Neutron yield n ∿ 2 x n̄. $N_s(s) =$
$N_s(m) = N_s(w) = 2$, same value for N_f's.
NE - 102 signal display as for Fig. 1. IEN
data.Radius of localized-source core r_o∿50μm.

$r_2 = 0.69$ for N_s(s+m), N_f(s+m); $r_3 = 0.36$ for N_s(s+m+w), N_f(s+m+w) (perfect correlation $r = 1$, no correlation $r = 0$; the probability P of obtaining three values of r as large as these purely by chance from N pairs of nonrelated N_s, N_f is (N = 42) respectively $P_1 < 10^{-3}$, $P_2 < 10^{-3}$, $P_3 = 0.02$; see Table V, p. 164, of Ref. 15). A correlation diagram for N_s(s+m+w), N_f(s+m w), the case with the smallest r, is shown in Fig. 4. From these data it becomes quite clear that a true correlation exists between N_s and N_f.

The correlation between N_s, or N_f, and the neutron yield n is less evident by a gross data presentation as in Fig. 5 a,b. A regression analysis of 109 shots gives $r \sim + 0.1$ for the pair n, N_s(s); $r \sim - 0.07$ for n, N_s(s+m); $r \sim + 0.07$ for n, N_s(s+m+w) (small $|r| < 0.01$ are also obtained for n, N_f). We can hardly consider significant these values of r because the probability to get the same value of r from pairs of unrelated (random) variables[15] is $P \gtrsim 0.1$. A nonlinear correlation for n, N_s can be detected by considering three groups of shots with a different neutron yield $n < \bar{n}$, $\bar{n} \lesssim n \lesssim 2\bar{n}$, $n > 2\bar{n}$ respectively. The result of our analysis on a series of 109 shots (same shots of Fig. 5a) is shown in the histograms of Fig. 6., where N vs. N_s is reported separately for each group. Nearly all N_s ($\gtrsim$ 90%) and most of N_m are observed within 50-70 ns from the onset of x-ray and neutron production. Medium h peaks and in a few shots a strong peak with h_s close to h_m can also be observed at a $\Delta t \sim$ 300-400 ns after onset. This Δt is the usual delay between emission time of hard x-rays ($\gtrsim$ 1 MeV) and a second period of activity ($\sim$ 100 ns long) with emission of relatively softer x-rays ($\sim$ 10 keV) and-in many shots-of neutrons. Shots with $\bar{n} < n < 2\bar{n}$ have $N_s = 2$ more frequently than other values; both groups of shots with low ($n < \bar{n}$) and high ($n > 2\bar{n}$) yields have instead $N_s = 1$. The strongest-by h and by peak area - x-ray burst in each discharge is always one of the initial peaks of hard x-rays and it has a FWHM with mean value $\sim$ 20 ns, 30 ns, 40 ns for the group of shots with $n < \bar{n}$, $\bar{n} < n < 2\bar{n}$, $n > 2\bar{n}$ respectively. In general a shot with many x-ray bursts N_s(s+m+w) has a lower x-ray emission in the whole discharge and a lower neutron yield than a shot with few but stronger x-ray bursts.

As a summary from Fig. 6, the table:

N peaked at when	N_s(s)	N_s(s+m)	N_s(s+m+w)
$10^{-1}\bar{n} \lesssim n \lesssim \bar{n}$	1	3	6
$\bar{n} \lesssim n \lesssim 2\bar{n}$	2	3	5
$2\bar{n} \lesssim n \lesssim 3.5\bar{n}$	1	2	4

indicates that the occurrence of many x-ray bursts can be competitive with n. Many x-ray bursts of low intensity seem to correspond to a particle-acceleration mechanism which becomes more effective also in accelerating positive ions (as it is indicated by the neutron yield) when a smaller number of stronger x-ray bursts is emitted.

3. INFRARED EMISSION

Two infrared detectors have been used:

(A) InSb crystal, area 0.42 mm x 0.37 mm, with a sensitivity $S(\lambda) = 8.1 \lambda/\lambda_{c.o.}$ Volt/watt-not affected by temperature-for incident radiation with λ smaller than the InSb cut-off wave length $\lambda_{c.o.} = 7$ μm (manufacturer calibration[16] with 500°K black body radiation and our calibration by He - Ne laser light); a Ge filter $\sim$ 3 mm thick eliminates radiation with $\lambda \gtrsim 1.8$ μm; response time of detector oscilloscope system $\sim$ 5 ns. A circular mirror focuses on the InSb crystal the image of a spherical plasma region of radius $\sim$ 1 cm at the center of the anode end. Defocussing effects of mirror and small detector area cut the intensity from other plasma regions. To reduce absorbtion the IR radiation reaches the detector by crossing a NaCl window in the wall of the plasma chamber. The IR-detector signal disappears by interposing a cardboard screen, that is, the small volume of the detector prevents response to x-rays. Other filters 2-3 mm thick (lucite, pyrex corning 7740, GE quartz 25) indicate that the bulk of the IR radiation has $\lambda \sim 2 - 4$ μm. Oscilloscope traces (Tektronix 7704) of InSb signals are reported in Fig. 7 a,b (detector at several angles θ from electrode axis-distance from plasma focus $\sim$ 160 cm); the IR-signal peak (80°) has a FWHM $\sim$ 40 ns with some unresolved structure: the slowly-rising precursor of the peak is due very probably to the gradual appearance of the emitting current sheath in the field of view of the detector; the peak at 0° has no precursor and a FWHM ∿ 80 ns. Our interpretation is that this larger FWHM at θ = 0° is due to a scissors effect of the current sheath closing on different points of the electrode axis over a time interval ∿ 80 ns. The power detected at IR-peak emission is ∿ 7 x 10³ watts; our data are consistent with an IR intensity somewhat higher at θ = 180° than at θ = 0°, 80°; refined measurements are necessary to discriminate effects due to a different source geometry by different directions.

(B) The second detector is an ITT vacuum photo-diode with an S-1 photocathode; a front surface mirror and a screen with a rectangular aperture (mask) were used for imaging various portions of the plasma on the photo-cathode. A short wave-length cutoff (at λ ≂ 1 μm) is obtained by Kodak gelatin filters (87 A, two, and 87 C). The pyrex glass containing the accelerator (detector outside) eliminates the radiation below λ ∿ 3000 Å so that only x-rays with ε ≳ 30 keV can eventually affect the infra-red detector below 1 μm. The photo-cathode sensitivity drops sharply for λ above 1.1 μm. In this way the response of the infra-red detector is sharply peaked at λ = 1.0 + 0.1 μm. A typical infrared signal from ITT photo-diode (θ ∿ 70°) is reported in Fig. 7c, mask aperture such that the image of a columnar region with diameter 1.5 cm, length 1.5 cm (in the discharge axial region, starting from anode end; axis along the electrode axis) is formed on the photo-cathode. The observed IR-peak with a width ∿ 40 ns, is not affected by hard x-rays when the photo-cathode surface is conveniently oriented parallel to the line stretching from this surface to the location of the radiation source. This infra-red peak and the first group of hard x-ray peaks occur on the same time interval. This is easily proved by using the infrared detector with a different orientation, e.g. at 180° (back side of accelerator; view along the hollow anode). The corresponding signal is the same as for Fig. 7c except that sharp peaks of hard x-rays emerge above the IR signal at the center of the broad IR-peak. The emitted power (λ ∿ 0.8-1.2 μm) at peak is ∿ 7 x 10³ watts in agreement with the InSb detector observation. The signal broadening due to the whole detector system is small (a δ-function input would give a ∿ 3 ns full-width signal). Since a single peak is observed we conclude that if a multiplicity of sources actually contribute to the infra-red peak, then the emission from each of

these sources should last longer than for a localized x-ray source, precluding time resolution. Infrared measurements by other laboratories (using plasma-focus devices as ours) have determined the non-thermal character of this infrared emission[3] essentially by plotting the intensity vs. λ⁻² in the interval 0.6 - 12 μm.

4. MICROWAVE OBSERVATIONS

Detectors for 3 cm and 10 cm microwaves have been located in different directions (70°, 80°, 180°) at a distance of 20 cm from the anode end. The video output was displaced on the scope (Fig. 8). Typically a first group of 2 - 5 microwave emission peaks, 3 - 10, ns full width (which are distributed on a 20 - 100 ns interval) is observed in each discharge. In some shots these peaks are followed after 200 - 250 ns (the same time separating the two groups of x-ray peaks) by a second group of microwave peaks. In other shots no large delay separates groups of microwave peaks. The horizontal component (i.e. parallel to the discharge axis) of the microwave electric field is usually larger by a factor ∿ 2 - 3 than the vertical component[2] (by a 90° rotation of the wave guide about its axis, the waveguide crystal detector, a P-N junction diode, can pick up either one of the two components). The microwave intensity is substantially the same in all directions (∿ 10 watts at peak intensity in the interval ∿ (1 ± 0.5)λ λ ∿ 3 cm; the intensity is smaller by a factor ∿ 3 - 4 for λ ∿ 10 cm). The details in the horizontal electric field signal are generally reproduced by the vertical field signal in the same direction[2]. Two typical "modes" of emission (say α,β) can tentatively be identified (these modes correspond respectively to the peaks reported in Fig. 8(a) and 8(c), α first peak, right). In the first mode, α, the emission peaks have (I) a smaller amplitude than in the second mode of emission; (II) occur earlier than β, at the same time of the hard x-ray emission; (III) a single peak is usually present in each discharge, with a FWHM ∿ 5 - 10 ns. The second mode of emission is usually characterized by (I) a remark-able regularity in the time-spacing be-tween peaks and (II) by a larger number of peaks (up to 10 - 12), FWHM ∿ 1 - 2 ns. Both modes of emission may occur in one discharge[2]. Usually the amplitude of the first emission mode becomes larger as

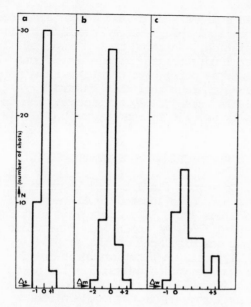

Fig. 3. a: N vs. $\Delta_s = N_s(s) - N_f(s)$.
b: N vs. $\Delta_m = N_s(m) - N_f(m)$. c: N vs.
$\Delta_w = N_s(w)$. IFN data. 42 shots.

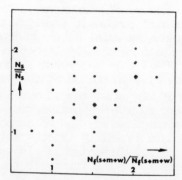

Fig. 4. Correlation diagram $N_s(s+m+w) \equiv N_s(s) + N_s(m) + N_s(w)$, $N_f(s+m+w)$ for same shots (42) of Fig. 3; mean value $\bar{N}_s(s+m+w) \simeq 6$, $\bar{N}_f(s+m+w) \simeq 4$. IEN data.

the neutron yield n increases. In some shots with large n the mode β is completely missing. Simultaneous time marking on the trace of the oscilloscopes is made by clipping the display beam in both scopes.

5. DISCUSSION AND CONCLUSIONS

Particle-acceleration by inductive fields during the decay of the plasma magnetic structure can consistently explain electron beaming (observed[1,5,6] via intensity anisotropy of x-ray localized sources) and D-ion beaming (via anisotropy of neutron-energy spectrum[7,8,14]) in the direction of the discharge axis. By this point of view the correlation among hard x-ray pulses, localized x-ray sources and

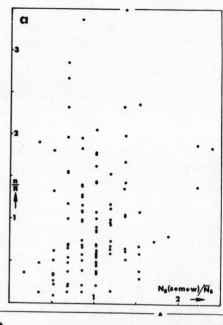

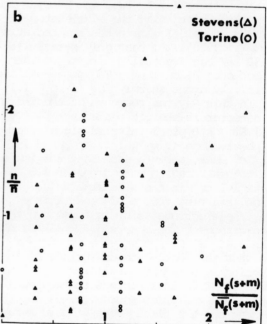

Fig. 5. a: Correlation diagram for neutron yield n, $N_s(s+m+w)$ from 109 shots (including series of 42 shots of Fig. 3,4; mean values $\bar{n} = 1.1 \times 10^8$, $\bar{N}_s(s+m+w) = 6$). IEN data.

b: $n, N_f(s+m)$ diagram from 95 shots (including 42 shots of Fig. 3,4 and 53 shots from SIT; here $N_f(s+m)$ is reported instead of $N_f(s+m+w)$ to check possible bearing of large value of r for $N_f(s+m)$, $N_s(s+m)$ on $n, N_f(s+m)$ correlation). Mean value $\bar{N}_f(s+m) \simeq 2.7$ for IEN data; $\bar{N}_f(s+m) = 3.0$, $\bar{n} = 1.0 \times 10^8$ for SIT data (sometime different conditions as pinhole diameters 30-75 μm, x-ray films K-RPR and K-PM3, Be fiters ∿ 15-50 μm thick have been adopted at IEN and at SIT; always K-NoS films at IFN),

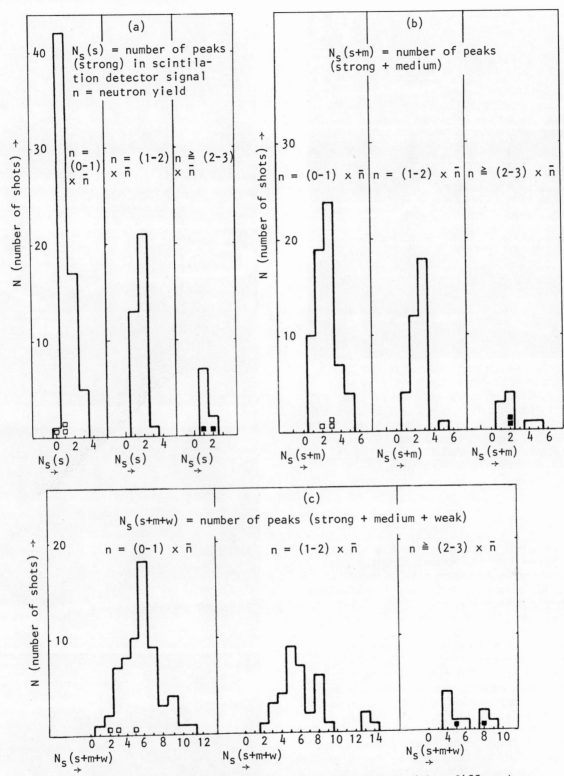

Fig. 6. N vs. N_S for three groups of shots (109 total) with a different n. Peaks with different h (s, s+m, s+m+w) have also been considered separately in (a), (b), and (c); shots with lowest yield (3) n ∿ 10^{-1} x $\bar{n}$ (◻) and highest yield (2) n ∿ 3.5 x $\bar{n}$ (■) are marked near N_S axis.

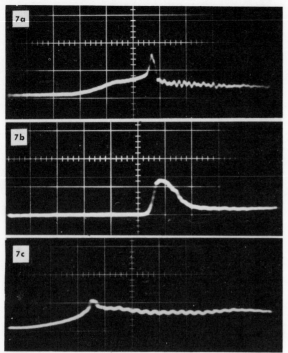

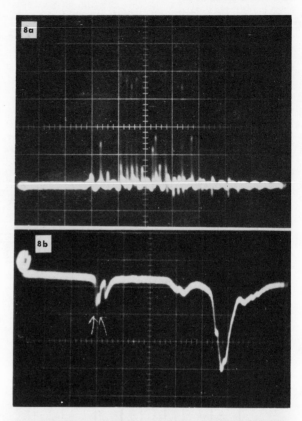

Fig. 7. IR detector signals: (a) InSb detector at 80° (n = 7.6 x 10⁷, 14 kV), (b) at 0° (n = 1.0 x 10⁸, 15 kV), (c) ITT photo diode at 70° (n = 2.3 x 10⁸, 15 kV); display 20 mV/cm in (a), (b), 50 mV/cm in (c), time increases at right 100 ns/cm. SIT data. Unresolved IR signal in (a) hints a double peak structure.

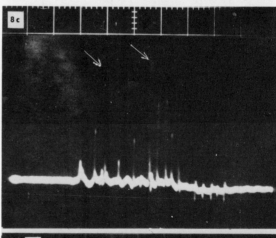

Fig. 8. (a) Microwave signal λ = 3 cm; (b) x-ray detector signal from the same low-neutron-yield shot n ∿ 7.6 x 10⁷; (c), (d) microwave and x-ray signals from high-yield shot n ∿ 7.1 x 10⁸. 0.1 V/cm in (a), (b), (c), 0.5 V/cm in (d) Tektronix 7704 display, 100 ns/cm, time increases at right; note lowering of zero-level signal due to ground-loop effect in late trace (c) and reflections of high microwave peaks, mode β, in (a), (c). Note different amplitudes of mode α , first peak at left in (a), (c). Microwave electric field // to discharge axis. SIT data. The broad x-ray peak∿ 480 ns after onset in (b) can be caused by recombination of Cu-ions near anode wall;second period of strong emission activity in the plasma occurs usually not later than∿ 250-400 ns after onset. A soft x-ray ($\varepsilon \lesssim$ 10 keV)detector was used here.

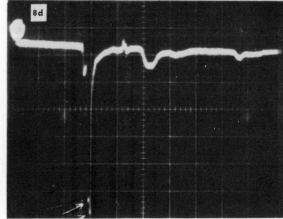

total neutron yield then indicates that the rate of decay of the magnetic structure - and so the inductive fields - can reach the highest values when the multiplicity of regions of decay do not exceed an optimum number (i.e. a fixed amount of magnetic energy is dissipated at the highest possible rate when dissipation occurs preferentially in a relatively small number of localized plasma regions)[17]. The same conclusion can be reached if fields due to charge separation accelerate particles, e.g., in the observed points of necking[5] of the axial plasma column. The peak intensity of the magnetic field $B \sim e\rho_{max}\bar{u}\pi r_o^2/r_o$ ($r_o \sim 50\mu m$) $\sim (1 - 2) \times 10^8 G$ was derived by estimating[5] the electron mean velocity $\bar{u}$ from the bremsstrahlung intensity of each localized x-ray source (with radius r_o) and by an independent determination of electron peak density $\rho_{max} \sim 10^{20}$ cm^{-3} (e = electron charge). The data of section 2 indicate that each x-ray localized source has a high-intensity component of hard x-rays ($\epsilon \gtrsim 30$ keV); this contributes to the film density as it is specified by the film exposure $R(\epsilon)$ with a much lower efficiency than $\epsilon \sim 2 - 10$ keV photons[5]. By using only x-ray film data to derive[5] the photon limiting-frequency $\nu_o = m\bar{u}^2/2h$ (h = Planck's constant, m = electron mass) the contribution of the hard component can be underestimated. The actual value of ν_o and consequently of $\bar{u}$, B can be somewhat higher, not smaller, than the values reported in Ref. 5,14, so that a fraction up to 2 - 10% of the total energy of the system can be stored as field energy $\pi r^2 \ell B^2$ in the volume of all localized x-ray sources. Use of screens and of focalization systems indicates that the most intense IR emission comes from the axial region of the plasma. Since IR and hard x-ray are simultaneously produced in the same plasma region their mechanisms of production can be interdependent. Microwave pulses are related in space and time with regions of high rate of decay of the magnetic structure. Pulsed mechanism of production and pulse polarization as it is reported in Ref. 2,14 can be explained in terms of particle beaming (mode α in particular) and/or in terms of rapid variations (mode β) of plasma local amperian current i, inside current loops with $i \gg I_{max}$.

The InSb detector was kindly given to us by Mr.J.D.Heaps of Honeywell Inc.,Minn.

*Work supported in part by AFOSR-U.S. , CNR-Italy,New Jersey Power and Light Co. and N.A.T.O. Scientific Affairs Division, Brussels,Belgium.
**On leave from Univ. of Buenos Aires.
† Now at Univ. of Calif.,Los Angeles.

REFERENCES

1. W. H. Bostick, V. Nardi, W. Prior, J. Plasma Phys. 8 (1972) p. 7.

2. W. H. Bostick, V. Nardi, W. Prior, Proc. 6-th Eur. Conf. Controlled Fusion and Plasma Phys., Moscow 1973; Vol. 2. p. 395. FWHM = full width half maximum.

3. H. Conrads, D. Gollwitzer, H. Schmidt, same as Ref. 2, Vol. 1, p. 367.

4. R. S. Post, Ph. D. Thesis 1973, Plasma Lab., Columbia Univ. Rep. No. 59, 1973, New York

5. W. H. Bostick, V. Nardi, W. Prior, Annals N. Y. Acad. of Sci., Vol. 251, Proc. Conf. Plasma Confinement and Relativistic Electron Beams, N. Y., March 1974, p. 2.
Same authors: Proc. Int. Conf. Physics in High Magnetic Fields, Grenoble, Sept. 1974, C.N.R.S. edit.

6. W. H. Bostick, V. Nardi, W. Prior, F. Rodriguez-Trelles: Proc. 2-nd Conf. Pulsed High-Beta Plasmas, Garching, July 1972, W. Lotz edit., p. 155.
Same authors: Proc. 5-th Eur. Conf. Controlled Fusion Plasma Phys., Grenoble 1972, C.F.A. edit., Vol. I, p. 70, Vol. 2, p. 239.

7. C. Patou, A. Simmonet, J. P. Watteau; Phys. Lett. 29A, 1, (1969).

8. J. H. Lee, L. P. Shomo, M. D. Williams, H. Hermandsdorfer; Phys. Fluids, 14, 2217 (1971).

9. V. S. Imshennik, N. V. Filippov, T. I. Filippova; Nuclear Fusion 13, 929 (1973).

10. A. Bernard , A. Coudeville, J. Durantet, A. Jolas, J. Launspach, J. de Mascureau, J. P. Watteau; Proc. 2-nd Conf. Pulsed High-Beta Plasmas, Garching 1972, W. Lotz edit. p. 147.

11. W. H. Bostick, L. Grunberger, V. Nardi, W. Prior; Proc. Int. Conf. Thermophys. Properties,Newton, Mass. 1970, American Assoc. Mech. Eng. p. 495.

12. N. J. Peacock, M. G. Hobby, P. D. Morgan: Proc. 4-th Conf. Plasma Phys. Controlled Nuclear Fusion Research, Madison 1971, I.A.E.A. edit. p. 537.

13. W. H. Bostick, V. Nardi, W. Prior: Prof. Conf. Dynamics of Ionized Gases, Tokyo 1971, Univ. of Tokyo Press, p. 375.

14. W. H. Bostick, V. Nardi, W. Prior: Proc. 5-th Conf. Plasma Phys. Controlled Nuclear Fusion Research, Tokyo 1974, in press.

15. H. D. Young: Statistical Treatment of Experimental Data; Mc Graw-Hill, New York 1962.

16. J. D. Heaps: Corporate Research Center, Honeywell Inc., Bloomington, Minnesota 55426.

17. This agrees also with findings of other laboratories: M. J. Bernstein (Aerospace), Proc. 4-th Conf. on Plasma Physics and Controlled Nuclear Fusion Research, Madison 1971; IAEA edit. 1971, Vol. 1, p. 572 (discussion of papers CN-28/D-3,D-4, D-5). R. L. Gullickson and R. H. Barlett (Lawrence Livermore Lab.), Advances in X-ray Analysis Vol. 18, Plenum Press, New York 1974. p. 186 (Proc. 23-rd Annual Conf. on Application of X-ray Analysis, Denver August 7-9, 1974).

APPLICATION OF THE RELATIVISTIC ELECTRON BEAMS ORIGINATING IN THE DISCHARGES OF PLASMA FOCUS TYPE FOR THE COMBINED LASER-REB PLASMA HEATING

V.A.Gribkov

Lebedev Physical Institute,Ac.Sci.,Moscow,USSR

ABSTRACT

Experimental investigations of the recent years allow to treat the discharge of plasma-focus-type (DPF) as an inductive storage with the switching time of the order of tens nsec. It results from the interferometry and X-ray measurements. The moment of the current sheath rupture coincides with the beginning of hard X-ray pulse. A considerable part of the bank energy (concentrated in the magnetic field near focus) converts into the electron beam energy with the mean energy of electrons of the order of IOO KeV after the disruption of the current. The interferometry allowed to investigate the dynamics of REB in DPF. The beam appeared to be divided into several filaments, and disturbed by "hose" instability and self-focusing inside the residual plasma. The X-ray pinhole pictures showed high plasma luminosity in the regions of beam focus, a neutron yield of the set-up had its maximum at this moment. All these data with anomalous scattering of diagnostic laser light in this region, high level of plasma noise with the frequency of the order of Langmuir frequency (by the factor of 10^6) and broad neutron spectra show us that an effective turbulent interaction of REB with plasma takes place in DPF. An experiment of particular interest for us involves the realization of laser initiated breakdown in the insulator and a combined laser-REB plasma heating. Analytical calculations indicate that when the REB energy is of the order of 10^4-10^6J and the laser pulse duration of 10^{-9}-10^{-10} sec, the laser energy of the order of 10^5J is required for the experiment.

Recent progress in research of DPF allows to treat the discharge of plasma focus-type[1] as an inductive storage with the switching time of the order of tens nsec. In fact, interferograms have shown (Fig.I) a disappearance of the interferometric fringes within the skin-layer just before the first compression phase of dense plasma focus (DPF) at some distance from the boundary of plasma and magnetic field. This phenomenon signifies that the diagnostic laser radiation has not passed through the objective aperture. This fact cannot be caused by such effects as laser radiation absorption, classical scattering in DPF plasma, refraction of diagnostic radiation on the density gradient, or insufficient temporal and spatial resolutions. The wave length of the plasma electron oscillations is:

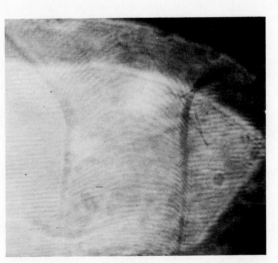

Fig. I

$$\lambda = \frac{2\pi}{k} = \frac{2\pi V}{\omega_{pe}} \gtrsim 5.10^{-5} cm,$$

where ω_{pe} = plasma frequency, derived from interferograms, and V = current velocity, derived from interferograms and current oscillograms. One can see that this wave length is more than half of the wave length of a ruby laser. This fact allows the diagnostic radiation to be scattered on these turbulent oscillations[2]. One can estimate the angle of radiation scattering and cross-section of this process[2] from the experiment geometry. These estimations lead to the conclusions that the cross-section of the anomalous scattering in the plasma within skin-layer is of 6 orders of magnitude higher than the classical one. On the other hand, the width of the skin-layer is of the order of c/ω_{pi} or even higher. The usual figure for anomalous conductivity of Z-pinch plasma is $\sigma_{anom.} = 10^{-2}\sigma_{cl.}$ This magnitude was used in the numerical calculations of the cumulation process[3], and was in a good agreement with the experimental data on the dynamics of the process. The analysis of the Rayleigh-Taylor instability dynamics has simultaneously shown that in the stabilization of short-wave modes of this instability the anomalous plasma conductivity plays the great role together with the plasma viscosity. The conductivity appeared to be of the same order of magnitude. In fact, the value of the minimum wave length of the instability (resulted from the equality of "inertial" increment of instability to the resistivity decrement) appeared to be of the order of experimental one for the magnitude of conductivity:

$$\lambda_{min} = \left(\frac{\pi}{2}\right)^{1/3} g^{-1/3} \sigma_{anom}^{-2/3} \simeq 1 cm \qquad (I)$$

where $g \simeq \dfrac{v_{T_i}^2}{R}$ (v_{T_i} – ion velocity, R – radius of magnetic field curvature), $\sigma_{anom} \simeq 10^{-2}\sigma_{cl.}$ – turbulent plasma conductivity. It's necessary to emphasize that the experimental results on the ability of plasma oscillations swinging, the noise magnitude, and its nature(Langmuir turbulence) are in a qualitative agreement with the theory[2,3,5]. The latter fact is confirmed by experiments with DPF working with different dopes to D_2(N_2 and Xe in our case). In fact, as it results from the theory of the faint electromagnetic wave scattering due to the turbulent plasma[2,5], the cross-section of scattering due to Langmuir oscillations is proportional to $\sqrt{\dfrac{M_i}{m_e}}$

whereas the cross-section of scattering due to ion-sound instability is proportional to $\sqrt{\dfrac{m_e}{M_i}}$ The comparison of the interferograms of N_2-doped discharges (Fig. I) with Xe-doped one (Fig.2) demonstrates an abrupt enhancement

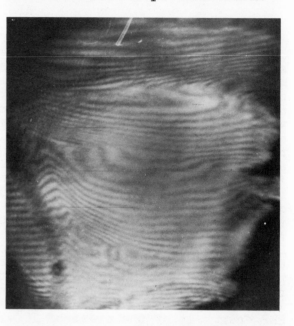

Fig. 2

of cross-section in the second case. The role of the impurities accumulated in DPF because of diffusion[4] up to the moment of the Ist compression is very essential for these phenomena. Subsequently, the turbulence level within the skin-layer apparently increased due to plasma cooling[6] resulting from the impurity ionization[4], whereas the pinch diameter and, consequently, the current velocity did not change practically. Unfortunately, plasma density is very high at this moment, so as it results from the above mentioned condition it is impossible to observe anomalous light scattering within the skin-layer (by means of a ruby laser with $\lambda = 0.694\mu$). But the experiments dealing with self-luminosity of the plasma in the wave range near plasma frequency have shown, that at that moment the luminosity power increased.

At that moment a large part of the bank energy is contained in the magnetic field near focus (as in the inductive storage). 30-40

nsec later than the first compression phase, both voltage and current oscillograms show a local "peculiarity" and at this very
moment the interferograms show an
"explosive-like" breach of the
sharp boundary between plasma and
magnetic field, i.e. a fast diffusion of the magnetic field[8] (Fig.3).

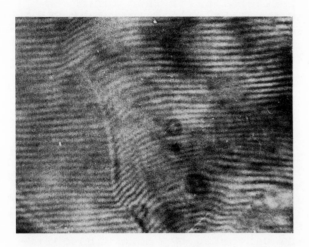

Fig. 3

An analysis of a great number of
interferograms have shown, that
such "an explosion" of the current
sheath usually began at one point
in the region of thickening, developed on the pinch surface due to
Rayleigh-Taylor instability, but
during the period of the order of
I.2 nsec it spread to the whole circumference of the pinch cross-section. We interpret this phenomenon
as the disruption of the conductivity current in DPF and the substitution of displacement current for
it. The absence of zero current on
the oscillogram can be explained
by the fact, that current is usually measured by Rogowski coil, i.e.
by magnetic field, which is equal
for both. The moment of the current
disruption coincides with the moment of the hard X-ray appearance
in DPF. This phenomenon indicates
the transformation of the magnetic field energy into the REB energy with an electron energy of the
order of 10^5 eV. A mechanism of
the REB origination in DPF cannot
be considered now as a fully understood one. Among several proposals for the explanation of the
phenomenon (an acceleration
due to quasi-stationary electric
fields induced during the current
sheath cumulation process;Cheren-
kov's acceleration by means of plas-

ma oscillations,etc.) we believe
the nearest to the reality is a
modification of Trubnikov's mechanism[9] . This model is capable of
explaining the REB existence during
the period of time while the process can be regarded as a two-dimensional one,i.e. during the period of the order of IO nsec.

It must be noted that usually
the beam originates not within
a turbulent zone (skin-layer) but
near its boundary inside the pinch
(Fig.4).This phenomenon can be

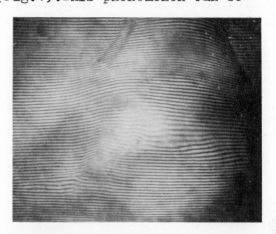

Fig. 4

explained by the fact (L.I.Rudakov)
that within the turbulent plasma
with the parameters of DPF -type
a magnetic insulation within the
turbulent skin-layer(which is larger than the Larmor radius) will
prevent REB from passing through
this zone. An interesting illustra-
tion of the fact is shown in Fig.5

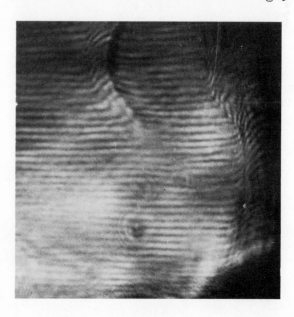

Fig. 5

where the turbulent zone has spread within the whole pinch. In this case an explosion of the current sheath takes place in that part of it, which is far from the pinch, and REB is injected into a region of a small magnetic field. It should be noted, that a calorimetry of the X-ray luminescence together with the analysis of the craters and the break-away of anode[10] have demonstrated the feasibility of existence of IO kJ REB in DPF of IOO kJ capacitor bank energy[11]. The interferograms allowed to investigate the REB dynamics in DPF. It is especially convenient with the Xe-doped regime in DPF functioning, since in this case the REB regions stand out against a background of plasma because of high anomalous scattering of laser light due to Langmuir turbulence (Fig. 6).

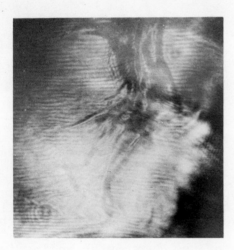

Fig. 6

During the initial period of REB formation the latter appeared to pass through the residual plasma to induce the backcurrent in plasma and to be slown down inside the anode. It is shown from the interferometry (Fig. 7), where one can see

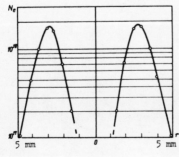

Fig. 7

an electron density decreasing near the beam axis and an increasing at the boundary[8]. This phenomenon is in qualitative agreement with the numerical calculations of the REB injection process into plasma /I2/. During this process the back-current heating takes place at plasma temperature of the order of several KeV[6]. Then during 20-50 nsec the self-focusing of REB takes place inside the plasma with the focusing length of the order of 1 cm and the focal spot less than 0.3 mm[11], and the focal spot goes from the anode into the plasma. This phenomenon looks like a method of REB focusing within the diode of high-power, pulsed relativistic electron accelerators[13]. It must be pointed out, that if in case of N_2 doping regime the REB is formed as a single filament, in case of Xe doping regime the beam, as a rule, is divided into several filaments (Weibel instability[12]) and distorted by "hose" instability. The focusing length and the "hose" instability increment are in a qualitative agreement with theory[14]. By this moment the hard X-ray luminosity of DPF is sharply decreased, but the neutron pulse has its maximum and the plasma "temperature" from X-ray measurements[15] and neutron spectrum[6] appears to be of the order of IO-20 KeV. All of these data together with the pin-hole pictures[4,16] show us a principal role of the beam effects in DPF. The investigations of plasma noises near plasma frequency result in the same conclusion[7]. It should be noted, that the main process of the transformation of REB energy into plasma energy takes place in DPF too late - only at the moment, when plasma density and beam electron energy are strongly diminished. So only a small part (several percent) of the beam energy converts into plasma energy.

In the light of above-stated it seems to be fruitfull to realize a combined laser-REB heating of plasma[17]. Let us consider the main idea of the experiment, not discussing its concrete scheme (we intend to verify several variants). First of all, the main role in the installation energetics must be played by REB, since it can be obtained with high efficiency.

Then, we shall use experimental data on slowing down length of REB originating in the discharges of Z-pinch-type, since a theory for such processes is now far from its comletion[13]. The analysis of the works[17] has shown that the beam heating in these installations began only after "preheating" of the plasma (usually, due to the backcurrent) In fact, it is not difficult to show that a stabilization of beam instability due to Coulomb collisions within plasma of solid state density disappears only at a temperature higher than I KeV. An approximation of the data on the REB slowing down length of these installations ($N_e = IO^{17} - IO^{21} cm^{-3}$) in the region of densities of the order of solid state density has shown that this length must be of the order of IO^{-5} cm. Then, from the same data it is shown that a diminishing of the pinch diameter increases (at least, up to some limits) an energy of the beam electrons, originating after the current cessation. The diminishing of a pinch size can also allow to increase the current and the magnettic field of the inductive storage of such a kind.

So, for our experiment we need plasma with $N_e \simeq IO^{23}$ cm^{-3}, preheating up to I KeV at the moment of the REB originating within Z-pinch. It can be fulfilled by means of the laser heating in the regime of the "heating wave". The solution of the proper task gives us the following relations for the required laser pulse duration τ_0 and for the laser flux density q_0 :

$$\tau_0 = 1.2 \cdot 10^{-29} z^2 \Lambda n_0 x_0^2 T_0^{-\frac{5}{2}} \quad (2)$$

$$q_0 \tau_0 = 2.5 \cdot 10^{-9} z n_0 x_0 T_0 \quad (3)$$

which are correct under the conditions:

$$\tau_0 \lesssim \tau_{ei} = \frac{10^{13} A T_0^{\frac{3}{2}}}{z^3 \Lambda n_0}$$

where x_0 - depth of the wave heated zone of the solid target, Λ - Coulomb log. From these relations it is shown, that in case of DT-target ($n_0 = 5 \cdot IO^{22} cm^{-3}$) the temperature $T_0 = I$ KeV can be distinguished within a layer of $x_0 = IO^{-5}$ cm thickness at $\tau_0 = IO^{-11}$ sec, and

$q_0 \tau_0 = IO^4$ j/cm^2. In case of CD_2-target the temperature $T_0 = 2$KeV can be distinguished with the same thickness layer at $\tau_0 \simeq IO^{-10}$sec and $q_0 \tau_0 \simeq IO^5$ j/cm^2.

At present time we have created a powerful 20-beam laser system with the parameters shown in Table I. We have under construction now a YAG-generator of O.I nsec pulse duration. It is shown that the laser can ensure the target preheating in the above mentioned regime.

Then, after the preheating, the REB must begin to slow down within this layer effectively and to heat plasma. As it is hard to expect that the REB flux density will be higher than the laser one, the subsequent heating must be in the regime with the corona gasdynamical motion. In case, when the electron conductivity takes place within the whole zone including the corona and REB slowing down zone, the hydrodynamic relations for plasma temperature and ion density look as follows:

$$T_{keV} = 1.4 \cdot 10^{-2} z^{\frac{1}{2}} A^{-\frac{1}{6}} (q_1 \tau_1)^{\frac{1}{3}} \quad (4)$$

$$n_i = 3 \cdot 10^{10} q_1^{\frac{1}{2}} t^{-\frac{1}{2}} z^{-\frac{9}{4}} A^{\frac{3}{4}} \quad (5)$$

where q_1, τ_1 = the REB flux density and pulse duration. From Eq. (4) and Eq. (5) one can estimate, for example, the temperature and the pressure of plasma near REB absorption zone within the target for the beam parameters of DPF beam-type. These are the following values: T=5KeV, $P \simeq IO^8$ atm.

It should be noted, that sharp density gradients which can arise in the plasma, can result in ionsound instability within the corona in case of $T_e \gg T_i$. So, we need to verify our results in view of the paper[18]. This work has shown that in such plasma a coefficient of thermal conduction changes by a factor of α : $\left(\frac{m}{M}\right)^{\frac{1}{2}} \lesssim \alpha \lesssim \left(\frac{m}{M}\right)^{\frac{1}{4}}$

It is shown that in this case we need a smaller laser system.

It should be kept in mind, that existence of a high power laser radiation can result in a high level of noises nonresonant to the REB-plasma system . As it is known[19], such noises can stabilize or distabilize the two-beam instability. So, for successful combines laser-REB heating we need an accurate analysis of the plasma instability excitation and stabilization conditions for this case.

At present time we have a fast capacitor bank of 10^5J, and with the characteristic time of the order of $1\,\mu$sec for the combined laser-REB heating experiment.

The author is thankful to O.N.Krokhin, G.V.Sklizkov, Yu.V. Afanasiev, N.V.Filippov for useful discussions.

TABLE I

τ_0 nsec	E_0 J	E_1 J	E_2 J	q_{01} W/cm^2	q_{02} W/cm^2	B_0 W/cm^2.ster
2	0.1	50	960	$9.6 \cdot 10^{15}$	$4 \cdot 10^{14}$	$2,2 \cdot 10^{15}$
20	0.5	110	2120	$2.1 \cdot 10^{15}$	$9 \cdot 10^{13}$	$5.1 \cdot 10^{14}$

Where E_0 - generator energy, E_1 - energy after the preamplifier system, E_2 - energy of the whole laser, q_{01}, q_{02} - laser flux densities in cases of 20 and 1 lens focusing correspondingly, B_0- laser brightness.

References

1. N,V.Filippov, T.I.Filippova, V.P.Vinogradov, Nuclear Fusion, Suppl.,part 2,p.577(1962).
2. V.N.Tsitovich, "Theory of Turbulent Plasma", Atomizdat,M.1971.
3. V.S.Imshennick, Preprint N 17, IAM Ac.Sci.USSR,1972.
4. V.A.Gribkov, O.N.Krokhin et al. 6th European Conf. on Contr.Fusion and Plasma Physics,v.I,p.375, M,1973.
5. V.V.Pustovalov, V.P.Silin, V.T. Tikhonchuk, Preprint N 183,LPI, Ac.Sci.USSR (1973).
6. Ch.Maisonnier et al. 4th Conf. on Plasma Phys. and Contr.Nucl-Fusion Res.,CN-28,D-1.D-2.Madison, Wisc., USA (1971).
7. R.F.Post, T.C.Marshall, Preprint,Columbia Univ.,N.Y.(1974)
8. V.A.Gribkov, V.M.Korzhavin, O.N.Krokhin et al. JETP Letts, 15. 329 (1972).
9. V.A.Gribkov, V.M.Korzhavin, O.N.Krokhin et al. 5th European Conf. on Contr.Fus. and Plasma Phys.,v.I,N 64,Grenoble,Fr.(1972).
10. S.V.Bazdenkov et.al.JETP Lettts, 18, I, II (1973).
11. V.A.Gribkov, O.N.Krokhin et. al.,JETP Letts, 18, II (1973).
12. R.Lee, M.Lampe, Phys.Rev. Letts, 31, 23, 1390 (1973).
13. G.Yonas, J.W.Poukey, K.R. Prestwich et al. Nuclear Fusion, 14, N 5 (1974).

14. A.A.Ivanov, L.I.Rudakov,JETP, 58, 4, 1332 (1970).
15. C.Patou, A.Simonnet, Note C.E.A., N 1189 (1969).
16. W.H.Bostick et al. 5th European Conf. on Contr.Fus. and Plasma Phys.,v.I, N 69, Grenoble Fr. (1972).
17. V.A.Gribkov, O.N.Krokhin et al. JETP Letts, 18, N 9 (1973).
18. E.V.Mishin, Doklady Ac.Sci., USSR 215, N 3, 565 (1974).
19. V.N.Tsitovich, V.D.Shapiro, Nuclear Fusion, 5, N 3, 228 (1965).

Inductive and Capacitive Storage Systems

SUPERCONDUCTIVITY, ENERGY STORAGE AND SWITCHING

H. L. Laquer
Los Alamos Scientific Laboratory
of the University of California*
Los Alamos, NM 87544

ABSTRACT

The phenomenon of superconductivity can contribute to the technology of energy storage and switching in two distinct ways. On one hand, the zero resistivity of the superconductor can produce essentially infinite time constants, so that an inductive storage system can be charged from very low power sources. On the other hand, the recovery of finite resistivity in a normal-going superconducting switch can take place in extremely short times, so that a system can be made to deliver energy at a very high power level. Topics reviewed include: physics of superconductivity, limits to switching speed of superconductors, physical and engineering properties of superconducting materials and assemblies, switching methods, load impedance considerations, refrigeration economics, limitations imposed by present day and near term technology, performance of existing and planned energy storage systems, and a comparison with some alternative methods of storing and switching energy.

1.0 INTRODUCTION

The phenomenon of superconductivity can be utilized in two quite distinct ways to provide solutions to some of the technical problems in the fields of inductive energy storage and of energy transfer or switching. On one hand, perfect conductivity or zero resistivity implies an essentially infinite time constant for the self discharge of a storage inductor, so that whatever energy is stored remains available indefinitely. On the other hand, with a properly designed device, it is possible to lose superconductivity i.e. recover normal resistivity very rapidly, so that one has a means of quickly transferring or "switching" the energy from the storage inductor into a load. This review article will be divided into three parts:
(A) The past or history of the relevant physics of superconductivity and switching,
(B) The present technology of superconducting materials and corresponding hardware developments, and
(C) The future large systems that are now at various stages of planning.

2.0 PHYSICS OF SUPERCONDUCTIVITY

The history of superconductivity can be marked by three concepts: perfect conductivity, perfect diamagnetism, and the existence of an energy gap. These ideas have formed our understanding of the phenomenon, in spite of the fact that there are important exceptions and limitations to all three of them.

2.1 RESISTIVITY

The electrical conductivity of all metals is caused by the mobility of some of the valence electrons, and is, in turn, limited by various scattering processes. For pure metallic elements the electrical resistivity, ρ decreases more or less linearly from room temperature (300 K) values of between 2 and 10 x 10^{-8} Ω-m to asymptotic low temperature (4 K) values of between 1 x 10^{-12} and 5 x 10^{-10} Ω-m. This temperature coefficient of resistivity results from the decrease in the scattering of the electrons by the atoms as the amplitude of the vibrations of the atoms within the crystal lattice decreases. The low temperature limit of the resistivity is set by scattering from physical or chemical impurities which reduce the perfection of the lattice. Thus the purer the metal the lower the limit, or the higher the residual resistance or resistivity ratio (Γ = ρ(300K)/ρ(4K). Resistance ratios well above 10 000 have been reported for ultrapure aluminum. It was the search for just these kinds of effects that caused H. Kamerlingh Onnes of the University of Leiden in Holland in 1911 to study the low temperature resistivity of mercury, an element

*Work performed under the auspices of the U.S.E.R.D.A.

that he could readily purify by
distillation, thus avoiding complex chemical and metallurgical procedures.[1]
Much to his surprise, he observed an abrupt
and complete disappearance of resistivity
at a critical transition temperature T_c,
Fig. 1, and after making certain that this
observation was not an experimental error,
Fig. 2, he systematically pushed his limit
of sensitivity from 2×10^{-11} Ω m to
5×10^{-18} Ω m. He also discovered superconducting transitions in lead and tin[3]
and later thallium[4] and indium[5]. All
of these are soft metals and are now classified as Type I superconductors. It was
with chagrin that Onnes discovered that in
all of these elements superconductivity is
destroyed not only by rather high current
densities[3] (critical currents, I_c), but
also by quite low applied magnetic fields[6]
(critical fields, H_c). Both vary essen-

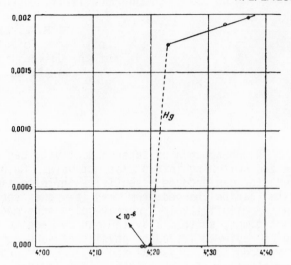

Fig. 2. _Details of the Superconducting
Transition in Mercury,_ from Onnes
1911.[2]

tially parabolically in the reduced temperature, e.g.

$$H_c(T) = \alpha \left[1 - (T/T_c)^2\right] \qquad (1)$$

The low values of the maximum $H_c(0)$, i.e.
H_c at T=0, of the order of 50 to at most
1000 gauss, squelched his earlier hopes of
using superconductors for the generation
of large magnetic fields without joule
losses. Temperature, currents or current
densities, and magnetic fields thus interact to define a region (phase diagram) for
the existence of the superconducting state;
outside this region the normal state pre-
vails, Fig. 3.

2.2 DIAMAGNETISM

Meissner's discovery in 1933[7] of
complete flux expulsion from a solid superconducting rod (or sphere) cooled through
T_c in a small transverse field, less than
H_c, justifies the description of superconductors as a distinct phase or state of
matter in the thermodynamic sense, i.e.
the diamagnetic, B=0, state is attained
independently of the path. It also explains the low critical fields of Type I
superconductors by the energy $(B^2/2\mu_0)$ required to exclude the magnetic field from
a unit volume of the bulk material, since
at least this amount of energy has to be
added to the thermodynamic free energy of
the metal by the "condensation" energy released in the transition from the normal
to the superconducting state.

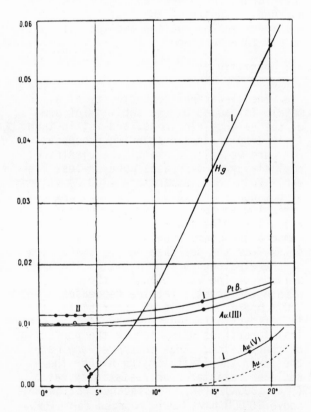

Fig. 1. _The Resistivity Ratio_ $\rho(273K)/\rho(T)$
as a Function of Temperature for
platinum, gold of different puri-
ties, and mercury taken from the
1911 publication of H. Kamerlingh
Onnes.[2]

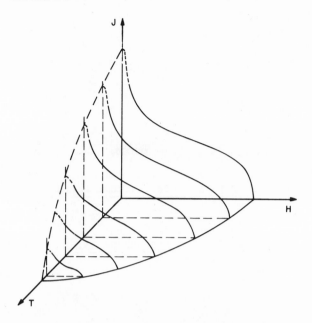

Fig. 3. *The J-H-T Phase Diagram, Schematic.* Hyperbolic shape of J-H curves is more characteristic for Type II superconductors; Type I have parabolic shapes on all axes.

The thermodynamic approach to superconductivity then allowed quantitative descriptions of the thermal and magnetic properties in terms of a phenomenological 2-fluid model[8] of normal and super electrons, but could not explain the reason for the phase change. The electrodynamics of superconductivity and in particular of the Meissner effect was described through the modifications of Maxwell's equations by F. and H. London,[9] who introduced the concept of a field penetration depth, λ. In such a surface layer very large (exponentially decreasing) current densities of the order of 10^{12} A/m^2 have to exist in order to provide the shielding currents needed for the Meissner phenomenon. Similarly, loss-less transport currents in Type I superconductors are pure surface currents and it was certainly clear by the end of the 1930's that superconductivity presents a nondecaying macroscopic quantum state in which a large piece of metal can behave like a single giant molecule.

2.3 ENERGY GAP

The third building stone in the theory of superconductivity is the existence and explanation of an energy gap. A gap has to exist in the allowed energy levels of the conduction electrons so that below T_c,

but still above the absolute zero, a single collision with the lattice cannot transfer enough energy to destroy the superconducting state. The microscopic theory of Bardeen, Cooper and Schrieffer (BCS)[10] now provides at least one physical model in which such a gap is created by an attractive pairing "condensation" of two electrons with opposite spin and angular momentum (Cooper pairs). This pairing or coupling is sufficient to overcome the omnipresent electrostatic or Coulomb repulsion and is mediated by phonons or sound waves in the crystal lattice. The gap decreases from zero at T_c to

$$2\Delta = 3.52 \ k_B \ T_c \qquad (2)$$

at the absolute zero, where k_B is the Boltzmann constant. However, there still is no theory available today that predicts the magnitude of the transition temperature of any material from first principles. The discovery of new and especially of high temperature superconductors remains in the realm of the inspired experimentalist.

2.4 TYPE II SUPERCONDUCTORS

It is perhaps understandable that much of the experimental research on superconductivity for the first 50 years should have been on sharpening and improving the precision as well as the thermal and magnetic reversibility of the transitions. These objectives were most easily obtained with ultrapure and single-crystal samples. As a result, there was a tendency to neglect and discount as "dirt" or inhomogeneity effects most of the observations on superconducting alloys and on some of the hard-to-purify transition elements, such as niobium and vanadium. All of these materials exhibit much steeper than usual critical field curves, the most outstanding being the Pb-Bi eutectic alloy found in 1930[11] to remain superconducting up to 1.9 T at 2 K. Again there was the promise of electromagnets without joule losses. However, it was not until 1961 when Kunzler reexamined the compound Nb_3Sn, whose very high transition temperature of 18.2 K had been discovered by B. T. Matthias in 1954,[12] and reported[13] current densities of 10^9 A/m^2 at 8.8 T and 4.2 K that the prospects and potential of superconducting electromagnets were finally taken seriously. Interestingly, Type II superconductors violate the two main precepts of classical superconductivity - perfect diamagnetism and zero resistivity.* Their outstanding feature, of course, lies in their much greater tolerance of magnetic fields, 2 to 50 Tesla or more. This is obtained by allowing the magnetic field to penetrate so that very little of the condensation energy is used up in trying to exclude the field (in the Meissner fashion). The field penetration, in turn, is made possible by a negative interfacial energy i.e. a gain rather than reduction in the thermodynamic free energy of the system when the material contains alternating normal and superconducting regions. Some of these ideas had been discussed much earlier by London[15] and Pippard[16] and most of what is now called the Ginzburg-Landau-Abrikosov-Gorkov or GLAG theory[17] was developed in the 1950's well before Type II behavior had become accepted as an experimental fact.[18] There are two relevant dimensions: the previously mentioned London penetration depth, λ, and the Pippard coherence length, ξ, which defines the distance over which the superconducting wave function can change by a significant amount (63% to 1/e). If $\xi > \lambda$, the interfacial energy is positive, and we have Type I behavior; if on the other hand $\xi < \lambda$, the interfacial energy is negative and the externally applied field starts to penetrate in individual flux quanta or vortex lines as soon as a lower critical value H_{c1} is exceeded. Bulk superconductivity persists and flux penetration is not complete, however, until an upper critical field, H_{c2}, is reached. The ratio of the two critical fields is given by

$$H_{c2}/H_{c1} = \lambda^2/[\xi^2 \ln(\lambda/\xi)] \qquad (3)$$

and can be as large as 1000. Figure 4 compares some of the properties of Type I and II superconductors.

2.5 FLUX PINNING

The magnetization curve, Fig 4 (right) of a pure or ideal Type II superconductor is reversible and there is nothing to maintain currents uniformly distributed throughout

---

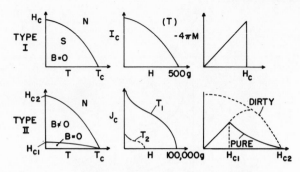

Fig. 4. *Comparison of Critical Field, Critical Current or Current Density, and Magnetization Curves for Type I (Top) and Type II (Bottom) Superconductors.*

the body of the material against the Lorentz force (JxB) which tends to push the current against the side of the conductor. Fortunately, however, from an applied point of view, ideal Type II materials are even harder to obtain than ideal Type I, and the usual Type II contains "dirt", such as grain boundaries and dislocations, which act as pinning sites to prevent the motion of the vortices under the influence of the Lorentz force and thereby maintain a uniform current density throughout the bulk of the conductor. This pinned current carrying state of the Type II superconductor is not a loss-less quantum state, but rather a metastable array of currents and flux lines. The critical current densities achieved in different wires, Fig. 5, depend on composition as much as on metallurgical factors and fabrication history, such as extrusion temperatures, amount of cold work, precipitation heat treatments and grain boundary growth. All of these factors need to be optimized and there can be a considerable payoff in terms of potentially higher current densities.

2.6 HYSTERETIC ENERGY LOSSES AND THE CRITICAL STATE

The main problem in properly utilizing high current density, Type II superconductors in an electromagnet stems from the fact that building up or changing the field and current in a magnet assembly will immediately induce critical currents in a

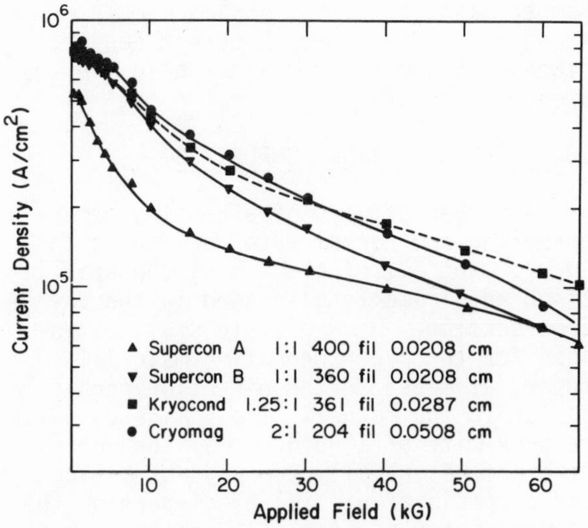

Fig. 5. *Critical Current Densities at 4 K for Some Commercial NbTi Wires.* Ratios of copper matrix to superconductor, number of filaments and wire diameters are listed.

surface layer of the conductor. The induced currents add algebraically to the transport currents and when their sum exceeds the permissible critical current density, the surface layer will expand into the body. During this flux motion or flux flow, work is done against the pinning sites thus dissipating energy within the conductor. Depending on the thermal balance, this process may raise the temperature and thereby lower the permissible current density. As long as the field and current penetration takes place isothermally and smoothly, the conductor is in the "critical state",[19] i.e. each region either carries the critical current density appropriate to the local field strength or it carries no current at all. If thermal run-away should occur we can have a "flux jump", the shielding structures collapse and the equivalent energy is deposited in the superconductor. Figure 6 schematically shows the induction of shielding currents and emphasizes the dilemma that in any complex current carrying structure the conditions $\rho=0$ and $\dot{B} \neq 0$ cannot be true rigorously and simultaneously. There are energy losses in a Type II superconductor during field and current changes and these losses have to be controlled and allowed for in the design of the superconductor as well as in the thermal and electromagnetic design of the overall magnet system.

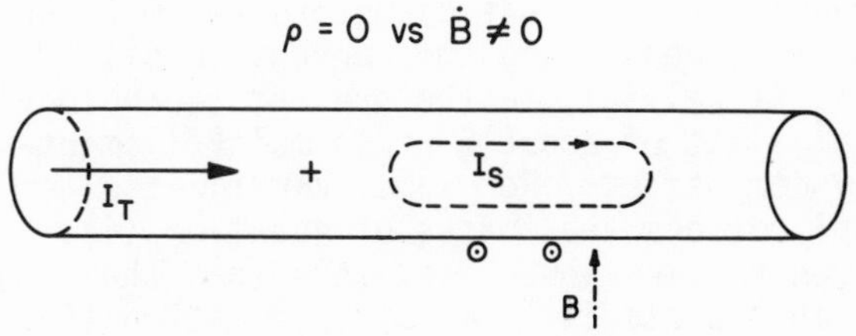

Fig. 6. *Induction of Shielding Currents in Superconducting Wire by Changing or Moving Magnetic Field (Schematic).*

2.7 STABILIZATION

The objectives of stabilization are to make a superconductor less sensitive to motion and charge or discharge rate effects, and to eliminate or recover from flux jumps. There are three complementary ways to achieve these goals: (a) reduce the intrinsic energy dissipation, (b) slow down the rate of energy release, and (c) allow the recooling of a current carrying assembly that may have been heated above T_c.

2.7.1 Intrinsic or Adiabatic Stabilization.

For a given field difference ΔB, the flux penetration process sketched in Fig. 6 will dissipate per unit volume of superconductor an amount of energy, Q_o/V, that is proportional to the critical current density, J_c, and to the superconductor thickness (in the direction of flux motion) or diameter, d,[19] thus

$$Q_o/V = \Delta B \cdot J_c \cdot d/4 \quad (J/m^3) \qquad (4)$$

This hysteretic energy dissipation is independent of the rate of field change. An adiabatic or intrinsically stable superconductor[20] is one in which the diameter or thickness of the superconducting elements (wires or filaments in a matrix, see below) has been reduced to the extent that an energy pulse equivalent to equation (4) can be absorbed in the specific heat S (per unit volume) of the superconductor itself without excessive temperature rise

$$d \leq d_1 = A(S/J_c)^{1/2} \qquad (5)$$

where

$$A = \left\{-3/\mu_o (dJ_c/dT)\right\}^{1/2} \qquad (6)$$

is constant for a given material in a given field. For NbTi at 4 K and 3 to 5 T equation (5) demands diameters below 25 or 50 μm, depending on the current density. This is well within the present technological limit of about 5 μm in multifilamentary composites. For some varying-field applications the losses of equation (4) often require lower diameters than the adiabatic stability limit of equation (5).

2.7.2 Dynamic Stabilization.

An alternative to reducing the total energy deposited adiabatically in a superconductor by flux motion is to reduce the _rate_ of flux motion or energy dissipation, so that thermal conduction can transport the energy to the surface (again without excessive temperature rise in the superconductor). This can be achieved by placing a good normal conductor of resistivity ρ in close proximity, so as to eddy current damp the flux motion through the superconductor of thermal conductivity k_s. The analysis of this problem[20] demands thicknesses

$$d \leq d_2 = (6\sqrt{2}/\pi)\left\{R(k_s/\rho)[(-1/J_C)(dJ_C/dt)]\right\}^{1/2} \quad (7)$$

where R is the area ratio of normal metal matrix to superconductor. The actual size limits for d_2 vary with ρ but are generally near those required by equation (5). However, the most important function of dynamic stabilization is to eddy current damp gross motion as well, and thus make a composite superconducting wire less sensitive to the effects of motion or vibration.

2.7.3 Cryostabilization.

A rather different way to guarantee system performance is cryostabilization[21] This consists of electrically paralleling the superconductor with sufficient normal conductor to carry the full current as a cryoconductor while the superconductor recools from an inadvertent thermal excursion, whatever the cause of that excursion. The analysis of this problem cannot be done in a general way since it requires a knowledge of the magnet structure, its anisotropic thermal conductivities, as well as the heat transfer coefficients and the total thermal capacity of the refrigeration system. There can be many degrees of cryostabilization depending on the confidence one has in the other stabilization methods. In one limit cryostabilization may be used only to the extent of allowing safe emergency shut down of a system in which the coolant has been lost. In the other extreme such as large bubble

chamber magnets,[22,23] enough normal copper is added to keep overall current densities between 0.5 and 1% of the actual J_C in the superconductor.

2.8 SUPERCONDUCTING COMPOSITES

A number of ways of arranging superconducting and normal metal are shown in Fig. 7. Not all of them meet the stabilization requirements discussed in the previous section. These dictate small dimensions for the superconducting material proper, at least in the direction perpendicular to the magnetic field lines, and/or the proximity of a good normal conductor. There are essentially two ways of making such a combination: (a) by dispersing the superconductor as fine filaments within a normally conducting usually OFHC copper matrix and (b) by backing a superconducting tape with a copper tape. The former is more or less isotropic, but the latter is not.

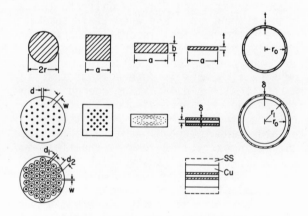

Fig. 7. _Some Configurations of Superconducting Wires and Composites._ Top row, pure superconductor: round wire, square wire, rectangular wire, tape, hollow tube. Middle row, simple two component composites: round wire with filaments in copper matrix, square wire, flattened rectangular wire, surface reacted or deposited tape, surface coated tube. Bottom row, multiple composites: coextruded mixed copper and copper nickel matrix, tape with added copper and stainless steel.

2.9 COUPLING AND EDDY CURRENT LOSSES

It is clear that the use of a good normal conductor in a superconducting composite will cause additional energy dissipation under ac or pulse conditions. However, even at slow charge, near dc conditions there can be problems with induced currents that are partly in the superconductor and partly in the low resistivity matrix and thus decay with time constants of minutes or hours. As indicated in Fig. 8 there is again the possibility of inducing shielding structures just as in the large single wire of Fig. 6. The energy associated with these structures, and released when J_c is exceeded, are proportional to the wire rather than the filament diameter. The remedy to this problem is transposition or twisting[20] as indicated in the lower half of Fig. 8. The induced currents now flow through the filaments in opposite directions within adjacent loops and cancel as far as stability is concerned up to a saturation field change rate.

Manufacturing technology limits the 360° twist pitch to about 8 wire diameters, a, and one now has a rate ($\dot{B}$ = dB/dt) dependent coupling loss Q_R in addition to the rate independent loss of Eq. (4). Thus[24]

$$Q_R/\lambda \, V_T \cong (3\pi/8) \, a^2 \cdot \Delta B \cdot \dot{B}/\rho(\lambda - \lambda^{3/2}) \qquad (8)$$

where λ = volume fraction of superconductor in composite

ρ = resistivity of normal matrix at operating temperature

V_T = total volume of conductor.

Equation (8) has a minimum value for λ=4/9. The only further means of reducing the coupling losses is to reduce the wire diameter or to increase the normal state resistivity, which, however, would lower the effectiveness of stabilization.

2.10 MIXED MATRIX CONDUCTORS

The obvious answer to having both high normal resistivity in the matrix for decoupling of the filaments and good copper for stabilization is to build a "mixed matrix" or triple composite conductor as indicated in Fig. 9. CuNi alloys, either 90/10 (90 wt% Cu-10 wt% Ni, $\rho_4 = \rho(4K)=1.7 \times 10^{-7}$ Ω-m) or 70/30 ($\rho_4 = 3.5 \times 10^{-7}$ Ω-m) are commonly used for decoupling. The coupling loss at maximum twist is now given by[24]

$$Q_R/\lambda_1 V_T = (3\pi/8) \, a^2 \cdot \Delta B \cdot \dot{B}/\rho_3 D \qquad (9)$$

where $D = \sqrt{\lambda_1(\lambda_1 + \lambda_2)} \cdot (1 - \sqrt{\lambda_1 + \lambda_2}) \qquad (10)$

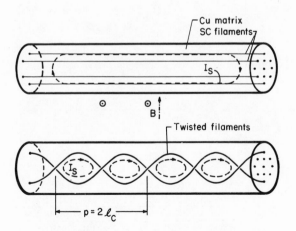

Fig. 8. *Induction of Shielding Currents in Straight and Twisted Multifilamentary Composites.* 360° twist pitch p is equal to two critical coupling lengths, ℓ_c.

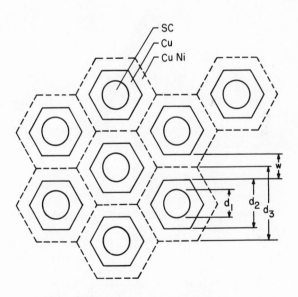

Fig. 9. *Design Layout for Mixed Matrix Composite Superconductor for AC Applications.* Each NbTi filament (d_1) is surrounded by copper (d_2) and decoupled by a web of high resistivity Cu-Ni alloy of width w.

and λ_1 = volume fraction of superconductor
   λ_2 = volume fraction of good normal
          conductor (Cu)
   λ_3 = volume fraction of resistive web
          (Cu Ni)
   ρ_3 = resistivity of resistive web
A normalized contour plot of equation (10)
is given in Fig. 10. Interestingly, there
is a maximum D, or minimum coupling, when
the optimally distributed resistive alloy
(λ_3) occupies 5/9 or 55.6% of the total
volume of the conductor;   that figure
is independent of the amount of stabilizer
(λ_2) used.

     Figure 11 gives a photomicrograph of
an early design mixed matrix superconductor
fabricated in 1972 with neither optimal
twisting nor optimal amount of alloy. We
have recently measured[24] the losses in
this material during single pulse field
discharges at various exponentially decaying rates and from various peak field values, as shown in Fig. 12. The initial
slope of the rate dependent losses is proportional to both ΔB and $\dot{B}$ as required by
Eqs.(4) and (9). The apparent saturation
is probably caused by the inevitable sample
heating and the resulting reduction in the
critical current density.

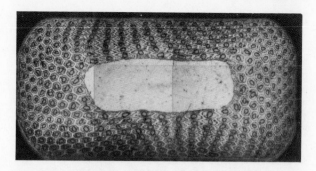

Fig. 11. *Mixed Matrix Superconductor
Manufactured in 1972.* Overall
dimensions are 2.0 x 4.0 mm,
360 filaments are twisted 20
turns/m, center region of Cu-Ni
contains no superconductor.

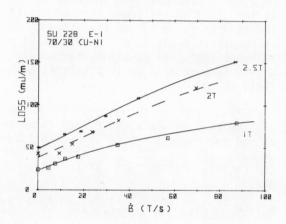

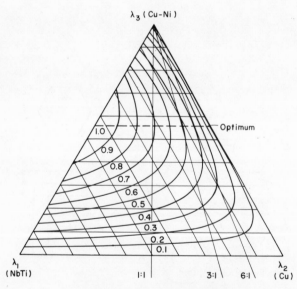

Fig. 10. *Contour Plot of Decoupling
D-Function Eq. 10*, normalized to
its maximum value of 0.148...
at $\lambda_1=4/9$, $\lambda_2=0$, $\lambda_3=5/9$. Note
that all lines of constant copper
to superconductor ratio (e.g.
1:1, 3:1, 6:1) have their maximum
at the same value of $\lambda_3=5/9$.

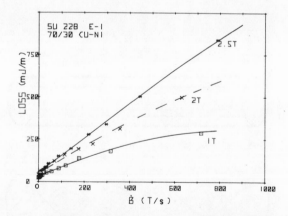

Fig. 12. *Energy Loss in Mixed Matrix Superconductor during Pulse Discharge
from Various Peak Field Values.*
Loss is given in mJ/m for conductor of Fig. 11, made with 70/30
Cu Ni. Identical conductor but
made with 90/10 Cu Ni shows correspondingly higher losses. Upper
plot has expanded scales.

2.11 SWITCHING

2.11.1 Switch Volume

The physics of switching might be considered as the opposite of stabilization. However, this does not mean that a normal-going superconducting switch does not need stabilization. On the contrary, a switch must be well stabilized in order to provide controlled switching on command, and this is made more difficult by the fact that a switch would be very bulky if it contained normal metal stabilizer. It was shown by Hake[25] at Los Alamos and also by Solé[26] at Limeil that the volume, V_{sc}, of superconductor in a switch, and thereby the materials costs, are governed by the peak power transferred or maximum voltage E_{max} times maximum current I_{max}, thus

$$\text{Cost} \propto V_{sc} = E_{max} \cdot I_{max} / J_c^2 \rho_N, \qquad (11)$$

where ρ_N is the normal state resistivity of the switch material (or assembly). Clearly, the highest normal state resistivity - i.e. no copper - and the highest self-field - or better yet, self-field compensated critical current densities- are wanted.

2.11.2 Switching Method and Latching

The phase diagram, Fig. 3, suggests that there are three ways of switching a superconductor by temperature, magnetic field or excess current. However, the transition is reversible which means that the switching conditions have to be maintained as long as needed. This can be difficult when the stored current decays. Alternatively the switch can be latched by delivering enough energy to it to first raise it to and then keep it and all associated matter above the transition temperature, T_c, as shown schematically in Fig. 13.

Thermal switching was used in an early study[27] but is bound to take tens of milliseconds for most switch geometries, except possibly with laser heating of thin films. Magnetic switching has been used in most of the French effort[26,28] and current switching has been developed in the Magnetic Energy Transfer and Storage (METS) program at Los Alamos.[29,30]

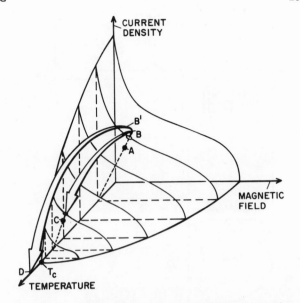

Fig. 13. *Dynamics of Switching and Latching.* Parts of current carrying superconducting switch originally at A may reach stable lower current superconducting state at C. Latching can only be guaranteed if entire switch reaches D.

A steady-state, magnetically-operated switch might conceivably be based on the rather difficult scale-up from a few hundred A to many kA of the niobium tape power cryotrons developed by Buchhold.[31] To be economical, such a switch must use a high current density or thin film superconducting material with a low upper critical field ($\sim$ 1T) or else be operated at temperatures above 4 K. For the more readily available conventional Type II conductors, pulsed-field switching appears more economical and, in terms of latching considerations and problems, there is very little difference between pulsed-field and pulsed-current switching.[32]

Current pulse switching is by far the simplest and most economic arrangement, provided one uses a circuit, Fig. 14, that guarantees delivery of all the trigger energy into the switch rather than into the load. Even then, there exists the possibility that the switch will hang for tens of milliseconds in a partially switched state, with exceedingly slow thermal propagation of the normal regions, and thus fail to transfer energy efficiently.

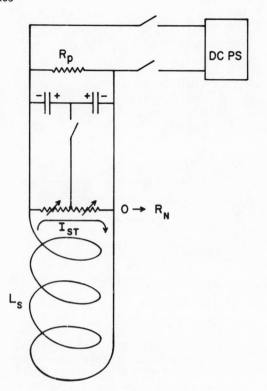

2.11.3 <u>Switching Times</u>
        Electromagnetic radiation of suf-
ficiently high frequency ν, so that the
energy

$$h\nu \geqq 2\Delta \cong 3.5 \; k_B T_c \; , \qquad (13)$$

will dissociate or break the condensed
Cooper pairs and thus switch a superconductor, where h = Planck's constant. However,
this frequency does not represent a limiting rate since higher frequencies will
simply deliver greater energies. The real
limitations are set by the time required
for communicating the switching signal
throughout the bulk of the switch material,
i.e. ultimately by magnetic and possibly
thermal diffusion times. It was shown in
our earlier studies[35] that, provided the
voltage was high enough, times as short as
100 nsec are attainable in current switching for 0.13-mm diam wires (Fig. 15) and
that these minimum times are a function of
the wire diameter (Fig. 16). It is very

Fig. 14. <u>Double Capacitor Switching Cir-
cuit</u> guarantees complete delivery
of capacitor energy into center
tapped switch, independent of
parallel impedances.

Wipf[33] has derived the conditions for a
partially normal current carrying superconductor with steady, more or less sinusoidal
temperature variations along its length.
We have obtained[30] an empirical relation
for the minimum voltage to guarantee the
development of full switch resistance during the initial trigger pulse:

$$E = 0.3 \; I_c \; R_N \; , \qquad (12)$$

i.e. the voltage has to be at least 30% of
the value required to drive the critical
current through the fully normal switch.
The minimum energy required for latching
is, of course, set by the enthalpy increment up to T_c for the switch and associated
matter, including potting material. Under
certain conditions the latching energy may
be abstracted from the storage circuit after the initial switch triggering. Some
of these considerations have been verified
and expanded upon in recent studies at
Karlsruhe.[34]

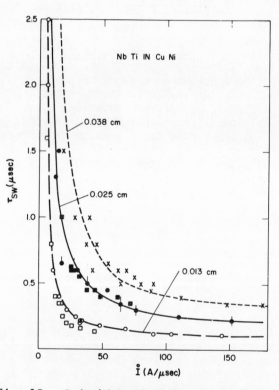

Fig. 15. <u>Switching Times in 10 cm Wire
Samples.</u> Note that in this experiment İ was also proportional to
the voltage on the switch element.

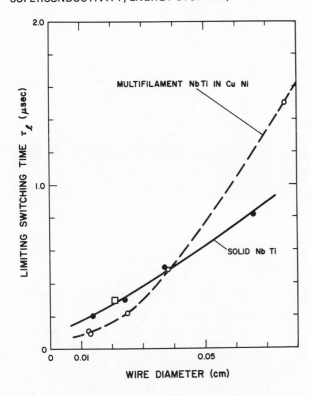

Fig. 16. *Limiting Switching Times as a Function of Wire Diameter* for solid and multifilamentary NbTi wires.

likely that high current thin film devices can be switched even faster than wires.* However, the practical limitations on switching speed usually arise from the time constants and jitter in the auxiliary switching circuit, i.e. the response of the triggered sparkgap, the inductance of the leads to the terminals of the switch, as well as the inductance of the switch itself. For magnetic switching there is an additional, hopefully reproducible delay due to the inductance of the field generating coil. The jitter or reproducibility problems become more severe, of course, when a number of similar units need to be activated simultaneously. As shown later, there is no difficulty in switching both halves of a 2 ohm 2000 A switch to full resistance simultaneously in 2 µs.

2.12 FLUX TRAPPING

There is an additional flux manipulating scheme that was discovered during early work on the behavior of Type II superconductors.[37,38] It combines elements of energy storage, switching and compression. A hollow cylinder, such as a Nb_3Sn sintered powder compact 25 mm o.d., 20 mm i.d. and 40 mm long is placed in a slowly rising magnetic field, Fig. 17. After initial shielding of the outside field, H_{ex}, from the inside, H_{in}, the field suddenly penetrates or flux jumps in about 100 µsec around 2T. The flux jump implosion step may be repeated 2 or 3 times before changing to flux creep at higher fields (lower J_c's), which, of course, is a beautiful indication of stable $\rho \neq 0$ behavior. On decreasing the external field the processes are reversed until at zero applied field 1 or 2T may be trapped in the cylinder. This type of experiment is more of a concept at the present time rather than a practical hardware scheme. There are thermal shock and control problems that should become less severe with thinner cylinders operating and triggered into flux jumping at current densities closer to their true J_c values.

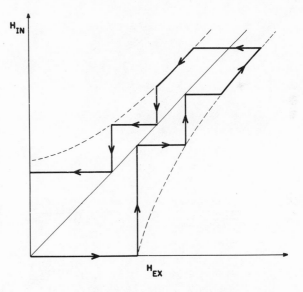

Fig. 17. *Flux Penetration into Hollow Cylinder.* (Tube Magnetization Experiment) Dashed lines are a measure of the critical current density in the cylinder walls.

---

*There is a large body of information on switching thin film superconductors at very small currents for computer applications.[36]

3.0 PRESENT HARDWARE AND SYSTEMS

3.1 CONDUCTORS

The requirements on the internal make-up of practical superconductors set by the various physics considerations discussed earlier are: finely divided superconductor, transposed or twisted within a matrix of a good normal cryoconductor, and with limited wire diameters and/or insulating regions of resistive alloys,depending on the intended B. Conductors with current ratings in the kiloampere range can be either monolithic, as shown in Fig. 11, or an assembly of many fine wires in the form of cables, braids, or spiral wraps, Fig. 18. Finer wires usually have higher J_c's, but are more subject to motion or vibration problems. Again, depending on the B requirements, the individual wires may be insulated either fully by organic or partly by inorganic insulation (such as copper oxide), or they may be bonded mechanically by a high resistance solder.[39]

The number of filaments in each wire can vary from a few hundred to at most a thousand, unless the manufacturer is willing to use multiple or sequential stacking and extrusions. In that case, filament numbers in excess of 20 000 can be obtained.[40]

The number of different superconducting materials that have been used since 1961 in commercial superconductors actually comprises a surprisingly small number of compositions. The properties of most of these are listed in Table I and clearly fall into two groups. First, there are the ductile, malleable alloys, NbTi and NbZr with cubic crystal structures, critical temperatures near 10 K and upper critical fields around 10T (at 4 K). Second, there are the brittle intermetallic compounds Nb_3Sn and Nb_3Ge, crystallizing in the A-15 or β-Wolfram crystal structure, with T_c's near 20 K and H_{c2}'s of 20 to 50 T. The highest transition temperature in bulk samples, > 20 K, has been seen on Nb_3Ge films recently prepared at Los Alamos.[41] These films carry self field critical current densities as high as 1.8×10^6 A/cm^2 at 14 K.[42] Their critical fields at 4 K could be in excess of 60 T.

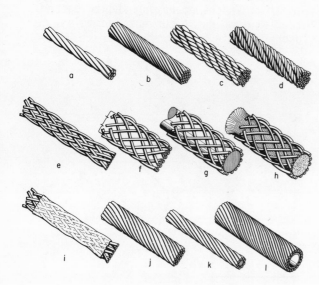

Fig. 18. *Transposed High Current Superconducting Assemblies.* (a)(b) Concentric lay stranded cable (hexagonally packed), (c)(d) rope lay stranded cables with proper and with incorrect twists, (e) flat braid, (f) flattened circular braid, (g)(h) circular braid on solid and flexible cores, (i) solder covered flat braid, (j) Roebel transposition, (k)(l) spiral wraps on solid wire and on hollow tube.

Table I

Properties of Commercially Important Type II Superconductors

	Atom Ratio	T_c (K)	$H_{c2}(4K)$ (T)	Cost ($/k·A·m·T)
NbZr	75/25	11.0	7.7	
NbTi	22/78	8.8±	9.2	
NbTi	36/64	9.3	11.6	0.072
Nb_3Sn	3/1	18.2	25	0.087
Nb_3Ge	3/1	22+	50+	?

The cubic alloys can be used over an extended range of compositions, they are easily fabricated by conventional means and behave like ordinary engineering materials. Their critical current densities can be optimized by cold working in combination with precipitation heat treating. The A-15 compounds must be of exact stoichiometric composition and require specialized fabrication procedures such as diffusion reactions or chemical vapor deposition. They degrade at tensile strains between 0.1 and 0.2%, which greatly limits their usefulness. For this reason and because it is readily stabilized in multifilamentary configurations, NbTi is the material of choice for most large scale engineering applications at the present time. The yield strength of these composites is a function of the strength and elastic modulus match of the individual components. As a first approximation one can use the appropriate volume averages with yield strengths of 50 to 100 MPa (7-15 000 psi) for copper, 700 to 1000 MPa (100-150 000 psi) for hard drawn NbTi, 430 MPa (62 000 psi) for hard 90/10 CuNi and 500 MPa (73 000 psi) for hard 70/30 CuNi.

Nb_3Sn is generally only available as tapes soldered to 0.025 mm thick copper tapes and sometimes strengthened additionally by stainless steel tapes (Fig. 7, bottom row). There is now much active development on multifilamentary Nb_3Sn conductors,[43] but the mechanical and engineering properties of such materials still remain to be evaluated.

Table I also lists some present costs in $/(kA·m·T) at 4 K. This unusual unit is convenient in field optimization analysis of energy storage inductors and can be justified as long as a given material is operated within the essentially hyperbolic regions of the critical current curves (Fig. 5). Powell[44] of the Brookhaven National Laboratory projects somewhat more optimistic costs for a mature fusion economy sometime after 2000 AD.

3.2 SUPERCONDUCTING SWITCHES

Current pulse switching (Sec. 2.11.2) avoids the costs and complexity of separate magnetic fields for switching and can be done effectively with the simple circuit of Fig. 14. Further subdivision is also possible, but it is necessary to minimize the inductance separately for each section of the switch. A coaxial switch geometry has been proposed by Mawardi.[45] This may well be the ultimate answer, but requires a considerable metallurgical fabrication development effort, is not readily scaled down to low currents and thus lacks the flexibility needed in the early developmental stages. For this reason, most of the Los Alamos effort[46] has been based on braided assemblies of multifilamentary NbTi wires. The best of the wires has been 0.2 mm diam with a 1.3:1 matrix ratio in 70/30 CuNi and with filament diameters of about 6μm. This material is adiabatically stable at self field critical current densities, but is very sensitive to motion and vibration and hence has to be potted fully and reliably. A loaded epoxy composition[47] has been found to be most satisfactory in conjunction with vacuum impregnation. A number of low inductance configurations have been constructed mostly from flat braids: spiral pancakes, stacked spiral pancakes, the classical Ayrton-Perry counterspiraling helical arrangement, and zig-zag accordion pleats in a solid slab of potting compound. A large Ayrton-Perry switch wound with two 87 wire flat braids is shown in Fig. 19, prior to final potting. This switch reached only 6000 A instead of the 10 000 A design figure, since the wire had a relatively large filament size (20μm) and the potting was not perfect. An obvious disadvantage of the Ayrton-Perry winding is the "useless" volume in the center of the tube which increases the needed dewar volume and costs. The zig-zag pleats have slightly higher inductances but fill space much more efficiently. A number of switch elements have been operated in parallel but it is more desirable to fabricate the switch from a single properly transposed conductor and thereby avoid problems of inductive mismatch and correspondingly inefficient current distributions. There is no difficulty in switching the two halves of a center tapped switch by the circuit of Fig. 14, in spite of the fact that the trigger currents add to the stored current in one half and subtract in the other half. Figure 20 gives the peak triggering currents measured in each half of a 2000 A braided Ayrton-Perry switch as a function of the stored current. The peaks were reached simultaneously within 2 μsec in the two halves, without any special efforts on minimizing the inductance of the triggering circuit.

Fig. 19. *Large Normal-Going Superconducting Switch with Noninductive
Ayrton-Perry Winding,* normal
state resistance 3.4 Ω, quenching current 6.1 kA.

3.3 ENERGY STORAGE COILS

A number of superconducting magnets
storing large amounts of energy incidental
to other uses have been constructed over
the last 10 years. Most of these are dc
bubble chamber magnets without need or
provision for fast energy removal or transfer. Figure 21 taken from a recent report
by Lubell[48] of the Oakridge National
Laboratory shows the gradual progress
since 1964 from a few MJ to almost 1000 MJ.
All of these coils represent large cryogenic and mechanical engineering efforts.
The addition of some data on pulsed magnets at the bottom of the graph indicates
that this technology is presently about

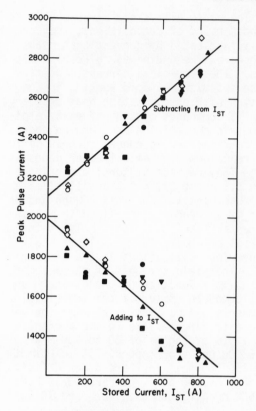

Fig. 20. *Peak Triggering Currents Delivered into each Half of Center
Tapped Ayrton-Perry Switch as a
Function of DC Current Stored
in System.*

three orders of magnitude below that for
dc magnets. Figure 22 is a nominally 300
kJ coil built in the Los Alamos METS program. This coil actually reached 12 500 A
and 375 kJ and, incidentally, represents
the highest current superconducting magnet
built anywhere. If the projected METS
goals are to be reached, there clearly has
to be an accelerated growth curve. Figure
23, also taken from Lubell,[48] gives the
costs for some of the large systems. The
slopes are empirical ones, because the
volume of a minimum conductor (Brooks coil)
solenoid varies as the energy to the 0.6
power, at fixed current density and packing
factor. The lower slopes on Lubell's graph
simply mean that relative costs have gone
down recently, or that a considerable
amount of development was paid for in the
early work, i.e. low energy dc coils should
now be cheaper. Even so, costs per joule
of stored energy decrease impressively from
$0.16/J at 1 MJ to $0.003/J at 1 GJ. By
comparison the materials and fabrication
costs for the bare 375 kJ METS coil were

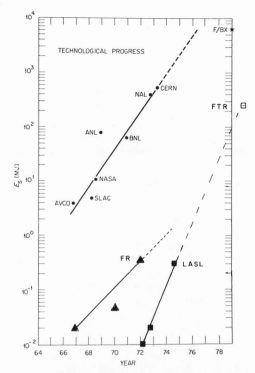

Fig. 21. *Technological Progress in Large DC and Pulsed Superconducting Coils.* DC coils from Lubell[48]; pulsed coils are French[26,28] (FR,▲) and LASL (■) work. Near term (1980) Tokamak and Theta-pinch objectives are indicated by F/BX (✱) and FTR (⊡).

about $0.12/J, not counting the in-house development, dewar, power supply and associated switch. The dewar and current leads present a similar cost. On the other hand, Wipf[49] has pointed out that at very high energy levels the amount of material and hence the cost will be directly proportional to the stored energy. At that point the economy of scale disappears and parallel systems are preferable.

3.4 REFRIGERATORS AND LIQUEFIERS

A major cost item in all cryogenic systems is represented by the liquefier or refrigerator. Not only do all superconducting components have to be cooled down, usually to 4 or 4.5 K, but any heat that leaks into the system (typically 0.2 to 0.3 watts/m^2 of dewar surface) or that is generated within the cryogenic environment has to be pumped to and discarded at room temperature. This involves the usual Carnot costs plus the inefficiencies of the cooling apparatus. The power, P_1, consumed by the compressors at room temperature, T_1, is

Fig. 22. *Experimental 375 kJ, 12.5 kA, 2.07 T Energy Storage Coil* designed for millisecond pulse discharges.

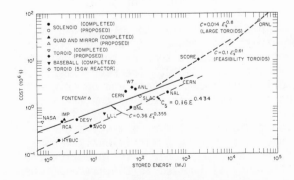

Fig. 23. *Costs of Large Superconducting Magnet Systems,* from Lubell[48] with line for solenoids, C_s, added.

related to the power, P_0, - the "frigiwatts" - removed from the low temperature, T_0, by

$$P_1 = P_0(T_1-T_0)/T_0 \; \eta, \qquad (14)$$

where η, is a measure of the mechanical and thermal efficiencies of the compressors, heat exchangers and expanders and can vary from less than 10% for small units to 25 or 30% for the largest refrigerators built to date. Thus it takes about 300 W of compressor power to refrigerate 1 W at 4 K. Strobridge[50] of the National Bureau of Standards at Boulder, CO has shown that the cost of large refrigerators is almost entirely determined by the compressor power (in kw), so that

$$Cost \; (\$) = 6000 \; P_1^{0.6} . \qquad (15)$$

Recent inflationary factors have increased the coefficient by about 25% from 6000 to 7500.

4.0 SYSTEMS CONSIDERATIONS

A complete systems study or parametric analysis can, of course, only be done in terms of the requirements of a specific system. However, some generalized statements can be made with regards to the range of applicability of various technological options and to the costs, both as a function of the stored energy and of the peak power output. Finally, these general statements can be illustrated by some of the large scale energy storage projects that are now being proposed.

4.1 TRANSFER CIRCUIT ANALYSIS

Figure 24 is a generalized schematic diagram of an inductive magnetic energy storage system. The power supply, PS, gradually builds up the required current, I_0, in a storage inductor, L_S, and a switch or variable impedance, X, breaks the current and transfers some of the stored energy into the load Z. The requirements of Z define the peak voltage and power characteristics of X as well as the transfer time constant, τ. If Z is resistive, very high energy transfer efficiencies can be achieved by making X sufficiently large. If, on the other hand, Z is inductive the efficiency is limited to 25%, at best when $Z=L_S$, unless a phase shifter such as a capacitor, C (or its equivalent in rotating machinery) is present in the circuit. In all cases, there are the additional

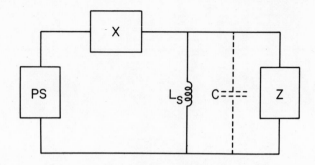

Fig. 24. *Generalized Inductive Energy Storage and Transfer Circuit.* Power supply, PS, is shown in series with switch or variable impedance transfer element, X, and storage inductor, L_S. Characteristics of load impedance, Z, determine size of optional transfer capacitor , C.

options: (a) of direct or transformer coupling of the load, Fig. 25c, with the possibilities of current and voltage transformation, (b) of having an energy diverting or voltage limiting resistor, R_p, in parallel with the switch, X, Fig. 25 b, c, (c) of inserting the power supply in series or in parallel in relation to the switch and storage coil assembly, and (d) of limiting peak voltages by modularizing the system with a sufficient number of ground points, Fig. 26, or with entirely separate units.

All of these trade-offs need to be considered in detail and we have changed our views on a number of them over the last few years.[51] At the present time and for a system with an inductive load, we favor:

1) Direct rather than transformer coupled load. The problem of multiple high current leads appears less formidable than the reactive forces during pulse discharge and the losses resulting from poor coupling coefficients in a transformer operating between cryogenic and room temperatures.

2) Series connected switch, power supply and storage coil.[30] This arrangement is safer and more convenient since it can be discharged quickly at any instant, whereas the parallel arrangement needs to be placed in the persistent mode and then disconnected from the low voltage charging

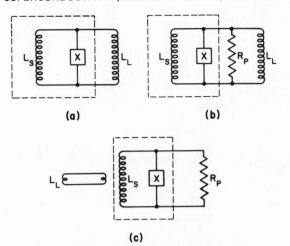

(a)   **(b)**

(c)

Fig. 25. *Some Transfer Circuit Options for Cryogenic Storage,* shown with inductive load, L_L, optional transfer resistor, R_p, and with direct (b) or transformer (c) coupling. Cryogenic region is enclosed in dashed lines.

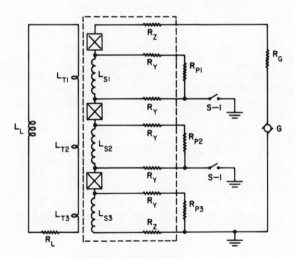

Fig. 26. *Modularized Energy Transfer System* with multiple ground points and three series connected coil-, L_S, and switch-, $\boxtimes$, modules. Resistances, R_Z, represent d.c. cryogenic high current leads from generator, G, with internal resistance, R_G. Pulse duty leads, R_Y, can have higher resistance than R_Z. Multiple winding transformer secondaries, L_T, and inductive load, L_L, are shown. R_L includes resistance of room temperature secondaries and load coil.

power supply. Also, there is then no need for superconducting links between coil and switch, at least for a storage unit with a single ground point.

3) A relatively small "protective" resistor, R_p, of the same impedance as X, Fig. 25b, which for a given circuit voltage implies merely a doubling of switch resistance and material costs. Further increases in X relative to R_p so as to reduce refrigeration costs, not only increase the materials costs, but also increase the required switching energy and are likely to introduce sizable voltage transients during the initial current transfer into R_p.

4) A transfer capacitor or its equivalent will ultimately be needed for reversible transfer. Its economics for improving transfer efficiency in intermediate size system is still under discussion.

4.2 SWITCH BURNOUT AND THERMAL RECOVERY

The amount of additional energy deposited in the switch during the transfer processes depends entirely on the relative circuit impedances and time constants and cannot be treated in a general way. Without R_p there is danger of burnout or at least annealing of the superconductor if the average current densities greatly exceed 10^9 A/m^2.[25]

Similarly, the rate of thermal recovery of a switch depends on the peak temperature attained during the transfer, as well as on the thermal environment, the amount of plastic insulating material, and the size and efficiency of any cooling channels. Typical recovery times in our switches have been between 10 and 200 seconds. If a system demands faster recovery or higher repetition rates than the 5 to 15 minutes considered for the METS system, it will be necessary to reduce the mass of the switch by eliminating the potting and probably to go to thin film switches or noncryogenic circuit breakers.

4.3 ALTERNATIVE SWITCHING METHODS

There exist, of course, a number of alternative switching methods, some discussed in these proceedings, that can be used in transferring energy from an inductive storage. Some of their advantages and drawbacks are listed here briefly for completeness:

1) <u>Fuses</u> are simple and inexpensive, but they are cumbersome in that they have to be replaced after every transfer,[52] although a prepackaged design greatly reduces the required time. Exact timing becomes a problem when currents have to be carried in a parallel mechanical contactor prior to energy transfer, especially with multiple elements. However, fuses do offer submicrosecond response times.

2) <u>Mechanical Breakers</u> include many commercially available devices that require mechanical motion for contact separation which, of course, will create an arc. Mechanical motion is limited to millisecond response at best; however, the time for the actual high voltage circuit interruption can be held to a few tens or hundreds of microseconds by the firing of a counter-pulse or commutating capacitor discharge to quench the arc[53]. Such methods have been used successfully[54] at 100 μsec transfer times and relatively high repetition rates (5 pps). Again, simultaneity is a problem when multiple elements are involved.

3) <u>Advanced Plasma Arc</u> devices[55] offer intriguing possibilities. When fully developed, they should be able to switch in times of a few tens of microseconds.

4) <u>Variable Inductance</u> devices[56,57] have rotating mechanical components and hence are limited to times longer than 1 msec or more likely 10 msec.

Clearly the primary <u>advantages</u> of the normal-going superconducting switch are that (a) in the resistive state it is ohmic with linear potential gradients, (b) the total voltage is only limited by surface breakdown and by the rest of the system, (there is no arc to extinguish), and (c) switching can be done in microseconds or less. The <u>disadvantages</u> are to be found in the complexity of cryogenic operations and in the refrigeration costs, especially when high pulse repetition rates are needed.

4.4   ENERGY STORAGE COIL

The decisions to be made in the design of the superconducting storage coil, some of which were implied earlier, are: type of conductor (i.e. NbTi), operating temperature, cooling method, current level and peak field value.

4.4.1 Coil Cooling

The prevailing practice of operating a large superconducting coil immersed in a "huge vat of boiling helium" at 4.2 K has been likened by H. H. Kolm[58] to the primitive cooling of early internal combustion engines with stationary sumps of water. Circulating cooling systems at supercritical pressures are certainly to be preferred in large systems, not only to reduce the helium inventory but also for a greater range of operating temperatures. So far, these ideas have been implemented in only two medium size systems.[59,60]

4.4.2 High Current Leads

The selection of kiloampere operating currents implies sizable additional heat leaks at the high current leads that bridge the temperature gap from the room temperature power supply to the cryogenic coil. With a liquid helium cooled magnet, a single well designed lead increases the boil-off by an amount equivalent to about 1 watt/kA. However, the actual refrigeration load is much greater, since the gas is returned to the refrigerator at room temperature after exchanging heat with the leads along their entire length. Various schemes for low temperature power supplies or "flux pumps" have been suggested,[61] but none has as yet been engineered for sizable power outputs at kiloampere levels. For these reasons it is unlikely that currents much greater than 25 kA will be used in superconducting coils in the near future.

4.4.3 Choice of Magnetic Field Level

For an energy storage coil, selection of the operating magnetic field presents an essentially free parameter, as distinct from ordinary magnets where it is usually governed entirely by physics considerations and preferably made as large as possible. It is easily shown that for a given amount of material with a hyperbolic J_c vs B curve the largest amount of energy would be stored in a coil operating at the lowest possible field, such as 1T or less. However, this simple view neglects the cost of the normal metal stabilizer, the size and costs of the dewar, and the corresponding refrigeration requirements. When these are taken into account, a broad optimum between 3 and 6T is obtained for NbTi at 4 K, with the exact value almost independent of the stored energy.[28,51]

4.5 LARGE SUPERCONDUCTING INDUCTIVE STORAGE SYSTEMS

A convenient way to view quickly the entire subject of energy storage and at the same time to intercompare various technological solutions and their ranges of applicability is through Fig. 27, the log (Power) vs log (Energy) plot first used by Solé[26] for historical comparison and projection. The two quantities are, related through the energy delivery time constants, τ, the 45° lines that are labeled 10^{-8} to 10^6 sec. The symbols mark the various superconducting experiments and proposals.

4.5.1 Classification

The storage units marked in Fig. 27 clearly fall into 4 areas or regions, depending on the intended delivery times.

A) Microsecond delivery of power into resistive loads such as flash lamps and lasers. This has included the French work[26,28] and also programs sponsored by the U.S. Air Force at Magnetic Corporation of America[54] and at Avco[62] that were based on earlier analyses by Stekly and co-workers.[63] A 100 kJ, 250 μsec coil has been built by MCA.[64] Unfortunately, there is as yet no generally available report on the very fine work done in these Air Force projects.

B) Millisecond delivery of power into inductive loads, such as the plasma compression coils for θ-pinch controlled thermonuclear (CTR) development. The goals of this METS effort are both sizable energies and very high peak powers. The θ-pinch fusion test reactor (FTR)[65] to be built in the early 1980's, will need 250 MJ of plasma compression energy delivered in 1 or 2 ms to a toroidal coil of 180 m circumference, Fig. 28. An energy transfer efficiency of 90% should be achievable with the aid of about 150 MJ of low voltage (10 kV) transfer capacitors, which are now conceived as having the dual function of also powering the experiment entirely during its early low energy tests. Since only a 15-minute repetition rate is required, the refrigeration costs inherent in normalgoing superconducting switches may be tolerated. Figure 29 is an example of one of the parametric comparisons that have been made for the plasma compression components of the FTR.[51]

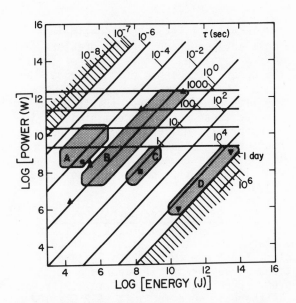

Fig. 27. *Classification of Energy Storage Systems* in terms of peak power, energy and delivery times.[66]

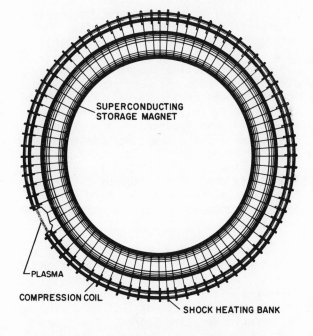

Fig. 28. *Sketch of 180 m Circumference θ-pinch Fusion Test Reactor.*

A full scale θ-pinch CTR power reactor[67] should operate at longer plasma compression times of about 10 ms, but the fields must be maintained for 30 ms. Every 10 seconds this reactor will require for plasma compression alone 178 MJ/m for a 351 m circumference, or a total of 6.3 x 10^10 J, while it only produces 101 MJ/m of usable thermal energy. Thus it will become extremely important to handle this large circulating power in a reversible manner, to reduce all losses to an absolute minimum, and to further develop low loss energy transfer methods.[57]

C) Second delivery to lower the short term energy demand placed on an electric utility. The Fermi National Accelerator Laboratory[68] has proposed such a device to limit average power consumption to 80 MW while energy is pumped into and out of the 500 GeV accelerator coils every 10 sec at peak rates of up to 300 MW. The unit will store 3.6 GJ, but only exchange 360 MJ with the accelerator.

D) Kilosecond delivery into a large utility system or grid with the objective of load leveling or peak shaving. The aim here is to provide an alternative to pumped hydroelectric units free of geographic and land use limitations. This will permit operating the most efficient base generating units, specifically nuclear reactors, at full power and avoid or reduce power generation from combustion turbines that burn expensive hydrocarbon fuels. The idea was first suggested by Ferrier[69] and there are active programs under the direction of Boom and Peterson at the University of Wisconsin[70] and of Hassenzahl and Keller at Los Alamos.[71] Andrianov et al.[72] have reported on a small experiment where 3-phase power was pushed back into the Moscow grid. Ultimately systems of this kind may store 30 GJ (8 MWh) to 100 TJ (30 GWh). To be economically practical the larger units will have to be built underground in rock to support the forces.

Clearly, energy transfer in the C and D regions could not be done by the types of switches discussed earlier, but will use solid state thyristor inverter-converter units. Present costs of such units including transformers are $30/kW. This means that peak power ratings can be increased with only a slight increase in capital costs. Power system stabilization becomes thus an additional benefit. Economic analyses of these and of transmission "credits"

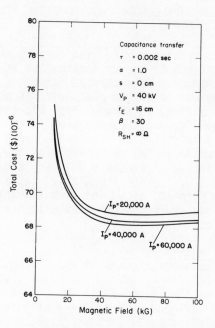

Fig. 29. *Parametric Study of Some METS System Options* from Rogers et al.(51)

from the utility point of view are discussed in the literature.[71]

4.5.2 Boundaries

Figure 27 also shows the boundaries or restraints imposed on inductive storage by near-term technological limitations as follows:

Transfer Times. Opening times of the most advanced high current breakers (except for fuses) set a lower limit of 100 ns at the upper left corner of the figure marked by the hatched lines perpendicular to the time constant lines. Distributed coil capacitance can produce a similar limit. Conversely, delivery times longer than 10 hours or 40 ks have no foreseeable economic use. This boundary is shown by the hatched lines at the lower right corner of the figure.

Unit Size and System Complexity. Although there are no a priori limits to the energy content of a single storage inductor, there are current and voltage limitations in any practical superconducting system. Even with an optimistic 100 kV and 25 kA there is a power limit of 2.5 x 10^9 W. Higher peak powers require a multiplicity of switches and grounding points. The numbers 1, 10, 100 and 1000 connected to the horizontal lines in the figure are a measure of that complexity, and 1000 switching units may well constitute a practical limit. On the other hand, below the line labeled "1", a single storage and switching unit will serve. An important consequence of being forced into a multiple unit system because of high peak power demands, can be the loss of much of the economy of scale. Thus costs per unit of stored energy will start to go up linearly with the number of units well before the strength of materials limit is encountered.

Charging Power. Total stored energy is limited by the amount of power, perhaps 1 GW, that can be drawn from a utility grid in 20 hours. This determines a vertical boundary of 10^{14} J at the right.

4.6 COMPARISON WITH ALTERNATIVE ENERGY STORAGE METHODS

4.6.1 Room Temperature Inductors

"Warm" or room temperature inductive storage is practical and can give sizable power amplification as long as the delivery time constant τ_1 is at least one or two orders of magnitude shorter than the self discharge time constant of the storage coil,

$$\tau_2 = L_S/R_S \qquad (16)$$

Walker and Early[73] at the University of Michigan slowly charged a warm aluminum coil to 525 kA and then transferred the energy into a 0.32 mΩ resistive load with a fuse. Similar methods continue to be useful in plasma research[52] and in ultra-high voltage experiments.[74] Carruthers[75] made a cost comparison between inductive and capacitive storage prior to the advent of high field superconductivity.

John Marshall[76] has been advocating large room temperature aluminum inductors as CTR power sources for some time. The inductors are charged directly from the power line through a rectifier at a power level, P_1, and after $3\tau_2$ the stored energy will have reached as its maximum practical value, 95% of

$$Q_2 = 1/2 \ P_1 \tau_2 \qquad (17)$$

He also notes that costs of a system are determined by the power supply (at $0.12/W) and the inductor. For a Brooks coil, mass and costs vary as $\tau_2{}^{3/2}$, independently of

the impedance or degree of subdivision. The total costs of such a unit, but without transfer elements are then (for manufactured aluminum at $4.4/kg)

$$\text{Cost (\$)} = 0.12 \ P_1 + 4200 \ \tau_2{}^{3/2} \qquad (18)$$

The ratio of power supply to Aℓ costs determines an optimum current density that is independent of the optimally stored energy, Q_2, and that for the assumptions of Eq. (18), a space factor of 0.8, and a resistivity of 2.83 x 10^{-6} Ω-cm amounts to 228 A/cm^2. The corresponding minimum costs are proportional to $Q_2{}^{0.6}$ and drop from $0.09/J at 1 MJ to $0.006/J at 1 GJ. A comparison with the costs of superconducting solenoids on Fig. 23 favors the latter above 30 MJ. Power supply limitations at perhaps 100 MW terminate the range of warm Aℓ coils at an energy of 10 GJ. Nevertheless, as long as weight and efficiency do not matter, they must be considered whenever the peak power requirement is above a line (not shown in Fig. 27) that would go from log (P) = 5.4 at 10^3 J to log (P)= 9.6 at 10^{10} J. This includes the entire A and B regions, except for the full size θ-pinch CTR reactor.

4.6.2 Capacitor Banks

Compared to inductive storage, capacitors are a mature technology. They remain the prime energy source for the entire left third of Fig. 27. Oil filled, paper dielectric capacitors are typically available in 3.3 kJ units or "cans" at $0.12/J (1.85 μf, 60 kV, 85% reversal, 0.1 μs transfer time) and in 8.5 kJ units at $0.03/J (170 μf, 10 kV, 25% reversal, 10 μs transfer time).[77] Thus, as usual, costs increase as relative power levels increase. Installed systems costs are higher by about a factor of 3 to 5. There is little economy of scale at larger energies and the system complexity or number of parallel units would appear as lines perpendicular to the energy axis on a display similar to Fig. 27. The 10 MJ Scyllac bank at Los Alamos[78] contains 3000 parallel capacitors and spark gaps that have to be fired simultaneously. It is unlikely that fast 60 kV systems will be pushed to higher energies; however a 10 kV 10 000 unit 150 MJ bank has been designed[77] with an estimated cost of $9.4 M or $0.06/J and thus competes for the near-term METS applications.

4.6.3 Electrolytic Batteries

Electrolytic batteries have been suggested for utility peak shaving, typically as 100 MWh (3.6×10^{11} J), 10 MW units located at substations. Such a unit using present day lead acid cells is estimated to cost[79] \$600/kW or \$60/kWh (for 10 hour peaking). A reduction in costs to \$30/kWh is expected with some near-term development. A project at the Argonne National Laboratory developing hot, 400°C, Li/S cells for storage and vehicle propulsion has a design goal[80] of \$15-25/kWh; however, more realistic near-term goals are given[81] as \$100/kWh to which \$10/kWh will have to be added for the inverter-converter, switchgear and installation. None of these batteries is intended for fast discharges and reversal from charge to discharge condition takes at least 1 sec. As far as is known, there are no active efforts to extend the classical work of Kapitza[82] on modifying the internal structure of electrolytic cells to improve their pulse discharge characteristics.

4.6.4 Flywheels.

Flywheels for energy storage have been rediscovered recently and made more attractive and safer through the use of modern anisotropic materials, such as fiber composites.[83] There are many proposed uses for propulsion as well as peak shaving. A relatively small 10 MWh (36 GJ), 3 MW unit to be located at substations is optimistically estimated[84] at only 10^{-5}/J and \$110/kW, which would be very competitive. However, the mere change to a 10 hour discharge would more than triple the cost per unit power to \$360/kW. The development of a constant frequency output AC generator driven from a variable speed shaft still presents a formidable engineering task.

The use of flywheels at discharge times of fractional seconds is much less attractive because of torque limitations.

4.7 COST COMPARISON

Table II summarizes and compares the cost estimates for most of the technological options that have been discussed. It is clear that in the A and B regions of Fig. 27 the objective must be cheap peak power, whereas in the D region it is low costs per unit of energy delivered per discharge cycle. Thus both \$/kW and \$/J

are given. In systems consisting of many identical components, such as capacitor banks, assembled costs are usually much higher than unit costs. This may also be true for highly subdivided inductive storage systems; hence great care must be exercised in comparing data from different sources. Varying discharge times also have an immediate effect on costs per kW. Operational convenience, weight, reliability and efficiency of energy usage are, of course, not included in this comparison.

We can conclude that on a capital cost basis superconductivity at the present time and for systems smaller than about 50 MJ competes only with some difficulty. However, it becomes more competitive and will have to be used at the very high energies where overall efficiency becomes paramount. There is some hope for lower costs in the long run, but only if experience and firmer cost factors can be obtained at medium sized units that will have to be subsidized. We have not shown detailed figures, but the situation is similar for superconducting switching, in that uses will have to be quite selective.

Although it is a fully developed science, as far as technology goes, superconductivity is still very young. To mature it will need nurturing, some time for growth, and informed and discriminating application.

TABLE II

Cost of Different Energy Storage Systems

	Energy (J)	Delivery Time	Capital Costs		
			Total ($)	$/Peak Power ($/kW)	$/Energy ($/J)
A. SUPERCONDUCTING STORAGE					
METS Coil (Fig. 22)	(375 k)*	---	45 k	---	(0.12)*
METS System (incl. switch)	75 k	2 ms	100 k	0.33	1.34
FTR Designs[51] (Plasma	208 M	1 ms	38.6 M	0.05	0.16
compression only)	208 M	2 ms	29.3 M	0.08	0.14
500 GeV Accelerator[68] (100 kWh - 300 MW pk)	360 M	4 s	20 M	67.	0.056
Univ. Wisconsin[70] (10 GWh - 1 GW)	36 T	10 h	430 M	430.	1.2×10^{-5}
LASL[71] (8.3 MWh - 25 MW)	30 G	0.3-3 h	6.1 M	245	2.0×10^{-4}
(278 MWh-30 MW)	1 T	9 h	52.1 M	1740.	5.2×10^{-5}
(11 GWh-1.4 GW)	39 T	8 h	640 M	470.	1.6×10^{-5}
B. OTHER SYSTEMS					
Capacitors 60 kV[78]	10 M	5 µs	7 M	3.5×10^{-3}	0.70
10 kV[77]	150 M	100 µs	9.4 M	6.3×10^{-3}	0.06
DC Supercond. Solenoids[48]	(1 M)	---	160 k	---	(0.16)
	(1 G)	---	3.2 M	---	(3.2×10^{-3})
Alum. Brooks Coils[76]	(1 M)	---	93 k	---	(0.09)
	(1 G)	---	5.9 M	---	(6×10^{-3})
Batteries, Lead Acid[79]	360 G	10 h		300	0.8×10^{-5}
(100 MWh) Li/S, now[81]	360 G	10 h		1100	3×10^{-5}
(10 MW) Li/S, goal[80]	360 G	10 h		200-350	$0.6-1 \times 10^{-5}$
Flywheel[84] (10 MWh - 3 MW)	36 G	3.3 h	325 k	110	0.9×10^{-5}

*Numbers in parentheses refer to energy stored in the system, all others are for energy delivered by the system.

1) H. Kamerling Onnes, Further experiments with liquid helium. C. On the change of electric resistance of pure metals at very low temperatures etc. IV. The resistance of pure mercury at helium temperatures, Leiden Comm. 120b (28 April 1911); V. The disappearance of the resistance of mercury ibid., 122b (27 May 1911).

2) H. K. Onnes, Sur les résistances électriques, Solvay Congress Nov. 1911, Leiden Comm. Supply No. 29.

3) H. K. Onnes, VII. The potential difference necessary for the electric current through mercury below 4°19 K, Leiden Comm. 133a (22 Feb. 1913); VIII.The sudden disappearance of the ordinary resistance of tin and the super-conductive state of lead, ibid., 133d (31 May 1913).

4) H. K. Onnes and W. Tuyn, X. Measurements concerning the electrical resistance of thallium in the temperature field of liquid helium, Leiden Comm. 160a (28 Oct. 1922).

5) W. Tuyn and H. K. Onnes, XII.Measurements concerning the electrical resistance of indium...., Leiden Comm. 167a (30 June 1923).

6) H. K. Onnes, Further experiments with liquid helium, I,....IX, The appearance of galvanic resistance in supra-conductors which are brought into a magnetic field at a threshold value of the field, Leiden Comm. 139f (28 Feb. 1914).

7) W. Meissner and R. Ochsenfeld, Ein neuer Effekt bei Eintritt der Supraleitfähigkeit, Naturwiss. 21, 787 (1933).

8) C. J. Gorter and H. Casimir, Zur Thermodynamik des supraleitenden Zustandes, Physik. Zeitschr. 35, 963 (1934).

9) F. and H. London, The electromagnetic equations of the supraconductor, Proc. Roy. Soc. A149, 71 (1935).

10) J. Bardeen, L. N. Cooper and J. R. Schrieffer,Theory of superconductivity, Phys. Rev. 108, 1175 (1957).

11) W. J. de Haas and J. Voogd, The influence of magnetic fields on supraconductors (Pb-Bi and other alloys), Leiden Comm. 208b (29 March 1930).

12) B. T. Matthias, T. H. Geballe, S. Geller and E. Corenzwit, Superconductivity of Nb_3Sn, Phys. Rev. 95, 1435 (1954).

13) J. E. Kunzler, E. Buehler, F. S. L. Hsu and J. H. Wernick, Superconductivity in Nb_3Sn at high current density in a magnetic field of 88 kgauss, Phys. Rev. Lett. 6, 89 (1961).

14) P. G. de Gennes, Superconductivity of Metals and Alloys, W. A. Benjamin Inc., New York, 1966, p. 265.

15) H. London, Phase-equilibrium of supraconductors in a magnetic field, Proc. Roy. Soc. 152, 650 (1935).

16) A. B. Pippard, The surface energies of superconductors, Proc. Cambridge Phil. Soc. 47, 617 (1951).

17) V. L. Ginzburg and L. D. Landau, On the theory of superconductivity, Zh. Eksper. Teor. Fiz. 2, 1064 (1950); A. A. Abrikosov, On the magnetic properties of superconductors of the second group, Soviet Physics JETP 5, 1174 (1957); L. P. Gor'kov, Theory of superconducting alloys in a strong magnetic field near the critical temperature, ibid., 10, 998 (1960).

18) B. B. Goodman, Type II superconductors, Rept. Progr. Phys. 29, 445 (1966).

19) a) C. P. Bean, Magnetization of hard superconductors Phys. Rev. Lett. 8, 250 (1962);
   b) H. London, Alternating current losses in superconductors of the second kind, Phys. Lett. 6, 162 (1963).

20) For a comprehensive treatment and summary of superconductor stabilization see: M. N. Wilson, C. R. Walters, J. D. Lewin and P. F. Smith, Experimental and theoretical studies of filamentary superconducting composites, Jour. Phys. D, 3, 1517 (1970).

21) A. R. Kantrowitz and Z. J. J. Stekly, A new principle for the construction of stabilized superconducting coils, Appl. Phys. Lett. 6, 56 (1965).

22) J. R. Purcell and H. Desportes, The NAL bubble chamber magnet, Proc. 1972 Appl. Supercond. Conf., IEEE Pub. No. 72 CHO 682-5-TABSC, p. 246.

23) G. Bogner, The development of large superconducting DC magnets in Europe, ibid., p. 214.

24) H. L. Laquer, Superconductors for millisecond pulse applications, to be published.

25) R. R. Hake, Single shot pulsed magnetic fields from inductive energy stores, Los Alamos Report LA-4617 MS, UC-20, TID 4500, March 1971, 44 pp.

26) J. Solé, Application des supraconducteurs à l'accumulation et à la libération de l'énergie électrique, Entropie 39, May-June, 21 (1971), and many CEA reports dating from 1967.

27) D. L. Ameen and P. R. Wiederhold, Fast-acting superconducting power switches, Rev. Sci. Inst. 35, 733 (1964).

28) J. P. Krebs, E. Santamaria, and J. Maldy, Superconducting devices for energy storage and switching, Proc. 4th Internat. Cryo. Eng. Conf. (IPC 1972), p. 172.

29) H. L. Laquer, J. D. G. Lindsay, E. M. Little and D. M. Weldon, Superconducting Magnetic Energy Storage and Transfer, Proc. 1972 Appl. Supercond. Conf., p. 98.

30) H. L. Laquer, J. D. G. Lindsay, E. M. Little, J. D. Rogers and D. M. Weldon, Design options and trade-offs in superconducting magnetic energy storage with irreversible switching, Technology of Controlled Thermonuclear Fusion Experiments and the Engineering Aspects of Fusion Reactors, E. Linn Draper, Jr., ed., USAEC CONF-721111, 1974, p. 177.

31) T. A. Buchhold, Superconductive power supply and its application for electrical flux pumping, Cryogenics 4, 212 (1964).

32) J. D. G. Lindsay and D. M. Weldon, personal communication.

33) S. L. Wipf and M. Soell, Flux flow properties of bare Nb-Ti wire, Proc. 4th Internat. Cryo. Eng. Conf. IPC Press 1972 p. 159.

34) K. Grawatsch, H. Köfler, P. Komarek, H. Kornmann and A. Ulbricht, Investigations for the development of superconducting power switches, IEEE Trans. Magnetics MAG-11, 586 (1975).

35) H. L. Laquer, D. B. Montgomery and D. M. Weldon, Superconductive energy storage and switching experiments, Proc. 13th Internat. Congr. Refrig. (IIR) 1971 p. 457.

36) a) J. W. Bremer, Superconductive Devices, McGraw-Hill, New York, 1962.
b) V. L. Newhouse, Superconducting Devices, Ch.22 in Superconductivity, R. D. Parks, ed., Marcel Dekker, NY, 1969.

37) Y. B. Kim, C. F. Hempstead and A. R. Strnad, Critical persistent currents in hard superconductors, Phys. Rev. Lett. 9, 306 (1962); Phys. Rev. 129, 528 (1963).

38) H. L. Laquer, Flux trapping and flux pumping with solenoidal superconductors, Proc. 11th Internat. Congr. Refrig. (IIR) 1963 p. 207.

39) A. D. McInturff, Superconductors for pulsed magnets, Proc. 1972 Appl. Supercond. Conf., p. 395.

40) R. A. Popley, D. J. Sambrook, C. R. Walters and M. N. Wilson, A new superconducting composite with low hysteresis loss, ibid., p. 516.

41) L. R. Newkirk, F. A. Valencia, A. L. Giorgi, E. G. Szklarz and T. C. Wallace Bulk superconductivity above 20 K in Nb_3Ge, IEEE Trans. Magnetics MAG-11, 221 (1975).

42) R. J. Bartlett, H. L. Laquer and R. D. Taylor, Critical currents in Nb_3Ge between 14 and 21 K, ibid., 405 (1975).

43) Ten papers at 1974 Appl. Supercond. Conf., ibid., pp. 231-302.

44) J. R. Powell, Design and economics of large DC fusion magnets, Proc. 1972 Appl. Supercond. Conf. p. 346.

45) O. K. Mawardi, Cryogenic switching systems for power transmission lines, U.S. Patent No. 3,384,762.

46) J. D. G. Lindsay et al, Development of a superconducting switch for magnetic energy storage systems, IEEE Trans. Magnetics MAG-11, 594 (1975)

47) Emerson & Cuming, Inc. Canton, MA Stycast 2850 FT and 2850 FT Blue, with catalyst 11.

48) M. S. Lubell, Superconducting toroidal magnets for fusion feasibility experiments and power reactors, Proc. 5th Internat. Cryo. Eng. Conf., to be published.

49) S. L. Wipf, personal communication.

50) T. R. Strowbridge, Refrigeration for superconducting and cryogenic systems, IEEE Trans. Nucl. Sci. NS-16, No. 3, 1104 (1969).

51) J. D. Rogers, B. L. Baker and D. M. Weldon, Parameter study of theta-pinch plasma physics reactor experiment, Proc. 5th Symp. on Engineering problems of Fusion Research,IEEE 73CH0843-3-NPS, 432 (1974).

52) J. N. DiMarco and L. C. Burkhardt, Characteristics of a magnetic energy storage system using exploding foils, J. Appl. Phys. 41, 3894 (1970).

53) A. N. Greenwood and T. H. Lee, Theory and application of the commutation principle for HVDC circuit breakers, IEEE Trans. on Power App. Syst. PAS 91, 1570 (1972).

54) E. J. Lucas, W. F. B. Punchard, R. J. Thome, R. L. Verga and J. M. Turner, Model coil tests for a pulsed superconducting magnet energy storage system, Proc. 1972 Appl. Supercond. Conf. p. 102.

55) M. A. Lutz, A Novel HVDC circuit interrupter for inductive energy storage systems, (cf. Ref. 30), USAEC CONF 721111, 1974, p. 298.

56) P. F. Smith and J. D. Lewin, Superconductive energy transfer systems, Particle Accel. 1, 155 (1970).

57) K. I. Thomassen, Reversible magnetic-energy transfer and storage systems, (cf. Ref. 30), USAEC CONF-721111, 1974 p. 208.

58) H. H. Kolm, The future of superconducting technology, Cryogenics 15, 63 (1975).

59) M. Morpurgo, The design of the superconducting magnet for the 'Omega' project, Particle Accel. 1, 255 (1970).

60) V. E. Keilin et al., Some problems of force-cooled superconducting magnet systems, Proc. 5th Internat. Cryo. Eng. Conf., to be published.

61) S. L. Wipf, The case for flux pumps and some of their problems, Proc. 1968 Summer Study on Superconducting Devices and Accelerators, Brookhaven National Laboratory Report BNL 50155 (C-55) April 1969, p. 632 and 689.

62) J. Teno, O. K. Sonju, L. M. Lontai et al, Development of a pulsed high energy inductive energy storage system, unpublished AVCO report.

63) E. J. Lucas, Z. J. J. Stekly, A. Foldes and D. Milton, Use of superconducting coils as energy storage elements in pulsed system operations, IEEE Trans. on Magnetics MAG-3, No. 3, 280 (1967).

64) Personal communication from Magnetics Corp. of America to John D. Rogers.

65) S. C. Burnett, W. R. Ellis and F. L. Ribe, Parameter study of a pulsed high-beta fusion reactor based on the theta pinch, (cf. Ref. 30), USAEC CONF-721111, 1974, p. 160.

66) H. L. Laquer, Superconducting magnetic energy storage, Cryogenics 15, 73 (1975).

67) R. A. Krakowski, F. L. Ribe, T. A. Coultas and A. J. Hatch, An engineering design study of a reference theta-pinch reactor (RTPR), Los Alamos Scientific Laboratory Report LA-5336, March 1974, 137 pp.

68) F. E. Mills, The Fermilab cryogenic energy storage system, IEEE Trans. Magnetics MAG-11, 489 (1975).

69) M. Ferrier, Stockage d'énergie dans un enroulement supraconducteur, Low Temperatures and Electric Power, Pergamon Press, Oxford, 1970, p. 425.

70) a) R. W. Boom and H. A. Peterson, Superconductive energy storage for power systems, IEEE Trans. on Magnetics MAG-8, No. 3, 701 (1972).
b) R. W. Boom, G. E. McIntosh, H. A. Peterson and W. C. Young, Superconducting Energy Storage,Adv. Cryo. Eng. 19, 117 (1974).

71) W. V. Hassenzahl, Will superconducting magnetic energy storage be used on electric utility systems? IEEE Trans. Magnetics MAG-11, 482 (1975).

72) V. V. Andrianov, V. B. Zenkevich et al, Discharge of a superconductor storage device into an inverter transformer, Soviet Physics-Doklady 16, 38 (1971).

73) R. C. Walker and H. C. Early, Half-megampere magnetic-energy-storage pulse generator, Rev. Sci. Inst. 29, 1020 (1958).

74) D. Kind, J. Salge, L. Schiweck and G. Newi, Explodierende Drähte zur Erzeugung von Megavolt-Impulsen in Hochspannungsprüf-kreisen, Elektrotechn. Zeitschr. A 92, 46 (1971).

LARGE SUPERCONDUCTING TOKAMAKS

D. N. Cornish
Euratom-UKAEA Ass. for Fusion Research
Culham Laboratory
Abingdon, U. K.

ABSTRACT

The magnetic field geometry for plasma containment receiving the most worldwide attention at the present time is the Tokamak System.

The main toroidal field in all existing machines and those at present being constructed is generated by an assembly of copper coils, often water cooled.

Design studies of power producing fusion reactors, however, show that this type of machine may only be economically viable if superconductors are used for this purpose.

The magnetic energy stored in the toroidal field of a 2500 MW(e) reactor will probably be greater than 10^{11} joules and even in the next generation of experiments designed to demonstrate the "feasibility" of fusion, will exceed 10^{10} joules. Due to the physical size and field strength of about 80 kG the forces and stresses will be much higher than ever before experienced. Pulsed fields for inducing the plasma current and stabilizing it will also react with the current in the coils giving rise to severe repetitive twisting moments.

Questions now being asked include: is it possible to use superconductors to meet all the technical and reliability conditions, can it be done economically and what development work must be done to bridge the gap between now and then?

These questions form the basis of the topics discussed in the paper.

SUPERCONDUCTIVE INDUCTOR-CONVERTER UNITS FOR PULSED POWER LOADS

H. A. Peterson,[*] N. Mohan,[*] W. C. Young,[**] R. W. Boom[†]
University of Wisconsin
Madison, Wisconsin 53706

ABSTRACT

Pulsed power loads of increasing magnitudes up to several hundred megawatts or more must be supplied in the near future. High energy physics research laboratory experiments and the longer range foreseeable needs for successful nuclear fusion reactors such as the University of Wisconsin Tokamak reactor designs are representative examples. Such large pulsed power demands are at best undesirable if not prohibitive, even for the largest electric power systems. Techniques for storing energy in superconductive inductors employing thyristorized converters are described. Circuits which serve to minimize both the pulsed power and corresponding reactive volt-ampere requirements from the three phase power system are presented. Exploitation of these circuits and related concepts from the control standpoint should provide a basis for designing power conditioning interface equipment to meet the challenging requirements of very large pulsed power loads looking to the future.

INTRODUCTION

Fusion reactors such as the University of Wisconsin Tokamak reactor, UW MAK-I,[3,4] -II,[5] according to present design thinking, will require repetitively pulsed power for the divertor and transformer coils with peak values of up to about 700 MW or more. These pulses are at best undesirable if not prohibitive, even for the largest electric power system.

Because the pulsed power is used to produce a magnetic field, the load is basically an inductor which might be supplied from a dc source, such as a Graetz bridge converter, Fig. 1, which serves as the interface with the conventional three phase power system. Such Superconductive Inductor-Converter or I-C units (for convenience) have been described elsewhere.[1,2,6,7] Figure 2 shows the distribution of the divertor coils, D1 to D4, and the transformer coils, T1 to T5, around the D shaped toroidal field magnets proposed for the UW MAK-I fusion reactor. The pulsed power requirements for these coils as a function of time are shown in Fig. 3.

Figure 3(a) shows the pulsed current and power requirements for the divertor coils. These are characterized by a charging time interval T_c, (in the range from 10 to 100 seconds), a hold interval T_h (in the range of minutes) and a discharge interval of T_d (similar to T_c). If there were no losses, all of the energy stored during the time interval T_c should be recovered during the period T_d after the hold period T_h. For this ideal no-loss case, the average power required from the power system would be zero, while the peak power required ($P_p = 1.0$) might be several hundred MW. Figure 3(b) shows the pulsed current and power requirements for the transformer coils. The slow change in transformer current during the burn time is used to maintain the necessary plasma current. Again for a no-loss case the average power required from the utility power system could be zero.

Figure 3(c) shows the pulsed current and power required from an I-C unit to reduce the power system requirements to a constant demand. Figure 4 shows how the divertor, transformer and I-C units with their converters and control systems would be interconnected with the three phase power system.

While this approach appears to be helpful in reducing or even eliminating the pulsed power demands made on a power system when supplying such loads, it does not provide a solution to the voltage "flicker" or voltage regulation problem. It is the purpose of this paper therefore

[*]Department of Electrical and Computer Engineering.
[**]Department of Engineering Mechanics.
[†]Departments of Nuclear Engineering, and Metallurgical and Mineral Engineering.

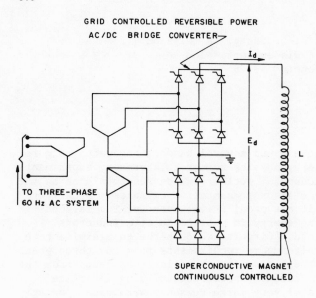

Fig. 1. Basic circuit elements for superconductive energy storage inductor-
converter or I-C unit for power system
applications.

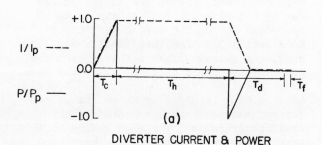

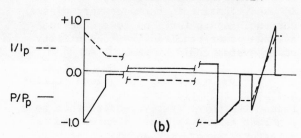

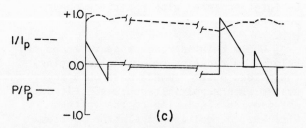

Fig. 3. UW MAK-I pulsed power requirement. Each quantity is normalized with
respect to its rating.

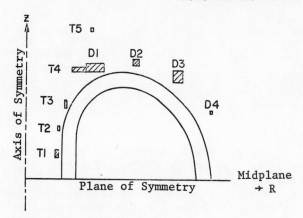

Fig. 2. Position of divertor and transformer coils.

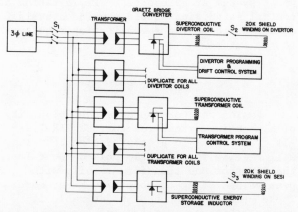

Fig. 4. Fusion pulsed power supply.

to present some circuit concepts and circuit arrangements which should be helpful
in satisfying very large pulsed power load
requirements imposing both excessive peak
power P and excessive peak kilovar Q
demands upon the power system source.

BASIC SYSTEM

The basic system which appears to
satisfy both of the foregoing requirements
is shown in block diagram form in Fig. 5.
In this basic system there is a large
superconductive inductor energy storage
unit. It is charged from the three phase
power system by means of the power source
converter. The combination of the inductor and this power source converter with
delay angle control make up an I-C unit
similar to that shown in Fig. 1. Energy
can be drawn from the three phase source
and stored in the superconductive-
inductor or alternately returned to the
source as desired by appropriate control
of the power source converter.

However, it is desired to provide one
or more (possibly several) pulsed power

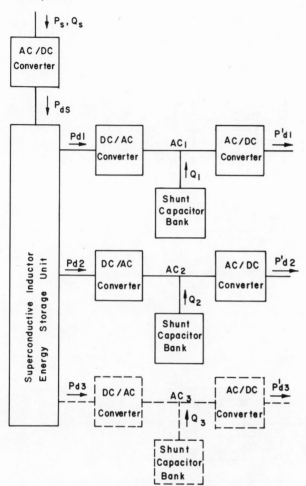

Power Source
3 Phase, 60 Hz.

Fig. 5. Block diagram representation of a
system to satisfy pulsed power requirements.

loads of large magnitude, and of arbitrary
shape and duration. This can be achieved
by drawing upon the energy stored in the
superconductive inductor, viewing it as a
constant current source and inserting into
the circuit a converter which is commutated
by a capacitor bank at a frequency preferably higher than 60 Hz (this frequency is
entirely independent of the 60 Hz source).
Thus, power P_{d1} can be made to flow out of
the storage unit provided we again convert
the AC_1 quantities to dc shown as P'_{d1}.
The capacitor bank simply provides the Q_1
required to achieve successful commutation
for both conversions. This double conversion sub-unit with terminals identified by
the quantities P_{d1} and P'_{d1} is power reversible in magnitude and direction so that if
the P'_{d1} load supplied is inductive such as
a fusion reactor load, that inductively
stored energy can in turn be returned to
the storage unit.

Clearly, if one sub unit can be made
to function in this fashion, a second
double conversion unit identified by
quantities P_{d2} and P'_{d2} can be added also.
Similarly others as needed such as the one
identified by P_{d3} and P'_{d3} can be added as
well. Net energy withdrawn, including
losses must come from the main source
converter. Indeed, in a broader sense,
the only need for the connection to the
60 Hz three phase source is to supply the
losses in the entire system. In turn, it
follows that as long as we design the
system so that the energy stored in the
storage inductor does not vary greatly
during normal operation, then the storage
inductor current will not vary greatly
either. If the power source converter
merely supplies losses, then it can be
operated simply as a rectifier. With
delay angle control and with constant
current on the dc side, P_s and Q_s are
essentially constant quantities while
operating with fixed delay angle. By
proper design, several pulsed loads of
arbitrary shape can be supplied and the
power system will be called upon only
for essentially average power at fixed
power factor free of individual pulses
or summary combinations thereof.

It will be appreciated that this
basic system in concept is the completely
electrical equivalent of a large flywheel
used for mechanical energy storage and in
turn supplying a plurality of pulsed
loads. The only need is a power source to
supply system losses and keep the energy
stored in the equivalent flywheel reasonably constant.

BASIC CIRCUITS

The basic circuit which permits hope
of realization for the foregoing system
objectives is shown in Fig. 6. Here it
is assumed that L_o is the superconductive
inductor energy storage unit. The power
source converter PSC provides the interface between the energy storage inductor
L_o and the source. From a practical
operating point of view, during normal
operation it may not be necessary for PSC
to be phase controlled. The converter
could be operated as a diode rectifier
interface with a current rating equal to
the current rating of the energy storage
inductor L_o. However, for complete flexibility, including return of stored energy
to the system, phase control should be
included. The PSC voltage rating should
be sufficient so that the PSC power rating
is sufficient to supply the average power

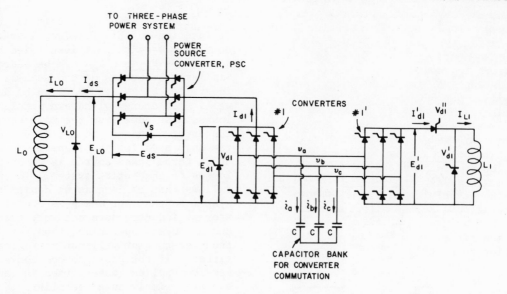

Fig. 6. Basic arrangement to supply pulsed power using three phase intermediate conversion.

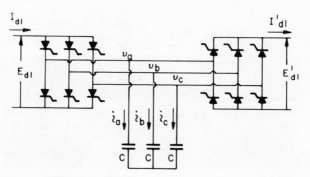

Fig. 7. Double conversion unit. Three phase, six pulse; dc/ac/dc.

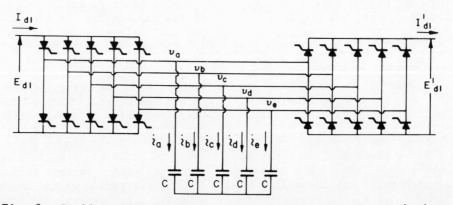

Fig. 8. Double conversion unit. Five phase, ten pulse; dc/ac/dc.

requirements corresponding to maximum pulsed power loads and corresponding losses in the system. For lesser pulsed load requirements and associated losses, the current I_{Lo} would tend to increase beyond rating. A limiting sensor could be employed to put PSC in bypass mode or reduced voltage mode until such time that I_{Lo} is once again within prescribed limits determined by design considerations. Delay angle control using thyristors would permit return of the energy stored in L_o to the power system source as desired, but would also be helpful in effecting bypass mode operation. Furthermore, phase control permits a degree of Q control which may be beneficial from a system voltage regulation point of view.[6]

Converters #1 and #1' together with the capacitor bank for commutation of both converters are a sub-unit for double conversion dc/ac/dc. If the pulsed load is purely inductive, L_1, as fusion reactor requirements might dictate, then power reversibility would be desirable so that energy stored in L_1 could subsequently be returned to the main energy storage circuit which includes L_o. Such a double conversion circuit permits reversibility under proper control. These circuits have some unusual properties which will be brought out in the next section.

DC/AC/DC CONVERTERS

The basic double conversion unit is shown in Fig. 7. It converts dc to ac to dc with reversibility under smooth continuous control. It may be contrasted with the now conventional ac to dc to ac conversion required for dc power transmission systems. This scheme, too, is reversible under smooth continuous control. The dc transmission scheme requires an ac system at each end to provide the means for commutation and the corresponding reactive volt-amperes or Q, as well as power P, in or out as the case may be. The dc/ac/dc converters require no Q at either end. The common ac terminals are connected to a capacitor bank, shown with three like elements each labelled C. This capacitor bank provides commutation and the corresponding Q for both converters. The frequency of this commutating bank is not narrowly restricted. It might well be in the range of 400 to 1200 Hz through proper design. It might well be variable over the operating load range. One big advantage of the higher frequency is the reduced physical size of the capacitor bank. A disadvantage is the increased number of commu-

tations per unit time which might impose more severe duty on thyristors. Increased capacitor losses must be carefully evaluated as well. Because energy stored in the capacitor bank is always small, power $P_{d1} = E_{d1}I_{d1}$ in at one set of terminals is nearly equal to power $P'_{d1} = E'_{d1}I_{d1}$ out at the other set, the small difference being due to losses in the capacitor bank and in the thyristor valves.

While the three phase ac terminals common to both converters provide a familiar Graetz bridge reference point, such dc/ac/dc units are not limited to three phases or six pulse designs. Multiphase operation (for smoother dc terminal voltage) can be obtained without any transformers needed to achieve phase shifting for more commutations per cycle of reference frequency. Consider a five phase system as in Fig. 8. This is a 10-pulse system, not easily achievable in the usual sense with the aid of transformers. Similarly a 7-phase 14-pulse system can be envisioned, or any number of phases provided that number is odd. As the number of phases n is increased, the number of capacitors C increases (since there is one capacitor in each phase) as does the number of thyristors. However, the product nC tends to remain constant for a fixed frequency and the average thyristor current is inversely proportional to n. The circuit arrangement for two dc/ac/dc converters is shown in Fig. 9 for a three phase intermediate conversion in each. It should be pointed out that the intermediate conversion frequencies need not be alike in the two; indeed the number of pulses need not be alike either.

COMPUTER STUDY RESULTS

Two such systems with three phase and five phase intermediate conversion (as shown in Fig. 7 and Fig. 8 respectively) were simulated on a digital computer and some representative results are presented.

SYSTEM CONSTANTS AND BASE QUANTITIES

In order to be able to generalize the conclusions to a wide range of voltage, current, and frequency values, the results are expressed in per unit (p.u.) of a set of arbitrary base system quantities.

      Base power   = 5 MW
      Base voltage   = 5 KV
      Base frequency = 60 Hz = 377 rad/sec

Therefore,

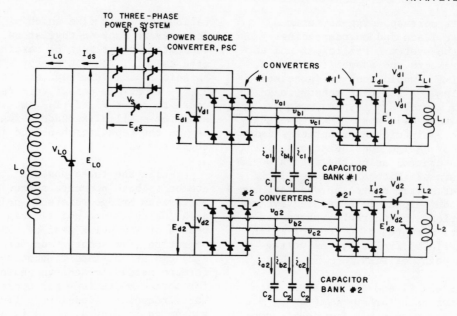

Fig. 9. An arrangement to satisfy more than one pulsed power load
using three phase, six pulse intermediate conversion.

Base current   = 1 KA
Base ohms   = 5 Ω
Base inductance   = Base ohms/
                    Base frequency
                    (in rad/sec)
                  $\cong$ .0133 H
Base capacitance = 1/(Base ohms
                    × Base frequency)
                  $\cong$ 530.5 µF

Now, for example, E_{d1} in p.u. = E_{d1} (in
volts)/Base voltage and L_1 in p.u. = L_1 (in
H)/Base inductance. The system constants
in p.u. of these arbitrarily selected base
quantities are given as follows (for three
phase as well as for five phase intermediate conversion):

L_0 = 750 p.u.
L_1 = 150 p.u.
 C = 0.518/n p.u.

(where n = 3, 5 for three and five phase
intermediate conversion, respectively).

I_{d1} = 1.0 p.u. and intermediate conversion frequency is 10.0 p.u. unless
otherwise specified.

THREE PHASE INTERMEDIATE CONVERSION

Figures 10 and 11 show the steady
state operating terminal quantities for a
three phase intermediate conversion scheme
as shown in Fig. 7. The converter 1'
operates with delay angle $\alpha_1^l = 0$ and may
consist of diodes instead of SCRs.

Figure 10 shows that the average
value of E_{d1}^l decreases as a function of
increasing frequency for different values
of I_{d1} from 1.0 p.u. down to 0.7 p.u. It
can be seen that a wide range of values
for E_{d1}^l are achievable by proper frequency
control.

Figure 11 gives the frequency at
which the system will have to be operated
if (E_{d1}^l) is to be 1.0 p.u. It shows that
the frequency of operation decreases with
the increasing value of I_{d1}^l.

Figure 12 shows the computer results
for the purely inductive load L_1.
Initially the current I_{d1} through inductor
L_0 is 1.0 p.u. and the converter 1 is in
the bypass mode (the bypass valves across
converter 1 and 1' are shown in Fig. 6).
The operation is resumed by initially
charging two of the three capacitors to
1.0 and -1.0 p.u. voltages (the third
capacitor is left uncharged) and then

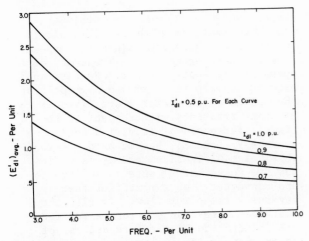

Fig. 10. Load voltage as a function of operating frequency with I_{d1} as a parameter and $I_{d1}' = 0.5$ p.u.

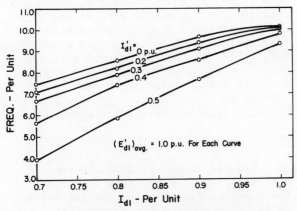

Fig. 11. Required operating frequency as a function of I_{d1} with I_{d1}' as a parameter for $(E_{d1}') = 1.0$ p.u.

firing the appropriate valves, at the same time blocking the grid pulse of the bypass valve. This transfers the current I_{d1} from the bypass valve to the convertor bridge and allows sufficient recovery time for the bypass vlave to regain control before it is subjected to forward voltage. The current I_{d1}' builds up slowly and the ac voltage v_a across the commutating capacitors exhibits a flat top (and bottom) and changes linearly with time from one extreme to the other. This is due to the very small value of I_{d1}' in relation to I_{d1}.

In Fig. 13, the system conditions are the same as in Fig. 12 except for the fact that initially I_{d1}' is equal to 0.5 p.u. and converter 1 and 1' both are initially in the bypass mode. The operation is resumed in the same manner as before (except for ±0.8 p.u. initial capacitor voltages instead of ±1.0 p.u.). It can be seen that E_{d1}' has much less ripple as com-

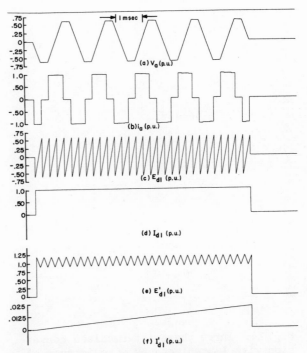

Fig. 12. Three phase intermediate conversion with initial I_{d1}' through L_1 equal to 0.

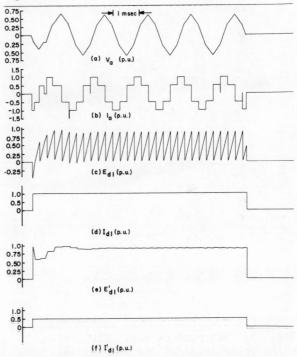

Fig. 13. Three phase intermediate conversion with initial I_{d1}' through L_1 equal to 0.5 p.u.

pared to the previous case. The harmonic content of E_{d1}' varies with the values of I_{d1} and I_{d1}'.

In Fig. 14, L_1 is replaced by a purely resistive load. I_{d1}' in steady

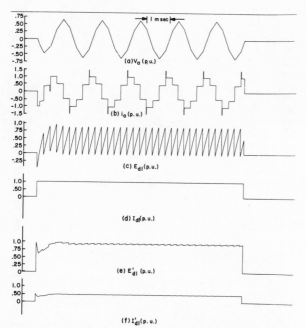

Fig. 14. Three phase intermediate conversion with L_1 replaced by a resistance of 2.0 p.u.

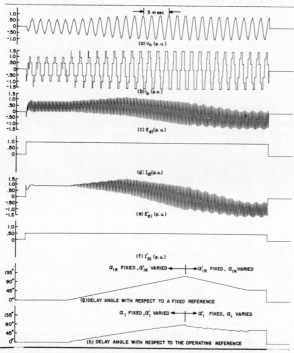

Fig. 15. Reversal of power flow using three phase intermediate conversion.

respectively and $\alpha_1' = 0$. To obtain the power reversal, delay angle α_1' is slowly increased by increasing α_{1R}' with respect to a fixed reference. The frequency of operation is kept constant. So far, no attempt is made to control α_1 except that the valves of converter 1 are fired at constant frequency with respect to the fixed reference. α_1' is varied until the average E_{d1} and E_{d1}' become zero. Beyond this point α_1' is fixed and α_1 is varied. Now the converter 1' acts as an inverter and converter 1 as a rectifier thereby reversing the power flow. It also shows that the energy can be transferred from one inductor to another, regardless of their current levels, by proper control of delay angles in converters 1 and 1'.

It should be pointed out that the peak values of the capacitor voltages and therefore instantaneous peak values of E_{d1} and E_{d1}' increase from a minimum with $\alpha_1'=0°$ to a maximum when α_1' is roughly equal to 90°. This, of course, is true at a constant frequency of operation. A simultaneous increase in frequency along with increasing α_1' could have kept the peak voltages from increasing in value.

FIVE PHASE INTERMEDIATE CONVERSION

Figures 16 and 17 show the results for five phase intermediate conversion (as shown in Fig. 8) with initial values of I_{d1}' through L_1 equal to zero and 0.5

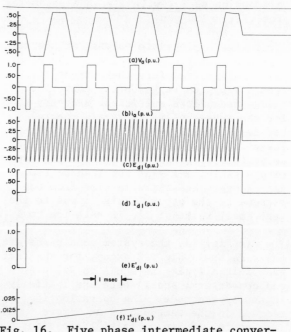

Fig. 16. Five phase intermediate conversion with initial I_{d1}' through L_1 equal to 0.

state is again roughly equal to 0.5 p.u. A comparison with Fig. 13 reveals that for the same value of I_{d1}', E_{d1}' is smoother for the inductive load as compared to the resistive load.

Figure 15 shows how the power transfer can be reversed in direction. Initially I_{d1} and I_{d1}' are 1.0 p.u. and 0.5 p.u.

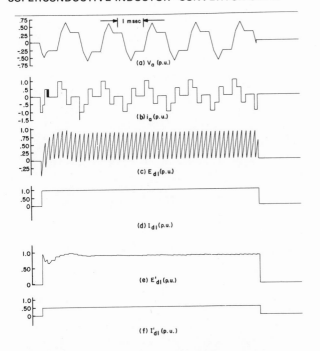

Fig. 17. Five phase intermediate conversion with initial I'_{d1} through L_1 equal to 0.5 p.u.

p.u., respectively. E'_{d1} shows the least harmonic content at no load (i.e., for $I'_{d1} = 0$).

Similarly, the results corresponding to Figs. 14 and 15 could be obtained for the five phase intermediate conversion.

SUMMARY AND CONCLUSIONS

In this paper, circuits are presented which minimize both the pulsed power P and the corresponding reactive volt-ampere Q requirements from the three phase power system when supplying very large pulsed power loads. These circuits permit transfer of energy with minimum loss from one inductor to another at a readily controllable rate, including reversibility of power flow. In the case of a resistive load, precise voltage regulation during pulsing is achievable; indeed, precise voltage regulation may be achieved even if the load is not purely resistive. A completely random or arbitrary load voltage can be generated as a function of time. Where the load voltage required can be prescribed as a function of time, such function can be generated also.

While the possibilities as well as the limitations of these circuits are not yet clearly established, it appears that greatest usefulness might be in the time range from a few milliseconds to many seconds or minutes. Thus, for fusion reactor requirements as foreseen by designers, these circuits may be useful. Other applications in this time frame in high energy physics research and other areas may be foreseen.

REFERENCES

1. R. W. Boom and H. A. Peterson, "Superconductive Energy Storage for Power Systems," IEEE Trans. on Magnetics, MAG-8, September 1973, pp. 701-703.

2. R. W. Boom, et al., "Superconducting Energy Storage", Advances in Cryogenics Engineering 19, 117 (1974).

3. B. Badger, et al., "UW-MAK-I, A Wisconsin Toroidal Fusion Reactor Design", UWFDM-68, Nuclear Engineering Department, University of Wisconsin, November 1973.

4. R. W. Boom, et al., "Pulsed Power Supply for the UW Tokamak Reactor", DCTR Power Supply and Energy Storage Review Meeting, March 5-7, 1974. AEC Hq., Germantown, Maryland. WASH 1310, UC-20.

5. R. W. Conn, et al., "Major Design Features of the Conceptual D-T Tokamak Power Reactor, UW MAK-II", Proc. 5th IAEA Conference on Plasma Physics and Controlled Nuclear Fusion Research, Tokyo, Japan, November 11-15, 1974.

6. "Wisconsin Superconductive Energy Storage Project". Feasibility Study Report, Vol. I, July 1974. Written by interdisciplinary team, College of Engineering (R. W. Boom, H. A. Peterson, W. C. Young, et al.).

7. H. A. Peterson and N. Mohan, "Superconductive Energy Storage Inductor-Converter Units for Power Systems", accepted for publication at IEEE 1975 Winter Power Meeting, New York, January 1975.

LOW INDUCTANCE ENERGY STORAGE AND SWITCHING FOR PLASMA PRODUCTION

Daniel N. Payton III, John G. Clark and William L. Baker
Air Force Weapons Laboratory
Kirtland AFB, New Mexico 87117

ABSTRACT

Most of the recent advances in pulse power have been driven by increased interest in power supplies to drive large pulsed plasma devices. Brief descriptions of such plasma devices as are being used in our laboratory are given. A more detailed description of the power supplies being developed is given with particular emphasis on design and switching.

INTRODUCTION

During the past decade research for producing short pulses of X-rays has been a significant factor in pushing the state-of-the-art in energy storage, compression, and switching. The techniques developed have been used in the past few years for the pulsed production of plasmas. The density and temperature of such pulsed plasmas are in general proportional to the power which can be delivered to the plasma. In this case, the plasma can be considered as the load in electrical engineering terms. The most successful of these experiments in producing hot, dense plasmas over the past few years have been those which were successfully engineered to match power delivery systems and the plasma loads.

In the present paper we wish to discuss several techniques being used within our laboratory to deliver power to pulsed plasmas that have been selected for research. These are principally hot, dense, high-Z plasmas produced to study their X-ray emissions. We have also done some power source development for field emission diodes for bremsstrahlung production recently. However, our main effort wih energy sources has been developing large, fast, high current capacitor banks. The switching used in such systems has varied from solid dielectric switches to pressurized rail gaps. The main thrust of this effort was principally directed toward developing high-current, low-inductance, simultaneously-triggered switches. Within the scope of this conference, our point of emphasis is the use of dynamic techniques to increase the power produced by the capacitor bank. Typical power systems capable of delivering power of the order of 10^{11} watts have been designed to deliver nearly 10^{13} watts of power radiated from the plasma.

In Section II we will expand briefly on the type of plasma devices used. In Section III we will discuss the energy sources and the switching techniques used.

PLASMA PHYSICS LOAD EXPERIMENTS

Several experiments are currently being pursued in our laboratory. The principal one of these is the SHIVA program. This device is a variation of a Z-pinch in which a fast current pulse is applied to a thin cylindrical foil stretched between two electrodes.[1] Half micron thick aluminum foils have been used with lengths varying from 1 to 2 centimeters and diameters varying from 10 to 16 centimeters. The various stages of the operation of the SHIVA experiment are shown schematically in Fig. 1. When the switches are closed the foil is heated through the vaporization state and becomes a low temperature plasma. It then expands to a thickness of approximately 0.5 cm.

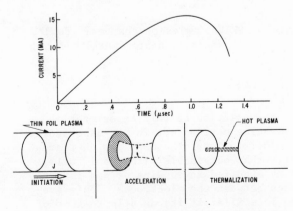

Fig. 1. SHIVA electromagnetic implosion operation.

As the current continues to rise through the load, the $\underline{J} \times \underline{B}$ forces drive the plasma toward the center accelerating it inward as energy is transferred from the energy source to the directed kinetic energy of the foil. Thus, we are delivering energy at state-of-the-art levels into the foil and concentrating it as inward directed kinetic energy. The major experimental problem observed was a restrike problem at the outer insulator. This problem has been reduced by taking care to prevent the UV emission from the plasma from producing photoionization of the insulator surface as shown in Fig. 2. As the plasma approaches the center, the energy source has reached its maximum current and the current starts to decrease. The foil collides with itself on the centerline converting directed kinetic energy into random thermal energy. As the thermalization from the directed kinetic energy takes place, the temperature of the plasma rises rapidly and thermal bremsstrahlung emission produces radiation. Thus, the power delivered to the plasma device has been transformed into an output power of radiation of approximately two orders of magnitude higher than the electrical power output of the capacitor bank.

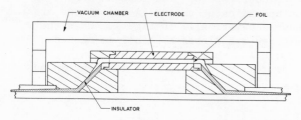

Fig. 2. SHIVA implosion chamber. Foil diameter is approximately 16 cm.

A variation of this experiment is also being carried out in our laboratory. In order to lower the mass per unit area below that obtainable with metal foils, we have puffed an annulus of gas into the space between the electrodes. This system is similar in its details to that of the foil SHIVA except that the implosion velocities and resultant temperatures are considerably higher.

These two imploding plasma experiments constitute the major portion of our present efforts in plasma physics, however, we have continued to pursue several other experimental efforts for development of plasma and X-ray diagnostics and as easily repeatable pulsed plasma sources. During the past few years an effort has been pursued in attempting to understand the radiation output of the dense plasma focus. These studies have included analysis of electron beams created within the focus,[2] neutron production by a D-T plasma,[3] and attempts at heating the dense plasma focus by external sources.[4,5,6,7] In the process of these studies extensive diagnostic measurements have been used to determine as many of the plasma parameters as possible.

Another area of research which has recently been completed in our laboratory was a study of laser-produced plasmas and laser-heated plasmas. A study of laser-driven instabilities in laser produced plasmas was the principal effort. High powered CO_2 and ruby lasers were used for these experiments.

Investigations into the collective acceleration of ions by relativistic electron beams is an area of continuing study. In these studies we are using the FX-100[8] and FX-25[9], Van de Graaff charged coaxial lines, to produce relativistic electron beams. These studies are an attempt to experimentally determine the feasibility of some recent theoretical models predicting an efficient, uniform ion beam acceleration by collective mechanisms.

An offshoot of the switching technology being developed is to use an extremely fast capacitor bank to directly charge diodes to produce bremsstrahlung radiation. These diodes machines will be discussed in detail elsewhere in this conference. The technology being developed in the design of these devices is certainly pushing forward state-of-the-art in power delivery systems. Experimental test modules are presently being developed at both Maxwell Laboratories Incorporated (MLI) and Physics International (PI) and have delivered direct pulses from a capacitor storage system through transmission lines to the diode. Without a pulse peaking capacitor, these devices have created voltage pulses on the diode of less than 30 ns rise time and 100 ns full width half maximum (fwhm) pulse time.

SHIVA POWER SOURCES

In order to supply the fast-rising
high-current power that is needed to
drive the SHIVA experiment, it has been
necessary to design a large, high-voltage
capacitor bank with a very low inductance.
This is necessary in order to obtain a
maximum current in one microsecond with
voltages in the hundreds of kilovolt
region and with energies sufficient to
drive the foil implosion to completion.
Initial experiments were conducted on an
available quarter megajoule capacitor
bank which used solid dielectric switches.
These experiments were carried out at
approximately 40 KV and lacked sufficient
power to produce hot plasmas by the SHIVA
concept. Following these early experiments, a capacitor bank was built which
operated at 45 to 50 KV storing one-
fourth of a megajoule of energy and
having a rise time of approximately 1.5
microseconds. This second bank also used
solid dielectric switches and is shown in
Fig. 3. Initial experiments were conducted approximately two years ago on
this bank and have demonstrated the technical feasibility of the initial stages,
that of initiation and implosion in the
SHIVA concept and have provided an excellent correlation of computational prediction and experimental data. The power
from this bank was not adequate to drive
the implosion hard enough to create a
plasma hot enough to clearly distinguish
the output radiation measurements of
plasma thermalization from the normal I^2R
heating of the imploded plasma column
itself. Thus, the need for a higher
power delivery capacitor bank was evident.

The most practical energy source for
these experiments in the laboratory
appears to be a one megajoule, 100 KV, 2
nanohenry capacitor bank. This capacitor
bank power supply is currently being
fabricated at MLI to the specifications
given by the Air Force Weapons Laboratory.
Extensive calculations of the SHIVA
dynamics have indicated such a power
supply can deliver sufficient energy to
the imploding foil plasma to achieve a
thermalization temperature significantly
different from background temperatures and
high enough to be able to be measured
accurately. This capacitor bank will consist of a parallel plate transmission line
in the shape of a cross as shown in Fig.
4. The top half of this capacitor bank
will be charged to plus 50 KV; the bottom
half will be charged to minus 50 KV. A
single 100 KV switch is used to connect
the upper and lower halves of the module
to the transmission plate. The load (not
shown in Fig. 4) is in the center and
consists of a one meter diameter center
section which is removable so that different experiments can be performed without tearing down and replacing the insulation and transmission plates on the entire
system.

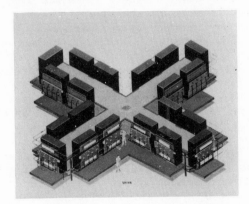

Fig. 4. Artist's concept of the one
         megajoule bank.

Fig. 3. Low inductance capacitor bank of
         Phase II.

Development of this megajoule bank
design has proceeded over the past year
in a modular concept. Initial studies
have indicated the desirability of using
individual modules which can be inter-
changed and removed for servicing. The
first developmental module of this type is
presently in our laboratory being used as
the energy source for the "puff gas"
system mentioned previously (see Fig. 5).
This energy source is a 55 kilojoule
module arranged in a plus-minus 50 kilovolt configuration. However, it differs

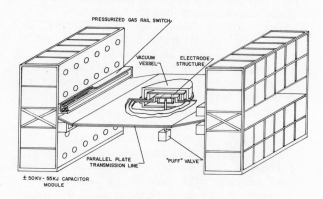

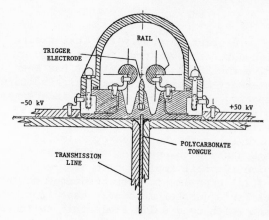

Fig. 5. Schematic of the puffed gas implosion system.

Fig. 6. Cross section of the 100 KV rail gap switch used on the one megajoule capacitor bank showing the polycarbonate tongue insert used to prevent tracking under the switch housing.

from that chosen in the final bank development in that it has two 50 KV switches per module as opposed to a single 100 KV switch. Development of these switches took place at PI, (see Fig. 6); more discussion of the design aspects and operating characteristics of such switches will be given in another paper at this conference.[10] In this prototype module a high voltage gasket was used which completely eliminates an insulator, such as mylar, extending out at the edges between the transmission lines. This was done to avoid tracking at the edge of the transmission lines and to provide a neat compact system. This attempt worked well when an adequate compression of the gasket could be maintained. However, the prototype development was not entirely successful because of difficulties in maintaining a uniform compression of the gasket under the 6-foot-long switch housings. Attempts at maintaining a uniform and tight insulation seal under these switches met with repeated failure and consequently, a different design was decided upon for the final bank development. The final design is shown schematically in Fig. 6. The capacitors have been inserted into the header plate above and below the transmission line with the 100 kilovolt switch between the headers. A polycarbonate tongue insert attached to the switch body extending between the transmission plate is used to avoid flash over under the switch. Anticipated final assembly and testing of the one megajoule capacitor bank should occur within the next few months.

The fact that the SHIVA concept becomes more efficient and easier to diagnose as the power is increased indicates that devices to supply more energy

at higher power than the present state-of-the-art for capacitor banks should be sought. Since the foil is being driven by magnetic forces, it is apparent that the increased power is needed to be obtained primarily through higher current rather than higher voltage. Consequently, one seeks a low voltage, but higher current, fast power supply as discussed earlier. Present pulse power technology suggests the use of explosive driven magnetic flux compression generators to drive the SHIVA load. In such a generator, flux compression is accomplished by use of a high explosive (HE) detonation to compress an ambient magnetic field inducing a higher current in the circuit. If one has an optimal matching, one can obtain efficient transfer of the energy from the explosive generator to the foil SHIVA. A schematic of such a generator is shown in Fig. 7 along with the electrical circuit at the bottom of the figure. In order to steepen the rise time of the explosive generator, a parallel inductance and a closing switch have been included. Initial research performed on the explosive generators utilized flat plate generator technology and has studied the coupling to the load. First experiments with a coupled load were performed recently but because of loss of instrumentation data no clear picture of how the SHIVA load performed is available. Future experiments are planned to evaluate the effectiveness of the transfer of energy to the foil load in comparison with those of the 1 megajoule bank in the laboratory. Beyond this level it is believed that one can reasonably extrapolate designs

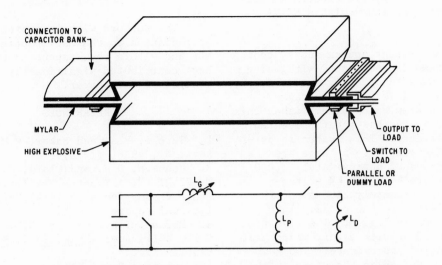

Fig. 7.   Schematic of the explosive driven magnetic flux
         compression current generator and the equivalent
         circuit.

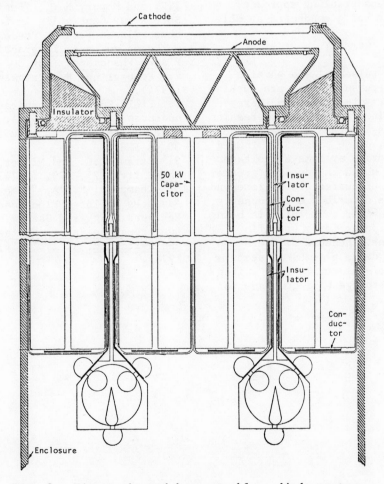

Fig. 8.   Direct charged bremsstrahlung diode system.

utilizing multi-megajoule electrical output explosive generators with a SHIVA device.

The final energy source and switching to be discussed is associated with using a capacitor bank to directly charge a bremsstrahlung diode at low voltages. One example system which accomplishes this is discussed in another paper presented at this conference. This system is performing as designed with a large area reflection diode of approximately 2,000 cm^2 being driven by a 120 KV voltage pulse with approximately 240 kamps of current delivered to the diode. This system is shown in Fig. 8. The rise time is less than 30 ns with a full width half maximum pulse of approximately 100 ns. Pinching is observed in the diode in the later phases of the pulse so that one appears to get blow off of the anode plasma sufficient to give slightly higher conversion efficiencies for the lower energy X-rays than are observed in a thick-target bremsstrahlung approach. A similar study is being conducted at MLI.

SUMMARY

Studies in pulsed plasma physics which are being pursued today are pushing forward the state-of-the-art in energy storage, switching, and power compression devices. The basic trend driven by the interest in hot, dense plasmas is toward large energy storage systems at voltages of about 100 KV which can produce greater than 20 megamps of current in microsecond pulses. We have described a 2 nanohenry, 1 megajoule capacitor bank which is being built to supply power for a specific plasma experiment. Because of increased interest in hot dense plasmas it seems that the state-of-the-art in fast switching and energy storage will continue to advance rapidly.

ACKNOWLEDGEMENT

The authors wish to express their appreciation to the many people at PI, MLI and within the Weapons Laboratory who have contributed significantly toward the content of this paper.

REFERENCES

1. Turchi, P.J. and Baker, W.M., J. Appl. Phys. 44, 4936, (1973).
2. Johnson, D.J. and Clark, J.G., AFWL-TR-72-31, "(U) Spectral Tailoring of a Dense Plasma Focus", (1972). classified internal report.
3. McCann, T.E., Payton, D.N. and Rogers, C.W., Bull. Amer. Phys. Soc., Ser. II, 17, 690 (1972).
4. Johnson, D.J., Ibid.
5. Mather, J.W., "Electron Beam and Dense Plasma Focus Interaction Heating Experiments", J. Appl. Phys. Vol. 44, No. 11, (1973).
6. Shattas, R., Stettler, J., Roberts, T.E. and Meyer, H.C., "Fusion Neutron and Soft X-ray Generation in Laser Assisted Dense Plasma Focus", Presented at the 2nd workshop on Laser Interaction and Related Plasma Phenomena, (Aug 30 - Sept 3, 1971).
7. Hoeberling, R., Private Communication, (1974).
8. Field Emission Corporation, model FX-100 operates at 6 MeV, 100 kA, 100 ns FWHM. Model FX-25 operates at 2.5 MeV, 25 kA, 30 ns FWHM.
9. Straw, D.C. and Miller, R.B., Appl. Phys. Ltrs. 25, 379, (1974).
10. Champney, P.D'A, "Some Recent Advances in Three Electrode, Field Enhanced Triggered Gas Switches", Inter-Conference on Energy Storage Compression, and Switching, Asti, Italy, (Nov 5 - 7, 1974).

THE DESIGN, FABRICATION, AND TESTING OF A
FIVE MEGAJOULE HOMOPOLAR MOTOR-GENERATOR

W.F. Weldon, M.D. Driga
H.H. Woodson, H.G. Rylander
University of Texas
Energy Storage Group
ENL 219
Depts. of Mech. and Electrical Engineering
Austin, Texas 78712, USA

ABSTRACT

The current and future generations of controlled thermonuclear fusion experiments require large amounts of pulsed energy for heating and confinement of plasma. Kinetic energy storage with direct conversion to electrical power (i.e. homopolar machines) seems to be the most economically attractive solution for meeting these requirements.

The University of Texas at Austin has a program intended to develop a design technology for homopolar machines to meet a broad spectrum of performance requirements in terms of stored energy and discharge times. The Energy Storage Group at the University of Texas at Austin has in the past ten months designed, fabricated, assembled and begun a thorough testing program on a second generation homopolar machine with a storage capacity of five megajoules. This machine, using room temperature field coils, solid electrical brushes, and hydrostatic bearings has been designed to deliver 42 volt pulses at current levels in excess of 150,000 amperes. The machine has been designed as a laboratory device with extremely stiff bearings, variable brush area as well as variable brush contact force, variable field strength for pulse shaping, and minicomputer controlled data acquisition, real time signature analysis and on line experiment control. A continuing program studying discharge characteristics, brush and rotor dynamics, machine losses, and system efficiencies is already underway and is currently funded through June, 1975. Funding for the project has been provided by the Atomic Energy Commission, the Texas Atomic Energy Research Foundation, and the Electric Power Research Institute.

INTRODUCTION

The next generation of thermonuclear fusion experiments currently under consideration requires large quantities of electric power for confinement and/or heating of the plasma. It is not feasible to take this energy directly from the electric power lines, and not economically feasible to construct new power generation or transmission facilities since the power is required only for short periods of time.

The last fact, however makes the use of energy storage systems which can slowly store energy from existing electrical distribution facilities and then deliver it to the experiment as rapidly as required, a practical alternative. In the past, both battery and capacitor banks have been used for this purpose, but the cost per joule of stored energy is rather high ($0.20 to $1.00 per joule). Additionally, as the size of either system increases, the complexity of the safety interlocks and switching gear required for safe operation increases substantially.

Several types of inertial energy storage devices have been proposed or constructed, but they usually consist of a separate motor, generator, and flywheel arranged along a shaft. In machines of this con-

figuration minimum discharge time is determined by the maximum torque which the shaft
is capable of transmitting. Additionally,
the individual motor and generator are
fairly expensive, especially since they
are usually conventional units originally
designed for continuous duty. The generator in this arrangement is often an alternator, thus requiring a high current rec-
tification system to deliver direct current to the experiment. The cost of such
units has been in excess of $0.20 per
joule.

The Faraday Disc or homopolar machine in the conventional configuration (axial magnetic field and radial current
flow, see Fig. 1.) can conveniently serve

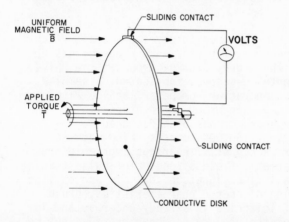

Fig. 1. Faraday Disc

as both motor and direct current generator
while the armature of the machine also
serves as the inertial store. Thus, the
forces which occur during motoring and
discharge are generated within the flywheel itself, essentially at the same
point where the energy is stored, eliminating the shaft torque limitation previously mentioned. The rotor itself is a right
circular cylinder of electrically conductive material, and the magnetic field
coil is simply a solenoid. Such units
have been built at the University of Texas at Austin for less than $0.01 per
joule in half megajoule and five megajoule
sizes and several design studies show that
larger units can be built for somewhat
less than this figure.

Prior to undertaking the five
megajoule homopolar machine project, the

Energy Storage Group at the University of
Texas at Austin had successfully completed
the design, fabrication, and testing of a
half megajoule homopolar motor generator.
This unit exceeded its design goals by
motoring to 6000 RPM in 100 seconds with
a current of 1200 amperes and discharging
in 7 seconds with a peak current of over
14,000 amperes.

The Energy Storage Group has also
completed design studies on homopolar
machines ranging in size from 50 to
2000 megajoules.

DESIGN PHILOSOPHY

Although the homopolar machine is simple in concept, many past examples have
employed extremely complicated solutions
(such as liquid metal brushes and superconducting field coils) in attempting to
overcome basic limitations. The design
philosophy of the Energy Storage Group
has been to strive for simplicity and
reliability; using conventional technology
in the development of the homopolar machine as a practical, low cost pulsed
power supply.

GENERAL LAYOUT

The general layout of the five megajoule homopolar motor-generator is shown
in Figs. 2 and 3. The 0.610 meter diameter,

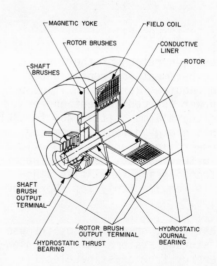

Fig. 2. 5 mj Homopolar Motor Generator

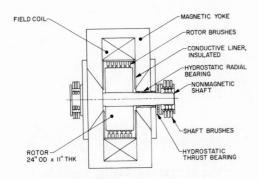

Fig. 3. 5 mj Homopolar Machine
Cross Section

0.279 meter thick steel rotor is shrunk
onto a 0.127 meter diameter, non-magnetic
shaft. This shaft is supported both axially and radially by hydrostatic bearings
which combine high stiffness and relatively low frictional losses. The bearing
housings are attached to the magnetic
yoke which also serves as a return path
for the axial magnetic field generated
by the field coil. The yoke is designed
to uniformly distribute the magnetic
field across the faces of the rotor.

ROTOR AND BEARINGS

The moment of inertia of the 0.610
meter diameter, 0.279 meter thick steel
rotor is given by:

$$I = \frac{\pi R_o^4 m a}{2} = 29.707 \text{ kg-m}^2 \qquad (1)$$

At a design speed of 584 radians per second the stored energy is:

$$KE = \tfrac{1}{2} I \omega^2 = 5.07 \times 10^6 \text{ joules} \qquad (2)$$

The stress at the periphery of the hole
in the center of the rotor is the critical factor which limits rotor speed. The
tangential and radial stresses due to
spin and shrink fit of 0.002 m/m are as
follows:

$$\sigma_\theta = \frac{(3 + \nu)\, m\omega^2}{8} \left[R_o^2 - \left(\frac{1+3\nu}{3+\nu}\right) R_i^2 \right] + \qquad (3)$$

$$\frac{E\delta}{2} \left(\frac{R_o^2 + R_i^2}{R_o^2} \right)$$

$$\sigma_r^2 = \frac{(3 + \nu) m \omega^2}{8} \left(R_o^2 - R_i^2 \right) - \qquad (4)$$

$$\frac{E\delta}{2} \left(\frac{R_o^2 - R_i^2}{R_o^2} \right)$$

Combining these equations gives the effective stress at any speed.

$$\sigma_{eff} = \sigma_\theta = \sigma_r = \frac{m\omega^2}{4} (1 - \nu)R_i^2 + E\delta$$

$$(\sigma_{eff})_{\omega=0} = 4.14 \times 10^8 \text{ nt/m}^2$$

$$(\sigma_{eff})_{\omega=584} = 4.16 \times 10^8 \text{ nt/m}^2$$

Thus the operating stress can be seen
to be nearly independent of rotational
speed. The limiting speed is reached when
$\sigma_r = 0$, i.e., when the slip fit is overcome by rotational strain. Setting Eq. 4
equal to zero and solving for gives:

$$\omega_{max} = \pm 2 \left[\frac{E\delta}{(3+\nu)R_o^2 m} \right]^{1/2} = 1174 \text{ rad/sec} \qquad (6)$$

The shrink fit of the disc on the rotor can
be seen to have a factor of safety of two
in terms of limiting speed vs. maximum
operating speed and it should be noted that
this represents a factor of safety of four
in terms of energy stored in the rotor assembly.

The radial bearings are four-pocket
orifice compensated hydrostatic bearings
supplied with oil at 4.14×10^6 nt/m².
The stiffness of each bearing is in excess
of 7.88×10^8 nt/m. Bearing stiffness is
kept high in order to maintain all natural
frequencies of the rotor-shaft-bearing
system above the maximum operating speed of
584 radians/second. Analysis of the shaft
indicates that it may be considered to be
rigid in comparison to the bearings.
For a worst case assumption, the upper and
lower bearing housings may be assumed to be
held together by two 19 mm diameter by
7.62 cm long bolts and two 19 mm diameter
by 15.24 cm long bolts. This assumption
gives an effective housing spring constant
of 2.32×10^9 nt/m. The effective spring
constant is:

$$K_{eff} = \frac{k_B k_h}{k_B + k_h} = 5.88 \times 10^8 \text{ nt/m} \qquad (7)$$

The first critical frequency can be
calculated by

$$\omega_{c1} = \left[\frac{2k_{eff}}{M_T} \right]^{1/2} = 1273 \text{ rad/sec} \qquad (8)$$

Gyroscopic effects cause a conical whirl whose frequency is given by

$$\omega_{cw} = \left[\frac{k_{eff}L^2}{2(I+I')}\right]^{1/2} = 1113 \text{ rad/sec} \qquad (9)$$

The thrust bearing is also a hydrostatic unit but is compensated by the use of flow control valves. It is designed to have a stiffness of 4.38×10^8 nt/m. Any misalignment in the bearing surfaces will produce a one cycle per revolution forcing function. The natural frequency of the rotor-shaft-bearing system in this mode is given by:

$$\omega_{ca} = \left[\frac{K_t}{M_t}\right]^{1/2} = 777 \text{ rad/sec} \qquad (10)$$

This critical frequency is only 33% above the maximum operating speed of the machine but detailed analysis as well as past experience with high flow rate hydrostatic bearings indicates that the internal damping is such that this will not be a severe vibration mode, even in the event that it is reached.

MAGNETIC CIRCUIT

The magnetic circuit of the five megajoule homopolar machine is designed to operate at a nominal flux density of 1.61 wb/m^2. This is the highest field attainable with moderate power losses in the field coil. The air gap at each end of the rotor is 0.0245 m. The number of ampere turns required to give the required flux density in the air gap is given by:

$$(Ni_{fc})_{air} = \frac{2gB}{\mu_0} = 65085 \text{ ampere turns} \qquad (11)$$

Additionally, the number of ampere turns required to produce the same flux density in the steel portion of the magnetic circuit (nominal length of $L_m = 1.7$ meters) is given by:

$$(Ni_{fc})_{steel} = L_m \cdot \overline{H} = 5950 \text{ ampere turns} \qquad (12)$$

$\overline{H}$ is determined from the magnetization curve for steel shown in Fig. 4. The total ampere turns required in the field

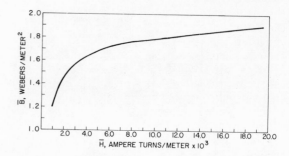

Fig. 4. Magnetization Curve for Steel

coil are:

$$(Ni_{fc})_{tot} = (Ni_{fc})_{air} + (Ni_{fc})_{steel} =$$

$$71035 \text{ ampere turns} \qquad (13)$$

Thus for a field coil of 260 turns, the current required is 273.2 amperes.

The field coil is wound of round copper tubing of 12.7 mm outer diameter and 1.24 mm wall thickness. Although electrolytic tough pitch copper is desirable because of its low electrical resistivity ($\rho = 1.71$ microhm-cm); deoxidized, high residual phosphorus copper ($\rho = 2.03$ microhm-cm) was used because of its ready availability. The nominal length of the coil is 762 meters. The resistance is given by:

$$R_{fc} = \frac{\rho L_{fc}}{A_t} = 0.343 \text{ ohms} \qquad (14)$$

Thus, the coil voltage and power requirements are:

$$P_{fc} = (i_{fc})^2_{tot} R_{fc} = 25.6 \text{ kilowatts} \qquad (15)$$

$$V_{fc} = (i_{fc})_{tot} R_{fc} = 93.7 \text{ volts} \qquad (16)$$

The field coil is divided into ten segments of 26 turns each and cooling water is supplied to each segment at 2.5 liters/min. and 20° C. The average exit temperature of the water is given by:

$$T_{out} = T_{in} + \frac{t P_{fc}}{C_p Q_m} = 34° \text{ C} \qquad (17)$$

This is certainly a conservative coil design in terms of temperature rise.

The 10 coils of the field winding of the machine can be connected in two groups of 5 coils, fed from two separate SCR power supplies and differentially controlled to serve as an electromagnetic thrust bearing. In this mode, the rotor is stabilized in the magnetic center of the machine against the axial electromagnetic forces tending to move the rotor towards the yoke wall. Previous experiments, made with the 0.5 mj. machine, led to a stable feedback system, controlling the two power supplies differentially in such a manner that the total ampere-turns remain constant. The position sensor has low inertia and strong derivative feedback to overcome the delay due to the relatively slow penetration of the magnetic field into the rotor.

This feedback system effectively controls the homopolar machine in the motoring mode. In the generating regime, (in the time of discharge) the feedback system remains connected. However, the rotor tends to be stabilized by the high currents in the compensating plates which, for the short time of discharge, play the role of a levitation system keeping the rotor carrying the high discharge current in the center of the air gap.

The magnetic yoke is designed so that the 1.61 wb/m^2 flux density in the air gap is maintained throughout the magnetic circuit. The manner in which this is accomplished may be seen more clearly by referring to Fig. 5. The area of the rotor penetrated by the uniform magnetic field in the air gap is determined as follows:

$$A_1 = \pi(R_o^2 - R_{im}^2) = 0.259 \text{ m}^2 \qquad (18)$$

The area of the yoke end plate penetrated by the total flux is

$$A_2 = 2\pi R_o t \qquad (19)$$

and in the outer ring the area is

$$A_3 = (d^2 + 2R_{oy}d)\pi. \qquad (20)$$

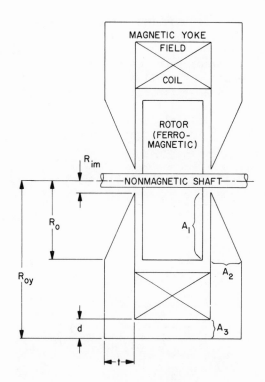

Fig. 5. Magnetic Circuit

But for uniform flux density:

$$A = A_2 = A_3,$$

or

$$t = \frac{A_1}{2\pi R_o} = 0.135 \text{ m} \qquad (21)$$

and

$$d = R_{oy} \pm \left[R_{oy}^2 - \frac{A_1}{\pi} \right]^{1/2} = 0.085 \text{ m} \qquad (22)$$

where R_{oy} = 0.673 meters for the 5 mj machine.

ELECTRICAL CHARACTERISTICS

If the magnetic flux density is assumed to be uniform across the faces of the rotor, the no load machine voltage is:

$$V_{nl} = \left[\omega B \frac{R_o^2 - R_{im}^2}{2} \right] = \frac{\omega \Phi}{2} \qquad (23)$$

Hall probes placed radially along the rotor

face indicate that the magnetic field in the air gap is uniform, even in the transient regime. At the maximum design speed of 584 rad/sec. and the rated magnetic flux density of 1.61 wb/m^2, the no-load voltage is approximately 42 volts.

The actual voltage across the machine terminals differs slightly for the two modes of operation.

(a) In the motoring mode:

$$V_m = \omega B \left(\frac{R_o^2 - R_{im}^2}{2} \right) + r_{tm} i_m =$$

$$\frac{\omega \Phi}{2\pi} + r_{tm} i_m \qquad (24)$$

(b) In the generate mode:

$$V_g = \omega B \left(\frac{R_o^2 - R_{im}^2}{2} \right) - r_{tg} i_g - L_r \frac{di_g}{dt} =$$

$$\frac{\omega \Phi}{2\pi} - r_{tg} i_g - L_r \frac{di_g}{dt} \qquad (25)$$

The motoring current is almost constant. The discharge current is varying in time and the role of the internal inductance of the rotor circuit, L_r (corresponding to the magnetic energy stored in the air-gap, compensating plates, and in the ferromagnetic body of the rotor) may be important. The values of r_{tm} and r_{tg} are different because the motoring regime occurs with only a small portion of the brushes on the rotor while the discharge occurs with all the brushes in contact - the friction and the wear of the brushes for the short period of time involved being negligible in comparison with the improvement in the electrical efficiency of the discharge. The influence of the resistance of the brushes can be seen if we compare it with the resistance of the rotor and compensating plates.

The electrical resistance of the rotor:

$$R_{rotor} = \frac{\ln(R_o/R_i)}{2\pi \sigma_{st} a} = 0.173 \times 10^{-6} \text{ ohms} \qquad (26)$$

The electrical resistance of the compensating plates

$$R_{c.p.} = \frac{\ln \frac{R_o}{R_i}}{4\pi \sigma_a d_{cp}} = 0.378 \times 10^{-6} \text{ ohms} \qquad (27)$$

The total resistance of the machine (measured) is less than 100×10^{-6} ohms including the brush system, rotor, compensating plates and leads. It is obvious that, as expected, the resistance of the brush contacts accounts for almost the entire resistance of the machine. In the near future, the external load will be replaced with a switch having a resistance of less than 5×10^{-5} ohms. In this way, the total resistance of the path for the discharge current will be less than 1.5×10^{-4} ohms. Then, the maximum value of the discharge current will be approximately

$$I_{max} = \frac{42 \text{ volts}}{1.5 \times 10^{-4} \text{ ohms}} = 280,000 \text{ amperes} \qquad (28)$$

Usually, the homopolar machine is considered (and analytically approached) as the condenser in an R,L,C circuit. Since a large amount of kinetic energy is converted into electrical energy at a relatively low voltage, the capacitance is very high. For the 5 mj. machine:

$$1/2 \ CV^2 = 5 \times 10^6 \text{ joules or}$$

$$C = 5750 \text{ farads} \qquad (29)$$

The parameter which describes the behavior of R,L,C circuits:

$$\lambda = \frac{1}{R} \left[\frac{L}{C} \right]^{\frac{1}{2}} = 6 \times 10^{-3} \qquad (30)$$

for the present circuit which indicates a highly damped circuit, while, for the improved discharge circuit λ will be only 0.02 when 0.5 is the value for the critically damped circuit.

The RC time constant, for the improved discharge circuit is:

$$\tau = 0.8625 \text{ seconds.} \qquad (31)$$

The expected variation in time of the discharge current is given in Fig. 6.

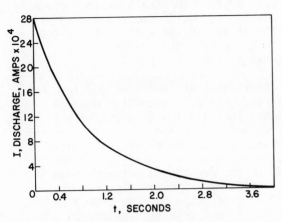

Fig. 6. Discharge Current As A Function of Time

The electromagnetic torque is given by the formula

$$T = \frac{\Phi i}{2\pi} \qquad (32)$$

assuming again that the magnetic flux density is uniform across the rotor. Then, for instance, the maximum braking torque occuring in the discharge time is

$$T_{max} = \frac{\Phi i_{max}}{2\pi} = 20{,}050 \text{ nt-m} \qquad (33)$$

BRUSHES

This machine uses solid brushes made of copper-graphite, chosen after intensive testing; because of their low voltage drop. Also friction, heating and wear of the brushes and the rotor were important considerations in our tests and our choice.

Cyclic D.C. machines have energy stored in the commutating coils which must be dissipated in the commutation period and leads to severe arcing. Homopolar machines do not experience this difficulty so that, for this reason, the current density at the brush contact can be increased to relatively high values while the operating conditions remain satisfactory. In the previous machine (0.5 mj.) discharge current densities exceeding 750 amps/cm^2 were

regularly reached without difficulties. In the 5 mj. machine, the contact area of the brushes is such that at the highest expected discharge current, the current density will be approximately 450 amps/cm^2 on the rotor and 1200 amps/cm^2 for the shaft brushes.

The arrangement of the rotor brushes comprises 6 separate tracks, each of them having 18 brushes of 6.45 cm^2 contact area. The shaft brush arrangement consists of 6 tracks of 6 brushes each. The contact force and number of brushes which make contact during either mode of operation can be pre-selected. The brushes are air retracted/spring extended such that in the event of a loss of air pressure all brushes will be lowered onto the rotor and shaft with full spring load.

CONTROL AND DATA ACQUISITION

The rapid response and large amount of energy involved in experiments with the five megajoule homopolar machine dictate the use of an automatic control system rather than the manual controls used on the half megajoule system. A small digital computer has been purchased to handle data acquisition and reduction as well as control and safety functions. The control duties assigned to the computer will be; (1) switching from motor to generate mode, (2) terminating an experiment in the event of a malfunction, (3) controlling the electromagnetic thrust bearing, and (4) shaping the discharge pulse by field and brush control. The computer is equipped with eight channels of analog to digital conversion and eight channels of digital to analog conversion to facilitate interfacing it with the machine.

Additionally, the computer will monitor and record the required data during the course of an experiment, perform the required data reduction, and plot the results in whatever form may be desired. It will also generate a permanent record of machine performance parameters after each run.

FABRICATION AND ASSEMBLY

The five megajoule homopolar machine was fabricated and assembled in a period of less than eight months. In order to do this, every effort was made to utilize readily available parts and materials wherever overall machine quality and performance would not be sacrificed by doing so.

The rotor was machined in the Mechanical Engineering Shop at the University of Texas at Austin from a billet of A.I.S.I. 4340 vacuum remelt steel.* Before final machining, the rotor was shrunk onto a shaft of type 304 stainless steel. The radial bearings were rough machined from stainless steel tubing and then babbitt liners were cast in place before the final machining.

Due to the unavailability of large stainless steel billets, the bearing housings were built up from three 7.6 cm thick stainless steel plates. These also were fabricated in the University's Mechanical Engineering Shop.

The magnetic yoke assembly, due to its large size, was the only part of the machine not fabricated at the University of Texas. The assembly was welded up from two 15.24 cm thick hot rolled steel end plates and a central ring rolled from 7.62 cm thick steel plate.** Next, bolted to the yoke and align bored as an assembly to assure proper bearing alignment.

The field coils were wound from 12.7 mm outside diameter copper tubing covered with a braided fiberglass sleeve. The ten segments of the final coil were then individually potted in glass reinforced epoxy. The winding and potting operations were carried out by undergraduate engineering students under the supervision of project personnel.

Electrical brushes were manufactured from a sintered copper-graphite material.*** The brushes are 2.54 cm square and 1.0 cm thick and are contoured to fit the rotor or shaft on which they run. They are silver soldered to 2.54 cm by 1.6 mm thick copper straps which act as electrical conductors and brush locating members. The brushes are forced down onto the mov-

* provided by Cameron Iron Works, Inc., Houston, Texas
** Fabrication by Astec Industries, Inc. Chattanooga, Tennessee
***Morganite Corp. CM=1S Homopolar Brush Material

ing surface by stainless steel coil springs. The brushes are lifted and the contact interface pressure is modulated by pneumatic cylinders attached to the back of each brush. The shaft brushes are able to utilize standard commercial air cylinders, but due to space limitations the rotor brush cylinders had to be fabricated at the University.

Additional preparations required for the five megajoule homopolar machine program included a high pressure oil supply and scavenging system for the bearings and a pressurized emergency oil reservoir to provide bearing lubrication in the event of a main oil pump or power failure. A pneumatic control system, unaffected by high magnetic fields, allows selection of any desired combination of brushes for motoring and/or generating as well as selection of the desired brush interface pressure for either mode. Also because of the high magnetic fields present in the machine, a fluidic tachometer has been designed and fabricated, as well as fluidic axial and radial rotor position sensors. Due to the unavailability of a suitable switch, an armature shorting switch capable of sustaining a current of 320,000 amperes for one second has been designed and is being fabricated for testing the generator.

TESTING PROGRAM

An extensive, year long, testing program, intended to thoroughly evaluate the new 5.0 megajoule homopolar motor-generator is currently underway at the University of Texas at Austin. The objectives of the program are to:
1) experimentally verify the basic machine constants,
2) determine machine efficiency in various modes of operation,
3) separate, investigate and attempt to minimize the various sources of mechanical and electrical losses,
4) determine the magnetic field distribution in all parts of the machine, under both steady state and transient conditions,
5) implement a signature analysis program to optimize operational

procedures and prevent failure conditions,

6) explore the operational limits of the various components of the machine and implement programs to extend those limits.

Shortly after completion of assembly of the 5.0 megajoule machine, a series of static tests were performed to verify certain design parameters. Oil flow rates and pressure drops in the radial and thrust bearings were verified. Radial bearing eccentricity was measured as was breakaway torque of the rotor. Breakaway torque was found to be higher than anticipated due to excessive drag from the thrust bearing oil seals. Provisions are now being made to relieve the pressure on those seals.

The magnetic field was measured with a radial array of Hall effect probes and the rise and decay times of the magnetic field were determined under various modes of switching. This information was required for the implementation of the magnetic thrust bearing control system as well as for several other schemes currently under investigation for switching the power output of the generator.

After the static tests, the machine ran satisfactorily the first time power was applied to the armature. A series of tests was then begun to generate data for the separation of losses calculations, classically known as a "retardation test". Essentially the machine is operated in four modes; first motoring up to a certain speed and then:

1) coasting down with all brushes retracted and magnetic field turned off (bearing and windage losses only)
2) coasting down with all brushes retracted and magnetic field on (introduces magnetic field induced losses.)
3) coasting down with motoring brushes down (brush losses.)
4) coasting down with generating brushes down (brush losses.)

The data generated in these tests allow the determination of the absolute as well as relative values of the various losses at all speeds, except those produced by eddy currents and electrical forces on the brush conductors. These

tests have already been run up to 86% of full speed and they verify the aforementioned seal drag problem, but all other losses seem to be in order.

The machine has also generated at various speeds up to 86% of full speed, into a rather high resistance external load since the new shorting circuit is not yet completed. Even with the high resistance loading circuit, current levels exceeding 139,000 amperes have been reached.

CONCLUSIONS

Although the testing and evaluation program on the 5.0 megajoule homopolar machine is just beginning, the preliminary results indicate that this machine will exceed its original performance goals by a clear margin. For example, the measured internal resistance of the machine is well below the value originally anticipated and continues to decline as the brushes wear in.

The Energy Storage Group at the University of Texas at Austin has shown that the disc type homopolar machine is a practical approach to the development of low cost pulsed power supplies where relatively low voltage, high current power is required or can be used. This type of system occupies less space, has a more flexible charge-discharge cycle, is less costly, and represents less of a safety hazard than competitive motor-generator, battery and capacitor systems. Investigations are now underway involving the application of this homopolar machine design technology to providing power for confinement systems of actual fusion experiments.

The Energy Storage Group wishes to express its gratitude to The U.S. Atomic Energy Commission, the Texas Atomic Energy Research Foundation, and the Electric Power Research Institute for this research.

LIST OF SYMBOLS

A_t = cross sectional area of field coil conductor (cm2)

$A_{1,2,3}$ = cross sectional area of magnetic yoke (m2) (see Fig. 5.)

a = rotor thickness (m)

$\overline{B}$ = magnetic flux density (wb/m^2)

C_p = heat capacity of water (watts/kg-$^\circ$C)

d_{cp} = thickness of compensating plate (cm)

E = Young's modulus (nt/m^2) (2.069 x 10^{11} nt/m^2 for steel)

g = air gap between end of rotor and magnetic yoke (m)

$\overline{H}$ = magnetic field strength (amp-meters)

I = moment of inertia of rotor about axis of rotation (kg-m^2)

I' = moment of inertia of rotor perpendicular to axis of rotation (kg-m^2)

i_a = armature current (amps)

i_{fc} = current in field coil (amps)

i_g = armature generating current (amps)

i_m = armature motoring current (amps)

K.E. = kinetic energy stored in rotor (joules)

k_B = radial bearing spring constant (nt/m)

k_{eff} = effective radial spring constant (nt/m)

k_h = radial bearing housing spring constant (nt/m)

k_t = thrust bearing spring constant (nt/m)

L = distance between radial bearings (m)

L_{fc} = length of field coil (m)

L_m = effective length of magnetic circuit in steel (m)

L_r = rotor inductance (henrys)

M_t = mass of rotor & shaft (kg)

m = mass density of rotor (kg/m^3) (7833 kg/m^3 for steel)

N = number of turns in field coil

P_{fc} = electrical power consumed in field coil (kw)

Q_m = total mass flow rate of field coil cooling water (kg/sec)

R_{fc} = resistance of field coil (ohms)

R_i = shaft radius (m)

R_{im} = radius of central hole in magnetic yoke (m)

R_o = outer radius of rotor (m)

r_{tg} = internal resistance of machine in generate mode (ohms)

r_{tm} = internal resistance of machine in motor mode (ohms)

T_{in} = inlet temperature of field coil cooling water ($^\circ$C)

T_{out} = outlet temperature of field coil cooling water ($^\circ$C)

t = time (sec)

V_{fc} = field coil voltage drop (volts)

v_{nl} = no load armature voltage (volts)

δ = shrink fit of rotor on shaft

μ_o = magnetic permeability of air (wb/amp-meter)

ν = Poisson's ratio (0.3 for steel)

ρ = electrical resistivity (ohm-cm)

σ_a = electrical conductivity of aluminum compensating plate (3.3 x 10^7 mhos/m)

σ_{eff} = effective combined stress at inner periphery of rotor (nt/m^2)

σ_r = radial stress at rotor inner diameter (nt/m^2)

σ_{st} = electrical conductivity of steel rotor (5 x 10^6 mhos/m)

σ_θ = tangential stress at rotor i.d. (nt/m^2)

Φ = magnetic flux (webers)

ω = rotor angular velocity (rad/sec)

ω_{ca} = axial critical frequency of rotor-shaft-bearing assembly (rad/sec)

ω_{cw} = critical whirl frequency of rotor-
      shaft-bearing assembly (rad/sec)

ω_{c1} = first critical frequency of rotor-
      shaft-bearing assembly (rad/sec)

A ONE MEGAJOULE FAST CONDENSER BANK FOR THE PLASMA FOCUS EXPERIMENT AT FRASCATI

C. Gourlan, H. Kroegler, Ch. Maisonnier
Laboratori Gas Ionizzati (Associazione EURATOM-CNEN), C.P. 65 - 00044
Frascati, Rome, Italy
and G. Villa
Ditta Passoni & Villa, Milano, Italy.

ABSTRACT

A 1 Megajoule fast condenser bank is being built at Frascati to energize a Plasma Focus experiment of large dimensions.

The condenser bank has the following characteristics: energy 1 MJ, voltage 40 kV, capacity 1250 microF, internal inductance 8 nH, number of condensers 96, repetition rate 1 shot every three minutes.

The construction of this modular condenser bank has been simplified to a large extent by using new techniques; spark-gaps built in the condensers, connection by cables with plug-in connectors, precollectors with built-in in R.C. matching elements, triggering by triax cables.

A rather unusual circuity (the spark-gaps are grounding the high voltage terminal of the condensers) eliminates the ringing of the charging and triggering elements at the frequency of the main discharge reducing the electric stress on these elements and the emission of electromagnetic noise.

The condenser bank is now completely assembled. Full energy tests are under way.

INTRODUCTION

Within the framework of the Joint European Programme on Plasma Focus, involving the Culham, Frascati, Juelich and Limeil laboratories, a 1 MJ Plasma Focus is being built, aiming in particular at assessing, the validity of the scaling laws. The bank is now completed and is being tested (Fig. 1). It will be connected to a plasma focus device of large dimensions in order to assess the validity of the scaling laws of neutron emission up to one Megajoule and to investigate better the mechanism of neutron production.

1. GENERAL DESCRIPTION OF THE BANK

1.1. Characteristics of the Bank

- Energy : 1 MJ
- Charging voltage: 40 kV
- External inductance of the bank: 8 nH
- Inductance of the load: 3 nH at beginning of the discharge
  50 nH at maximum of the current.
- Maximum current: 6/7 MA
- Voltage reverse: 40%
- Repetition rate: one shot

Fig. 1: 1 Megajoule Bank Plasma Focus
E, Experiment; R, Liquid Resistor; S, Sub-Master-Gap

every three minutes
- Requested life: 50,000 shots
  at full voltage.

1.2. General Lay-out (Fig. 2)

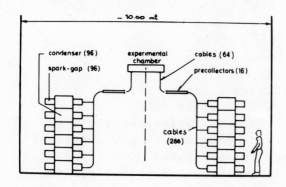

Fig. 2: General Lay-out

The condenser bank has the shape of a cylindrical shell with an opening of one meter width to give access to the inside.

The dimensions are:
External diameter: 8 m
Internal diameter: 5.5 m
Height   : 3 m

This compact structure provides:
- a short length of connections between the bank and the discharge chamber;
- enough room inside for auxiliary instrumentation for the experiment,
- an economic transversal dimension for the building which has heavy walls and ceiling 1 meter of concrete because of radiation hazard (the device is expected to yield more than 10^{13} neutrons of 2.45 MeV per discharge). Strong metallic holders bolted on the condensers make them able to be assembled in column without the need of a metallic structure.

1.3. The Electric Circuit (Fig. 3)

The bank is made of 16 columns of 6 condensers each with built-in spark-gaps. The condensers are connected by coaxial cables to a precollector made of 16 radial sectors. Each sector of the precollector has R.C. matching elements and is con-

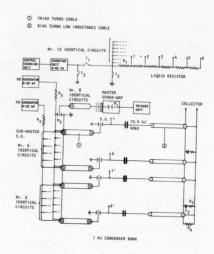

Fig. 3: Electrical Diagramme

nected to the experiment by four coaxial cables.

The trigger circuit includes one master trigger unit and 8 submaster trigger units. Each submaster triggers two columns of condensers through triaxial cables. The main charging generator is connected to the condenser through 16 electrolytic resistors. A return charging current resistor is connected in parallel to the experiment in order to provide:
- a return circuit for the current during charging and in case of emergency discharge of the bank
- a damping parallel circuit for the discharge in case of accidental firing of the spark-gaps when no breakdown is possible in the experiment.

2. DESCRIPTION OF THE ELEMENTS

2.1. Capacitors (Fig. 4)

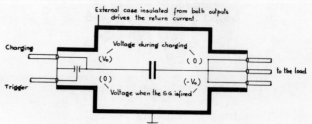

Fig. 4: Connections to the capacitors

Each capacitor includes:
- a capacitive element,
- a spark-gap
- a plug type output for coaxial
  cables.

2.1.1. Capacitive Element.

The capacitive element is made of 160 packages, 40 kV, of the overlapping winding type connected in parallel by welding to the two insulated outputs (Fig. 5). The capacitor has the following characteristics:

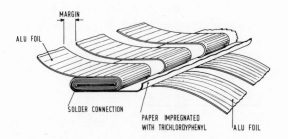

Fig. 5: Sketch of one capacitive element

- charging voltage: 40 kV
- capacity   : 13 μF
- energy   : 10.4 kJ
- internal inductance: 50 nH
- total inductance (capacitor +
  spark-gap + plug): 120 nH
- insulant: paper + trichlordyphe-
  nyl (synthetic oil)
- voltage gradient: 67 kV/mm
- two outputs insulated.

2.1.2. Spark-gap (Fig. 6).

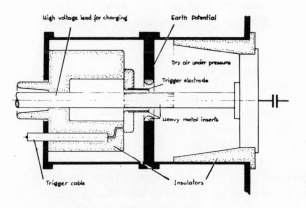

Fig. 6: Spark-gap

One output of the capacitor is connected to a built-in pressurized spark-gap of the field distortion type: the pressure of dry air for a working voltage of 40 kV is about 2 kg/cm^2. The high voltage electrode of the gap is connected directly to an output of the capacitor and goes through the spark-gap providing a charging lead for the capacitor, the opposite output of the capacitor being connected to ground during charging through the circuit including the coaxial cables, the precollector and the return charging resistor. The trigger electrode is a sharp edged ring coaxial with the high voltage electrode. The ground electrode is a plate connected directly to the case of the capacitor. The active parts of the spark-gap have heavy metal inserts.

2.1.3. Plug-type output for coaxial cables.

The opposite output of the capacitor has special sliding elastic contacts of "multicontact" type in which the connectors of three coaxial cables going to the precollector are plugged. The elastic elements of these contacts have a venitian blind configuration and provide a good contact between two concentric metallic cylinders.

2.1.4. Advantages of the two insulated outputs.

The capacitor with two insulated outputs works in the following way: the capacitor is charged at -40 kV; when the gap is fired, the high voltage electrode is grounded and an opposite voltage of +40 kV appears at the output towards the experiment. This type of circuit has the following advantages:
- maintenance: the spark-gap with
  their connections are located at
  the rear of the condenser bank, in
  a very accessible zone free of
  cables;
- safety: the supply and trigger
  circuits are grounded when the

spark-gaps are fired and are not
varying in voltage with the main
discharge: in this way the risks
of accidental breakdown as well
as the level of spurious signals
are considerably decreased.

2.2. Coaxial Cables

288 coaxial cables with double
braid conductors connect the capacitors to the precollectors: they
are 4.5 meters long with an inductance of about 100 nH/m. Both ends
of the cables are terminated by special connectors fitting with the
"multicontact" elements. Special
care has been taken to increase the
dielectric strength against superfi-
cial discharge.

64 similar cables of shorter length
(2.20 m) are used to connect the experiment to the precollectors.

2.3. Precollectors with R.C. Matching System (Fig. 7)

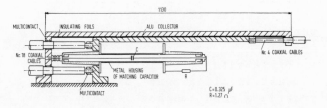

Fig. 7: Precollector with RC matching system

Each sector of the precollector
is made of two thick plates of aluminium insulated by polythylene foils
4 mm thick: at one extremity are connected 18 cables coming from a column
of condensors and, at the opposite
one, 4 cables going to the experiment. They are paralleled by a R.C.
system made of a 0.3 μF condenser
and a 1.27 Ω resistor matching the
18 cables. This R.C. system is built
in the precollector to reduce parasitic inductance.

2.4. Dummy Load

A dummy load with well-known characteristics is mounted on the bank

in order to make the acceptance
tests. This load has to simulate
approximatively the plasma focus
at the time of maximum current:
this means an inductance of about
60 nH, a voltage reversal of about
40% and enough heat capacity to
absorb 1 MJ. It is made of 8 elements of coaxial stainless steel
tubes, 3 meters long. In each
element, inner and outer tubes are
welded together at one end; they
are connected to a ring collector
at the other end. Insulation is
made by a polyethylene foil wrapped
on the inner tube; cooling is
provided by water circulation
along the tubes.

2.5. Trigger System (Fig. 8)

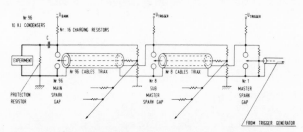

Fig. 8: Trigger system 1MJ bank experiment
plasma focus

The trigger system includes:
- a small Marx generator triggered
  by a 100 V signal with an output
  of 30 kV;
- a master unit triggering 8 sub-
  master units; each of these triggering 12 capacitor spark-gaps;
  masters, submasters and capacitor spark-gaps are of the same
  type;
- connections made by triaxial
  cables.

The essential components of the
trigger system are the triaxial cables operating as Blumlein circuits:
the external conductor is at ground
potential; the intermediate conductor, before firing, is charged at
+ 30 kV. The inner conductor is
connected at one end to the trigger

electrode of the gap to be triggered and at the other end, through
a decoupling resistor of 1 k Ω , to
a potential divider at a voltage of
60% of that of the bank. The master
and submaster units are essentially
spark-gaps which short to ground the
intermediate conductor of the triaxial cables: a negative pulse of
-30 kV propagates along the cable,
finds an open circuit at the trigger electrode, doubles in amplitude
and makes the spark-gap fire with a
jitter less than 5 ns with a forma-
tive time lag inferior than 10 ns.

2.6. Charging Resistors

The charging resistors are 16
tubes made of fiber glass/epoxy with
a copper sulfate solution and copper
electrodes, divided in series in 6
sections of 4 Ω each connected to
the 6 capacitors of a column. Each
section is able in emergency to
absorb 1 MJ. This could happen if
the first capacitor of a column
breaks down at the end of charging;
the energy liberated in the failing
capacitor would then be only its own
energy (10 MJ), which is easily contained by the case of the condenser.

2.7. Protection Resistor

The protection resistor (1 Ω)
connected in parallel with the experiment, provides as said before,
a return circuit for the charging
current of the bank and an emergency energy dump if the experiment
fails to breakdown. This would
happen if the plasma focus experimental chamber was under good
vacuum at the time of firing. It is
made of a copper electrode in a
solution of copper sulfate contained in a copper tank. Its heat capacity is such that one megajoule
makes its temperature rise by only
a few degrees.

2.8. Auxiliaries

The bank is charged by a main
60 kV/0.8 A DC HV supply, with a
present constant current controlled by thyristors. The trigger circuits are charged by a 50 kV/50 mA
H.V. generator, the voltage of
which can be applied permanently
or when a present fraction of the
bank voltage is reached.

2.9. The Compressed Air System

A compressed dry air circuit
(dew-point-20°) supplies the spark-
-gaps and the pneumatic valves. At
the inlet of the circuit is a bal-
last tank of 500 liters at 10 kg/
/cm^2. From this reserve, the different elements are supplied by
pressure reducers. The pressure
in the main spark-gap is supplied
by a low impedance line containing
at the input a high conductance
pressure reducer which can be monitored from the control room in
order to adjust the pressure at
the working voltage; the exhaust
is made by a common valve, the
aperture of which is triggered
from the control room. After every
shot, the spark-gaps are flushed
by 20 liters each of air at atmospheric pressure.

2.10. Control Circuit

At the output of each condenser
is placed a magnetic loop, sending
a signal of about 50 V to a gated
monitor which indicates whether all
the spark-gaps have been fired simultaneously within 40 ns.

The "Failure Detector" makes it
possible to identify those gaps
which are not working in the right
way and the operator can intervene
with a preventive maintenance.

3. CONTROL AND ACCEPTANCE TESTS

In the firm, the packages of the
capacitors are tested individually
to check their ionization level.

Then the packages are welded and
after impregnation, the capacitors are tested at 50 kV D.C. and
8 kV A.C. (50 Hz). On the site,
the spark-gaps mounted on the capacitors are tested at atmospheric
pressure: their self-breakdown voltage must be between 23 kV and 24
kV.

The main cables have been tested
under pulsed conditions: a ringing
discharge of 55 kV amplitude with
an overshot of 15 kV, 50 nsec at
the beginning; 5% failed during
the tests (4000 shots). Then the
individual column has been tested
at full voltage on a dummy load for
40 discharges. During these tests,
the trigger system has been pulsed
for over one thousand discharge
without failure.

When all the bank was connected
to the dummy load, spikes with
voltage amplitude twice the charging voltage appeared across the
load although the main cables were
well matched; additional damping
resistors are being installed
before the final tests: 100 con-
secutive shots at full voltage.

DEVELOPMENT OF A RELIABLE, LOW-COST, ENERGY-STORAGE CAPACITOR FOR LASER PUMPING*

John R. Hutzler
Aerovox Industries, Inc.
740 Belleville Avenue
New Bedford, Mass. 02741

William L. Gagnon
Lawrence Livermore Laboratory, University of California
Livermore, California 94550

ABSTRACT

The 14.5-μF, 20-kV energy discharge capacitor developed in the 1950's far exceeds the present needs of laser fusion research. This unit was designed to operate under conditions of large voltage reversal and high peak currents. In glass-laser systems, voltage reversal occurs only during occasional faults, and peak currents are below 1000 A. It is relevant, therefore, to pursue lower-cost, higher energy-density capacitors, which more closely fit the requirements of laser fusion research. Several companies have designed and produced prototype units that meet the cost criteria of 5 to 6¢/J and energy density of 5 to 6 kJ/unit.

INTRODUCTION

Large glass-laser systems are characterized by low conversion efficiency of electrical energy to light energy. This means that a substantial portion of the total cost for a laser-fusion research facility goes to energy-storage capacitors. These capacitors are a well-developed technology[1-5] and have been extensively applied in the Sherwood and CTR programs over the last 20 years. However, their applications require high peak currents and large voltage reversals, which is not the case in laser-fusion research. In large glass-laser systems, the capacitors normally see no voltage reversal and only moderate peak currents. Also, the required operating life of a laser-fusion facility is significantly shorter than for Sherwood or CTR applications. With the capacitor industry, LLL is working to develop energy-storage capacitors more specifically suited to the needs of laser-fusion research.

DESIGN AND COST CONSIDERATIONS OF ENERGY-STORAGE CAPACITORS

For laser fusion research, it is desirable to reduce the cost of capacitors while still maintaining the reliability essential to successfully operating the laser facility. The raw materials used in a capacitor represent approximately 50% of its procurement cost. Many of these materials have shown extraordinary inflationary trends over the past five years. All the materials used in the capacitor, therefore, have been reviewed to determine their design and reliability merits along with their cost effectiveness.

The capacitor used in laser-fusion research has a rating of 14.5 μF at 20 kV. In 1970, the Naval Research Laboratory procured 368 capacitors rated at 14.0 μF, 20 kV for which there is reliability and extended-life test data.[4] The NRL capacitor is described in design A in Table 1. Manufacturing loss records, acceptance tests, customer reports and extended-life tests indicate that it is reliable. Its cost, based on October 1974 prices, is $0.16/J.

Design B in Table 1 describes the capacitor first supplied by Aerovox† to the LLL for use in the Laser Fusion Program. It is identical to design A except for a slightly higher capacitance (14.5 vs 14.0 μF) and the use of two

---

*Work was performed under the auspices of the U.S. Energy Research & Development Administration.

†Reference to a company or product name does not imply approval or recommendation of the product by the University of California or the U.S. Energy Research & Development Administration to the exclusion of others that may be suitable.

Table 1. Design, cost, and performance characteristics of energy-storage capacitors.

	Capacitor design					
	A	B	C	D	E	F
Energy, J	2800	2900	2900	2900	6000	5800
Capacitance, μF	14.0	14.5	14.5	14.5	30	29
Voltage, kV	20	20	20	20	20	20
Cost, $/J[a]	0.160	0.151	0.104	0.102	0.058	0.062
Oscillatory life, pulses	>100,000	>100,000	75,000	~18,000	<100	>1,000
Nonoscillatory life, pulses	—[b]	—[b]	—[b]	—[b]	>175,000	>175,000
Bushing type	Epoxy	Ceramic	Molded	Molded	Molded	Molded
No. of bushings	1	2	2	2	2	2
No. of capacitors in series	4	4	4	2	2	2
Dielectric stress, V/mil	2,000	2,000	2,083	2,041	2,564	2,564
Paper weight, lb	56	60	53	56	74	72
Foil weight, lb	14	15	12	7	10	17
Oil weight, lb	42	43	43	47	52	51

[a]All costs are based on October 1974 prices.

[b]No data.

ceramic bushings instead of one vacuum-cast epoxy bushing. The life and reliability of the two designs are equivalent, but the cost of B was 6% lower, primarily due to the higher cost of the epoxy bushing.

Although design B was approved for the Laser Fusion Program, the cost was still higher than desired. A major cost-reduction program consisting of the following modifications was instituted.

IMPREGNANT

The preferred impregnant is castor oil, a vegetable product subject to weather and other conditions that affect its price from time to time. Over the past several years, the price of capacitor-grade castor oil has increased as much as 300%. To offset this price increase and reduce handling costs within the plant, a bulk storage system for castor oil was implemented.

ELECTRODE

The electrodes were reduced in thickness, thereby reducing the weight and cost of aluminum foil.

DIELECTRIC

Probably the most fruitful approach to reducing capacitor costs is in selecting the kraft paper used as the dielectric. In Fig. 1, paper cost versus thickness is plotted. The cheapest paper is 0.0007-in. thick, with a very sharp rise in cost as paper becomes thinner. However, the quality of sheet formation improves as the paper becomes thinner, since thinner sheets have a higher dielectric strength.

The formula for capacitance of a parallel plate capacitor is $C = KA/T$, where K is the dielectric constant, A is the area of the electrodes, and T is the thickness of the dielectric. Reducing the thickness of the dielectric reduces the weight of paper used and increases the capacitance. Since a higher capacitance is not required, it is maintained at a constant value by reducing the area of the foil and paper. This double reduction in paper weight is proportional to the square of the paper-thickness reduction. The reduction in foil weight is directly proportional to the reduction in foil area.

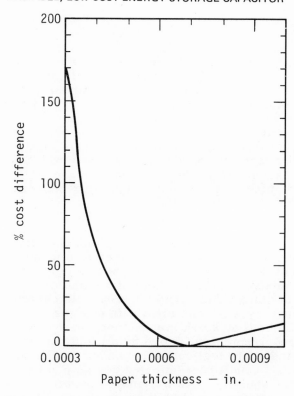

Fig. 1. Cost vs thickness of kraft paper used as the dielectric.

Table 1 shows the weights of these components with different dielectric stresses. Since the cost of paper varies with the thickness of the sheet, care must be exercised in choosing the makeup of the dielectric pad. The ideal solution is to reduce the dielectric and choose multiple sheets of paper close to 0.0007-in. thick. Much of the savings of design C over B in Table 1 comes from a change in the dielectric pad: B used five sheets of 0.0005-in. paper, while C used four sheets of 0.0006-in. paper.

The limitation on dielectric reduction is an increase in voltage stress and a loss of life. The life of a capacitor is inversely proportional to the fifth power of the applied voltage[6] and is reduced by one-half for each 10°C increase in operating temperature.[7,8] Life tests to demonstrate compliance with the specifications as shown in Table 1 are essential before approving a dielectric reduction.

BUSHINGS

The ceramic bushing with solderable metal inserts is costly to manufacture and assemble. A lower-cost bushing made by injection molding of a thermoplastic

resin was designed, tested, and approved for capacitors used in the LLL laser program.

RESULTS OF DESIGN ECONOMIES

Design C of Table 1 incorporated most of the design changes mentioned above. Bulk-storage vacuum tanks were installed for the castor oil, thinner foil electrodes were used, the paper dielectric was reduced by using fewer sheets of thicker paper, and injection-molded thermoplastic resin bushings were used. These changes brought about a significant decrease in capacitor cost when compared with design A or B. While the oscillatory life was also reduced somewhat, from more than 100,000 to 75,000 pulses, this was still more than adequate for the design life of the laser.

Design D (Table 1) is similar to C, but multiple sheets of thicker paper were used by reducing the number of capacitor sections in series from four to two, which raised the voltage on each section from 5,000 to 10,000 V. The savings on this design were minimal, while a very significant reduction in life was noted. Design D was not approved for the Laser Fusion Program.

Testing and analysis of prototype operating banks showed that the most probable fault was an exploding flash-lamp. This results in a 1500-Hz ringing condition for the capacitor, with only a slight increase in current above normal operating levels. Consequently, the required number of oscillatory test pulses was reduced to 1000.

This reduction permitted a major redesign of the capacitor. In design E, the total dielectric thickness was reduced, increasing the voltage stress; the paper efficiency was increased by using two capacitor sections in series, a wider width of paper and foil, and thicker sheets of paper (see Fig. 2). As a result a 44% cost improvement was obtained over design D and a 63% improvement over A. The life under operational conditions was satisfactory at more than 175,000 non-oscillatory pulses. But oscillatory life, which represents fault conditions, was not satisfactory. Dissection analysis after an oscillatory-life test indicated that failure was caused by the rupture of the foil, or the connection between the aluminum foil and the copper conductors in the base of

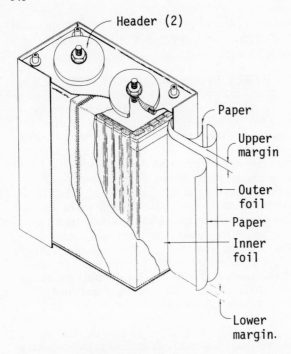

Fig. 2. Energy-storage capacitor
(design E).

the capacitor where the two series sections
were connected.

The pertinent parameters of the
various designs were reviewed and the

data considered are shown in Table 2.
The terminal current density for design D
was 104,527 A/in.^{2} of available aluminum-
foil terminal area. Not only was there a
very high current density at the terminal
foil interface, but also the electromagnetic force associated with the current
flow was reversed when the two groups of
capacitors were connected. The high current flow, and the concentration of electromagnetic force due to the reversal of
current flow, were believed to have
ruptured the foil and/or the interconnection,
which caused the failure.

A new design, F, was manufactured,
using the same paper dielectric thickness
but thicker aluminum foil with a current
density of 59,153 A/in.^{2} of available
aluminum-foil terminal area. Its current
density was consistent with designs C and
D, both of which met the oscillatory life
requirements of more than 1,000 pulses.
And a life test conducted under oscillatory
fault conditions indicated design F was
satisfactory. Its life under operating
conditions would be the same as for E,
since the same paper thickness and voltage
stress were used. The thicker foil added
a 7% increase in the cost of the capacitor
over design E, however. Design F is
currently being further tested.

Table 2. Critical capacitor parameters.

	Capacitor design				
	C	D	D_1[a]	E	F
Energy, J	2,900	2,900	4,350	6,000	5,800
Capacitance, μF	14.5	14.5	14.5	30	29
Voltage, kV	20	20	25	20	20
No. of capacitors in series	4	2	2	2	2
Voltage stress, V/mil	2,083	2,040	2,551	2,564	2,564
Foil thickness, mils	0.22	0.22	0.22	0.25	0.35
Foil termination area, in.^{2}	1.6	0.83	0.83	1.21	1.66
Peak current oscillatory discharge, A[b]	51,790	51,790	64,738	126,478	98,194
Current density of termination, A/in.^{2}	32,369	67,398	77,998	104,527	59,153
Peak current (critically damped discharge), A	427	427	533	881	854
Current density of termination, A/in.^{2}	216	514	643	728	514

[a]D_1 is the same design as D, but operated at 25 instead of 20 kV.

[b]Oscillatory discharge was 33 kHz with 85% voltage reversal.

CAPACITOR RELIABILITY AND QUALITY

RELIABILITY

Capacitors, like other electrical components, show three distinct failure rates. A typical failure rate curve is shown in Fig. 3. The high failure rate during the early portion of life is due to defects in the raw materials and mechanical damage during manufacture. The ensuing much lower failure rate is the rate for which the capacitors were designed. The increasing failure rate during the latter stages of life is due to the wearout of materials. The end-to-life and design failure rates can be controlled by changing specifications; the early failure rate can be controlled by a stringent quality-controlled manufacturing operation plus a subsequent power burn-in.

QUALITY CONTROL

A quality-controlled manufacturing operation is essential to produce large quantities of reliable energy-storage capacitors.[2] A dossier is compiled for each unit consisting of batch identifications for all materials and personnel assignments for each manufacturing step. Incoming materials are inspected to certify that they conform to specifications. Inspection continues during each stage of manufacture. More important than the inspection, however, is the attitude and capabilities of the production personnel. If they are well trained and motivated, there will be very few defects for the inspector to find. What defects are found are communicated back to the production personnel and rarely occur again. Many of the problems here are associated with new employees who are just learning their job.

After manufacture, each capacitor is tested for capacitance, power factor, voltage withstand, and insulation resistance. All capacitors are also leak tested for a minimum of 18 hours in a 65°C oven. Finally, a power burn-in is performed on the capacitor to simulate the operational use or the fault conditions of the circuit. Any unit that fails during the manufacturing or testing operations is dissected and analyzed for cause of failure. The cause is then communicated to the production worker responsible for the failure, or to the vendor who supplied the defective material, and corrective action is taken.

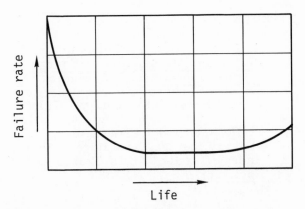

Fig. 3. Typical failure-rate curve for a capacitor.

ACCEPTANCE TESTING

To assure that all capacitors will meet operational requirements and specifications, acceptance testing of large production lots has been explored. One approach is to perform a power burn-in of 1,000 to 4,000 pulses on every capacitor; any units that fail are replaced by the vendor. This approach is expensive and time consuming to the customer, however. Another approach is to perform a life test of 10,000 to 20,000 pulses under accelerated conditions on a random sample of the production lot.[9] If the sample passes the life test, the entire production lot is accepted; if it fails, the lot is rejected. Capacitors procured by NRL and LLL have used the lot-acceptance life-test procedure successfully.

CONCLUSION

The combination of adequate design, quality-controlled manufacture with a power burn-in, and acceptance testing will produce a reliable energy-storage capacitor. The data presented in this paper show that reliable, low-cost, energy-storage capacitors are available, but successful production requires close communication between the user and the manufacturer.

REFERENCES

1. E. L. Kemp, Considerations in the Design of Energy Storage Capacitor Banks, Los Alamos Scientific Laboratory, N. Mex., Rept. LA-2530 (June 1961).

REFERENCES (Continued)

2. G. P. Borcourt, Problems in the Design and Manufacture of Energy Storage Capacitors, Los Alamos Scientific Laboratory, N. Mex., Rept. LA-142-MS (January 1970).

3. P. Hoffman and J. Ferrante, "Energy Storage Capacitors of High Energy Density," IEEE Trans. Nucl. Sci. NS-18 (4), 235 (August 1971).

4. J. R. Hutzler et al, "Reliable Capacitors for Large Energy Storage Facilities," Proc. Symp. Eng. Problems Fusion Res., 5th, Princeton, November 1973 (IEEE Pub. No. 73CH0843-3NPS).

5. E. L. Kemp, "The Study of Capacitive Energy Storage for a Theta-Pinch PTR Compression Coil," in Proc. Symp. Eng. Problems Fusion Res., 5th, Princeton, November 1973 (IEEE Pub. No. 73CH0843-3NPS).

6. M. Brotherton, Capacitors, Their Use in Electronic Circuits (D. Van Nostrand Company, Inc., N.J., 1946), p. 49

7. M. Brotherton, Capacitors, Their Use in Electronic Circuits (D. Van Nostrand Company, Inc., N.J., 1946), p. 48.

8. "Capacitors, Selection and Use of," in MIL-STD-198C (December 1971), p. 11.

9. M. P. Young et al, "Procurement and Testing of Capacitors for Large Energy Storage Facilities," J. NRL Prog. (September 1973), p. 72.

DESIGN AND TEST RESULTS OF A NEW TYPE OF VERY HIGH VOLTAGE CAPACITORS FOR LOW INDUCTANCE APPLICATIONS

J. Cortella, J. Jouys, J. Raboisson,
J. Leborn, and J. J. Wavre
C.E.A. - Valduc, France
and
E. Haefely
Saint Luis, France; Basel, Switzerland

ABSTRACT

This paper will describe new 10 kJ, 240 kV low inductance capacitors which are designed for fast discharge applications where large peak currents and large voltage reversal are required.

These high energy capacitors belong to a 12 MV Marx generator comprising 50 stages of 240 kV each housed in an oil-filled tank. The Marx generator feeds a "Blumlein" line which produces flash X rays.

These capacitors are immersed in oil and for maintenance reasons a high reliability as well as a long life expectancy are also required.

A description of the special design including both mechanical and electrical considerations necessary to meet the tight specifications will be given.

In the last part of the paper we describe the results of the measurements performed on these capacitors.

THE MAGNETIC ENERGY STORAGE SYSTEM USED IN ZT-1

L. C. Burkhardt, R. Dike, J. N. Di Marco, J. A. Phillips, R. Haarman, A. E. Schofield
Los Alamos Scientific Laboratory
University of California
P. O. Box 1663
Los Alamos, New Mexico 87544

ABSTRACT

The ZT-1 toroidal pinch experiment is driven by a magnetic energy storage system that uses four fuses to interrupt the current. One fuse package can interrupt 800 kA, develop 70 kV peak, and absorb 30 kJ. Each quadrant is driven by a fuse to increase the total voltage around the torus to 280 kV and the total inductively stored energy is 100 kJ. Approximately 10% agreement is obtained with a numerical calculation used to predict the fuse behavior. Inductive coupling of 14 nH between the four fuses eliminates the jitter of ~ 0.1 μs, observed with a single fuse.

ZT-1 is a toroidal pinch experiment[1] that utilizes magnetic field programming to achieve a high-β stable plasma. Plasma confinement is achieved by the magnetic field generated from a toroidal current in the plasma. Stability results from the high magnetic field shear, which develops when the toroidal magnetic field is programmed to reverse in direction at the outside of the pinch. An example of a stable magnetic field profile is shown in Fig. 1.

The size of the experiment can best be described by the dimension of the discharge tube which is 0.103 m I.D. and has a major diameter of 0.76 m.

The power supply used to drive the toroidal current must have specific characteristics in order to avoid wall breakdown at the surface of the discharge tube after the pinch has formed. It is also required that the power supply be capable of transferring current into the plasma in a very short time, i.e., on the order of an Alfven wave transit time across the discharge tube (~ 0.5 μs).

To avoid wall breakdown after the pinch has formed, the current should remain approximately constant, or from another point of view, the voltage along the discharge tube should be low. From our experience breakdown does not occur for electric fields less than 30 kV/m. A high voltage, low inductance, capacitor bank could satisfy the first requirement of rapid current transfer, but would require crowbar switches to remove the voltage from the discharge tube. Since such a system has low inductance, the resulting L/R decay of the current would be too short for the purposes of this experiment. A better way to meet the specifications is to use a power supply that has constant current characteristics. Two supplies that approximate this requirement are shown in Fig. 2, along with a schematic representation of the electrical system of the experiment.

To drive the toroidal current in the plasma the aluminum primary is divided into quadrants, which are electrically connected in parallel. A magnetic core is used to couple the aluminum primary and the plasma secondary.

The electrical circuits displayed in Fig. 2 are representations of two

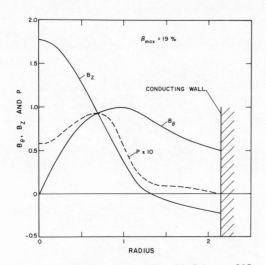

Fig. 1. Stable magnetic field profile

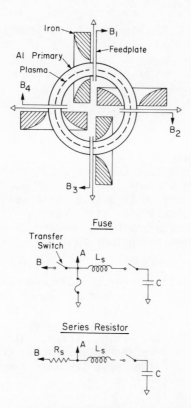

Fig. 2. Schematic diagrams of the electrical system of the experiment and power supplies.

alternatives used to drive each quadrant of the torus. There are four positions represented by "A" which are electrically coupled together with ∼ 14 nH in order to equalize the voltage around the torus. This coupling "locks" the fuses together, thereby avoiding the timing jitter of ∼ 0.1 μs observed with a single fuse. More will be said about this point later. L_s, the storage inductor, is really one inductor that is common to all quadrants, but for these equivalent circuits it is considered to be made up of four parallel inductors. Its magnitude can be varied from 40 to 160 nH by means of adjustable rods at the base of the machine.

First, consider the series resistor circuit. The amplitude of the current flow is controlled by making R_s greater than the characteristic impedance of the circuit. L_s is made as small as possible to make the L/R rise time compatible with experimental requirements. The series resistor is made from 304 stainless steel folded into a parallel plate transmission line to reduce its inductance. The stainless steel is immersed in an oil bath to control its temperature. Its resistance is typically 0.25 Ω.

The fuse or magnetic energy storage circuit shown at the top of Fig. 2 uses a copper fuse to interrupt the current flowing of the storage inductance L_s, which is now increased to its maximum value of 160 nH.

The transfer switches are pressurized, low inductance, spark gap switches which are preset to breakdown at a specified voltage by varying the pressure in the gap. A small pulsed arc in one electrode generates photoelectrons to reduce the statistical variation of the gap breakdown voltage. The switches are not triggered; simultaneity of breakdown is achieved by the transformer effect which automatically will overvolt any gap that delays in breaking down.

The fuse[2] is copper foil 25 μm .025 mm thick, 90 mm wide and 180 mm long, folded at the midplane to form a parallel plate transmission line. ABS insulation separates the folded copper which is immersed in 100 μm glass beads and the entire assembly is molded into a package. This fuse will develop 70 kV and interrupt 800 kA, and can absorb 30 kJ without damage. Figure 3 shows the fuse package.

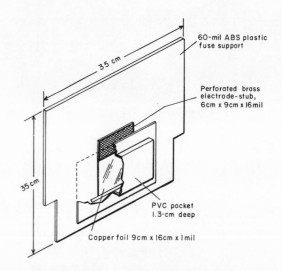

Fig. 3. Fuse package

Both circuits are driven by the same capacitor bank, operating with the same stored energy. Considered on the basis of all four quadrants the total capacitor bank is 600 μF, typically charged to 20 kV giving a stored energy of 120 kJ.

A photograph of the machine is shown in Fig. 4.

Fig. 4. The ZT-1 machine

The capacitors connect to the coaxial base of the machine. The adjustable storage inductance is below the floor level in this figure; adjustment is achieved by changing the inner diameter of the coax. Connection between the storage inductance and the fuses is accomplished by the "wave" sections which also provide the low inductance coupling between fuses. The fuse is located just inside the junction between the feed plates radial to the torus and the wave section. The transfer switch is located between the fuse and the torus in the extension that is perpendicular to the radial feed plates. The torus, transformer iron and vacuum system can be seen beyond the transfer switch.

Figure 5 shows a summary of the parameters of the machine which also indicates the advantage of fuse operation.

	SERIES RESISTOR	FUSE	
R	0.4 to 0.2	--	Ω
L_s	140	300	nH
I_ϕ	35 to 70	180 to 220	kA
$\dot{I}_\phi$	0.06 to 0.12	0.8 to 1.2	MA/μs
C	150	150	μF
V_{BANK}	20	20	kV

Fig. 5. Summary of machine parameters for the series resistor and fuse mode.

The remainder of this paper will be devoted to the behavior of the fuse used with this experiment.

To predict the behavior of the four-fuse system a model is used to describe the fuse dynamics. From our experiments[3] the resistivity of the copper fuse, as a function of the internal energy, follows published[4] values up to the point at which vaporization starts; at higher energy densities an analytic fit is made to the data. This information is used in a numerical code to predict the fuse behavior and also, the effect of the coupling between the fuses to overcome the observed jitter of ~ 0.1 μs.

Figure 6 shows a comparison between experiment[3] and calculation for a single fuse.

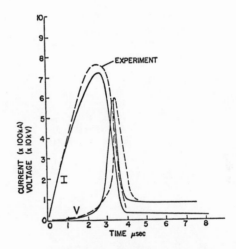

Fig. 6. Comparison between numerical calculation and experiment for single fuse without current transfer to the load.

To numerically determine the effect of coupling, an effective jitter is created by using fuses with different cross sections, and without electrical coupling. Figure 7 shows the results of the calculation and an effective jitter of 1.5 μs in the voltage that occurs without coupling. S_1 and S_2 refer to the fuse cross sectional area. The fuse having the larger area takes longer to interrupt the current as would be expected from the "action integral" equation.

Figure 8 shows the effect of coupling the fuses together with a 14 nH inductor. The two voltage traces cannot be resolved on this graph. The coupling can be observed by noting the respective currents, I_2 is decreased when compared to the uncoupled case.

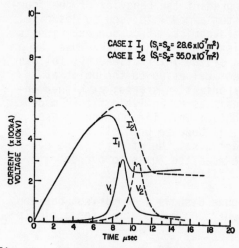

Fig. 7.   Display of the effective jitter
            induced in the numerical calcula-
            tion by means of different fuse
            cross section.

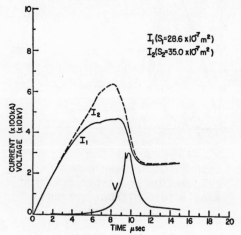

Fig. 8.   Numerical results using 14 nH
            coupling between fuses to
            eliminate timing fitter.

This coupling is also verified experimentally in that no discernible voltage unbalance between the quadrants is
measured. Figure 9 shows the measured
pinch current and voltage for the fuse
mode of operation. The sweep time is
2 μs per division, the peak current is
209 kA, and the peak voltage for this
discharge is 40 kV. Maximum energy in
the storage inductor for this shot is
100 kJ.

SUMMARY

        For those applications requiring a
constant current power supply a magnetic
energy storage system, using fuses to

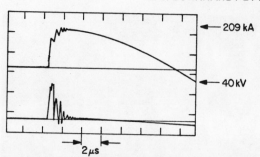

Fig. 9.   Experimentally obtained pinch,
            current and voltage for the fuse
            mode of operation.

interrupt the current, can be a practical
solution. It enhances by almost an order
of magnitude both the rate of rise and
magnitude of current that can be obtained
when compared to the series resistor
technique. To increase the power transfer above that for a single fuse a parallel arrangement of fuses is shown to be
a feasible technique provided sufficient
electrical coupling is designed into the
system.

REFERENCES

1] Burkhardt, L. C., et. al., Proc. 2nd
    Topical Conf. on Pulsed High-Beta
    Plasmas, 1972 (Max-Planck Institut
    fur Plasmaphysik, Garching 1972) 33.

2] R. A. Haarman, R. S. Dike, and M. J.
    Hollen, Fifth Symp. on Eng. Problems
    of Fusion Research, Princeton, Nov.
    (1973) pg. 459.

3] J. N. Di Marco, L. C. Burkhardt,
    Journal of Appl. Phys. 41, 3894
    (1970).

4] Smithsonian Physical Tables, Vol. 120
    Ninth Revised Edition, pgs. 155 & 393,
    Published by the Smithsonian
    Institution (1956).

INDUCTIVE ENERGY STORAGE WITH SHORT-CIRCUIT GENERATORS

C. Cortese, C. De Bernochi, E. Tessitore
I E N (+)
Torino - Italy

ABSTRACT

A scheme has been developed for testing the convenience of inductive energy stora ge to power a plasma-focus device. In our scheme the storage inductor can be charg- ed up to ~0.1 MJ or ~2.3 MJ by using two phases of two different synchronous altern- ators. The inductor-charging circuit is closed when the alternating voltage is cross- ing zero. A convenient synchronization circuit for switch-time controlling was built and successfully tested on a scale model (~0.1 kJ). The same synchronization circuit can be used for any large scale system.

INTRODUCTION

Our inductive-energy-storage scheme is devised to operate a plasma-focus (PF) at an energy level which can be orders-of- -magnitude higher than that of today plas- ma-focus experiments at IEN with a 12 kJ capacitor bank at 18 kV. Inductive energy seems particularly convenient for repetiti ve pulsed operations of a plasma-focus within our research program. For order of magnitude estimates we have used the parameters of a short-circuit generator of IEN, a Brown Boveri (BB) with 23 MJ and of a larger one, an English Electric (EE)° with a 167 MJ rotating group. Principal characteristics of those two short-circuit generators are reported in Table 1. The storage inductor (SI) is charged directly by the a.c. power given by the short-cir cuit generator (G). The driving motor (brings the alternator at the rated angul- ar velocity ω) is disconnected when an output voltage of about 4 kV (for BB) ap- pears at the stator terminals. The sto- rage inductor is then charged. With a ro tor kinetic energy much larger than the magnetic energy which is absorbed by SI, ω decreases only by a small percent age during each charge of SI. We can have so repetitive charges of SI without connecting further the driving motor to

the network. Efficiency need in energy transfer sets limits in the value L_{SI} of the inductance of the storage inductor. On the other hand to have a maximum of energy in SI we have to connect SI to G when the voltage $V(t) = V_{max} \sin(\omega t + \psi)$ of G vanishes. This can be done by us- ing a convenient synchronization circuit in the design of fast switches. We have built this synchronization circuit (SC). The main purpose of this paper was the designing and testing of SC. To test this part of our scheme we have inserted it in a scale model of inductive-storage- -and-load system. As a model load we have inserted the coaxial accelerator we normally use for our plasma-focus experiments [1]. Since the synchroniza- tion circuit is only a timing device, its design does not require substantial varia tions with the increasing of the charging currents and so it may be used for eve- ry a.c. inductive storage scheme.

SYNCHRONIZATION CIRCUIT. SCALE MODEL.

Energy can be transferred with a maximum of efficiency from the genera- tor to SI if the impedance of storage in- ductor and that of G have the same value. The internal reactance of our alternators

---

(+) Istituto Elettrotecnico Nazionale Galileo Ferraris
Corso Massimo d'Azeglio 42 - I-10125 Torino

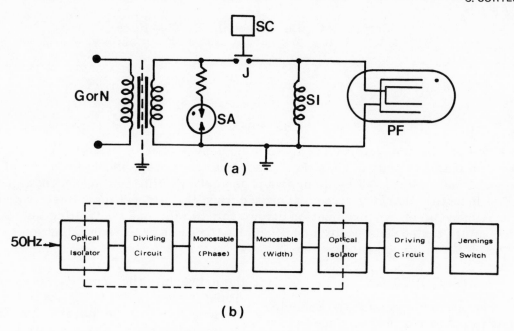

Fig. 1.   Inductive energy-storage model circuit
a) Electrical circuit
b) Synchronization circuit

with cylindrical rotors, is equal to twice the subtransient reactance x''_d (Ω/phase) (we use a two-phases connection) [2]. Our choice for L_{SI} therefore will be $L_{SI} = 2x''_d/\omega$ (by neglecting the total circuit resistances). If V_{rms} is the voltage between the two phases of the alternator the steady rms current in the inductor is:

$$I_{rms} = V_{rms}/(2x''_d + x_{SI}) = V_{rms}/4x''_d$$

($I_{peak} = \sqrt{2}\ I_{rms}$). By connecting the inductor with G at the time the voltage of G is crossing zero ($\psi = 0°$) the peak value of the current in the inductor (half period later) (i. e. at the time in which G is disconnected and L_{SI} starts discharge on the load) is: $I_{peak}(0°) = 2\sqrt{2}\ I_{rms}$ (by closing at $\psi = 90°$, $I_{peak}(90°) = I_{peak}(0°)/2$, i. e. the storage energy at peak current in the inductor would be four times smaller). To store the maximum energy in the SI, it is necessary to build a special circuit for activating the closure of the make-switch when $V(t) = 0$ and to open it again when the current in SI is at peak value. Both operations must be done as quikly as possible and (during the second step) by reducing arc losses for maximum energy

transfer. Our model was built as in Fig. 1a by using a 24 mH air coil as storage inductor, a 220/220 V, 1000 VA insulation transformer, a Jennings vacuum switch (J) (type R1, peak voltage 50 kV, rated current 50 A_{rms}), to close and open the charging circuit. This is driven by the synchronization circuit (SC) and a plasma coaxial accelerator as discharge system (center electrode diameter 3 cm, external electrode diameter 10 cm, electrode length 14 cm, D_2 pressure ~ 2 torr). In operating our small scale system we have used a 220 V network (N) ($\omega/2\pi = 50$ Hz) instead of a large generator G. As it is shown in Fig. 1b an optical isolator powered by an a.c. voltage gives a synchronization signal free from electric noise to a frequency dividing circuit which sends a pulse to a monostable every 100 periods of the a.c. voltage and at the correct time with respect to the sinusoidal voltage. This pulse triggers the second monostable which provides a signal of suitable width (~ 10 ms). This pulse is sent to another optical isolator which controls the driving circuit (power amplifier) for supplying the Jennings switch coil.

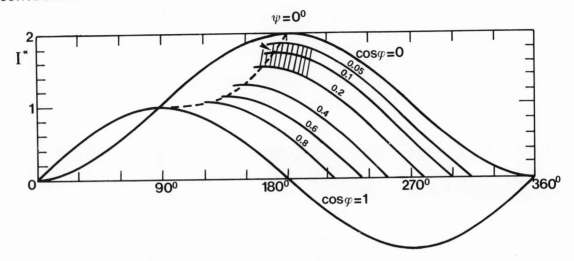

Fig. 2.   Variation of the charging current as function of the circuit resistance in the
case $\psi = 0°$. The arrow marks the range of values of $\cos\varphi$ for our system.

The entire digital circuit is built
with MOS integrated circuits to achieve
a better noise immunity. In this way
the Jennings switch is closed at 2 sec
intervals for 10 ms starting exactly at
zero voltage. The fast rise time ($\gtrsim 10\mu s$)
and the high peak value of the voltage
(~ 10-20 kV) which is developed across
the inductor at the switch-opening time
were enough to start a discharge in the
gun (the surge arrester (SA) at the tran
sformer secondary winding is necessary
to prevent a disruptive overvoltage on
it). Those values are essentially the sa
me as for plasma-focus operations. In
our small scale test the current value
in the coaxial accelerator ($\sim 10^2$A) is in
stead substantially smaller(by a factor
$\sim 10^{-3}$). For these small values of the
current we have not observed plasma-fo
cus formation. The luminous current
sheath between the electrodes is localized (during most of the discharge time)
between the electrodes at the breech of
the gun or, in a different discharge, at
the other end of the electrodes. This is
the maximum allowed current in our J.
For a test at substantially larger current (e.g. 10^4-10^5A) the Jennings
switch should be replaced by a second
coaxial accelerator. The coupling of
two plasma guns in a scheme where
each gun can perform as a switch for

the circuit of the other gun was proposed
and analyzed by Belan at al.[3]. Minor
changes in the driving circuit are necessary for the largest power that can be
supplied by the generators (see Table 1).
These variations in the driving circuit
depend on the element which is used to
replace J. By using a second coaxial accelerator it is convenient to use the same
driving circuit that triggers the discharge in our 12 kJ plasma-focus experiment.
The performance of SC is not affected by
variations of ω .

NUMERICAL ESTIMATES

The resistance (R) in the whole circuit decreases the peak current, $\cos\varphi$
increases with R ($\varphi = \mathrm{arctg}\,\omega L/R$). Fig.
2 shows: $I^* = I_{peak}(0 \leqslant \cos\varphi \leqslant 1)/I_{peak}(\cos\varphi = 1)$.
Our estimate gives $\cos\varphi \simeq 0.1$ and so the
current in the SI is $I_{peak} \simeq 1.7\sqrt{2}\ I_{rms}$
(see Fig. 2). For our system $R = R_G + R_E$
where R_G is the internal resistance of
the generator (R_G 3% of x''_d), R_E is the
resistance of the SI and connections. Obviously R lowers the peak energy value
of SI and would reduce the shots frequency in a repetitive discharge system because ω is also decreased. In the case
$x_{SI} = 2\,x''_d$ the maximum energy that can
be stored during a half period of V(t) (clo
sure at $\psi = 0°$) can be calculated by using for the storage energy W_{SI} the equa-

Table 1. Characteristic parameters of two short-circuit generators

			BB	EE
Rated power	P	(MVA)	5.5	60
Short-circuit power		(MVA)	60	1600
Speed		(rpm)	1000	3000
Frequency	$\omega/2\pi$	(Hz)	50	50
Phases			3	3
Rated voltage	V_{rms}	(kV)	4	11/6.35
Stator connections			Star	Star/delta
Rated current	I_{rms}	(kA)	0.8	3.15/5.45
Short circuit current		(kA)	8.7	84/146
Subtransient reactance	$x''_d{}^*$	(per units)	0.066	0.034
Rotor pd^2		(kg. m^2)	16800	13500
Rotating group energy	W	(MJ)	23	167
Driving motor		(kW)	600	1100

tion[4]:

$$W_{SI} = \frac{3.24 \cdot P}{8 \cdot \omega \cdot x''_d{}^*} \qquad (1)$$

where $x''_d{}^* = x''_d \cdot P/V^2_{rms}$ is the subtransient reactance (per unit). By using equation (1) and values of P, $x''_d{}^*$ from Table 1 we have for generator BB: $x''_d =$ = 0.192 Ω, $x_{SI} = 2x''_d = 0.384\,\Omega$, $L_{SI} =$ = 1.22 mH and $W_{SI} = 0.1$ MJ. For the generator EE we have: $x''_d = 0.0687\,\Omega$, $x_{SI} = 0.1374\,\Omega$, $L_{SI} = 0.44$ mH, $W_{SI} =$ = 2.3 MJ. For a circuit with the SI directly connected to G as Fig. 1a the values of the current in SI are below useful values for a plasma-focus operation: $I_{peak} = 13$ kA (for BB), $I_{peak} = 102$ kA (for EE). To increase the SI charge current, and so the discharge current in the gun (as it is necessary for a plasma focus experiment), we can use the insulation transformer itself as step-down transformer and at the same time as SI, while the gun is directly connected to its secondary. In this way the secondary is shorted during the charge time by a convenient switch; at its opening time the storage energy will be transferred to the gun. By using the generator BB with the step-down transformer normally employed for short-circuit test (4000/350 V, 3 MVA, percent impedance 2%) we can obtain up to $I_{peak}(\psi = 0°) =$ = 0.2 MA for charge (and gun discharge)

currents. For the EE generator, by using a convenient step-down transformer, I_{peak} ($\psi = 0°$) $\simeq 2$ MA. In the case of the BB, of course, a better transformer can give substantially higher values of I_{peak}.

CONCLUSION

We have built and successfully tested, on a 0.1 kJ scale model, a synchronization circuit for an a.c. inductive storage energy system. A critical analysis reveals that such a circuit is useful also in the case of a larger systems that make use of the short-circuit generators.

AKNOWLEDGMENT

We thank V. Nardi for suggesting this work and for helpful advices.

Work supported in part by CNR.

REFERENCES

1) W. H. Bostick et al., in this same Proc.
2) Standard Handbook for Electrical Engineers, McGraw Hill, (1957).
3) N. V. Belan et al., Soviet Physics - Technical Physics 18 51 (1973).
4) G. A. Sipailov et al., Elektrichestvo 1 52 (1972).

° This generator at the moment is disassembled at IEN.

HIGH MAGNETIC FIELD CRYOGENIC COIL FOR THE FRASCATI TOKAMAK TRANSFORMER

Andreani, R. and Lovisetto, L.
Laboratori Gas Ionizzati (Ass. CNEN/EURATOM), C.P. 65 - 00044 Frascati,
Rome, Italy

ABSTRACT

The Frascati Tokamak transformer is made up of a system of coils magnetically linked with the toroidal plasma column of the machine.

The main element of this setup is a 1.2 m long solenoid, inner diameter 0.3 m, outer diameter 0.7 m, designed at Frascati and already built.

The flux density on the solenoid centerline, at a 31 kA current, is 17 Tesla. The energy stored in the transformer 30 MJ.

The current variation produced opening a commutating switch generates a maximum voltage of 40 kV across the transformer, of this about 28 kV across the solenoid.

Ionization and heating of the plasma are produced by the corresponding emf induced along the plasma column.

Liquid nitrogen cooling is employed in order to reduce the energy dissipation during the current build-up.

1. INTRODUCTION

Tokamak machines for thermonuclear fusion research are intended to obtain thermonuclear reactions in a dense hot deuterium plasma contained in a toroidal vacuum chamber.

The plasma confinement is assured by the magnetic field produced by large toroidal magnets while the plasma breakdown and heating is obtained transferring energy to the plasma from sets of circular windings (Tokamak transformer), magnetically linked with the plasma itself (see Fig. 1).

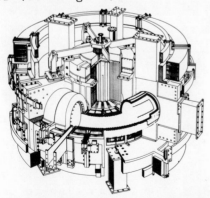

Fig. 1. Frascati Tokamak simplified layout.

In the case of the Frascati To

kamak, the operation of the transformer is the following.

A current rising to 31 kA is fed into the windings by a 21 MW power supply (motor-flywheel-generator set and diode rectifiers).The energy stored is 30 MJ. A commutation system, described in paper[1] forces the transformer current I through a resistance R producing a large overvoltage RI (40 kV) across the transformer (see Fig. 2). The

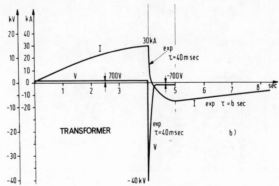

Fig. 2. Transformer current and voltage waveshapes.

current and the voltage decay thereafter with the circuit L/R time constant (40 msec). The corresponding secondary voltage, which is of about 100 V peak, ignites the plas ma column and generates the plasma current. The operation continues reversing the power supply polarity and feeding more energy to the transformer and, through the transformer, to the plasma.

2. COIL DESIGN

2.1. Parameter Values

The results of the large work of analysis and optimization performed; by the Frascati design team are reproduced in Fig. 3 and 4,

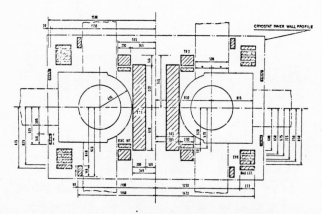

Fig. 3.   Transformer layout.

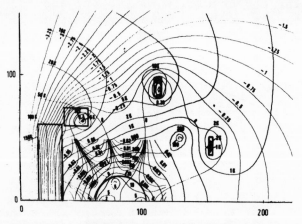

Fig. 4.   Magnetic flux lines and field intensity profiles.

which show respectively the transformer layout and the magnetic flux lines and field intensity profiles.

What came out, was that the item most stressed was the central coil.

The total required magnetic flux through it, 3,5 V sec at peak current, had an available cross section of only 71 cm diameter. The corresponding maximum magnetic field was of 170 kG. An air core solution was consequently chosen. The current density was so high: 87 A/mm^2 that, in order to limit the energy dissipation and the power required, it was decided to cool the transformer windings at liquid nitrogen temperature to take advantage of the factor 7.3 reduction in copper resistivity with respect to the ambient temperature value. Another constraint was due to the skin effect problem, important during the current decay phase, which limited the radial conductor thickness to a maximum of 1 cm.

The main coil parameters are collected in Table I.

2.2. Type of Structure and Constructional Details

Due to the scarcity of available volume, we were bound to consider a radially self-supporting copper structure.

The coil is essentially a large solenoid made up of 18 cylindrical layers of helically wound conductors (see Fig. 5). The elementary conductor is composed of 12 specially built enameled conductors (3 x x 10 mm^2) in parallel. The material chosen is ETP hard drawn copper. Hard drawn copper has a yield stress of 27 kg/mm^2 with 0,2% permanent elongation at 20°C($\ast$). The enamel coating has shown also to perform satisfactorily at 77°K.

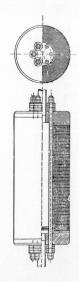

Fig. 5.   Coil.

($\ast$) Performance improves at LN$_2$ temperature.

Table I. Coil Parameters

Magnetic flux stored	3.5 V sec
Current (at 3.5 V sec)	31 kA
Power supply voltage	700 V
Maximum voltage (at commutation time)	28 kV
Maximum voltage to ground	14 kV
Magnetic field on axis	170 kG
Number of turns	540
Current density (at 31 kA)	87 A/mm^2
Resistance (at 77°K)	5.9 m
Elementary conductor	3x10 mm^2
Number of elementary conductors in parallel per turn	12
Weight	2.7 ton

The conductors move from one layer to the next one at both ends of the coil in groups of three in four positions 90° apart. Each conductor has only one brased junction near one of the ends of the solenoid and the different junctions are scattered along the circumference. The coolant flows in between layers through channels created by glass reinforced epoxy spacers.The electrical insulation is assured by 4 mylar sheets. (see Fig. 8 par. 2.3) The space factor results 0.8.

Both ends of each conductor, at the beginning and at the end of the solenoid are held into position by a thick disc of glass reinforced epoxy. The coil is therefore self supporting with regard to the radial forces.

In the axial direction, the coil is tightened by 6,42 mm dia. stainless steel bolts. The bolts transfer a force of 7 ton/ea. to the copper, through the glass reinforced epoxy discs and a series of 90° glass reinforced epoxy sectors (see Fig. 6) the retaining structure includes series of cup springs installed between the bolts and the discs.

The coil and the bolts can expand and contract independently over the entire temperature range without releasing the pressure on the copper conductors or damaging the bolts.

2.3. Mechanical and Electrical Stresses

Mechanical stresses in the winding have been evaluated assuming no cooperation between the successive layers.

The relations emploied are:

$$\sigma_r = j t B ; \quad \sigma_t = j r B ; \qquad (1)$$

$$B = \frac{a_2 - r}{a_2 - a_1} B_o ; \qquad (2)$$

$$\sigma_{t \, max} = \frac{3}{4\pi} \frac{a_2^2}{a_2^3 - a_1^3} \overline{J} \times \Phi ; \qquad (3)$$

$$u = r \mathcal{E}_t ; \qquad \mathcal{E}_t = \frac{1}{E} (\sigma_t - \nu \sigma_1) \qquad (4)$$

where:

σ_r, σ_t radial and circumferential stresses

t   radial thickness of the elementary conductor

j   current density

r   radius

a_1, a_2 inner and outer radii

B   magnetic field value on the centerline of each elementary conductor

B_o magnetic field on the coil centerline

Φ total magnetic flux

u   radial displacement

$\mathcal{E}_t$ tangential strain

E   elastic modulus

ν Poisson's ratio

All the parameters are expressed in the MKS system of units. The values of σ_t, σ_r, σ_v vertical stress, and B are indicated in Fig. 7. From the electrical point of view, the voltage applied to the coil, at the commutation time, is ±14 kV to ground maximum. A detailed drawing of the insulation is

Fig. 6. Coil terminal disc and 90° sectors.

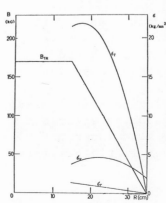

Fig. 7. Mechanical stresses and magnetic field in the coil

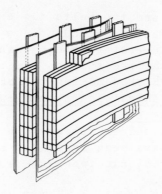

Fig. 8.   Coil insulation detail.

reproduced in Fig. 8. The maximum
voltage between two successive layers during operation, can be assumed to be 6.2 kV considering the
possibility of feeding the transformer with a capacitor bank provid
ing a very fast rise time. In this
case, a 100% voltage reflection
could occur.

Tests of the insulation have
been performed on a full scale model of two adjacent layers and on a
1 : 2 scale model of the coil. At
such a voltage, the electric field
in the gaseous nitrogen is 44 kV/cm
and the electrical rigidity is
clearly exceeded. One has therefore
to rely on the resistance of the
mylar sheets to punch through.

A sinusoidal voltage wave at
industrial frequency, of 8 kV rms
value, has been applied to the first
model for half an hour totalling
about 180.000 cycles. The absorbed
current has been kept under control
and no sign of deterioration has
been observed.

The 1 : 2 scale model of the
coil has undergone   10.000 volta
ge shots: 16 kV total, 4 kV between
adjacent layers without any sign of
incipient failure.

The insulating materials used:
enamel, glass reinforced epoxy and
mylar sheets have also shown to
behave very well at liquid nitrogen
temperature.

3. MODELS

Apart from few models of particular items to study special problems like insulation, mechanical
characteristics etc., a 1:2 scale
model of the coil has been built
and tested mechanically and electri
cally. The model underwent 21 shots
at current between 29 and 34 kA and
22 at 35 kA without any detectable
sign of deterioration. The model was
also tested electrically (see par.
2.3.)

4. COIL CONSTRUCTION

The coil construction has
been already completed by Indelve,
Padova following the Frascati design. Fig. 9 shows the coil under

Fig. 9.   The coil under construction

construction. Voltage tests have
been already performed. Satisfactory liquid N_2 cooling has also been
demonstrated. Full power tests will
be possible as soon as the power
supply will be available.

REFERENCE

1
   R. Andreani, L. Lovisetto,
   G. Vittadini: Int. Conf. on
   "Energy Storage, Compression, and
   Switching", Torino, November,
   5 - 7 1974.

COMPUTER CONTROLLED FLYWHEEL TYPE MOTOR GENERATOR FOR FUSION RESEARCH

A. Miyahara and E. Bannai[*]
Institute of Plasma Physics, Nagoya University, Nagoya, Japan
*Public Facilities Application Engineering Department,
Tokyo Shibaura Electric Company, Tokyo, Japan

ABSTRACT

A 125MVA/100MW flywheel type motor generator is prepared for JIPP-T-2 stellarator. Special character of this generator are with control algorithm to excite the load coil current in the shortest rise time and with voltage and current feedback loops to keep the load coil current constant. Small pilot plant of 100 kVA generator with approximately same circuit time constants was prepared to simulate the control system. The operation is satisfactory by this scheme.

INTRODUCTION

Fast rise of coil current to reduce the heat dissipation during initial phase and constant current regulation of toroidal coil, are important techniques for large fusion devices because time constant of coil becomes longer with increasing coil inner diameter[1,2]. To develop this technique, the application of computer controlled system to 125MVA/100MW flywheel type motor generator has been done, which was prepared for power supply of JIPP-T-2 stellarator[3]. The consideration of time constants along control loop is essential and we set up a small pilot plant of 100 kVA generator to investigate this situation previsely.

DESCRIPTIONS OF 100MVA/100MW GENERATOR

PARAMETERS AND STRUCTURES

The main generator has a structure similar to that used for turbine generators, and is completely enclosed. Two cooling devices are located on top of the stator frame of the generator and the generator is of the revolving field type. This generator was designed and manufactured by Tokyo Shibaura Electric Company[4], and has the specifications as Table 1.

The stator windings consists of two independent three-phase star-connected windings having 30° phase. This arrangement has been adopted to avoid overheating the rotor surface by higher frequency harmonic currents generated at the time of converting the generator output into DC.

Table 1. Specification of 125MVA/100MW flywheel type motor generator.

Rotor weight	24,000 kg
Stator weight	56,000 kg
Flywheel weight	50,000 kg
Rotor current	5,000 A
Rotor input voltage	1,600 V
Rotor speed	3,600 rpm
Number of phases	6
Peak voltage per phase	4,500 V
Normal voltage per phase	3,000 V
Peak power injected into rotor	11,200 kVA
Peak power delivered by motor	100 MW
GD^2 of flywheel (including generator and motor rotor GD^2)	34.5 ton-m^2
Repetition rate at full power	1 impulsion every 4 min.

The rotor material determines the mechanical strength of a generator. In order to avoid fatigue caused by repeated torque at a low frequency characteristic of this generator, a $Ni-M_O-V$ alloy was solid-forged. The speed of the rotor is expected to drop by about 20% after each power output. Therefore, the characteristic frequency of the entire rotating system is designed not to fall in the range of 100 to 80% of the natural oscillation.

In order to have most effectively a maximum possible moment of inertia to provide rotational energy to the generator, the highest quality shaft material obtainable at present (Ni-Cr-Mo-V alloy forgings) was used for the flywheel.

CONTROL SYSTEM

Since it is required to reach the full current in 1 sec, regardless of the field time constant of ∿l sec and the load time constant of ∿l sec, the field voltage is required to be 10 times the no-load rating and the terminal voltage is required to one and a half times of the rated voltage. Thus the field current is controlled by the computer, which also maintains 1 second flat-top without overshoot or undershoot.

The analog device, which measures the generator terminal voltage, field current, field voltage and coil current and converts these values into voltage signals via transducers, consists of a field voltage control device with an overvoltage and overcurrent protection device and a pulse generator which converts the output pulse of the voltage regulator into pulses. Signals from the computer, control the generator output by controlling the thyristor gate of the field through this analog device.

An optimum control computer program is stored in the computer as a control algorithm. This program computes the control signals to produce the coil current in the shortest time. Because of restrictions arising from the generator and coil structure, it is fairly complicated program. The method of optimization is achieved by Pontrijagin's maximum principle. It solves the problem by means of linear programming utilizing a feature where the result of the maximum principle for a continuous system is closely approximated by a discrete state equation with a short sampling cycle. The result is then converted into the Bung-Bung form. The generator is accurately described in terms of Park's equation, taking into consideration the damper windings and magnetic saturation. Therefore, if a desired value is merely given to the computer initially, the control signals will be computed according to this value and the coil current can be controlled accurately.

EXPERIMENT BY 100 KVA GENERATOR

100 kVA generator having characteristics of 3 phase-60Hz-220V-262A-0.8pF was employed to simulate the control system of large flywheel type motor generator.

Generator parameters of T_{do}', X_d and X_q are 1.1 sec, 0.77 PU and 0.405 PU respectively. This generator was connected through rectifier to load coil having L/R of 0.7 sec. Both generator output voltage and load coil current were picked up by individual sensor element and were fed back to CPU field excitation control system. Coil current and armature voltage waveforms were shown in Fig.1.

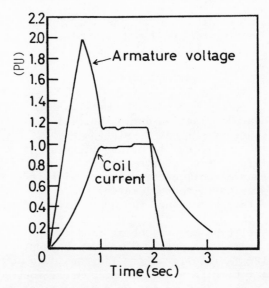

Fig.1. Coil current and armature voltage waveforms.

CONCLUSION

The main motivation to build computer controlled flywheel type generator is to control the toroidal coil current of Tokamak parallel with poloidal current for compression of plasma column. To achieve this object, further investigation of plasma current control is necessary. Besides above mentioned problems, the use of a computer is indispensable for this facility in order to respond properly to start up and stopping of the generator and to accidents. A supervisory program to monitor the condition of the entire system, a logging program and a start-and-stop program for the generator are stored in the computer. All operations, controls and checks, as well as proper procedures in case of accidents, are automatically performed, thus eliminating the need for trained personnel.

REFERENCES

1. B. Oswald: IPP 4/96 Nov. 1971
2. K. Matsuura: IPPJ-DT-36, Feb. 1973
   (in Japanese)
3. K. Miyamoto: IPPJ-205, Nov. 1974
4. A. Miyahara et al: Toshiba Review 94
   (1974) 7

Collective Effects

PULSED FUSION

J.G.Linhart

ABSTRACT

Basic conditions for pulsed thermonuclear fusion systems are examined. Simplified Lawson and trigger criteria are derived for magnetically and inertially confined plasmas. Two radically different proposals for a fusion reactor are described. In the first a plasma whose density is of the order of 10^{19} ions/cm^3 is supposed to be confined by a slow massive liner, in the second a plasma of super-solid density is generated by a liner whose speed is about 10^8 cm/sec. A remark on economy of such single-shot devices is made.

INTRODUCTION

It was appreciated very early on (late 1920's) that exothermic nuclear reactions may exist by either combining two light nuclei (fusion reactions) or by splitting a very heavy nucleus into two or more parts (fission reactions). In either case the energies available in such reactions are of the order of Mev per nucleon, which is six orders of magnitude more than that available in chemical reactions per atom. The fusion reactions require a close enough approach of two light nuclei for the nuclear, cohesive forces to become operative.
This means in all cases a partial mutual penetration of the nuclei's coulomb potential barrier. The energy W liberated in a fusion reaction can be expressed as

$$W = \delta m.c^2 \qquad (1)$$

where δm is the mass-defect of the reaction. The extreme example of a fusion reaction is the particle-antiparticle annihilation where $\delta m = 2\,m$, m being the mass of the particle.
The requirement of sufficiently close collisions for nuclear fusion between light nuclei implies high collisional energies or, in an ensemble of many interacting particles, high temperatures.

Such extreme conditions of matter, necessary for fusion, did not hold much hope for man to be able to imitate stars and liberate thermonuclear energy in usable quantities.

We all know how this feeling of an impossible task was swept away by the advent of the A-bomb which during its explosion produced conditions of matter not unlike those in the stars. The gambit of using the A-bomb as a match for a thermonuclear reaction was taken up and solved by E.Teller. The solution of the "match" problem was not a straightforward one because the fission bomb does not really generate high enough temperatures to light the fusion process in deuterium (the only fusion fuel then available) directly. This can be easily appreciated if we consider that a complete combustion of a given thermally isolated U or Pu mass puts most of the fission energy into the almost one hundred electrons per nucleus which works out to be less than 1 Mev/electron. The electrons radiate this energy, filling the available radiation modes of a black body until a thermodynamic equilibrium is reached. It is easy to show that the temperature corresponding to this equilibrium is about 10^8 K and that it is reached in a time $\tau < 1$ nsec. This is not long enough, and after τ the temperature is not high enough to ignite deuterium.

The first release of man-produced thermonuclear energy occured in 1952. Subsequently the H-bombs were perfected, made smaller, the A-bomb trigger used more efficiently and later it was suggested that these explosive devices could be used in peaceful applications, such as blasting out canals, mining and perhaps provide even propulsion for space-ships. Many of the aspects of such applications were worked out in the project Plowshare. Most of these proposals remain on paper for fear of radioactive pollution.

In order to show how difficult it is trigger a selfsustaining fusion detonation let us derive, in a simplified way, the trigger criterion for a D-T and D plasma.

THE TRIGGER-CRITERION

Let us consider first an infinite D-T medium, whose density is n_o and temperature $T \ll 10^8 °K$. Let energy W_o be deposited in a small spherical volume of the medium, heating it thus to $T \gg 10^8 °K$. The expansion of this volume generates a spherical diverging wave. The radiation emitted by the hot plasma can be considered lost to the surrounding cold medium and, therefore, the only pressure taken into consideration will be the kinetic pressure of a hot plasma. As long as the energy losses, as well as the energy gain through fusion reactions are small with respect to W the diverging wave is described well by the Taylor's self-similar solution[1] for the $\gamma = 5/3$. The speed of the shock front is then

$$\dot{r}_o = \sqrt{\frac{32}{3} \frac{k}{M} T_o} \qquad (2)$$

and the thermal energy is $W_{th} = \frac{9}{13} W_o$

The temperature T_o behind the shock-front decreases with expansion, partly owing to the expansion of the hot medium, partly owing to the shock-heating of new material.

This leads to a mean temperature-drop

$$\langle \dot{T} \rangle = -3 T_o \frac{\dot{r}_o}{r_o} \qquad (3)$$

Let us now calculate the fusion power $\dot{W}_f$ for a 50/50 D-T mixture. Let us assume that $\dot{W}_f$ is that carried by the α particles and deposited entirely within r_o. Only 9/13 of this energy is converted into W_{th} and, therefore, the mean rise of temperature due to fusion is

$$\langle \dot{T}_F \rangle = 0.7.10^{10} n_o \langle \sigma v_o \rangle \qquad (4)$$

It is clear that the $\langle \dot{T} \rangle$ is initially higher than $\langle \dot{T}_f \rangle$.* If at some r these were to be equal, the temperature T and r_o will stop decreasing and as $T_o \propto 1/r_o$ whereas T_f is independent of r_o it is evident that after the moment when

$$\langle \dot{T} \rangle + \langle \dot{T}_F \rangle = 0 \qquad (5)$$

the T_o can only start growing. This corresponds to the spreading of the fusion reaction until a new distribution, corresponding to a true detonation-wave regime is established.[2] If the condition (5) is not satisfied for any r_o, the temperature T_o will drop monotonically below the ignition temperature T_L of the D-T reaction ($T_L \sim 4.10^7 °K$). We shall, therefore, take Eq.(5) as a trigger-criterion and using Eqs.(3) and (4) we get

$$n_o r_o = 6.5.10^{-6} \frac{T_o^{3/2}}{\langle \sigma v \rangle_o} \qquad (6)$$

---

* It is not correct to compare the "mean" temperature rise and drop. However, it can be shown not to cause a large error.

The minimum of the function on the right-hand side of Eq.(6) is about 0.55×10^{28} for $T_o = 1.5.10^8 °K$. This gives

$$n_o r_o = 3.5.10^{22} \qquad (7)$$

The corresponding energy W_o is

$$W_o = \frac{384\pi}{5} k n_o T_o r_o^3 \int_0^1 \frac{n}{4n_o} \frac{T}{T_o} x^2 dx \,, \quad x = \frac{r}{r_o}.$$

The minimization w.r. to T_o should be done for W and not in Eq.(6). The resulting error is small and has been taken into account in Eq.8ab. The value of the integral is approximately 0.09. Finally

$$W_o = 700 \left(\frac{n_s}{n_o}\right)^2 \cdots (MJ) \qquad (8a)$$

$$r_o = 0.67 \frac{n_s}{n_o} \cdots (cm) \qquad (8b)$$

where $n_s = \frac{1}{2}10^{23}$ el/cm^3 corresponding to solid density.

In case that a solid density D-T plasma is precompressed by a strong shock we have $n_o = 4n_s$ and, therefore, $W_o = 43.7$ MJ and $r_o = 1.6$ mm.

We shall call this criterion the "optimistic trigger-criterion" because in its derivation we have not taken into account the following physical processes, all of which will make triggering more difficult :

1) energy loss due to heat condition ahead of shock,
2) energy deposition outside the fusion zone (i.e. in plasma whose $T < 4.10^7 °K$),
3) T_i being always inferior to T_e,
4) the departure from Taylor's self-similar distribution,
5) the undeveloped tail of the Maxwellian ion-distribution,
6) the radiation loss.

When the effect of the first two processes is included in the analysis, the values of W_o and r_o are found to be considerably higher than the optimistic ones of Eq.(8 a,b). More accurate computations using a hydrodynamic code[3] suggest as a practical compromise

$$W_o \simeq 1000 \left(\frac{n_s}{n_o}\right)^2 \cdots MJ \qquad (9a)$$

$$r_o \simeq 0.75 \frac{n_s}{n_o} \cdots cm \qquad (9b)$$

The same type of analysis can be carried out for pure deuterium giving

$$n_o r_o \sim 5.10^{23} \quad \text{at} \quad T \sim 1.5.10^9 \ (^\circ K)$$

which implies a trigger-criterion

$$W_o \simeq 3.10^6 \left(\frac{n_s}{n_o}\right)^2 \dots \ (MJ) \qquad (10a)$$

and

$$r_o \simeq 10 \frac{n_s}{n_o} \quad \dots \ (cm) \qquad (10b)$$

Ex.: for a deuterium plasma precompressed by a strong shock we get $n_o = 4n_s$ and $W_o = 40$ (tons TNT), $r_o = 2.5$ cm – which implies an A-bomb as a trigger.

In the case of deuterium it is clear that the charged products of the reaction will be mostly deposited within r_o as well as a good part of neutrons. The range of neutrons is

$$\lambda_n < \frac{10^{24}}{n} \ , \ 4n_o > n > 0$$

and, therefore, making use of Eq.10b, we have

$$\frac{\lambda_n}{r_o} < 2$$

Suggesting that about 20% of the neutrons will deposit their energy within the hot spherical region. Note that the λ_n/r_o relationship is independent of n_o and, therefore, of precompression.

Let us note in passing that the initial trigger-energy must in all cases be dissipated within the region whose dimension is r_o within a time

$$\tau < \frac{r_o}{v_s}$$

where

$$v_s = \dot{r}_o \sim 2.5.10^8 \ cm/sec.$$

This implies for the DT triggers times

$$\tau < 3 \frac{n_s}{n_o} \ (nsec)$$

and powers of

$$\dot{W} > 2.10^{17} \frac{n_s}{n_o} \ (Watts)$$

Of course, A-bombs can easily provide such powers - but in the 1950's nothing else could and consequently most scientists dreaming about CTR including the fathers of the H-bomb agreed that triggering super-dense fuels was out of question and all focused their efforts on magnetic confinement.

MAGNETIC CONFINEMENT

We are familiar with the history of the search for a satisfactory magnetic insulation of plasma. First simple Z-pinches and dipole magnetic bottles, then θ-pinches and sophisticated bottles of the non-zero B-minimum type, a development of toroidal bottles, particularly Tokomaks. All this research has been continuously frustrated and further stimulated by the discovery and suppression of more and more refined instabilities. Confined plasma is a system far from thermodynamic equilibrium, which is expressed by density and temperature gradients and by the departure of the particle velocity-distribution from a Maxwellian one.

Back in the 1940's Bohm, working then on calutrons, proposed a formula for the diffusion of plasmas across a magnetic field. The formula resembles that of Fermat's last theorem - in that its derivation is lost in the past and attempts at its rigorous justification have not been entirely satisfactory. However, in many circumstances, when the MHD instabilities have been made inocuous, plasma diffuses at speeds not inconsistent with the Bohm diffusion.

For this reason let us use the Bohm plasma loss as a yardstick and say that a particular plasma is lost at a rate of f-times smaller than that corresponding to Bohm's formula.

Let us first write down the Lawson energy criterion for magnetically confined reactors.

The simplest criterion for a thermonuclear reactor states that the fusion energy output W_f must be large enough to recharge the source of the system. This can be written as

$$\epsilon_c \, \epsilon_p (\dot{W}_F \tau + W_p) \geqslant W_p \qquad (11)$$

where : ϵ_p is the efficiency of conversion of the source-energy into the energy W_p of a DT plasma, ϵ_c that of heat in electricity.

Taking into account only bremsstrahlung loss and putting $\epsilon_c = 1/3$ we obtain a curve in the n,T diagram (Fig.1). A typical entry-point in the reactor region is then characterized by

$$n\tau = 10^{14} \ , \ T = 10^8 \ ^\circ K$$

As it is likely that $\epsilon_p \ll 1$ we have (taking always $\epsilon_c = 1/3$)

$$1/3 \, \epsilon_p \dot{W}_F \tau \geqslant W_p$$

giving for $T = 10^8$ an approximate relation

$$n\tau \geqslant \epsilon_p^{-1} . 10^{14} \qquad (12)$$

Next let us consider a cylindrical plasma of radius r_o confined by a magnetic field B within a cylindrical tube of radius R. Let us further assume no end-loss such as may be the case of a toroidal plasma. If the loss of plasma is f-times smaller than Bohm's, then the particle life-time in the tube whose radius is r is

$$\tau = 10^{-4} \, \gamma \frac{r^2 B}{T} \tag{13}$$

Using Eq.(12) we get for the line-density N

$$N = \pi r^2 n = \frac{\pi}{\gamma \epsilon_p} 10^{26} B^{-1} \quad \text{(ions/cm)} \tag{14}$$

The pressure balance can be written as

$$\beta B^2 = 16\pi n k T \tag{15}$$

and using Eqs (12), (13), (15) and taking $T = 10^8 \, °K$,
we have

$$r \simeq 10^{10} B^{-3/2} (\epsilon_p \beta \gamma)^{-1/2} \tag{16}$$

The radius of the wall (R) is determined
by the considerations of wall-dissipation
W_w. Assuming $\dot{W}_w = 500 \, W/cm^2$ and using
Eqs (13,14) and (16) we have

$$\dot{W}_w = \frac{3 N k T}{2\pi R \tau} \simeq \frac{5 \cdot 10^{10} \beta B}{\gamma R}$$

from which

$$R \simeq 10 \, B \frac{\beta}{\gamma} \tag{17}*$$

There are obviously two ways to
satisfy the reactor criterion. The first
is to improve the magnetic confinement –
which essentially means making γ as
large as possible (a consideration must
also be given to radiation) whilst choosing a practicable B, e.g. 200 KG.

The second is to accept a bad γ
but counter it by increasing B.

Let us say a few words about the
first approach. From Eqs.(16) and (17) it follows that reactors of the Tokomak
plus Stellarator type will have mange-
able dimensions only if $\gamma > 1000$
(assuming $\beta \sim 5\%$). Experimentally and
for relatively low temperatures
($T_i < 1$ keV) γ's of 30 were achieved.
If nature proves cooperative and a
further order of magnitude improvement in
γ can be achieved at Ti > 10 (keV) then
this type of reactor will become feasible –
but not necessarily economic. Such γ's may
perhaps be achieved using some feedback mechanisms to minimize microinstability.

We shall now discuss the approach
to a fusion reactor by using strong magnetic fields.

THE HIGH MAGNETIC FIELD CONFINEMENT

One of the benefits of relatively
short confinement times is the feasibility
to produce and to confine high β plasmas.
Confinement times of miliseconds rather than
seconds also permit us the use of high pulsed

---

* As $R > r$ it follows that $B > 4 \left(\gamma/\epsilon_p \beta^3 \right)^{1/5} \cdots (kG)$
For $\gamma = 1000$, $\epsilon_p = 0.1$ and $\beta^3 = 10^{-4}$ we get
$B_{min} = 160$ KG.

magnetic fields – perhaps as high as 300 (KG).
Also the problems of heat dissipation at the
wall become rather different.
Putting B = 360 (KG), $\epsilon_p = 0.1$, $\beta = 1$ into
Eq.(16) we get

$$\gamma > \frac{2 \cdot 10^4}{r^2} \tag{18}$$

if $r = \frac{1}{2}$ m is admitted as technologically
feasible then $\gamma > 8$. If R is not much
larger than r the wall-dissipation will
be, of course, much higher than 500 watts
per cm^2. It has been suggested that in
these devices one would use wall-sweating
as an essential part of the cooling process.

In order to decrease the reactor
dimension still further one would have to
use magnetic fields in excess of $\frac{1}{2}$ M Gauss.
which implies the destruction or at least
a distortion of the coil (magnet) system.
In such a case we may exploit the ideas
on magnetic flux-compression by a converging liner which is formed and discarded
each time a hot plasma is formed, compressed and dissipated. Experimentally liner –
compressed magnetic fields up to 10^7 Gauss
were produced. Velikhov in USSR and
Robson in USA envisage to use this approach
at a level of about 4 M Gauss. At this level
the maximum plasma density is about 10^{19}
ions/cm^3 and the inertial confinement time
τ_i by the liner is shorter than the Bohm
diffusion-time. Let us describe a system
of this type[5].

A massive theta-pinch coil encloses a thin metallic liner which in
turn encloses a warm plasma. A buffer
magnetic field separates the plasma from
the liner, as in a conventional theta-
pinch. The outer coil is energized from
an external source, producing a field of
$100 \div 200$ kilogauss on the outside of
the liner. The coil is strong enough to
withstand the field but the liner collapses radially inwards, gaining kinetic
energy from the driving field and reaching speeds of 10^5 cm sec^{-1}. At the point
of maximum collapse the energy is transferred to thermal energy of the plasma
and potential energy of the buffer field.
If the system is long enough that end
losses may be neglected, the confinement
time of the compressed plasma is determined by the inertia of the surrounding
liner.

One-dimensional calculations show
that $n\tau_i \propto B_o r_o \rho^{1/2}$ where B_o is the peak
magnetic field, r_o the minimum plasma
radius and ρ is the density of the
liner-material. With a heavy liner such
as lead, break-even is obtained with
$B_o = 4$ M Gauss and $r_o \sim \frac{1}{2}$ cm. However, in order

to avoid MHD instabilities in a theta-pinch the field-lines must be kept straight which means that plasma can flow freely out of the ends. End-losses must be included in the energy balance, and it can be shown that to minimize the total energy invested in the system, the end-loss time should be 3 times the inertial confinement time. The necessary length of the system is $\propto B_o^{-2}$ and with $B_o = 4$ M Gauss a theta-pinch feasibility experiment needs to be 30 meters long.

There is no fundamental reason why a cylindrical liner cannot be driven to a speed of 10^6 cm/sec, in which case field of about $20 \div 30$ M Gauss could be achieved. At this field-level the length of an θ-pinch system will be only about $\frac{1}{2}$ meter and reacting assemblies at an energy level of 100 MJ can be envisaged.

An interesting variation of this scheme consists in using $B \ll \sqrt{16\pi n k T}$, i.e. operating at $\beta \gg 1$. In such a case, the magnetic field reduces the heat-flow to the liner, the pressure of the hot plasma being transmitted to the cold wall by particle collisions. The magnetic field provides, therefore, an energy confinement, the plasma pressure is uniform and is born directly by the liner-wall. Computations show[6] that the radiation loss and the heat losses can be tolerated provided the liner-speed is in excess of $2 \cdot 5$ cm/μ sec. Even if one were able to generate much higher fields than, let us say, $B = 100$ MG, it is doubtful whether one should strive for a purely magnetic confinement. The plasma densities in this case start to approach the solid state densities and one may, once again, consider the purely inertial confinement related to the trigger-criterion. This has been suggested by several groups, partly because since the 1950's there has been a development of such very high-power devices as lasers and electron beams and, therefore, the high power requirements implicit in the trigger criterion is no longer an unsurmountable barrier.

MICROEXPLOSIONS

The release of fusion energy in a plasma whose density is of the order of solid density or higher has the character of a micro-explosion. If such explosions did not cause damage or expenditure of costly material, then generation of fusion power by such a method would become possible provided the reactor criterion (from Eq.21, taking $\tau = \frac{r}{v}$, $v = 1 \cdot 5 \cdot 10^8$)

$$n \, \mathscr{X} \, r \geq 1 \cdot 5 \cdot 10^{22} \, \epsilon_p^{-1} \quad (19)$$

would be satisfied, where n is the plasma density, $\mathscr{X}$ the inertial confinement factor (tamping factor)*, r the plasma dimension, and ϵ_p the efficiency of transformation of stored energy W_s into plasma energy W_p. We have assumed here that the efficiency ϵ_c of conversion of the thermal energy into electrical energy is 30% and the plasma temperature $T = 10^8$ °K.

The minimum plasma energy is, therefore, given by

$$W_p \sim \frac{7}{(\epsilon_p \mathscr{X})^3} \left(\frac{n_s}{n}\right)^2 \cdots \begin{array}{l}\text{spherical} \\ \text{geometry}\end{array} \; (MJ) \cdots (20a)$$

$$W_p \sim \frac{28}{(\epsilon_p \mathscr{X})^2} \frac{n_s}{n} \cdots (MJ/_{cm}) \begin{array}{l}\text{cylindrical} \\ \text{geometry}\end{array} \quad (20b)$$

within a plasma whose radius is

$$r \simeq \frac{0 \cdot 2}{\mathscr{X} \epsilon_p} \frac{n_s}{n} \quad (21)$$

for both geometries (n_s is the solid state density). The value of the stored energy W_s is, by definition, $W_s = \epsilon_p^{-1} W_p$ and in the case of spherical geometry it follows from Eq.(20) that

$$W_s \propto \epsilon_p^{-4} \quad (22)$$

It is suspected that in most cases ϵ_p will be small, probably of the order of 0.1, which implies a very large value of W_s.

In the case of such small ϵ_p it would seem advantageous to attempt to start a fusion detonation in a volume of fusion fuel so small as to still permit spreading of the fusion explosion.

This leads to the concept of a trigger where W_p is supplied with the intention of producing a detonation in a reservoir of a DT fuel close to the trigger, where it is hoped that the price of the fusion energy W_f released will, above some still not too large W_f, compensate for the cost of the damage per shot. Provided such a W_f exists and $W_f \gg W_p$, consideration of ϵ_p is superfluous and one has just to be able to provide the trigger energy W_p, which is given by the extension of Eq.(15a)

$$W_p \sim \frac{1000}{\mathscr{X}^3} \left(\frac{n_s}{n}\right)^2 \cdots (MJ) \begin{array}{l}\text{spherical} \\ \text{geometry}\end{array}$$

$$(23ab)$$

$$W_p \sim \frac{700}{\mathscr{X}^2} \frac{n_s}{n} \cdots (MJ/_{cm}) \begin{array}{l}\text{cylindrical} \\ \text{geometry}\end{array}$$

within a radius $r = 0 \cdot 75 \, \mathscr{X}^{-1} n_s/n$ (cm) for both geometries.

---

* Defined as the ratio of plasma life-time to a free-expansion time of the same plasma.

Which of the two schemes is more favourable from the point of view of smaller W_s, keeping always the total energy of explosion W_f constant, follows from Eqs. (20) and (23).

If the trigger scheme were to be preferred then $W_{st} < W_{sr}$ where the subscripts t and r, respectively, denote trigger and reactor. For the spherical case, we obtain $1000 < \dfrac{7}{\epsilon_p^3}$ or

$$\epsilon_p < 0.19 \tag{24}$$

For $\epsilon_p < 0.19$, the trigger-scheme will become more feasible and the required source energy will be

$$W_{st} = \left(\frac{\epsilon_p}{0.19}\right)^3 W_{sr} \tag{25}$$

Thus, e.g. in case of $\epsilon_p = 0.05$, which could well be a realistic estimate, W_{st} will be two orders of magnitude lower than W_{sr}. It would seem that, unless other problems in the trigger scheme arise, the trigger approach will allow a much smaller W_s to be used.

However, the values of W_p are very large for both triggers and reactors and the only obvious way of making them low is by making

a) $\dfrac{n}{n_s} \gg 1$ i.e. by pre-compressing the fusion fuel

b) $\varkappa \gg 1$ i.e. by confining the plasma inertially (tamping)

Calculations have been made[7] on the compression of dense D-T targets up to $n/n_s \sim 10^4$ by a powerful spherical laser beam. The extremely high pressures required to bring about such a compression ($p > 10^{10}$ atm.) are supposed to be generated not by the radiation pressure of the laser, but by the reaction $\rho_j v_j^2$ of the jet produced by the absorption of laser energy near the surface of the target.

The generation of high pressures and, consequently, of $n/n_s \gg 1$ can be also achieved during the impact of a hypervelocity projectile (liner) on a D-T target. This pressure is of the order of ρv^2 where ρ is the density of v and velocity of the projectile before the impact[8].

If one wished to produce a spherically symmetrical impact by a spherical liner on a spherical D-T target it can be shown that the magnitude of radiation from the liner and the radial heat loss from the D-T core precludes the use of any other liner-material except deuterium or, perhaps, lithium, although if the liner completely explodes before the maximum compression is reached,

much denser liner-materials may be used.

If one is prepared to use cylindrical liners, it is possible to use magnetic fields in order to cut down the radial heat-loss and, thus, one should be able to use heavy liners and achieve inertial confinement with $\varkappa > 1$. On the other hand, the liner must be, at least, $\varkappa r$ long in order that the end loss does not invalidate the utility of the inertial confinement factor.

If one wishes to compress the D-T target to $n/n_s \gg 1$ it is necessary to avoid excessive shock formation during the liner-impact. This can be achieved in a "soft" impact. Whereas such a soft impact may, perhaps, be achieved in the laser-project by tailoring the output of the laser (i.e. providing an initially soft piston), in the D-T compression by liners the same effect can be achieved by having a suitably soft target. This is particularly feasible if a combination of a magnetic field and D-T medium is used as a target[9] or if a liner shell is compressing a D-T shell[10].

There is another possibility of achieving $n > n_s$ and this is by means of a super-pinch. A super-pinch is, to a certain extent, an analogy of a neutron star in which the gravitational pressure overcame all material pressures until most of the star-material is condensed to a nuclear density. In a super-pinch the magnetic field of a Z-pinch type must be able to collapse the plasma to densities $n > n_s$. This is possible only if the plasma pressure $p = 2nkT$ during the radial compression rises slower than the magnetic pressure i.e.

$$p < p_m = \frac{B_\varphi^2}{8\pi} = \frac{I^2}{2\pi r^2}$$

In case of a cylindrical adiabatic compression we have

$$p = p_1 \left(\frac{r_1}{r}\right)^{10/3}$$

where index 1 corresponds to an initial situation. It is clear that p grows faster than p_m as r decreases. Consequently in order to obtain high final densities one must start with a $p_1 \ll p_{1m}$. Such a situation would, however, lead to a fast radial implosion in which shocks would be generated and p would rise faster than in a purely adiabatic compression. In order to avoid the shock-formation one must insert an additional radial force p_r such that $p + p_r = p_m$ at all times and remove the p_r as the collapse proceeds. This can be achieved in the axial-flow pinches in which case p_r is the mechanical resistance of an internal, conical electrode.

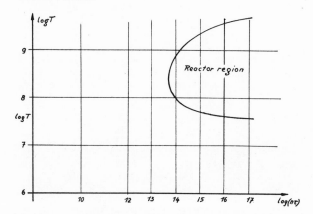

Fig. 1   Simple Lawson's criterion for 50/50 DT and $\varepsilon_c = \frac{1}{3}$

The initial suggestion of the axial-flow pinch has been made by Morozov[11] who intended to generate plasma densities of the order of $10^{18} - 10^{19}$. Experiments in which initial plasma densities $n_i \sim \frac{1}{2} 10^{16}$ are compressed to $n_{max} \sim \frac{1}{2} 10^{18}$ have been described by Morozov et al. recently[12].

It is clearly important to extend these experiments to initial densities in the range $10^{16} - 10^{17}$ so that final densities of 10^{19} can be attained. Morozov's experiments were plagued by the presence of impurities – a problem which should become less troublesome as the plasma density increases. Consequently the experiments planned at Camen represent an important step towards the realisation of a superpinch at even higher densities, eventually possibly at $n_{max} > n_s$.

ECONOMICS

The proponents of quasi-stationary reactors have delt in great detail with the question of competitive chances of these machines when compared to fossil-fuel or fission type power-stations. Similar analyses have been performed for high-β, non-selfdestructive reactors. Let us mention here briefly a general criterion for the economy of single-shot devices. It can be written, assuming realistic values for KW hour-cost, as

$$\frac{C}{Y} < 2 \qquad (26)$$

where Y is the energy-output per shot in tons of TNT and C is the cost of making a shot in US dollars[11]. Taking some of the laser-proposals in which the trigger energy $W_o = 100$ KJ and $\varepsilon_p \sim 1\%$ we must have $Y \gtrsim 10$(MJ) i.e. 2 kg TNT. Then

$$C < 0·4 \text{ (cents)}$$

Of course not all the fusion reactors have to compete with power-stations. Teller suggested recently that if a fusion device could be used in space transport, no ordinary economic rules would hinder its development.

A particularly attractive application of the microexplosions would be to a rocket-drive for large, long-range spaceships. Such an energy-source would be able to provide jet-speeds an order of magnitude higher than those furnished by chemical rocket-fuels, improving thus drastically the pay-load to initial load ratio. In this way the old dream of the mastodontic "Orion" project could become true.

REFERENCES

1. G.I. Taylor, Proc.Roy.Soc., A 201, p.159 and p.175 (1950).

2. R. A. Gross, Proc.Varenna School on High Energy-Densities (July 1969).

3. J . P. Somon, Nucl.Fusion, 12, p.461 (1971).

4. J.G.Linhart, Plasma Physics, p.303, (EURATOM 1969)

5. A.E.Robson, Proc.Confer.Elst.and Elmag. Confinement, (New York, March 1974).

6. C.Rioux and C Jablon, to be published in Nuclear Fusion.

7. J. Nuckolls et al., Nature 239, p.139 (1972).

8. J.G.Linhart, Nuclear Fusion, 13, p.321 (1973).

9. H.Knoepfel in "Physics of solids in intense mag.fields", p.467(Plenum Press 1969).

10. K.Brueckner, Proc.Erice School on pulsed fusion, (Sept.1974).

11. A.I. Morozov, Sov.Tech.Phys., 12, p. 1580 (1968).

12. A.K. Vinogradova et al., Sov.Tech. Phys., 18, p.1604 (1974).

13. V. Gilinsky and H.Hubbard, Nucl.Fusion, 8 p.69 (1968).

LASER FUSION: CAPITAL COST OF INERTIAL CONFINEMENT

Ray E. Kidder

Lawrence Livermore Laboratory, University of California
Livermore, California 94550

ABSTRACT

In the context of laser-induced fusion of solid pellets, a quadratic relation between peak laser power and the inertial confinement parameter ρR is derived and discussed. This relation is combined with the linear relation between laser system cost and peak output power to obtain an estimate of the capital cost of inertial confinement.

INTRODUCTION

In the context of laser-driven fusion, a laser pellet-compressor needs to accomplish two tasks: to bring the thermonuclear fuel to its ignition temperature T_b; and to provide sufficient inertial confinement, measured by the ρR-product achieved in the fuel, so that substantial fuel burn-up takes place. (ρ and R denote the density and radius of the fuel at peak compression.) Provided the maximum achievable efficiency of fuel burn-up $\varepsilon_{max} \ll 1$, both ε_{max} and the Lawson confinement parameter $n\tau$ are directly proportional to ρR according to the relations

$$n\tau = \rho R/4M_i c_s, \qquad (1)$$

$$\varepsilon_{max} \le \rho R/\{(\rho R)^* + \rho R\}, \qquad (2)$$

$$(\rho R)^* = \{8M_i c_s/<\sigma_{DT}v>\}_{min}$$

$$= 7 \text{ g/cm}^2, \qquad (3)$$

where M_i is the average ion mass, $<\sigma_{DT}v>$ is the Maxwell-averaged DT reaction cross section, and c_s is the isothermal speed of sound in the fuel at peak compression.

The laser pulse energy W_L required to raise a mass M $(= 4\pi\rho R^3/3)$ of thermonuclear fuel (DT) to its ignition temperature T_b can be written

$$W_L \text{(joules)} \simeq 10^9 (\rho R)^3 T_b/\varepsilon_H \eta^2, \qquad (4)$$

where ε_H is the efficiency of conversion of laser pulse energy into heat in the fuel, η is the ratio of the fuel density ρ to its normal solid density ρ_o $(= 0.2 \text{ g/cm}^3)$, T_b is in kilovolts, and ρR is in units of grams per square centimeter. Taking T_b to be 10 keV and ε_H to be 0.03 as typical values, Eq. (4) becomes

$$W_L \simeq 4 \times 10^{11} (\rho R)^3/\eta^2. \qquad (5)$$

We note that the laser pulse energy required to achieve ignition temperature at a specified value of ρR decreases as the square of the fuel compression η, this being the reason why extremely high fuel compression is required. If for example ρR is to exceed 0.3 g/cm^2, the minimum value for which spherically divergent thermonuclear propagation can occur, then η must exceed 10^3 if the laser pulse energy is not to exceed 10 kilojoules.

We shall see that a more direct relation exists between the laser pulse power P_L and the inertial confinement parameter ρR, at least in the case of isentropic compression, than exists between the laser pulse energy W_L and ρR as expressed by Eq. (4). This result is particularly useful, because the size and cost of a laser pellet-compressor is also more directly related to its peak power capability than to its output pulse energy.

RELATION BETWEEN CAPITAL COST AND OUTPUT POWER OF LASER-COMPRESSION SYSTEMS

The cost C of a large multibeam laser pellet-compression system is proportional to its rated peak optical output power P_L, because both the total cost and output power are proportional to the number of beams of given aperture, i.e.,

$$C(M\$) = \alpha P_L \text{(terawatts)}. \qquad (6)$$

The maximum useful power achievable in each beam is limited by nonlinear wavefront distortion accumulated in the laser medium and other optical elements in the beam, and is presently limited to less than 3×10^9 W/cm^2 of beam aperture. The maximum beam aperture is limited by superfluorescence and the disproportionate cost of large-aperture optics.

The coefficient α is currently estimated to be 0.7 M\$/TW, based on the \$17 M cost (excluding building) of the 25 TW neodymium-glass Laser Facility SHIVA now under construction at the Lawrence Livermore Laboratory. It is interesting to note that 25 TW is also the approximate output power of the most powerful of present pulsed electron beam machines, the AURORA Facility at the Harry Diamond Laboratory, White Oak, Maryland.

RELATION BETWEEN ρR AND PEAK OPTICAL POWER

It has been shown[1] that the mechanical power P_M required to homogeneously and isentropically compress an ideal ($\gamma = 5/3$) gas is proportional to the square of the ρR-product achieved by the compression. This result can be obtained very simply by making use of the property of such compressions that the internal energy W_i of the gas being compressed,

$$W_i = 2\pi\rho R^3 c_s^2, \qquad (7)$$

doubles in a time τ proportional to the sound transit time

$$\tau \propto R/c_s. \qquad (8)$$

The mechanical power supplied is then given by the proportionalities

$$P_M \propto W_i/\tau \propto \rho R^2 c_s^3 = (c_s^3/\rho)(\rho R)^2. \qquad (9)$$

However, the factor (c_s^3/ρ) appearing on the right of Eq. (9) is constant along an isentrope, so that

$$P_M \propto (\rho R)^2. \qquad (10)$$

If, in addition, we assume that the efficiency with which laser energy can be converted into compressive work does not depend significantly on the ρR-product achieved, we may write the proportionality above in terms of laser power P_L as

$$P_L \propto (\rho R)^2. \qquad (11)$$

This result has indeed been derived from more detailed considerations elsewhere, based on a model of self-regulating pellet ablation by hot electrons of the pellet corona, and is supported by detailed computer calculations of the isentropic compression of solid, spherical fuel pellets by the absorption of laser light. These computer calculations also provide the value of the coefficient of proportionality

$$P_L (\text{terawatts}) = \beta\{\rho R(g/cm^2)\}^2 \qquad (12)$$

$$\beta = 300 \text{ TW}/(g/cm^2)^2. \qquad (13)$$

It must be stressed that the simple relation expressed by Eq. (12) is based on assuming the pellet corona to be a collision-dominated, quiescent plasma, an assumption that may well be false unless the wavelength of the laser light is quite short. (A wavelength of 0.265 μm was employed in the computer calculations of Ref. 1.) An alternative model treated by Rudakov[2] considers the corona to be a collisionless, turbulent plasma. This latter model may be more realistic for pellet compression with longer wavelength laser radiation, and according to Rudakov provides less efficient coupling between the laser beam and the compressed pellet core.

CAPITAL COST OF INERTIAL CONFINEMENT

Combining the results of Eq. (6) and Eq. (12), we arrive at the following estimate for the capital cost of a laser pellet-compression system having a given ρR rating:

$$C(M\$) = \alpha\beta\{\rho R(g/cm^2)\}^2, \qquad (14)$$

$$\alpha\beta = (0.7)(300) \sim 200 \text{ M\$}(g/cm^2)^2. \qquad (15)$$

Values of estimated capital cost, together with optical power and fuel burn-up efficiency, are listed in Table 1 for selected values of ρR. A ρR-product of 3 g/cm^2 is thought to be required in the application of laser fusion to the production of electric power[3].

SUMMARY AND CONCLUSIONS

A significant property of the results listed in Table 1 is the quadratic increase of system cost with inertial confinement to be achieved. This property is based on a theoretical relation between peak laser power P_L and inertial confinement ρR that applies to the special case of nonturbulent, homogeneous, isentropic pellet compression, and which is as yet untested by experiment.

It is expected that inertial confinement will be largely determined by peak laser power, though perhaps not in accordance with the simple relation we have presented. We believe that an experimental investigation of this important relationship should be undertaken.

Table 1. Capital cost, optical power, and fuel burn-up versus compressor ρR-rating.

Compressor ρR-rating (g/cm^2)	Fuel burn-up (ε) (%)	Optical power (P) (TW)	Capital cost (C) (M\$)
0.3	4	30	20
1.0	13	300	200
3.0	30	3000	2000

REFERENCES

*This work was performed under the auspices of the U.S. Environmental Research and Development Administration.

1. R.E. Kidder, Nucl. Fusion 14, 797 (1974).

2. L.I. Rudakov, JETP Lett. 19, 376 (1974).

3. J.H. Nuckolls, Laser Interaction and Related Plasma Phenomena, Schwarz and Hora, Eds. (Plenum Press, New York, 1974), p. 399.

ACCELERATION OF ELECTRONS BY AN ELECTRODYNAMIC SPACE-CHARGE EFFECT

Walter R. Raudorf

Physics Department, Sir George Williams University
1445 de Maisonneuve Blvd. West
Montreal, Quebec, Canada H3G 1M8

ABSTRACT

The acceleration method is based on the transfer of the energy of a stream of electrons to the electrons at the front of the stream by virtue of a collective space-charge effect. The apparatus used to demonstrate this effect comprises a cathode at one end of a 214 cm long drift tube, a tubular decelerating electrode at the other end, and solenoidal coils around the grounded drift tube. Under the influence of the magnetic field, produced by these coils, the electrons drawn from the cathode form a cylindrical beam advancing in a screwlike manner towards the electrically insulated decelerator which is negative with respect to the cathode. The increasing magnetic flux and the retarding electric field there stop the arriving electrons. The running on of the succeeding electrons causes the charge density at the front of the beam to grow and the potential to drop until a virtual cathode is formed whose potential is far below that of the decelerator. Electrons ahead of the virtual cathode are accelerated into and through the decelerator. The 2.5 m long instrument, operated with a 20 kV, 350 mA beam, accelerated electrons to energies up to 14 MeV. The energy transfer efficiency was about 1.7 %.

INTRODUCTION

Heese and Raudorf[1] described the construction and operation of a low cost linear accelerator for electrons referred to as "electronic ram" (ER). The acceleration is accomplished by an electrodynamic space-charge effect, referred to as "ram effect" (RE).

In this article design and performance of the ER will be discussed, and the mechanism of the acceleration process in terms of an equivalent circuit explained. Results of theory and experiment will be presented and possible improvements of the machine suggested.

DESIGN

Referring to Fig. 1, the present ER is axially symmetric and consists of the following components: an electron gun at one end of a 214 cm long stainless steel tube (drift tube) of 1.6 cm i.d., and a 6 cm long tubular electrode (decelerator) of the same inner diameter at the other end. The decelerator is electrically insulated from the 30 cm long stainless steel envelope of 10 cm i.d. The drift tube, the housing of the

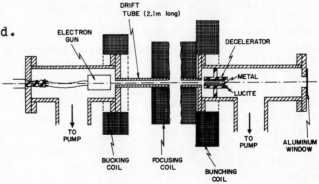

Fig. 1. Schematic diagram of the electronic ram.

electron gun as well as the envelope of the decelerator are grounded during operation.

The dominant feature of the
design is the employment of a static axially symmetric magnetic
field. This field, generated by
solenoids around the electron gun
(bucking coil),the drift tube (focusing coil) and the decelerator
(bunching coil), is uniform within
the drift tube, vanishes in the
vicinity of the cathode, reaches
its highest flux density in the
region of the decelerator, and
falls off toward the aluminum disk
(exit window) which terminates the
system.

The electron gun is of a modified Pierce design, using a directly heated tungsten filament. The
bucking coil around the housing
provides magnetic beam focusing if
necessary. The gun is capable of
projecting a 350 mA cylindrical
beam of about 1 cm diameter at a
cathode voltage of 20 kV relative
to ground. To avoid overheating
the cathode was not continuously
operated but square pulsed at a repetition rate of 100 Hz. The focusing coils around the drift tube
are capable of producing flux densities up to 2000 G. The bunching
coil serves to raise the flux density sharply in front of the decelerator to 2500 G. Independent
power supplies energize the various
solenoids.

The decelerator is firmly fitted into a lucite tubing which is
rigidly aligned with the drift section by a three-legged "spider" of
polystyrene. The spacing between
drift tube and decelerator is .5 cm.
During operation stray electrons
charge up the decelerator negatively w.r.t. ground. The ER is terminated by a stainless steel plate
with an aperture of 4 cm diameter.
This aperture is hermetically sealed with an aluminum disk (173 mg/cm^2)
so that electrons of energies above
524 keV can escape. Standard flan-
ges and O-ring seals are used in
joining the various sections vacuum tightly together.

OPERATION

Under the influence of the
applied magnetic field the electrons projected by the gun form,
as they enter the drift tube, a
cylindrical beam which rotates about its axis as it advances[2] .

The angular velocity of the
electrons in the beam is proportional to the flux density. As
this is uniform within the drift
tube, the angular velocity is the
same for all the electrons.

The space-charge repulsive
forces are just balanced by magnetic focusing forces so that the
cross section of the beam remains
unchanged. The space-charge density within the beam is proportional to the square of the flux density and hence uniform. The axial
velocity of the electrons is also
uniform and the same for all electrons. In sum, the electrons in
the drift tube form a cylindrical
beam which rotates about its axis
as a unit and advances much in the
manner as a screw.

As the beam travels towards
the decelerator, electronic charge
is displaced from the wall of the
drift tube to ground at the same
rate as electronic charge flows
into the drift space. In other
words, the convective beam current
is continued in the form of displacement current in accordance
with the principle of continuity.
At the end of the drift space the
beam encounters a retarding electric field due to the negatively
charged decelerator and a rapidly increasing magnetic flux density. The retarding electric field
reduces the axial speed of the
arriving electrons, the increasing magnetic flux density, generated by the bunching coil, brings
about complete conversion of axial
to angular velocity of the foremost electrons. The Lorentz force
on the other hand causes a compression of the space-charge in
that portion of the beam which re-
volves without advancing, and consequently a potential drop there.

A potential barrier develops against which the succeeding electrons run. Both space-charge density and potential barrier grow until the beam is stopped. At this instant the potential within the front section of the beam attains a minimum. The cross sectional area in which the minimum occurs takes on the role of a virtual cathode whose potential is far below that of the decelerator.

The electric field beyond the virtual cathode, generated at the expense of the energy associated with the entire beam draws electrons from the virtual cathode and accelerates them into and through the decelerator.

The induction lines emerging from the inner surface of the envelope around the decelerator section terminate at the front of the beam and do not penetrate because of the high space-charge density. This screening effect of the outer electrons is in fact the cause of the development of the virtual cathode.

The electrons on which the induction lines end, constitute the charge beyond the virtual cathode in the region of the positive potential gradient. These electrons are drawn out and become accelerated at the expense of the electric field energy . The electrons passing through the decelerator traverse a region of decreasing flux density. Consequently they gain axial velocity at the expense of angular velocity. Electrons of sufficient energy pass through the exit window. Simultaneously with the emission of electrons into the decelerator the virtual cathode begins to recede within the beam in direction of decreasing magnetic flux density until it disperses radially. The stopped beam starts moving again and the whole process is repeated.

As pointed out in the first paper on the subject[3], the drift section assumes the role of an inductance, the decelerator section that of a capacitance. The equivalent circuit is therefore very simple, consisting of an inductor in series with a capacitor across which an ideal switch periodically opens and closes. The period during which the switch is closed is equal to the time during which the beam advances in the drift tube. The stopping of the beam corresponds to the opening of the switch, and the subsequent development of the virtual cathode to the charging up of the capacitor.

The beam area in which the virtual cathode arises is equivalent to the capacitor plate which is being charged up, the grounded envelope around the virtual cathode to the other plate. Incidently the larger diameter of the envelope there assists in setting up the virtual cathode[4] .

The field between virtual cathode and envelope corresponds to the field between the plates, and the discharge of the virtual cathode to field emission between the plates of the capacitor.

For given operating conditions and geometry of the ER, inductance and capacitance of the equivalent circuit can be computed, and hence the energy level $\hat{V}$ to which the charge of the capacitor will be raised.

MEASUREMENTS AND RESULTS

All measurements and tests were carried out under the following operating conditions. The amplitude of the square pulses generated by the pulsing system was 20 kV, the pulse width 10 μs, and the pulse repetition rate 100 Hz; the axial flux density within the drift tube close to 1340 G and in front of the decelerator 2500 G. The amplitude of the beam current was 350 mA, and the pressure inside the envelope .2 μTorr.

To check on the occurence of the RE a hollow cylindrical collector electrode was mounted coaxially with respect to the beam on the terminating plate of the ER, replacing temporarily the exit window.

A typical oscillogram of the output is shown in Fig. 2. Unfortunately the speed of the sweep was too fast for taking single sweep photographs; the exposure time had actually to be a multiple of the sweep period. The traces in the oscillogram, therefore, correspond to several successive pulse trains.

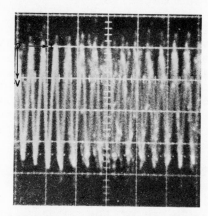

FIG. 2. Output pulses across 15 Ω. Time scale 0.02 μsec/cm, sensitivity 0.5 V/cm (1 cm=1 large div).

The rise time of each pulse which is equal to the bunching period, is about 5 ns.

As the potential within the bunched portion of the beam drops below that of the decelerator, the front electrons start to move towards the collector. To account in the equivalent circuit for the collector current during the bunching period, the capacitor has to be shunted by a "bleeder" resistor.

At the end of the bunching period the electric field due to the non-uniform charge distribution within the beam will accelerate electrons backwards causing the charge compression to move in direction of decreasing flux density. Finally when the interelectronic repulsive forces are no longer balanced by the magnetic focusing forces, the charge compression dissolves radially. With it vanishes the potential barrier that prevented the beam from advancing.

To obtain information about the current pulses outside the ER a series of measurements was taken using the following experimental set-up. Three collimating lead disks were mounted in front of the exit window. The disk closest to the window had a circular aperture of 1.9 cm diameter, coaxial w.r.t. the beam; the other two 3 and 6 cm, respectively, away from the first one had concentric holes of only .6 cm diameter. In front of them, at an axial distance of about 5 cm was an electromagnet in such a position that the collimated beam had to pass through a 6.35 mm air gap between the circular pole pieces of 6.5 cm diameter.

At a distance of about 60 cm from the air gap in the direction of the beam, was a Si(Li) detector with 2 mm of silicon and an active area of 50 mm^2 perpendicular to the beam. The detector was biased with 500 V and connected to a charge sensitive pre-amplifier of gain 1, the output of which was fed into the oscilloscope.

The number of electrons impinging per second on the detector was estimated to be at least 10^9. In view of this the detecting system could not be expected to resolve the output of the ER into distinct pulses produced by individual electrons. In fact it performed the operation of integration of the signals as shown in Fig. 3. The oscillogram was taken after insertion of an aluminum absorber (1564 mg/cm^2) between detector and collimators. Taking the stopping power of the window and path in air into account, electrons of energies above 3.8 MeV could reach the Si(Li) detector. The time scale is 10 μs/div, with the time increasing to the right, and the vertical scale is 50 mV/div.

The contribution of X-rays to the detector output was checked by energizing the electromagnet to deflect the electron beam. As anticipated it proved to be relatively small.

The oscillogram shows qualitatively the growth of electronic charge as function of time. Of

physical significance is the charging up process which appears to
be short relative to the decay of
charge. Particularly interesting
is the rate of growth of charge
as this is equal to the mean value
of the current into the detector.

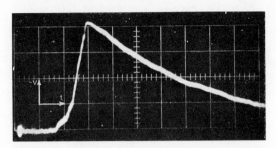

FIG. 3. Si(Li)-Detector output. Energies of incident electrons
above 3.8 MeV. Time scale 10 μsec/cm, sensitivity 50 mV/cm
(1 cm = 1 large div), single sweep.

ENERGY MEASUREMENTS

For determining the ener-
gies of the electrons in the extracted beam two fundamentally different methods were used.

The first one was based on
magnetic deflection. The electron
beam emerging from the exit window was collimated by three lead
diaphragms and passed through the
air gap of an analyzing electromagnet which produced in the air
gap a homogeneous horizontal magnetic field, transverse to the
motion of the electrons. The detector was mounted such that the
electrons could enter the detector
window only if their original direction of travel was rotated
through a certain angle by the
transverse magnetic field in the
air gap. The knowledge of this
angle, the flux density, and geometry allowed one to compute the
energies of the electrons striking
the detector.

The second method made use
of the penetrating power of the
electrons. The experimental arrangement was basically the same
as for the deflection method, except that the detector was aligned
with the axis of the system. Aluminum disks, calibrated in mg/cm²,
were placed between window and
lead collimators. The minimum

thickness of absorber, just preventing the electrons from reaching the detector, was then determined. The analyzing magnetic
field was briefly applied during
each measurement to check on the
background of the output.

The results of the energy
measurements obtained by either
method agreed surprisingly well.
under operating conditions as
stated above, the energies of the
fastest electrons in the beam
were found to be 14 MeV ± 10 %.

ENERGY DISTRIBUTION

Initially the familiar method
of pulse height analysis was employed to investigate the energy spectrum of the accelerated electrons.
When it was found, however, that
the time constant of the detecting
circuit exceeded by far the risetime of the incoming pulses, this
convenient method had to be discarded. Instead a simpler but much
less instructive procedure was used.

The experimental arrangement
was the same as that for testing
the maximum range of the accelerated electrons in aluminum. With
the ER in operation the range was
gradually increased from zero to
5400 mg/cm² by inserting more and
more absorbers between exit window
and collimators until the scope
trace could no longer be analyzed.
The type of oscillograms obtained
for the various ranges is shown
in Fig. 3. The increase of voltage
with time displayed on the scope is
proportional to the growth of the
incident charge. The slope of the
voltage rise is therefore proportional to the rate of growth of charge,
that is proportional to the incoming current. Hence relative changes of the slope will be equal to
the corresponding changes of current. If, for example, the lengthening of the path in aluminum causes
a drop of the current by a factor
10, then the slope of the voltage
rise will also decrease by a factor
10. From the scope pictures taken
at different ranges the corresponding slope values can be readily
established.

With the aid of these data the currents for the various ranges can be calculated, if the current for one particular range is known. The path in the window and air is characterized by the range $R = 350$ mg/cm^2. For the sake of convenience it was decided to determine for this particular range the current which passed to the detector.

The time average of the current passing through the exit window was found to be 10^{-10} A, resulting from current pulses which occur with a frequency of 10^8 Hz (Fig. 2)during time intervals of 10 μs. Since the cathode happened to be pulsed 100 times per second, there are 100 such intervals per second.

If I_p denotes the amplitude of each pulse, then the mean value of the current during each of the 10 μs intervals is about $I_p/2$. The charge landing per second on the collector amounts to

$$I_p \times 10^{-5} \times 100/2 = I_p \times 10^{-3}/2 \text{ C/s.}$$

Equating this to the average current of 10^{-10} A gives $I_p = 2 \times 10^{-7}$ A, and for the mean value of the current during the 10 μs intervals 10^{-7} A. Using this value and the slope data obtained from the oscillograms, the currents corresponding to the various ranges in aluminum employed in this investigation could be calculated.

The empirically found connection between range and energy allows one to relate these currents to the energies of the accelerated electrons. The graph in Fig. 4 in which the average current during the 10 μs interval is plotted vs energy, presents summarily the results of a series of measurements.

Effects of scattering by the absorber are not accounted for in this plot, but this does not mean that they are negligibly small.

All relevant results reported in the preceeding sections were obtained by averaging over the results of several sets of measurements. The sets were repeated under identical operating conditions, but usually not on the same day. Often weeks or even months intervened between tests of the same nature. Nevertheless the deviations from the mean values presented here never exceeded 10 %.

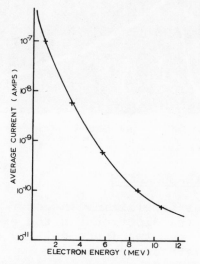

FIG. 4. Average external beam current vs electron energy.

THEORY

The efficient operation of the ER requires a uniform beam in which the axial velocity of all the electrons is the same. Such a beam can be produced by projecting a cylindrical stream of electrons of predetermined velocity into a static and homogeneous longitudinal magnetic field which is coaxial with the stream. Referring to the original article[3] on the ER the equations of motion of an electron in the beam can be found by means of the standard Lagrangian scheme. Together with the energy equation they determine the properties of the beam.

From the preservation of the generalized momentum it follows that all electrons in the beam turn round with the same angular velocity (Larmor precession)

$$\omega_L = \tfrac{1}{2} k B_0 \quad (\text{M.K.S. units}$$

are used), provided the cathode is free of magnetic flux.

$k = e/m =$ ratio of charge to mass

of an electron, B_o = flux density of the focusing magnetic field.

The forces on the electrons in the beam must be in equilibrium, hence the second time derivatives of the electron coordinates must vanish. On the other hand Poisson's equation must be satisfied subject to the prescribed boundary conditions fixed by the conducting envelope. Accounting for these conditions one arrives at a uniform axial velocity

$$v_o = 5.93 \times 10^7 \, V_o^{1/2}(1-\alpha^2)^{1/2} \text{cm/s} \quad (1)$$

and a uniform space-charge density

$$\rho_o = \epsilon_o k B_o^2/2$$

V_o being the voltage between tube and cathode, b the radius of the beam, R the inner radius of the drift tube, and $\alpha = B_o/B_1$, B_1 is the limit field defined by

$$B_1 = 6.65 \, V_o^{1/2}/b(1 + 2\ln \tfrac{R}{b})^{1/2} \text{ G} \quad (2)$$

with b in cm and V_o in volts.

The total current is given by

$$I_o = 6.58 \times 10^{-5}\alpha^2(1-\alpha^2)^{1/2} V_o^{3/2}$$
$$\times (1+2\ln \tfrac{R}{b})^{-1} \text{ A} \quad (3)$$

No current can pass down the drift tube if $\alpha \geq 1$.

Two fundamental laws govern the operation of the ER, namely the principle of continuity and the principle of conservation of energy. The first law is usually expressed in the form of the continuity equation

$$\text{div}(\rho \vec{v} + \partial \vec{D}/\partial t) = 0, \quad (4)$$

where $\vec{D}$ is the electric induction vector. Applying the continuity equation to the discharge process within the ER one finds that the electronic current entering the drift tube is continued in the form of displacement current, and that during the bunching period Δt the space-charge distribution within the beam becomes non-uniform.

The energy principle is employed in the form of Poynting's theorem

$$\text{div } \vec{S} = -\rho \vec{v} \cdot \vec{E} - dw/dt$$

It states that the energy streaming into unit volume per unit time is partially used to accelerate the electrons and partially to increase the electromagnetic energy.
$\vec{S}$ = Poynting's vector,
w = electromagnetic energy density,
$\vec{E}$ = electric field intensity.

Integrating Eq. (4) over the space inside the grounded envelope of the ER, and over the transit time $t_o = l_o/v_o$, l_o being the length of the drift tube, yields

$$W_b(t_o) = I_o V_o t_o - W_e(t_o) \quad (5)$$

$W_b(t_o)$ = energy of the beam at $t=t_o$
$W_e(t_o)$ = energy of the electric field within the drift tube at $t=t_o$
$I_o V_o t_o$ = energy supplied by the pulser.
In particular

$$W_e(t_o) = b^4 \pi \rho_o^2 l_o(1+4\ln \tfrac{R}{b})/16\epsilon_o \text{ J}$$

and
$$W_b(t_o) = V_o^2 \frac{1.11 \times 10^{-10}\alpha^2 l_o}{1+2\ln R/b}$$
$$\times (1- \tfrac{\alpha^2}{4} \frac{1+4\ln(R/b)}{1+2\ln(R/b)}) \quad \text{J} \quad (6)$$

The integration from t_o to $t_o + \Delta t$ is based on the assumption that a virtual cathode, i.e. potential minimum develops at the front of the beam in a plane normal to the axis of the system when the beam is decelerated.

During the deceleration of the beam the magnetic field associated with the beam current collapses and induces an axial electric field. This field, however, appears in the region beyond the virtual cathode only; within the beam it is compensated by the electric field due to the non-uniform space-charge distribution caused by the run-on of the electrons.

The potential distribution on either side of the virtual cathode is proportional to the four-thirds power of distance[5]. This is just the way in which the potential must vary in order that the radial field component $E_r = 0$. For this reason there is field flux emerging only from the beam front at $t = t_o + \Delta t$. This flux determines the charge Δq in front of the virtual cathode. The charge becomes the object upon which the electric field, generated by the total charge of the beam, acts.

Applying the divergence theorem to a Gaussian "pillbox" of vanishingly small height which encloses the virtual cathode, one finds that $\Delta q = \hat{V} C$, where $\hat{V}$ is the potential of the virtual cathode w.r.t. ground, and C a quantity which depends on the geometry of the decelerator section. It is convenient to consider C as the capacitance of the equivalent LC-series circuit.

In terms of the components of the equivalent circuit, the beam energy can be expressed as $W_b = I_o^2 L /2$, where L represents the inductance of the drift space.

If W_b is fully passed on to the front electrons, then the following relations are readily obtained by integrating the differential equation

$$d(I^2 L) + d(V^2 C) = 0$$

in the usual way, assuming that the output pulses are in good approximation sinusoidal,

$$\hat{V} = I_o (L/C)^{1/2}, \Delta q = I_o (LC)^{1/2}, \text{ and}$$
$$\Delta t = \pi (LC)^{1/2}.$$

Actually, however, only a fraction of the beam energy is likely to be transferred to the front electrons. The above relations, therefore, have to be replaced by the more general equations

$$\hat{V} = (2 \eta W_b / C)^{1/2} \qquad (7)$$

$$\Delta q = (2 \eta W_b C)^{1/2} \qquad (8)$$

$$\Delta t = \pi (LC)^{1/2} \qquad (9)$$

The ratio $\eta = \hat{V}^2 C / 2 W_b \leq 1$ defines the efficiency of the energy transfer.

APPLICATION

The present ER was operated under the following conditions:

Cathode voltage $V_o = 20$ kV
Beam current $I_o = .35$ A
Diameter of beam $2b = 1.0$ cm
Diameter of drift tube $2R = 1.6$ cm
Length of drift tube $l_o = 214$ cm

The density of the focusing field was slightly less than that of the limiting field B_1 which according to Eq. (2) was 1.34 kG.

From Eq. (3) one obtains

$$\alpha^2 (1 - \alpha^2)^{1/2} = 3.8 \times 10^{-3} \approx (1 - \alpha^2)^{1/2},$$

Eq. (1) gives $v_o = 3.2 \times 10^7$ cm/s,

hence $t_o = l_o / v_o = 6.8 \times 10^{-6}$ s,

and $I_o V_o t_o = 4.8 \times 10^{-2}$ J.

From Eq. (6) one gets $W_b = 3 \times 10^{-2}$ J,

therefore $L = 2 W_b / I_o^2 = 0.5$ H.

To compute the capacitance accurately is a rather tedious task It is, however, readily obtained from Eq. (9) if L and Δt are known. Fig. 2 shows a rise time of the pulses of 5 ns. Since this rise time is the same as the bunching period, it follows from Eq. (9) that

$$C = (\Delta t / \pi)^2 / L = 5 \times 10^{-18} \text{ F.}$$

Inserting this value in Eqs. (7) and (8) yields

$$\hat{V} = 110 \text{ MV and } \Delta q = 5.5 \times 10^{-10} \text{C.}$$

The actual value of $\hat{V}$ turned out to be 14 MV, hence

$$\eta = (14/110)^2 = 1.7 \% \text{ and}$$

$$\Delta q = 7 \times 10^{-11} \text{ C.}$$

The charge, however, which actually reached the collector was roughly 10 times larger as can be seen from Fig. 2. This discrepancy is due to the fact that the discharge of C during the bunching period was not taken into account.

DISCUSSION

The calculation of the energy-transfer efficiency η assumes an ideal beam, that is a beam which behaves exactly in accordance with the theory. The obtained value for the beam energy W_b, therefore, should be considered as an upper limit and consequently that for η as a lower limit. In fact the energy transfer efficiency may lie well above 1.7 %, particularly if also the energy losses in form of electromagnetic radiation are taken into account which are caused by the acceleration and deceleration of electrons.

The operation of the ER could probably be significantly improved by replacing the present electron gun by a gun which has at least a 10 times larger perveance, and which is equipped with an indirectly heated equipotential cathode instead of a tungsten filament. The fact that the acceleration process is due to a collective space-charge effect suggests to operate the ER with a low voltage-high density beam rather than with a high voltage-low density beam.

Both high beam stability and efficient operation can be expected if the ratio α happens to be .82. For this value the beam current attains a maximum at a given cathode voltage V_0. For example
α = .82, V_0= 10 kV, I_0= 13 A, B_0= 650 G, B_1= 800 G, R/b = 1.6.

A further improvement of the ER is likely to result from the application of a stronger and faster rising bunching field. Unfortunately the bunching coil in use was not designed for generating a flux density above 2500 G.

CONCLUSION

As far as the demonstration of the RE is concerned both design and construction of the present machine proved to be successful but neither of them should be considered as being optimal. Further experimentation will be necessary to find out how the device should be modified in order to improve its performance, in particular to increase the output currents.

To achieve very high energies of the output electrons two or more ER's may be arranged in series. The beam emerging from the decelerator of the first ER would serve as the input to the next one in series.

Apart from simplicity and low cost the acceleration method has the advantage of being applicable not only to electrons but also to ions.

Concluding it may be mentioned that T.G.Mihran and B.R. Andal [6] observed almost 10 years ago acceleration of electrons by a space-charge effect. More recently a team of researchers at the Maryland University [7] also reported the observation of the RE.

REFERENCES

[1] N.R. Heese and Raudorf, Rev. Sci. Instr. 43, 1594 (1972).

[2] L. Brillouin, Phys. Rev. 41, 260 (945).

[3] W.R. Raudorf, Wireless Engr. 28, 215 (1951).

[4] A. Coutu, N.R. Heese, and Raudorf, J. Appl. Phys. 42, 4504 (1971).

[5] L.P. Smith and P.L. Hartmann, J.Appl. Phys. 11, 225 (1940).

[6] T.G.Mihran and B.K. Andal, IEEE Trans. ED-12, 208 (1965).

[7] W.W. Destler et al., Proc. IX th International Conference on High Energy Accelerators, Stanford 1974.

HYDRODYNAMIC AND OPTICAL VIEWPOINT ON
PARTICLE BEAMS

J. D. Lawson
Rutherford High Energy Laboratory
Chilton, U. K.

ABSTRACT

A particle beam may be thought of as a drifting hot with finite temperature and pressure or as a bundle of optical rays whose collective properties are expressed in terms of the Liouville invariant emittance; the physical meaning of the paraxial envelope equation including self fields will be discussed from both points of view.

SUPER - STRONG FOCUSING OF AN INTENSE RELATIVISTIC

ELECTRON BEAM BY AN ANNULAR LASER BEAM

F. Winterberg
Desert Research Institute
University of Nevada System
Reno, Nevada 89507

ABSTRACT

It is shown that by the nonlinear transverse radiation pressure within a plasma of a pulsed convergent annular high power laser beam it is possible to focus an intense relativistic electron beam down to a radius of $\sim 10^{-4}$ cm. The transverse radiation pressure results from the dielectric property of a plasma in conjunction with the phenomena of the self-focusing of intense laser light. The tightly focused electron beams make possible the release of thermonuclear energy by micro-explosions. The components for such a system consisting of a $\sim 10^4$ Joule IR laser and a $\sim 10^6$ Ampere relativistic electron beams are already available.

In concepts of controlled thermonuclear energy release by micro-explosions initiated with intense relativistic electron beams one of the principal problems is the focusing of these beams onto a very small area. It had been shown some time ago[1] that the focusing of an intense relativistic electron beam with beam currents in excess of $\sim 10^6$ Ampere, down to a diameter of $\sim 10^{-3}$ cm promises the release of thermonuclear energy by micro-explosions with remarkably small beam energy inputs. The tightly focused beam would have to be projected in dense thermonuclear material which has to be in the liquid or solid state preferably but would not be required to be compressed to high densities as in the laser implosion fusion scheme by Nuckolls et al.[2]. The beam would then form a very narrow pinch channel with a very strong magnetic field of the order $\sim 10^8$ Gauss. This very strong magnetic field by action of magnetic confinement would
a) delay the thermal expansion of the narrow pinch channel
b) reduce the electronic heat conduction losses to the surrounding dense plasma in the radial direction and
c) confine the charged fusion products to within the pinch channel and hence reacting region

which is the condition for a thermonuclear detonation.

Although the question of stability for such a tightly confined electron beam within dense material is still unresolved, there are at least three reasons favoring a higher degree of stability than in an ordinary pinch discharge: 1) The pinch would be enclosed by a plasma with solid state density acting against hydrodynamic instabilities[3]. 2) Since the current would be carried by relativistic electrons higher stability is expected due to radiation damping. 3) Since the beam propagating freely in dense matter can only establish a high magnetic field after repelling its return current by the Weibel instability, a process which is very fast and much shorter than the beam discharge time, the high current filament will be surrounded by a annular return current sheet which is likely to enhance magnetohydrodynamic stability.

In fact, with intense relativistic electron beams, stable pinches for the time scale of the beam discharge time of $\sim 10^{-8}$ sec and with a diameter of $\sim 10^{-1}$ cm have been observed within the high voltage diode by the attachment of a small guide electrode onto the

center of the cathode surface facing the anode[4,5]. However, attempts
by this method to focus the beam
down to a smaller diameter have so
far been unsuccessful or at least
inconclusive. We will give here the
likely reason for this failure and
propose a way for focusing the beam
down to substantially smaller diameters.

A beam current I (Ampere) focused down to a radius r (cm) will
have a selfmagnetic field $H = 0.2$
I/r with a magnetic stress $H^2/4\pi =$
$(10^{-2}/\pi)(I/r)^2$. Since the guide
electrodes to focus the beam have a
finite tensile strength σ it is obvious that in order to prevent their
mechanical destruction by the magnetic forces of the beam one must require that $H^2/4\pi < \sigma$. A typical value for the tensile strength of a
solid is $\sigma \sim 10^{10}$ dyn/cm² resulting
in $H \lesssim 3 \times 10^5$ Gauss or $I/r \lesssim$
$10\sqrt{\pi\sigma} \simeq 1.8 \times 10^6$ Ampere/cm. This
shows that the concentration of a
current in the megampere range by a
solid guide electrode down to a diameter much less than ~ 1 cm, as it
is contemplated for thermonuclear
micro-explosions, will be difficult
to achieve. The limit of 3×10^5
Gauss is also in good agreement
with the experimentally reported maximum beam focusing of $\sim 10^7$ Ampere/
cm² so far achieved by the technique
using a solid guide electrode with
$\sim 10^5$ Ampere beams focused down to
a mm² resulting in a selfmagnetic
field of $\sim 4 \times 10^5$ Gauss. With high
tensile strength whiskers for which
$\sigma \simeq 10^{11}$ dyn/cm² it may be possible
to reach $I/r \sim 6 \times 10^6$ Ampere/cm,
which would permit to focus a 10^6
Ampere beam down to $r \sim 2$ mm. This
however, is still a long way off
from the desired goal to focus a
beam down to $\sim 10^{-3}$ cm.

In order to overcome these limitations we propose here the following technique explained in Fig. 1
and Fig. 2. Fig. 1 shows an axial
cross-section of a relativistic
electron beam projected into a thermonuclear target. As seen in Fig. 2
the beam is first discharged into
the target by a guide electrode.
But also any other method producing
an initially well collimated beam
would serve the same purpose. The
target is grounded to the anode so
that the beam is focused within the

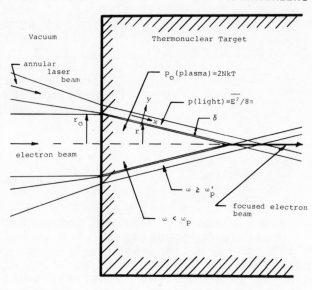

Fig. 1. The focusing of an intense relativistic electron beam by a conical
annular laser beam

diode space. The target itself may
be prepared in form of a solid piece of thermonuclear material, for
example TD, which upon interaction
with the beam is transformed into a
plasma. A short time before the
beam is projected onto the target
an annular conical convergent high
intensity pulsed laser beam is shot
into the target as shown in Fig. 1
and Fig. 2. In Fig. 2 one can see
how this can be done by an annular
laser beam, which after propagating
along the cathode is focused onto
the target by an annular lens positioned around the end of the cathode. The laser frequency ω shall be
chosen such that $\omega < \omega_p$ where ω_p is
the plasma frequency of the target
material. This condition is met by
a high intensity IR laser as for
example a CO_2 or a HF chemical laser. The annular laser beam projected onto the thermonuclear target
shall have at its impact a width
which is close the diffraction limit. At the high contemplated intensities this annular laser beam
will penetrate into the target material and by action of optical
self-focusing will form a narrow
convergent annular plasma channel
of width δ. This channel by thermal
expansion will then form an annular
plasma region with plasma frequency

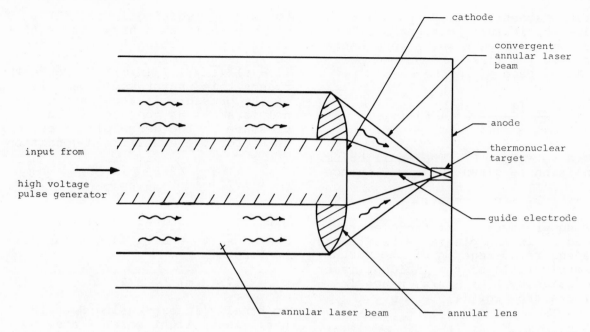

Fig. 2.   The bombardment and focusing of the thermonuclear target within
the diode space by the electron beam with the annular laser beam

ω_p' for which $\omega_p' \gtrsim \omega < \omega_p$ making it optically transparent and permitting the laser pulse to propagate within this channel. After this channel has become optically transparent no further appreciable heating and hence thermal expansion will take place such that from there on ω_p' will remain just slightly smaller than ω. The time to form this channel is of the order ℓ/v_s where ℓ is the length of the channel and v_s the velocity of the shock wave generated by the laser pulse propagating into the target. For $\ell \simeq 1$ cm and $v_s \simeq 10^7$ cm/sec this time would be $\sim 10^{-7}$ sec. However, one may also form the annular channel by a short wave length laser prepulse with $\omega_0 \gtrsim \omega_p$ prior to the main pulse propagating through the thusly formed channel region in which then due to thermal expansion $\omega \gtrsim \omega_p'$. Both the electron beam and the annular laser beam will have upon impact on the target the same initial radius r_0.

The laser radiation confined within the annular convergent channel will create a transverse radia-

tion pressure which can be easily computed from Maxwell's stress tensor in a dielectric medium[6,7], with the total force density acting on the plasma given by

$$\underline{f} = \nabla \cdot \left[\sigma_{ik} + \frac{\varepsilon - 1}{4\pi} E_i E_k - p\delta_{ik} \right]$$

$$+ \frac{\varepsilon - 1}{4\pi c} \frac{\partial}{\partial t} \underline{E} \times \underline{H} \, , \qquad (1)$$

where σ_{ik} is the Maxwell stress tensor in vacuum and p the total pressure within the plasma. The dielectric property of the plasma is given by

$$\varepsilon = 1 - \frac{\omega_p^2}{\omega^2 + \nu^2} \left(1 + i \, \frac{\nu}{\omega} \right) \qquad , \quad (2)$$

where ν is the electron collision frequency. If the propagation of the annular light beam is along an x-axis and the direction perpendicular to it along a y-axis of a local cartesian coordinate system (see Fig. 1), then we can distinguish between two polarizations for which $E_x = E_z = H_x = H_y = 0$ and $E_x = E_y = H_x = H_z = 0$. For an electromagnetic wave propagating in a dielectric substance one has for the time average $H^2 = \varepsilon E^2$ and hence for both the first and second polarization case

$$f_y = -\frac{\partial}{\partial y}\left[\frac{1-\varepsilon}{8\pi}\overline{E^2} + p\right] \qquad . \quad (3)$$

where $\overline{E_y^2} = \overline{E_z^2} \equiv \overline{E^2}$. The equilibrium condition is given by $f_y = 0$, hence

$$(1 - \varepsilon)\frac{\overline{E^2}}{8\pi} + p = \text{const.} \qquad (4)$$

The square average of the electric field $\overline{E^2}$ in the plasma is related to the square average $\overline{E_o^2}$ of the vacuum electric field of the incident laser radiation by $\overline{E^2} = \overline{E_o^2}/\sqrt{\varepsilon}$, hence it follows from eq. (4)

$$\frac{1-\varepsilon}{\sqrt{\varepsilon}}\frac{\overline{E_o^2}}{8\pi} + p = \text{const.} \qquad (5)$$

In the dense plasma region where $\omega < \omega_p$ the laser light intensity decreases rapidly over the distance $\sim c/\omega_p$, therefore $\overline{E^2} \simeq 0$ and $p = p_o$ where p_o is the maximum hydrostatic pressure within the plasma, therefore const. $= p_o$ and thus

$$\frac{1-\varepsilon}{\sqrt{\varepsilon}}\frac{\overline{E_o^2}}{8\pi} + p = p_o \qquad . \quad (6)$$

In the region of the annular laser channel one has $\omega \gtrsim \omega_p'$. For $\nu = 0$ and $\omega = \omega_p'$ one would have (from eq. (2) putting $\omega_p = \omega_p'$) $\varepsilon = 0$, and $\overline{E^2} = \infty$. The radiation pressure would therefore diverge. However, if ν is finite $\overline{E^2}$ remains finite. At the contemplated laser intensities the collision frequency is not determined by the classical electron-ion collision frequency which normally is very small compared to ω_p and which would lead to very large radiation pressures but rather by the growth rate of the oscillating two-stream instability[8], and which approximately is given by $0.7\,\omega_{pi} = 0.7\,(m/M)^{1/2}\omega_p \simeq \nu$, where ω_{pi} is the ion plasma frequency and m/M the electron ion mass ratio[9]. Recognizing that $\nu/\omega_p \ll 1$ one obtains from eq. (2) the approximation

$$\varepsilon \simeq 1 - \frac{\omega_p^2}{\omega^2}\left[1 + 0.7i\left(\frac{m}{M}\right)^{1/2}\frac{\omega_p}{\omega}\right] \quad . \quad (7)$$

For $\omega \simeq \omega_p'$ (putting in eq. (7) $\omega_p = \omega_p'$) one would then have with sufficient approximation $1 - \varepsilon \simeq 1$ and $\text{Re}|\sqrt{\varepsilon}| = 0.6\,(m/M)^{1/4}$, hence $(1-\varepsilon)/\sqrt{\varepsilon} \simeq 1.7\,(M/m)^{1/4}$. For a TD plasma this factor would be $\simeq 14$. If the pressure in the laser channel is small as compared to the pressure in the dense plasma one obtains for the equilibrium condition

$$p_o = 1.7\left(\frac{M}{m}\right)^{1/4}\frac{\overline{E_o^2}}{8\pi} \simeq \frac{\overline{E^2}}{8\pi} \qquad . \quad (8)$$

With the average Poynting vector $S = c\,\overline{E \times H}/4\pi = c\,\overline{E_o^2}/4\pi$ one can write for this

$$p_o = 1.7\,(M/m)^{1/4}\,S/2c \qquad , \quad (9)$$

or finally in expressing S by the total laser light power P and the area A onto which the light is focused one has $S = P/A$, and hence

$$p_o = 1.7\,(M/m)^{1/4}\,P/2Ac \qquad . \quad (10)$$

The pressure p_o in the plasma is composed of essentially three parts 1) the pressure of the plasma ions and electrons, 2) the pressure of the relativistic electrons in the beam and 3) the magnetic pressure of the selfmagnetic field generated by the current of the intense electron beam. For a beam propagating within the diode there exists a strong axial electric field. As a consequence of this strong axial electric field the beam can propagate within the diode in a background plasma neutralizing the beam electric field but not the beam current and hence beam magnetic field. The current is thereby equal to the total beam current and which can be much larger than the Alfvén current. These conclusions, although not yet fully understood, are in good experimental agreement with the observed beam pinching inside the diode. As a consequence the forces resulting from the pressure of the beam electrons and beam magnetic field are in equilibrium. One can therefore put $p_o = 2NkT$, where $2NkT$ is the pressure of the plasma ions and electrons. In order to obtain beam pinching, the beam magnetic field has to be trapped within the target plasma. This is possible to happen if the target was initially in a cold low conducting state such that it can trap the beam magnetic field prior to the transformation of the target by beam heating into a highly conducting plasma[1]. The beam magnetic field and hence the beam itself are thereafter confined within the thusly formed plasma. Therefore, if the electron beam propagates into the plasma confined by the convergent annular light beam, it too will be confined to the plasma and hence focused down to a small diameter as it approaches the vertex point of the conical light beam. The maximum beam focusing is thereby determined by the width δ of the self-focused laser radiation. Experimentally values as small as $\delta \sim 10^{-4}$ cm have been observed. However, how far down the beam can be actually focused depends on the fulfilment of the confinement condition eq. (10) which now reads

$$1.7 \ (M/m)^{1/4} \ P/2Ac = 2NkT \qquad . \quad (11)$$

For a plasma radius and hence beam radius r somewhere within the converging annular light beam channel of width δ one has $A \simeq 2\pi r\delta$. For a target material with $N = 5 \times 10^{22}$ cm^{-3} (i.e. solid TD) and expressing kT in eV one has

$$P = 4.3 \times 10^{21} \ Tr\delta \qquad . \quad (12)$$

The plasma temperature in the converging channel is likely to increase as the beam approaches its highest concentration. The most important mechanism to heat the plasma is by the two-stream instability. However, if sufficient collisions take place in between the beam electrons and the background plasma, the growth of the two-stream instability can be suppressed. This could be done by increasing the scattering cross section with the addition of high-Z impurities in that part of the plasma where the focusing takes place. The admixture of high-Z material has the additional advantage of slowing down the growth rate of the oszillatting two-stream instability thus making the effective collision frequency ν entering eq. (2) smaller and hence $\overline{E^2}$ larger leading to an increased transverse radiation pressure $\overline{E^2}/8\pi = E_0^2/8\pi\sqrt{\varepsilon}$. In case of the electron beam what remains is target heating by classical stopping

power, a process which is not very efficient. It therefore, may be possible to keep the plasma temperature in the focusind region below $\sim$ 100 eV ($\sim 10^6$ °K). This in addition is also supported by the fact that above 100 eV the radiation losses of the plasma would become significant if the plasma is optically opaque which can happen by the admixture of high-Z material. Therefore, the thermonuclear material perhaps may best be placed from and behind the region of highest beam focusing. The plasma in the convergent, the focusing process causing region, may thereby consist of other than thermonuclear material with a Z-value high enough to keep the plasma optical opaque but not too high as to cause too much energy loss of the beam by classical stopping power. Further detailed calculations are needed to clarify this point.

Assuming beam focusing down to $r \sim 10^{-3}$ cm with $\delta \sim 10^{-4}$ cm would require $P = 4.3 \times 10^{16}$ erg/sec, which could be accomplished by a laser pulse of 43 J = 4.3×10^8 erg to be delivered in $\sim 10^{-8}$ sec and which would correspond to a plasma and radiation pressure of 1.6×10^{13} dyn/cm^2 $\simeq$ 16 Mb.

In order to keep the focused electron beam stable against hydrodynamic instabilities one has to require that

$$\frac{\overline{H^2}}{8\pi} < \frac{\overline{E^2}}{8\pi} = 14 \ \frac{\overline{E_0^2}}{8\pi} \qquad , \quad (13)$$

where $H = 0.2 \ I/r$ is the beam magnetic field. Condition (13) therefore reads

$$\frac{10^{-2}}{2\pi} \left(\frac{I}{r}\right)^2 \leq 14 \ \frac{P}{2Ac} \qquad . \quad (14)$$

For $r = 10^{-3}$ cm, $I = 2 \times 10^6$ Ampere and $\delta = 10^{-4}$ cm one obtains $P \gtrsim 1.7 \times 10^{19}$ erg/sec which could be achieved with a 1.7×10^4 J laser pulse in 10^{-8} sec. In this case the radiation pressure would be $\sim 6.4 \times 10^{15}$ dyn/cm^2 = 6400 Mb.

After the beam has been focused down it may be projected into dense plasma placed just behind the anode and may from there on propagate as a freely drifting current neutralized beam. In this case then, because of the fast growing Weibel instability, the return current in the dense plasma may be expelled from the beam resulting in localized very strong magnetic fields as they are desirable for electron beam induced thermonuclear reactions. It thus seems likely that with the proposed method thermonuclear reactions not only can be initiated in the diode space but also in the drift space behind the diode filled with dense plasma.

REFERENCES

1) F. Winterberg, Nuclear Fusion 12 353 (1972).
2) J. Nuckolls et al., Nature 239 139 (1972).
3) M. Lampe et al., Bull. Am. Phys. Soc. 17 1029 (1972).
4) W.H. Bennett et al., Apl. Phys. Letters 19 444 (1971).
5) I.M. Vitkovitsky, Bull. Am. Phys. Soc. 18 1331 (1973).
6) L.D. Landau and E.M. Lifshitz, Electrodynamics of Continuous Media; Addison Wesley Publishing Company p. 242 (1960).
7) H. Hora, Physics of Fluids 12 182 (1969).
8) P.K. Kaw and J.M. Dawson, Physics of Fluids 12 2586 (1969).
9) O. Buneman, Phys. Rev. 115 503 (1959).

OPTICAL PULSE COMPRESSION

A. J. Glass
Lawrence Livermore Laboratory, University of California
Livermore, California 94550

ABSTRACT

The interest in using large lasers to achieve a very short and intense pulse for generating fusion plasma has provided a strong impetus to re-examine the possibilities of optical pulse compression at high energy.

Pulse compression allows one to generate pulses of long duration (minimizing damage problems) and subsequently compress optical pulses to achieve the short pulse duration required for specific applications. The ideal device for carrying out this program has not been developed.

Of the two approaches considered in this report, the Gires-Tournois approach is limited by the fact that the bandwidth and compression are intimately related, so that the group delay dispersion times the square of the bandwidth is about unity for all simple Gires-Tournois interferometers. The Treacy grating pair does not suffer from this limitation, but is inefficient because diffraction generally occurs in several orders and is limited by the problem of optical damage to the grating surfaces themselves.

Nonlinear and parametric processes have been explored. Some pulse compression has been achieved by these techniques; however, they are generally difficult to control and are not very efficient.

INTRODUCTION

The subject of this paper is the storage, compression, and switching of energy in the form of electromagnetic waves of optical frequency, that is to say, light waves. We particularly emphasize the concept of optical pulse compression.[1,2]

In any pulse compression scheme the bandwidth of the system limits the degree of compression. The compression of pulses of energy that travels in the form of waves is therefore limited by the frequency of the carrier. The advent of lasers, sources of coherent light, induces us to examine the possibility of performing pulse compression using light waves as carriers. The objective is, of course, to compress optical energy in time.

There are four principal reasons for compressing optical pulses and, as shown below, they are primarily associated with very high-power lasers:

- The principal failure mode of high-power lasers is the destruction of the laser elements by the interaction of the intense electric field of the light wave with the nominally transparent medium.[3] A way of reducing this self-damage phenomenon is to stretch out the optical pulse and thereby lower the peak field at a given energy density. For efficiency, one would like to extract as much of the stored energy from the laser as possible. This requires that the laser be saturated; that is to say, that the energy density extracted from the laser be greater than the saturation energy density characteristic of the laser medium. However, in achieving this high energy density in a short pulse, we must be careful not to exceed the damage threshold of the laser medium.

- In many applications, for example laser fusion, a short pulse duration is required and we quickly find that we are unable to achieve efficient energy extraction consistent with the required pulse duration. Again, a solution to this impasse is to generate a long pulse in which the energy can be efficiently and safely extracted from the laser amplifiers, and subsequently compress the pulse.

- The original motivation for optical pulse compression, and the only application to which this concept has been applied, is to correct for the self-phase modulation induced in large laser amplifiers by the passage of the intense light and thereby to achieve bandwidth-limited pulses of extremely short duration.

- Finally, some applications require selecting a small fraction of a pulse and compressing it in time to have a synchronous probe for diagnostic purposes. This is particularly true in experiments like laser fusion.

PULSE COMPRESSION SCHEMES

There are certain common features of all pulse compression schemes. First of all, the pulse to be compressed must be coded; i.e., each temporal element of the pulse must be tagged or identified in some way. Four possible ways to code a pulse are as follows: associating each element of time with its separate frequency; splitting the pulse into a number of elements, each of which propagates at a different angle or in a different direction; associating each element of the pulse with a different polarization state; and coding the pulse by using its intensity variation if the application allows intensity-dependent effects for pulse compression.

Once the coding is impressed on the pulse, a selective or dispersive group delay element is introduced into the system. This is an element through which the time for the passage of the pulse depends on the coding. Finally, to achieve ideal compression of the pulse, we arrange the coding and the dispersive element so that all pieces of the pulse emerge from the dispersive element at the same time.

DISCRETE CODING SYSTEM

In Fig. 1 we see one example of a pulse compressor. Here we have illustrated an angular or direction-sensitive coding scheme. The initial pulse is divided into a number of discrete elements and, by means of beam splitters, each of these elements is propagated along a slightly different path through the

amplifier. The dispersive element that is introduced into this system is illustrated as a stacker in Fig. 1. When the pulse elements emerge from the amplifier, a system of beam splitters is used to recombine the separate pieces of the original pulse along a single propagation path. The delay times in the stacker are arranged to cause the various elements of the pulse emerging from the amplifier at different times, to emerge from the stacker simultaneously.

The disadvantage of a discrete coding system is that the compression ratio is greatly limited. The shortest pulse that can be achieved is no shorter than the duration of the individual elements. Also, beam splitters are very vulnerable to optical damage at high energy.

NONLINEAR PULSE COMPRESSION SCHEME

An example of a nonlinear pulse compression scheme is seen in Fig. 2. Here a short pulse, identified as the signal, is injected on one side into one end of a parametric amplifier and into the other end is injected a long pulse identified as the pump pulse. The parametric medium is chosen in such a way that the two pulses interact and the energy from the pump pulse is converted to the optical frequency of the signal pulse. The pulse which emerges is a highly amplified short pulse having the duration of the original injected signal, but carrying the energy of the long pump pulse. An example of such a device is the backward traveling Raman amplifier in which the parametric process is stimulated Raman scattering.[4]

The disadvantages of such a system are the intrinsic difficulty of controlling nonlinear processes and the requirement for a complicated system of beam splitters and dichroic elements for bringing the

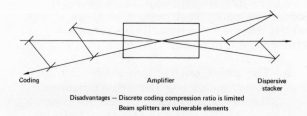

Coding   Amplifier   Dispersive
                                                                                        stacker

Disadvantages — Discrete coding compression ratio is limited
           Beam splitters are vulnerable elements

Fig. 1. Angular stacker.

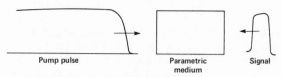

Example: Backward traveling Raman amplifier

Fig. 2. Parametric converter.

pump and signal pulses into the medium from opposite ends and separating them after they have emerged from the parametric medium.

CONTINUOUS FREQUENCY CODING

The most commonly used technique of optical pulse compression is based on continuous frequency coding of the optical pulse. There is a slightly different value of the carrier frequency associated with each time element of the pulse. The optical frequency varies monotonically as a function of time through the pulse. This is known as chirping.

The chirp can be achieved by one of two methods, it can be self-generated by the nonlinear interaction of the pulse with the optical medium itself or it can be externally generated by means of an electro-optic shutter.[5,6] The requirements on a dispersive delay line for optical pulse compression are precisely the same as the requirements for a high-resolution optical spectrometer. Two types of devices are commonly used. One is a totally reflecting Fabry-Perot resonator, first proposed by Gires and Tournois.[7] The other is a pair of diffraction gratings, first proposed by Treacy.[8] We shall discuss the special features of each of these devices.

We assume that the compressor is a loss-free device and therefore associated only with a frequency-dependent phase shift. If we inject an optical electric field E_0 at frequency ω then the field emerging from the device at frequency ω, is identified as $E_1(\omega)$. We write the output wave in the form

$$E_1(\omega) = E_0(\omega) \exp\left(i \, \Phi(\omega)\right), \quad (1)$$

where $\Phi(\omega)$ is the phase shift in the device. As long as the chirp bandwidth is much less than the optical frequency, writing $\omega = \omega_0 + \nu$, we find

$$\Phi(\omega) = \Phi(\omega_0) + Q_1 \nu + Q_2 \nu^2/2 + \dots, \quad (2)$$

where the group delay $Q_1 = (d\Phi/d\omega)$ at $\omega = \omega_0$, and the group delay dispersion $Q_2 = (d^2 \Phi/d \omega^2)$ at $\omega = \omega_0$.

If we neglect the effects of bandwidth, we can think of the optical pulse entering the compressor as made up of a series of packets of energy, each impressed on a carrier wave of a slightly different frequency. Each of these packets moves through the compressor at a slightly different speed, experiencing a slightly different delay. In this approximation, the output intensity is given by

$$I_1(t_1) = \int dt_0 \, I_0(t_0) \, \delta(t - t_0 + Q_1 + Q_2 \nu). \quad (3)$$

Here we assume that each group travels independently in time.

For a linear chirp, the frequency offset $\nu = \beta t_0$ and

$$I_1(t_1) = \int d\nu \, I_0(\nu) \, \delta\left(\nu - \frac{\beta(t_1 - Q_1)}{1 + \beta Q_2}\right). \quad (4)$$

If we examine the δ function carefully we see that the best compression can be achieved in those circumstances in which the denominator of the second term, $1 + \beta Q_2$, vanishes. Under these circumstances, the entire pulse will emerge at the time $t_1 = Q_1$, the average group delay through the compressor. The optimal compression criteria is thus given by the relation $\beta Q_2 - 1$.

In the more exact Fourier analysis of the problem, we express the chirped input pulse as

$$E_0(t_0) = A_0(t_0) \exp\left(i \beta \, t_0^2/2\right). \quad (5)$$

The output pulse takes the form

$$E_1(t) = \frac{1}{2\pi} \exp\left(i \, \Phi(\omega_0)\right) \int dt_0 \, A_0(t_0)$$

$$\times \exp(i \beta t_0/2) \int d\nu \exp(i \, Q_2 \, \nu^2/2)$$

$$\times \exp\left(i\nu(t - t_0 + Q_1)\right), \quad (6)$$

where the frequency integral is taken over the range of frequencies for which Q_2 is sensibly constant. A careful analysis of the pulse compression by this approach leads to the not surprising conclusion that the final pulse duration that can be achieved in any pulse compression device is limited to the inverse of the bandwidth, where the bandwidth is determined either by the range over which the frequency is chirped or the range over which the group delay dispersion can be treated as constant.

From these simple considerations we can obtain a figure of merit for any compression scheme. We identify T_0 as the initial duration of the input pulse and T_1 as the final duration of the compressed pulse. As we have just seen, the final duration T_1 is given by the reciprocal of the compressor bandwidth that, for a linear chirped pulse, is simply the reciprocal of the product of the chirp coefficient β and the initial pulse duration T_0.

To obtain optimal compression, the group delay dispersion must be equal to $-1/\beta$, which from the previous analysis is simply $-T_0 T_1$. This expression is incomplete, since if $Q_2 = 0$, no compression results, and $T_0 = T_1$. We therefore write the optimal compression condition in the form $Q_2 \sim -(T_0 - T_1)T_1$. Multiplying this expression by the square of the bandwidth, we arrive at a dimensionless compression ratio given by

$C = (T_0/T_1) - 1$, for optimal design. For high compression, $T_0 \gg T_1$ and C is just the compression ratio. Let us now discuss the dimensionless compression ratio C for the Gires-Tournois and grating pair compressions.

Gires-Tournois Approach. The Gires-Tournois interferometer has been extensively discussed in the literature.[2,7] It consists of a fully reflecting, Fabry-Perot interferometer. The group delay dispersion in this device is given by

$$Q_2 = -t_o^2 \frac{(1 - r^2)}{4r} \frac{\sin \omega t_o}{\left[\left(\frac{1 + r^2}{2r}\right) - \cos \omega t_o\right]^2},$$

(7)

The back surface is assumed to be 100% reflecting. As we raise the reflectivity of the front surface of the device the maximum value of Q_2 increases. However, it turns out that the bandwidth decreases

as we raise the finesse of the interferometer in such a way that the dimensionless compression ratio C is always approximately equal to unity, independent of variation of the design parameters. Thus to achieve significant compression using the Gires-Tournois approach, a large number of interferometers are required. This leads to significant problems in the alignment of the device.

In Fig. 3 we see the functional form of the frequency dependence of the dispersive group delay, Q_2. Here, for each value of the amplitude reflectivity r, Q_2 is normalized to its maximum value. As the reflectivity is increased, the maximum value increases, but as we can see from Fig. 3, the response curve becomes narrower in frequency. Numerical evaluation of these cases shows that the product of Q_2 times the square of the bandwidth is approximately 1, independent of the value of reflectivity. Here the bandwidth is defined as the frequency range over which the dispersive group delay Q_2 differs from its maximum value by no more than a factor of 2.

Treacy Approach. Treacy has proposed a pulse compression scheme based on a pair of diffraction gratings.[8] Light is incident on the first diffraction grating blazed for efficient diffraction in a given order. The second grating is adjusted to recollimate the light, so that all frequencies emerge on parallel paths. A diagram of the grating pair is shown in Fig. 4. The group delay dispersion is given by

$$Q_2 = \frac{B t_o}{c} \left[(\omega t_o \cos \gamma)^2 + 2\omega t_o \sin \gamma - 1\right]^{-3/2}, \quad (8)$$

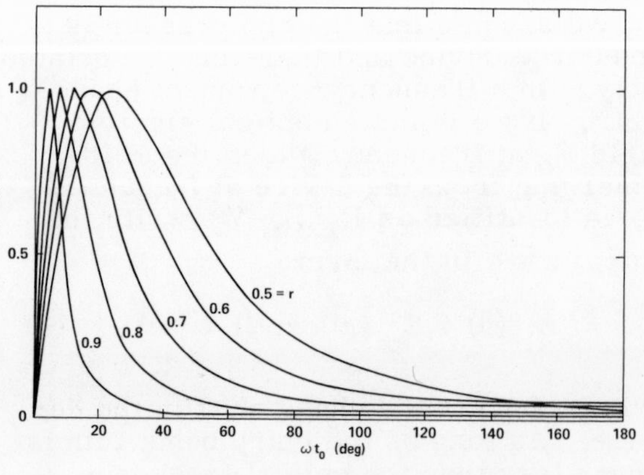

Fig. 3. Gires-Tournois response.

where the parameters are defined in Fig. 4.

Unlike the Gires-Tournois device, the grating pair is not limited in bandwidth. The group delay dispersion, Q_2, can be increased independently of the bandwidth by the simple expedient of increasing the separation between the gratings. The grating pair is a strongly dispersive device, however. Figure 5 is a graph of Q_2/Bt_0 vs ωt_0, plotted on a logarithmic scale to show how strongly the group delay dispersion Q_2 varies as a function of frequency. The dimensionless quantity, Q_2/Bt_0, is effectively the group delay dispersion normalized in units involving the grating separation and grating spacing, and ωt_0 is the product of the carrier frequency times the characteristic time. The different values of the angle of incidence are represented by the different curves in Fig. 5. Despite the strong frequency dependence of the group delay dispersion in this device, the bandwidth limitation is not severe. Instead compression is limited by the range over which the pulse can be chirped and ultimately by the problem of overlapping orders of diffraction in the gratings themselves.

The grating pair has been used for the compression of ultrashort optical pulses. Pulses with initial durations of a few picoseconds have been compressed to durations as short as 10^{-13} sec to what seemed to be bandwidth-limited pulses. Pulse compressions in the approximate range of 5 to 10 have been achieved. These were, however, pulses of low intensity and the measurements were carried out in a qualitative fashion.

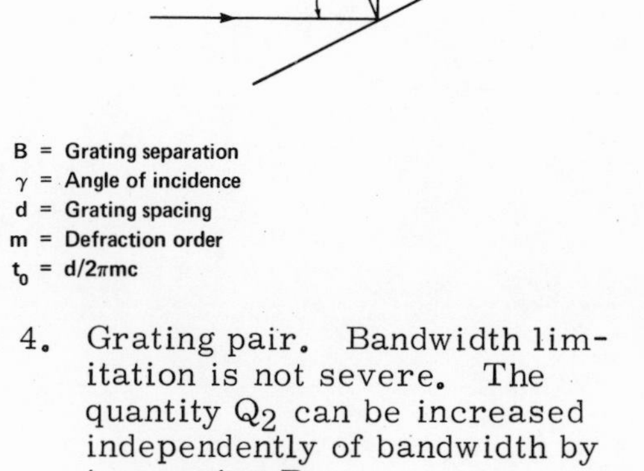

SUMMARY

The compression of optical pulses is particularly suited to applications associated with high-power lasers. In this cursory survey of optical pulse compression, we examined the methods of pulse compression and the features common to all schemes. Of the schemes available, we emphasized the Gires-Tournois and the Treacy approaches, pointing out the advantages and disadvantages of each.

ACKNOWLEDGMENTS

This work was performed under the auspices of the U.S. Energy Research & Development Administration. It is a great pleasure to acknowledge fruitful discussions of the problems of pulse compression with Robert Fisher of the Los Alamos Scientific Laboratory and Roy Bastian of the Perkin-Elmer Corporation.

REFERENCES

1. J. A. Giordmaine, M. A. Duguay, and J. W. Hansen, IEEE J. Quantum Electron. QE-4, 252 (1968).

2. A. Lauberau and D. von der Linde, Z. Naturforsch. 25a, 1626 (1970).

$$Q_2 = \frac{Bt_0}{C} [(\omega t_0 \cos\gamma)^2 + 2\omega t_0 \sin\gamma - 1]^{-3/2}$$

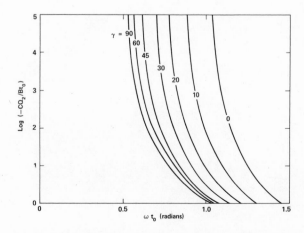

B = Grating separation
γ = Angle of incidence
d = Grating spacing
m = Defraction order
t_0 = d/2πmc

Fig. 4. Grating pair. Bandwidth limitation is not severe. The quantity Q_2 can be increased independently of bandwidth by increasing B.

Fig. 5. Dependence of Q_2 on frequency.

3. A. J. Glass and A. H. Guenther, Laser Interaction and Related Phenomena, H. J. Schwartz and H. Hora, Eds. (Plenium Press, New York, 1974).

4. M. Maier, W. Kaiser, and J. A. Giordmaine, Phys. Rev. 177, 580 (1969).

5. R. A. Fisher and W. K. Bischel, Bull. Am. Phys. Soc. 18, 1586 (1973); Appl. Phys. Lett. 24, 468 (1974).

6. E. B. Treacy, Phys. Lett. A28, 34 (1968).

7. F. Gires and P. Tournois, C. R. Acad. Sci. 258, 6112 (1964).

8. E. B. Treacy, IEEE J. Quantum. Electron. QE-5, 454 (1969).

DISPERSIVE TEMPORAL COMPRESSION OF THE ENERGY IN LASER PULSES: A REVIEW*

Robert A. Fisher
University of California
Los Alamos Scientific Laboratory
Los Alamos, NM   87544

ABSTRACT

Analagous to chirp radar compression schemes, the dispersive temporal compression of "chirped" light pulses has been discussed for many years now. Here we briefly review developments of these optical pulse compression techniques. This scheme can provide pulses so short that they would otherwise be unavailable, and such a scheme has the potential for improving the efficiency of short-pulse laser amplifier chains.

This meeting is concerned with the storage and compression of energy. A pulse of light is a form of energy and the power of a given light pulse can be increased if the energy in that pulse is compressed into a shorter time span. Although there are many ways to change the duration of a pulse, we review here the technique which involves the passage of "chirped" pulses through dispersive materials.

To first understand the concept of a "chirped" pulse, consider Fig. 1. The top half of this figure shows the time history of a pulse. The horizontal axis denotes time which is increasing to the right. The left-hand side of the pulse is then the "early" part and the right-hand side is the "late" part of the pulse. The bottom half of Fig. 1 shows the plot of the instantaneous frequency versus time and, in this example, the frequency increases linearly with time. This pulse is said to have a positive linear chirp. More complicated chirps correspond to more complex instantaneous frequency curves.

A pulse can acquire a chirp through many mechanisms; singled out here are the most popular techniques. A material with dispersion can chirp a pulse; this is accomplished since different colors travel at different group velocities. A short pulse with its high color content then emerges longer and chirped. Externally driven electro-optic modulators can chirp a pulse because the material index of refraction can be programmed to vary with time. Thus the crests and valleys of the light-pulse electric field do not remain at constant separation, and

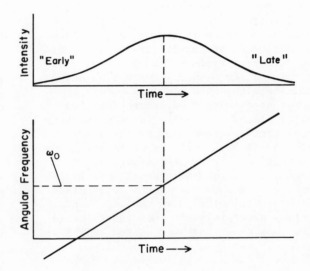

Fig. 1. A chirped pulse.

the pulse develops a frequency modulation while maintaining its same shape. A material with a nonlinear index of refraction can also chirp a pulse. In this case the index of refraction of the material is written as $n = n_0 + n_I \cdot I$, where I is the intensity, and n_I is the nonlinear index coefficient. Since at any point in the sample the pulse intensity varies in time, the index also varies in time and modulates the frequency. This chirping technique is difficult because of the lack of external control, and because other competing nonlinear effects (such as self-focusing and Raman scattering) may significantly influence the results.

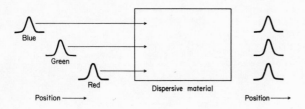

Fig. 2. Demonstration of the dispersive
         compression of a chirped pulse.

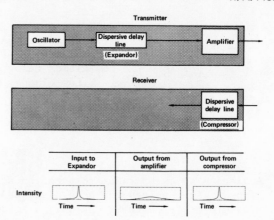

Fig. 3. Chirp radar schematic.

Figure 2 depicts how a linearly
chirped pulse might subsequently be compressed in a dispersive material. We
note that a material with dispersion has
different group velocities for different
color pulses. We consider for the sake
of discussion that the chirped pulse is
made up of three different color pulses
which arrive at the dispersive material
at different times. The left-hand side
of Fig. 2 represents a chirped pulse,
and the right-hand side represents the
compressed pulse. We assume that the
early part of the pulse is red, the central part is green, and the trailing
portion is blue. If the different group
velocities in the material are just
right, all of the colors will emerge in
unison. Thus we see that a long pulse
with a particular chirp can be compressed
into a short pulse if the dispersion in
the compressor has been chosen to appropriately compensate for the chirp.
Figure 2 can also be looked at "backwards," indicating that a short pulse,
when passed through a dispersive material,
will come out longer and chirped. The
compression scheme we are discussing
here has two parts: the first is preparing the pulse so that it has a chirp,
and the second takes advantage of the
chirp by passing it through something
with the proper dispersion. The pulse is
then temporally compressed.

The interest in pulse compression
started some time ago with the development of "chirp radar".[1] This scheme
allowed much stronger radar pulses to
be transmitted from a given radar set
by making a very long chirped pulse and
then compressing it only after reflection from the target. This process is
shown in Fig. 3. The top shaded area is
the transmitter and the bottom shaded
area is the receiver. We see that,
prior to amplification, the oscillator

pulse is made longer (and chirped) in a
dispersive delay line so that is is weaker
when it goes through the amplifier. The
return pulse is then compressed in a dispersive delay line with the opposite dispersion. If it were not for the expansion
prior to amplification, the amplifier
would have burned out. Thus both increased
range and resolution could be achieved
through the use of chirp radar.

Interest in optical pulse compression
grew after Gires and Tournois presented
in 1964 an interferometer which has adjustable dispersive properties.[2] This
interferometer is shown in Fig. 4. It is

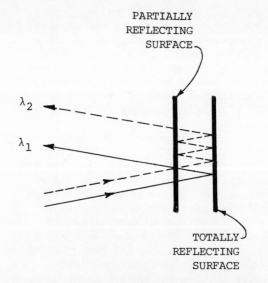

Fig. 4. The interferometer of Gires and
         Tournois. Note that different
         colors have different delays.

a variation of the Fabry-Perot interferometer in which the back mirror is a 100%
reflecting surface while the front mirror
is a partially reflecting surface. All
parallel rays which enter the interferometer leave parallel but different colors
are delayed inside the device for different
amounts of time. This is a very compact
way of putting a lot of dispersive delay
into a small volume. The difference in
transit time through this interferometer
for different colors is a result of the
interference between partial waves in the
device. This interferometer was successfully used in 1969 by Duguay and Hansen to
compress externally chirped helium-neon
laser pulses.[3] Another dispersive device
used for compressing chirped pulses was
invented in 1968 by Treacy,[4] and is shown
in Fig. 5. This is a pair of gratings
where the two grooved surfaces are facing
each other. Again, all rays which enter
parallel leave the grating pair parallel,
but different colors travel through different paths. Treacy used this grating
pair to study the chirp naturally occurring on the pulses emitted from a mode-
locked neodymium:glass laser. His work
was a major stimulus for optical pulse
compression studies.

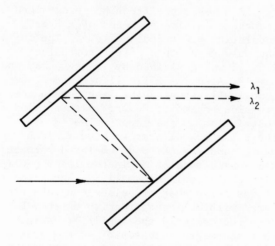

Fig. 5. The grating pair of E. B. Treacy.
Note that different colors travel along different paths.

In 1969, Fisher, Kelley, and Gustafson showed by calculation that passage
through a nonlinear material could put a
chirp on a pulse which could be subsequently compressed in a grating pair.[5]
This scheme is shown in Fig. 6. Required

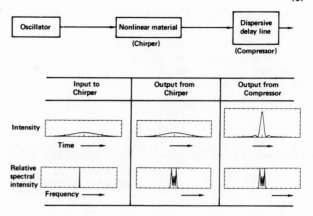

Fig. 6. The optical pulse compression
scheme of Fisher, Kelley, and
Gustafson.

here is the passage of an unmodulated
pulse through a nonlinear material, and
then the passage of the chirped pulse
through a suitably dispersive material
or structure. The top three curves in
this figure indicate the intensity history of the pulse in various stages of the
system. Both the input to the chirper
and the output from the chirper are long
pulses, while the pulse emerges shorter
after passage through the compressor.
The bottom row of curves depicts the
pulse spectrum (or "frequency content")
at various points in the system. The
input pulse to the chirper is a long pulse
having a correspondingly narrow spectrum,
but the effect of propagation in the nonlinear material puts a chirp on the pulse
and dramatically increases its frequency
content, as can be seen in the center
curve. Since the dispersive delay line
is a linear nonattenuating device, the
spectrum emerging from the compressor is
identical to that which the pulse had
when it entered the compressor. The uncertainty principle puts a lower limit on
the duration that a pulse with a particular spectral content would have, and the
increase in color content in the nonlinear
material gave the pulse enough spectral
width to be subsequently temporally
shortened.

Also in 1969, Laubereau demonstrated
that this two-step compression scheme can
work.[6] His use of ultrashort pulses necessitated the detection technique known
as two-photon-fluorescence (TPF),[7] which
unfortunately can lead to misleading

results.[8] The following year, Laubereau and Von der Linde published a long paper on the subject of pulse compression by the same two-step process.[9] Here they show how the compression effect is reduced because of deviations from the optimally compressing dispersive delay setting. In 1972, Reintjes, Carman and Shimizu reported[10] the study of transient self-focusing in liquids, and the probe pulse used by them was first compressed by a grating pair. In this case the experimenters relied upon the naturally occurring chirp in mode-locked Nd:glass laser pulses. It does not appear that direct measurements of the compression were made.

In 1973, Eckardt, Lee and Bradford modeled the mode-locked Nd:glass laser[11] and showed that only the first few pulses in the train would possess a good compressible chirp; the rest of the pulses were predicted to come out "scrambled". This also strengthened the arguments of Ref. 8. At the same time, Fisher and Bischel showed by calculation that dispersion of the chirper material must be properly accounted for in the two-step scheme.[12] The chirper material dispersion has the wrong sign to compress the chirped pulse as it evolves, and therefore the second step (the dispersive grating pair) must have more dispersion than had been previously expected.

In 1974, Fisher and Bischel showed by calculation that the detrimental problems of self-focusing in a Nd:glass amplifier chain could be significantly reduced by passing a longer (and weaker) pulse through the amplifier chain and then later compressing the pulse.[13] This scheme is similar to that of Ref. 5, and the crucial point is that the laser amplifier chain itself is found to be sufficiently nonlinear to put a compressible chirp on an uncertainty-limited pulse. Thus the scheme is also depicted by Fig. 6 if one substitutes an amplifier chain for the nonlinear material as the chirper. It was shown that if the pulse peak intensity averaged 2 GW/cm^2 along a 2-m glass chain, a 1000-ps duration pulse could subsequently be compressed to 125 ps in a series of Gires-Tournois interferometers. Such a scheme has relatively good stability against input temporal or spatial amplitude fluctuations, but it is sensitive to jitter in the oscillator center frequency and it requires numerous

expensive Gires-Tournois interferometers. Because the nonlinear propagation in the chain puts quite a chirp upon an unmodulated pulse, it is essential to understand the scheme for unmodulated input pulses before considering more complicated inputs. Also in 1974, Drexhage and Eisenthal reported the development of subpicosecond structure on picosecond pulses when thin flats were placed within a mode-locked Nd:glass laser oscillator.[14] Although the flats do have dispersive properties similar to the Gires-Tournois interferometer, the production of these subpicosecond features is yet unexplained. Again in 1974, Grischkowsky[15] repeated the experiment of Ref. 3 using a dispersive metallic vapor as the compressor for electro-optically modulated dye laser pulses. This is possible because the dye laser can be tuned within the vicinity of the Rubidium line. Also in 1974, Ippen, Shank and Gustafson achieved self-phase modulation in CS_2-filled fibers in the absence of either self-focusing or Raman scattering.[16] To date this appears to be the most controlled demonstration of the self-induced modulation effect.

In 1975, Loy demonstrated that an unmodulated pulse could be compressed if it is passed through an absorbing gas with its resonance frequency programmed to change in time by means of a Stark modulating electrical pulse.[17] The formal equivalence of this effect to that of Ref. 15 was described by Grischkowsky and Loy.[18] Also in 1975, Elliott and Fisher numerically studied the laser-chain compression scheme of Ref. 13 with a relaxed restriction on the input pulse.[19] Here the pulse was allowed to have a prechirp on it, and the chirp accumulated through self-phase modulation in the laser amplifier chain could be small by comparison. It was shown that if pulses could be prepared with a linear prechirp on them before they entered the amplifier chain, that bandwidth excesses of 40 could lead to single interferometer compressions of 1.5 - 2.0. These calculations took into account the role of nonlinearity and dispersion in the amplifier chain. Results indicate that in comparison to Ref. 13 the scheme's sensitivity to center frequency jitter in the oscillator pulse is reduced, and that sensitivity to uncorrectible beam divergence is also reduced.

In conclusion, experiments in which the role of self-phase modulation is important in producing the chirp upon the pulse are not yet reliable, and competing nonlinear effects must be avoided. For practical reasons, large laser systems cannot enjoy the luxury of compression in molecular absorbers because of the lack of coincidences between laser frequencies and properly absorbing candidates. It is, however, hopeful that a compression scheme can aid the large and expensive efforts which are presently in progress to make multi-kilojoule subnanosecond pulsed laser systems.

REFERENCES

1. J. R. Klauder, A. C. Price, S. Darling-ton, and W. J. Albersheim, Bell Syst. Tech. J. 39, 745 (1960); C. E. Cook, Proc. IRE 48, 310 (1960).

2. F. Gires and P. Tournois, C. R. Acad. Sci. (Paris) 258, 6112 (1964).

3. M. A. Duguay and J. W. Hansen, Appl. Phys. Lett. 14, 14 (1969).

4. E. B. Treacy, IEEE J. Quantum Electron. QE-5, 454 (1969) and references therein.

5. R. A. Fisher, P. L. Kelley, and T. K. Gustafson, Appl. Phys. Lett. 14, 140 (1969).

6. A. Laubereau, Phys. Lett. 29A, 539 (1969).

7. J. A. Giordmaine, P. M. Rentzepis, S. L. Shapiro, and K. W. Wecht, Appl. Phys. Lett. 11, 216 (1967).

8. R. A. Fisher and J. A. Fleck, Jr., Appl. Phys. Lett. 15, 287 (1969).

9. A. Laubereau and D. Von der Linde, Z. Naturforsch 25A, 1626 (1970).

10. J. Reintjes and R. L. Carman, Phys. Rev. Lett. 28, 1697 (1972); J. Reintjes, R. L. Carman and F. Shimizu, Phys. Rev. A 8, 1486 (1973).

11. R. C. Eckardt, C. H. Lee, and J. N. Bradford, Opto-Electron. 6, 67 (1973).

12. R. A. Fisher and W. K. Bischel, Appl. Phys. Lett. 23, 661 (1973).

13. R. A. Fisher and W. K. Bischel, Appl. Phys. Lett. 24, 468 (1974); IEEE J. Quantum Electron. QE-11, 46 (1975).

14. K. H. Drexhage and K. B. Eisenthal, J. Appl. Phys. 45, 2614 (1974).

15. D. Grischkowsky, Appl. Phys. Lett. 25, 566 (1974).

16. E. P. Ippen, C. V. Shank, and T. K. Gustafson, Appl. Phys. Lett. 24, 190 (1974).

17. M. M. T. Loy, Appl. Phys. Lett. 26, 99 (1975).

18. D. Grischkowsky and M. M. T. Loy, Appl. Phys. Lett., to be published.

19. C. J. Elliot and R. A. Fisher, Los Alamos Scientific Laboratory LA-UR-75-22.

*Work performed under the auspices of the Energy Research and Development Administration.

LOW 3.3 KV THRESHOLD 100 KW COAXIAL N$_2$

LASER WITH A RESONATOR

Heinz Fischer, R. Girnus, and F. Ruehl
Darmstadt, Germany

ABSTRACT

Increased laser power with a very low threshold is obtained with a 1500 pf pulsed coaxial discharge applying a 22% reflection sapphire plate. The pulse width is increased from 3 ns to 6 ns by the resonator. The discharge is pulsed by a high current rise "Nanolite" pulser. The output is 80 kv with a 14 kv pulse voltage. The beam divergence amounts to approximately 1 mrad only. The energy conversion factor is increased to 0.35%.

Switches

SOME ADVANCES IN HIGH POWER, HIGH dI/dt, SEMICONDUCTOR SWITCHES

Derrick J. Page
Westinghouse Research Laboratories
Pittsburgh, Pennsylvania 15235 U.S.A.

ABSTRACT

The design and operation of modern high power thyristors is outlined and some of the limitations of present designs are detailed. Two new methods of turning on power semiconductor switches which overcome these limitations are described. The first method involves the use of special emitter shunt patterns which are used in conjunction with laterally directed junction charging current to achieve a large area of turn-on. In this mode of operation the device is turned on by a high rate of change of anode voltage. The second method employs a neodymium doped YAG laser to produce, instantaneously, a high density of electrons and holes within the base regions of the device. Both of these methods enable rates of rise of current in power semiconductor switches that are significantly higher than those available in conventional thyristors. Rates of rise of 20,000A/μsec, rising to 3000 amperes, have been achieved to date using the laser triggering technique. These devices are ideal for series operation due to the isolated nature of the triggering source. Series strings of ten laser-fired devices have been operated successfully to date. Due to the extremely high switching rate, the dynamic voltage equalization networks customarily used to protect high voltage series strings of power devices can be significantly reduced in complexity. High-power thyristors with fast turn-on capability are likely to be applied in circuits requiring the switching of large currents and voltages with moderate duty cycle.

INTRODUCTION

In recent years the power handling capability of power semiconductor devices, particularly thyristors and rectifiers, has increased significantly. With the availability of large diameter silicon crystals, large area devices are being developed. At the present time, commercial devices are available with areas of nearly 20 cm^2 and devices with over 40 cm^2 are at the development stage. Such devices will be capable of conducting several thousand amperes continuously.

Advances in processing techniques have allowed the voltage rating of power devices to increase and blocking voltages of 3KV and over are now available. However, increasing the blocking voltage is at the expense of forward conduction efficiency and turn-off time.

In applications where power devices are required to switch at rates much higher than line frequencies, the switching performance of the devices is important. By reducing the turn-off time of thyristors, it is possible to signifi-cantly reduce the cost of the auxiliary circuit components and improve the circuit efficiency. Advances in reducing the turn-off time have been accomplished by improvements in processing techniques, by controlling the carrier lifetime and by the use of special electrode designs. Fast thyristors are now available with less than 10 μsec turn-off time and devices with turn-off times of less than 5 μsec are in development. However, there is a compromise between turn-off speed and forward conduction efficiency.

Another important parameter is the turn-on time which again is significant in high frequency applications. By employing special electrode designs, it has been possible to significantly reduce the turn-on switching losses and to obtain devices which can tolerate a rapid rise in the conduction current. However, there are limits to what can be achieved by electrode design improvements. This paper addresses itself to the turn-on time problem. The operation of a thyristor is first described and is followed by a description of other techniques that can significantly improve the turn-on performance.

Fig. 1.   A selection of power semiconductor
         devices showing the growth in size
         over the past decade.

If the current is large enough the voltage
drop laterally within the p-base region
underneath the n+ emitter will be sufficient to forward bias the p-n+ junction
and electrons will be injected into the
base region at the inside edge of the
cathode. These electrons will diffuse
towards the depleted region and then be
carried by drift within the depleted
region due to the high electric field.
Upon reaching the other side of the depleted region the electrons will forward
bias the anode p-n junction, causing the
injection of holes into the n-base region.
These holes will traverse the device in
a manner similar to the electrons and result in further forward bias of the cathode p-n+ junction. This will result in

Figure 1 illustrates the growth in
size of power devices over recent years.
The thyristor on the left, developed about
ten years ago, is rated at 80A and 1400V.
Next is a thyristor, introduced about
1965, with a rating of up to 300A and
1500V. Then comes the 550A, 2000V thyristor introduced in 1968. Next to it is
a 50mm device rated at 1000A and 1400V or
600A and 3000V that was made available in
1972. The large device on the right is a
2000A, 2000V experimental device that is
not yet commercially available.

THYRISTOR OPERATION

When a thyristor is turned on, it is
not possible, with present day devices, to
allow the current through the device to
rise at an unlimited rate. The reason for
this is that only a small area of the device is initially turned on and the result-
tant power dissipation in the "on" or conducting region must be kept down to safe
limits. This is best explained by referring to Fig. 2 which shows the cross
section of a conventional thyristor. Let
us assume that the device is in the off
state and the anode contact is at some
high positive potential relative to the
cathode contact. The central p-n junction
is reverse biased and a depletion region,
shown cross-hatched in Fig. 2 exists. The
device can be turned on by applying a
small positive signal to the gate electrode,
relative to the cathode. The resultant gate
current will flow along the path indicated
from the gate to the cathode electrode.

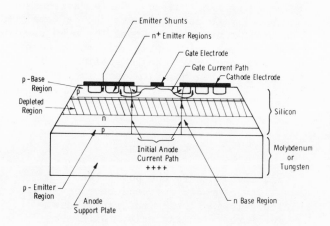

Fig. 2.   Cross section of a conventional
         thyristor (not to scale) illustra-
         ting the turn on process.

further electron injection from the cathode
and hence further hole injection. The net
result will be double injection of charge
into the base regions and the voltage
across the depletion region will collapse.
The turn-on process has now commenced and
current flows at the inside edge region
of the cathode. This results in anode
current flowing in this region as indicated in Fig. 2. This "on region" will
then spread at about forty microns per
microsecond until the entire cathode region
is conducting. However, the anode current
must be limited by the external circuit
until substantial spreading of the "on
region" has taken place.

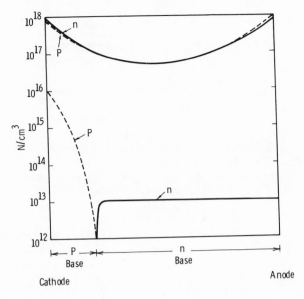

Fig. 3. Carrier distribution within a
thyristor in the conducting state
and in the unbiased state.

When the turn-on process is completed,
a plasma of electrons and holes exists in
the base regions of the device with a density distribution as indicated by the upper
curves in Fig. 3. The lower curves indicate the free carrier density distribution
in the device in the unbiased state.

The conducting region in a thyristor
can be observed by imaging the infrared
radiation emitted from the recombination
of electrons and holes within the on region in the device. This is done by
etching an array of small holes in the
cathode electrode to allow the radiation
to leave the device. This is then observed
using an image converter tube. By gating
the imaging system, the turn-on process in
a thyristor can be examined in detail. A
schematic of the system used is shown in
Fig. 4.

Figure 5 shows the on region in a
thyristor 12 μsec after initiation of turn
on and carrying 140A and 50 μsec after
initiation of turn on and carrying 350A and
finally 120 μsec after turn on and carrying
400A. Note here the small fraction of the
total device area that is initially
turned on.

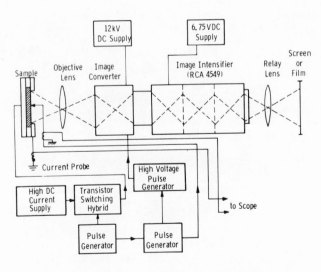

Fig. 4. Schematic diagram of the apparatus used to image the infrared
radiation from the carrier recombination within the thyristor.
This indicates the regions of the
thyristor that are in conduction
at a given point in time.

Fig. 5. Images of the infrared radiation
emitted from a conventional thyristor taken at various times
after the initiation of turn on.
This shows how the spreading of
the conduction region takes place
from the center of the device.

Figure 6 shows a device that failed
by allowing the current to rise too rapidly
during the initial turn on. The power dissipation was sufficient to melt the silicon.

One way of improving the device is to
increase the inside perimeter of the cathode electrode so that the initial on region
will be larger and hence the current can be
allowed to rise more rapidly. However, if

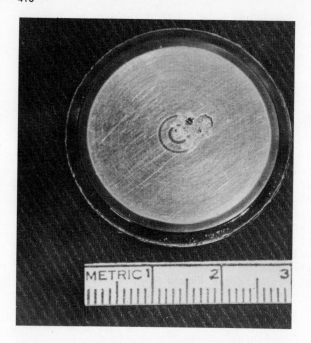

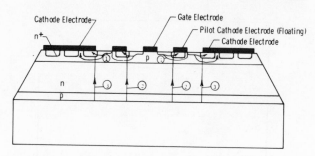

Fig. 7.   Cross section of a thyristor incorporating an amplifying gate
structure. Here the sequence of
turn on is illustrated, starting
with a gate current labeled 1,
resulting in a pilot anode current
2 which in turn results in a main
anode current 3.

Fig. 6.   A thyristor chip that has failed
during turn on by allowing the
current to rise too rapidly before
a sufficiently large area of the
device was in conduction. The
current density was sufficiently
large to melt the silicon.

electrode and the main cathode can be designed with an interdigitated pattern,
giving the desired large cathode perimeter.
The current can be allowed to rise fairly
rapidly in such devices, typically at the
rate of about 1000A/µsec. An interdigitated amplifying gate thyristor is shown in
Fig. 8.

this is done the gate current to fire the
device will become excessively large because of the lower lateral resistance of
the p-base region. This can be compensated
by incorporating an amplifying gate. A
cross section of a device with this feature
is shown in Fig. 7. Here a small thyristor
is integrated into the center of the
thyristor. This is the n^+ emitter region
beneath the floating pilot cathode electrode. The turn on of the device is similar to the first case discussed and a gate
current is made to flow from the gate electrode to the cathode (labeled 1). The
geometry of the n emitter regions is adjusted so that the emitter beneath the
floating cathode becomes forward biased before the main emitter region. The gate
current then causes injection from the
pilot cathode and this area turns on and
results in pilot cathode current (labeled
2). This current then supplements the
original gate current and assists in turning on the main cathode region, resulting
in the flow of main cathode current
(labeled 3). The floating pilot cathode

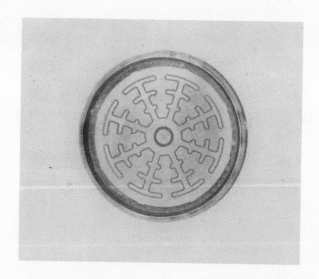

Fig. 8.   A thyristor chip incorporating an
amplifying gate and interdigitated
cathode electrode.

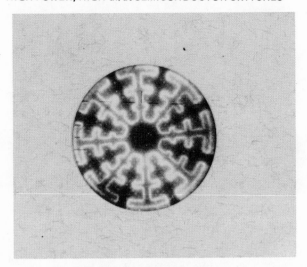

Fig. 9.   Infrared radiation emitted from a
thyristor chip having an interdigitated electrode structure and
showing the large area in conduction only 10 μsec after turn on.

Figure 9 shows the infrared radiation
emitted from such a thyristor chip. This
was measured only 10 μsec after turn on
and at an anode current level of 1000A.
Note the large area in conduction at this
point compared to that for a more conventional thyristor shown in Fig. 5.

There are two other schemes for turn
on that allow much more rapid turn on.
These are described in the following
sections.

REVERSE SWITCHING RECTIFIER

The first structure is shown in cross
section in Fig. 10 and is referred to as
the RSR (Reverse Switching Rectifier).
This is a two-terminal device and operates
as follows: Let the anode electrode be at
a high positive potential. Most of the
n and p base will be depleted of charge because the voltage is dropped across the reverse biased center p-n junction. In this
state the device can be regarded as a capacitor. If now a positive voltage pulse is
applied to the anode, charge will flow
within the device to further charge the
capacitor. This current will flow in the
manner indicated in Fig. 10 beneath each
of the n emitter regions and hence to the
cathode electrode via the emitter shunts.
If this capacitive current is large enough,
it will create a lateral voltage drop suf-

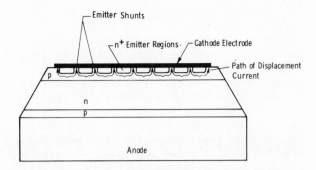

Fig. 10.   Cross section of a reverse conducting thyristor showing the
displacement current paths when
the anode is subjected to a positive step of voltage. This current causes the emitter junctions
to become forward biased and turn
on takes place over a large area
of the device.

ficient to forward bias the emitter p-n
junctions. The emitter regions will become
forward biased at a matrix of points inter-
stitial to the emitter shunt pattern. In
this manner, a large area of the device can
be turned on rapidly. Devices of this
type are well suited for series stacking
because it is not necessary to produce the
isolated trigger circuitry that is required for conventional thyristor stacks.

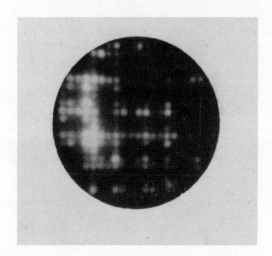

Fig. 11.   Image of the infrared radiation
emitted from an RSR device. This
picture was taken only one microsecond after turn on. The large
area in conduction should be
noted.

Devices of this type have achieved rates of rise of current up to 4000 A/μsec. Figure 11 shows an infrared picture of an RSR device. Note here the relatively large area initially turned on after only one microsecond. The current at this point is 600 amperes. The nonuniformity in Fig. 11 is attributed to the fact that in order to perform this infrared imaging measurement, the cathode electrode had to be removed in certain regions which upset the uniformity of the resistive paths for the capacitive current.

LIGHT ACTIVATED SEMICONDUCTOR SWITCH

The second structure is shown schematically in Fig. 12. This again is a two-terminal pnpn structure. If a positive potential is applied to the anode electrode, the center p-n junction will be reverse biased and carriers will be depleted from the central region and the device will be in the forward blocking (off) state. In all the devices described earlier, turn on involved various means of obtaining injection of electrons and holes from the cathode and anode to achieve the plasma charge density shown in Fig. 3. This is a relatively slow mechanism because it involves diffusion processes which are inherently slow. There is another, much faster, method of creating a plasma of electrons and holes in the base regions and that is by optical absorption. By illuminating the silicon with light, photons can be absorbed and create electron-hole pairs. This process is extremely rapid and can take place within the depletion region and a plasma created without having to rely on diffusion of charge from the electrode regions.

In practice, the infrared radiation is obtained from a neodymium doped YAG laser and has a wavelength of 1.06 microns. Such radiation is closely matched to the band gap of silicon and results in efficient conversion of photons to electron hole pairs. To obtain efficient optical coupling to the silicon, the polarized radiation is introduced into the silicon at the Brewster angle.

Using this technique, the plasma of electrons and holes that normally exist in the base regions of a thyristor in conduction can be instantaneously produced. The area of turn on can be large and transit time limitations to establish the plasma by injection from the cathode and anode emitters are greatly reduced. Fur-

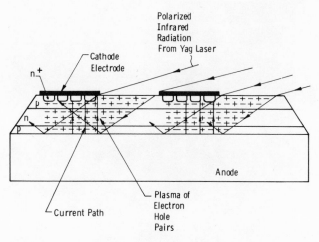

Fig. 12.   Schematic cross section of a light activated semiconductor switch.

thermore, series stacking is greatly simplified by the isolated nature of the trigger system. An optical system for stacking devices to produce a high voltage switch is shown in Fig. 13. Here, a single laser beam is split up into a number of beams and each is fed to a semiconductor device. The scheme ensures simultaneity of switching. Because the switching process is so rapid, a minimum of equalization network is necessary to ensure equal voltage division by the devices. We have demonstrated stacks of devices with only shunting resistors instead of the usual RC networks.

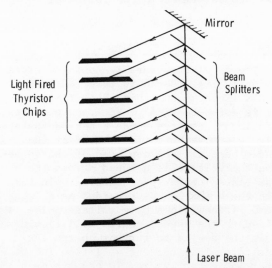

Fig. 13.   An optical scheme employed to turn on a series stack of thyristors. The beam splitters are designed to equally divide the laser beam.

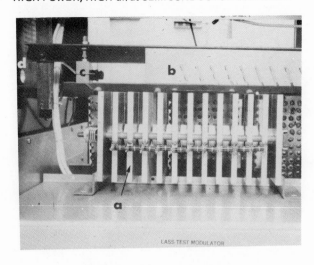

Fig. 14. A stack of ten light activated
semiconductor switches complete
with a beam splitter system.

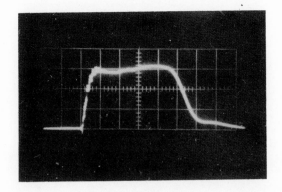

Fig. 15. Current waveform obtained through
a switch consisting of ten laser-
fired thyristors in series. The
vertical current scale is 600A/
division and the horizontal time
scale is 1.0µsec/division.

A stack of ten devices with the beam
splitter system is shown in Fig. 14. The
laser beam enters through hole 'd' and
through telescope c to the beam splitters
labeled 'b'. The light then is split into
ten equal beams and reflected onto the
stack of ten thyristors which are located
at the centers of the heat sink plates
'a'. This stack has been operated at
4000A/µsec with a peak current of 1800
amperes at a 60 Hz repetition rate. The
power limit was set by the circuit and not
by the devices. Figure 15 shows the current waveform obtained by discharging a
lumped delay line through a series string
of ten light activated semiconductor
switches. This demonstrates the extremely
fast switching performance that this technique can yield. Each thyristor was capable of blocking 1200 volts.

Single devices have been operated at
20,000A/µsec, rising to 3000 amperes at
low repetition frequencies. In this case
it was necessary to arrange the device in
a co-axial system to reduce the circuit
inductance and a capacitor was discharged
through the device.

CONCLUSIONS

This paper has shown that improvement
in dI/dt performance of thyristors can be
achieved by interdigitation of the cathode
electrode. However, by using displacement
current turn on or by light activated turn
on, considerable improvement in switching
speed is possible. In addition, these
techniques are ideal for series stacking
as devices incorporating these functions
are two-terminal and do not require gate
isolation. Although dI/dt values as high
as 20,000 A/µsec have been measured at low
duty cycle, considerable development effort
will be required before devices having this
kind of performance become commercially
available. However, the developments to
date indicate that, with suitable developmental effort, such devices can be made a
practical reality.

ACKNOWLEDGMENTS

The author would like to express his
thanks to many members of Westinghouse
Research Laboratories for supplying material for this paper. In particular, to
Mr. J. S. Roberts and Dr. P. F. Pittman
for information on light-fired thyristor
performance and to Drs. H. Yamasaki and
M. H. Hanes for providing information and
furnishing data on the infrared imaging
measurements. Thanks are also due to the
U.S. Navy for support in the area of light
triggered thyristors under contract
No. N00039-71-C-0228.

FRASCATI TOKAMAK TRANSFORMER SWITCHING SYSTEM

Andreani, R., Lovisetto, L.
Laboratori Gas Ionizzati (Ass. CNEN/EURATOM) C.P. 65, 00044 Frascati,
Rome, Italy
Vittadini, G.
T.I.B.B., Milan, Italy

ABSTRACT

Plasma ionization and heating, in the Frascati Tokamak, is obtained
generating an emf along the plasma column, by switching the dc current
flowing in the Tokamak transformer. 30 kA flowing in the 60 mH transformer
inductance must be commutated into a resistance to generate 40 kV across
the transformer itself.

Studies and tests to solve this problem have been conducted, on different types of breakers, in cooperation between Tecnomasio Italiano Brown
Boveri, Milan and Laboratori Gas Ionizzati, Frascati.

Satisfactory results have finally been obtained using a DLF commercial air blast breaker in a chopper type circuit.

A capacitor bank in parallel to the breaker is discharged immediately
after the contacts separation and the arc in the switching element is
estinguished at the first current zero. A saturable reactance in series
with the breaker reduces the current decay rate to allow sufficient deionization time.

1. INTRODUCTION

Tokamak machines appear to be,
at the moment, one of the most prom
ising ways to achieve the controlled thermonuclear fusion of the
hydrogen isotopes.

In these machines, typically,
a toroidal column of plasma in
which a current of the order of
many hundred thousands ampere is
flowing, is kept in geometrical e-
quilibrium by the interaction of
the current itself with a large toroidal magnetic field. The plasma
current is also used to heat up the
plasma and bring the plasma temperature in the range of many millions
of degree centigrade where the ther
monuclear fusion reactions take
place.

The plasma current is generated
and sustained using the plasma as
the secondary of a large set of
windings, the Tokamak transformer.
A large dc power supply feeds the
transformer and an emf on the plasma is generated switching the trans

former current through a resistance
and producing a large dI/dt.

The problem of commutating the
large transformer currents and hold
ing off the high voltages generated,
has been studied and solved, in the
case of the Frascati Tokamak, by a
joint effort made by the Laboratori
Gas Ionizzati - CNEN - Frascati and
the Tecnomasio Italiano Brown Boveri - Milano.

2. MACHINE PARAMETERS

The Frascati Tokamak, represented in Fig. 1, has two main sets

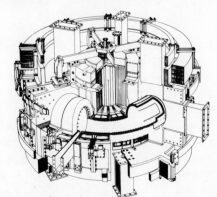

Fig. 1. Frascati Tokamak.

of windings: the toroidal magnet, producing a field of 10 T at 40 kA, and the transformer. The magnet and the transformer are cooled at liquid nitrogen temperature: 77° k to gain a factor 7 on resistance and power with respect to room temperature values. Two motor-flywheel generator sets with rectifiers deliver respectively 120 MW to the magnet and 21 MW to the transformer.

The transformer current and voltage waveshapes are reproduced in Fig. 2. Peak current is 30 kA,

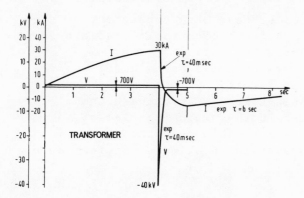

Fig. 2.   Transformer voltage and current waveshapes.

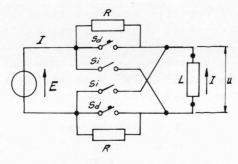

Fig. 3. Simplified equivalent circuit schematic.

peak voltage 40 kV. Fig. 3 shows the simplified equivalent schematic of the circuit where L is the transformer inductance and R the commutation resistance. At commutation time, Sd opens and the transformer current I is switched through the resistances R generating a voltage across L given by U = E - 2RI. E being rather small (700 V), U ≃ 2RI = = 40 kV. The corresponding voltage peak, along the plasma centerline

ionizes the gas. U decreases exponentially with time constant T = = L/2R. When U = 700 V, Si closes feeding the transformer in reverse to keep the plasma current flowing.

3.BREAKER PERFORMANCE

To evaluate the possibility of using real breakers for the Sd function, the equivalent circuit of Fig. 4 has been used, replacing the

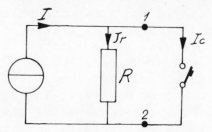

Fig. 4. Commutation equivalent circuit schematic.

transformer inductance with the current generator I.

The arc characteristics of existing breakers are already known so that their performance can be checked comparing the resistance characteristic with the arc characteristic as in Fig. 5 which shows

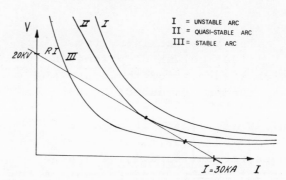

Fig. 5.   Arc characteristic versus resistance characteristic

the three possibilities. Tests have been performed at CESI in Milan on airblast and magnetic deionization breakers to measure the voltage limit RI which is respectively: 1300 V for the airblast breaker 3600 V for the magnetic deionization   breaker

In both cases therefore, a great number of interrupting chambers should be connected in series to provide the voltage required and the cost and reliability of operation would be very unfavourable.

Unfortunately, in all the existing ac breakers, high values of arc voltage at large currents are highly undesirable because they imply large energy dissipation and large contact wearing. Moreover dc breakers are generally produced only for voltages up to 3 kV and their arc voltage cannot be much larger than that to avoid overvoltages on the networks during the normal operation.

We considered the possibility of developing something new for this purpose, but we realized that the cost would have been extremely high and also the performance not safely quaranteed so that finally we decided to try to open the circuit producing an artificial current zero.

This method is already widely used in thyristor switching circuits where a condenser is used to produce one or more current zero trough the thyristor and so clear the circuit.

5. CIRCUIT DIAGRAM

The circuit used is represented in Fig. 6. The Sd breaker is an

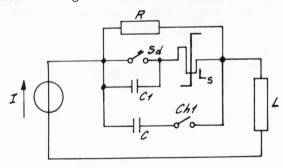

Fig. 6. Commutation system test circuit schematic.

air blast DLF type. One chamber per pole is used.

The circuit operation is the following.

At commutation time, the contacts of Sd separate and a steady arc, sustained by the current I, is,

established between them. When the contacts are fully away (few msec afterwards), Ch 1 closes and the precharged capacitor bank C is discharged through Sd generating an oscillatory current. The saturable reactor L_S, largely saturated at the current I, unsaturates close to the current zero and slows down the current rate of decay to allow more deionization time to the arc. When the current crosses the zero line, the arc extinguishes and a recovery voltage appears across the breaker. The rate of rise of such a voltage is determined by the condenser bank C_1 and the saturable reactor L_S. The performance of the circuit has been analyzed using computational methods.

The arc in the breaker has been represented, by two equations:

$$V = \frac{A}{i} + V_o ;$$

$$\frac{dg}{dt} = \frac{G - g}{\tau} ;$$

The first is the stationary arc characteristic, in the second, G is the conductivity of a stationary arc with the same current, g is the transient conductivity at time t and τ is the so called arc time constant.

The saturable reactor has been represented considering an hysteresis loop made up of rectilinear portions.

The performance of the circuit has been studied solving numerically the related system of equations to determine the correct value of the different components. The critical element has shown up to be the size of the capacitor bank C.

6. PERFORMANCE TESTS

To check the computation results, two series of tests have been performed at CESI Milan, using a large rectifier bank to get the required 30 kA current.

The inductance used was quite smaller than the transformer one but this produced only a faster voltage decrement after the commutation.

The other circuit elements were very close to the final values required.

During the first series of tests, the size of the capacitor bank C was only slightly larger than the minimum computed value. The results have shown a 50% probability of satisfactory performance. This is quite reasonable considering that the arc interruption, like many other phenomena in electrical engineering, has a statistical behaviour, while the computational method used cannot account for the momentary variations which occur in the behaviour of complex systems and produce the statistical distribution in the performance itself.

In the second series of tests, the size of the capacitor bank has been doubled and the performance has been completely satisfactory. 17 tests have been performed at full current; 30 kA, and full voltage: 20 kV, without replacing the contacts. The contacts were still in reasonable conditions after the 17 tests. The tests could not be continued for lack of available time. Fig. 7 shows typical test

current and voltage waveshapes. The fast current reversal for few micro seconds after the first zero crossing, is determined by the skin effect in the saturable reactor which was made of 0.3 mm thick iron sheets.

The frequency of oscillation of the condenser bank C with the saturable reactor L_S is of about 2000 Hz.

7. CONSTRUCTION STAGE

The construction of the circuit elements is presently under way in the TIBB machine shops.

The schematic of the circuit is represented in Fig. 8. Two poles

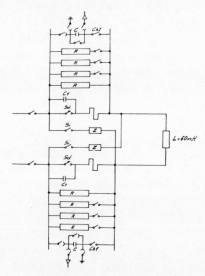

Fig. 8.   Commutation system circuit
schematic.

of a DLF breaker are used, one on each side of the circuit. A safety switch with a very short operating time, less than 10 msec, is provided, in parallel with each interrupting chamber, to short the main breaker in case of misbehaviour and sustained arc. This switch, identical to the Ch switches used to vary the number of resistances R in parallel during the discharge, has been built especially for this application. The entire system is automatically controlled using a programmable timer which operates the different components following a programmed sequence.

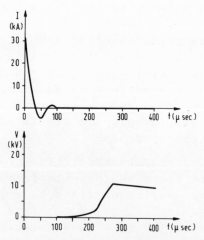

Fig. 7.   Typical test voltage and
current traces reproduced
from oscilloscope pictures.

SWITCHING PROBLEMS IN CROWBAR AND

POWER CROWBAR SYSTEMS

G. Klement and H. Wedler
Max-Planck-Institut für Plasmaphysik
Garching, West Germany

ABSTRACT

Comparing different schemes of energy storage such as capacitive
storage, inductive storage, inertial storage and battery, it can easily
be shown that for the submillisecond range the capacitive storage is advantageous because of its high power density and its low cost per MW. From
an examination of the cost for the various capacitor banks at Garching it
follows that its costs rise with the square root of the rise time[1][2].
Considering also commercial low-cost, high-density capacitors, we find the
costs converging at 10 DPf/joule for a risetime of approximately 0.5 msec.
A reliable crowbar-switch is essential for an economically reasonable employment of a capacitive storage to generate a quasi-stationary magnetic
field with a short rise time. Switching problems are crucial for any
storage system of that type. Besides starting switches - which have been
extensively described in literature - we will restrict ourselves to compare
and discuss the various crowbar switches which have been developed at
Garching[3][4][5]. First to mention is the ferrite decoupled spark gap
developed early in 1963 and extensively incorporated in many experiments.
Because the current taken over is somewhat delayed in this switch we initiated in 1968 the development of power crowbar spark gaps with simultaneous break down. We will discuss this feature in detail and present
experimental results. Lately it was found that the laser-triggered power
crowbar spark gap is very promising. Experimental results with spark gaps
with pressurized argon show that the power of presently available laser
is sufficient even to switch big systems. This is also true with a view
of the rather simple and powerful CO_2 laser. From the resistance-current
characteristics of the different crowbar switches it follows that - for
long current pulses - in addition to the fast switching spark gaps one
has to employ also low resistive mechanical closing switches. We will
report on the development of a free from thumbing and therefore high load
mechanical closing switch[6].

SIMPLE, SOLID DIELECTRIC START SWITCH*

W. C. Nunnally,[†] M. Kristiansen, and M. O. Hagler
Department of Electrical Engineering
Texas Tech University
Lubbock, Texas 79409

ABSTRACT

A field-distortion, solid dielectric switch of the type introduced by Martin[1] at A. W. R. E. and later extensively investigated by Dokopoulos[2] at Jülich is used as a start switch for a fast theta pinch system. Previous solid dielectric switches of this type required extensive electrode and switch package preparation as well as a fast trigger system to provide repeatable, multiple-arc channels during switching. In contrast, this switch was designed to facilitate ease of operation, switch package preparation and replacement. Repeatable performance is obtained with pitted and scarred electrodes and a relatively slow trigger system. The switch, 20 cm wide, produces an average of four arc channels, holds off approximately 110 kV and switches 70 kV with a maximum current of 500 kA. The inductance of the switch, including the connecting stripline inductance of 11 nH, is approximately 12 nH. The theta pinch circuit resistance is 13 mΩ, most of which is due to the four, parallel, Tobe Deutschmann ESC 249 capacitors. A very low switch resistance of 1-2 mΩ is indicated. The trigger system, a compact, photoelectrically triggered Marx circuit, produces a 80 kV, 50 joule pulse with a rise time of 50 ns. The average delay from Marx trigger initiation is .1 μs with an average jitter of .05 μs. The switching action is performed under oil to decrease switching noise. The oil also prevents high voltage tracking problems and thus reduces the switch package size. All of the current carrying joints in the switch are reinforced by a hydraulic jack that holds the switch plates together during switching.

INTRODUCTION

The multiple arc, solid dielectric switch is a simple method of switching large currents at high voltages with very low inductance. Such a switch which is simple to operate and performs reliably with pitted electrodes and a relatively slow trigger rise time was designed and developed for use in the theta pinch system shown in Fig. 1.

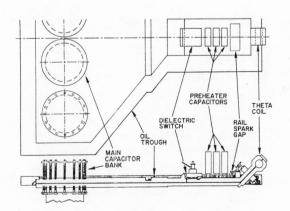

Fig. 1. Theta Pinch System.

OPERATION

The dielectric switch operates similar to a three electrode spark gap. A simple model is shown in Fig. 2.

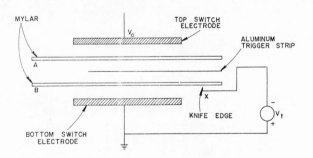

Fig. 2. Model of Dielectric Switch.

The top switch electrode is held at the source voltage, V_o, and the bottom electrode is grounded. Initially, the trigger strip voltage is determined by capacitive voltage division depending upon the Mylar insulation thickness of layer A and B in Fig. 2. The switching sequence begins when a negative trigger pulse with a finite rise time is applied to the knife edge in Fig. 2. When the dielectric strength of switch package layer A is exceeded, between the knife edge and the trigger strip, a small arc occurs at point X. The trigger strip voltage changes to the potential of the knife edge with a rise time much less than the trigger system rise time. The fast change in the trigger strip potential distorts the electric field at the edge of the trigger strip. A rise time of greater than 4 kV/ns is necessary for multiple arcs to occur between the trigger strip and the top electrode.[2] This action causes the trigger strip voltage to change very fast to the top electrode potential. This again distorts the electric field and causes multiple arcs in switch package layer B, which completes the switching action. This method was first suggested by Martin[1] at A.W.R.E. and later extensively investigated by Dokopoulos[2] at Jülich.

A schematic of the theta pinch circuit is shown in Fig. 3.

The electrode and circuit inductance of the switch were approximated by using the inductance per unit length of a stripline or[3]

$$L = \frac{\mu_o d}{W},$$

where d is the separation and W is the width. The inductance of one arc channel was approximated from the inductance of a straight round wire. This inductance is given by[4]

$$L_A = .002\ell \left[2.3 \log_{10} \left(\frac{4\ell}{r} \right) - .75 \right] \mu H,$$

where ℓ is the arc length in cm and r is the arc radius.

The switch is 20 cm wide. The electrodes are 7.5 cm in length with a separation of .115 cm. The electrode inductance is then .6 nH. The average number of arcs is four. The initial arc inductance for four arcs, assuming an arc radius of 10^{-3} cm and a length of .115 cm, is .25 nH. The connecting stripline inductance is 11 nH. Thus the total inductance is about 12 nH, most of which is due to the connecting stripline.

CONSTRUCTION

The dielectric switch is constructed as illustrated in Fig. 4. The stainless steel switch electrodes are mounted directly in the aluminum stripline leading to the theta coil. The switch electrodes become pitted and scarred after many switchings. They are shown in Fig. 5 after 1000 switchings. Figure 4a illustrates the low inductance, high current hinge joint designed to ease switch package replacement. A hydraulic jack holds the switch together during switching. The jack, set to a pressure of 105 kg/cm[2], reinforces the current joints between electrodes and stripline and at the hinge.

The switch package is a very important part of the construction. The switching region is submerged in transformer oil to prevent tracking and thus

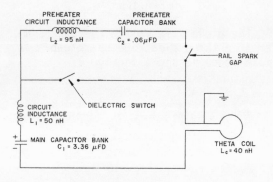

Fig. 3. Theta Pinch Schematic.

Capacitor C_1 is initially pulse charged to V_o. For preheating, the rail spark gap is closed and capacitor C_2 charges through the theta coil from C_1. The maximum voltage across the dielectric switch is approximately

$$V_m = V_o \left(\frac{2C_1}{C_1 + C_2} - \frac{L_2}{L_1 + L_2 + L_c} \right),$$

or approximately 1.5 V_o. The dielectric switch must thus hold off about 1.5 times the switching voltage during the preheating transient.

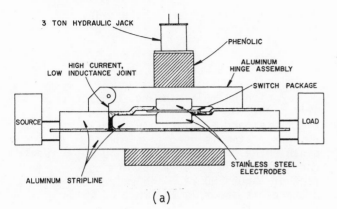

(a)

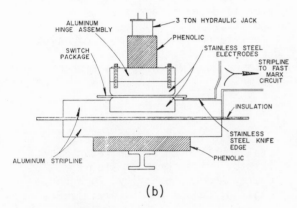

(b)

Fig. 4. Construction of Dielectric
Switch

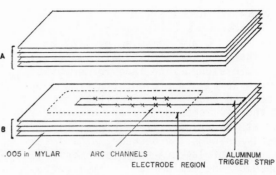

Fig. 6. Switch Package

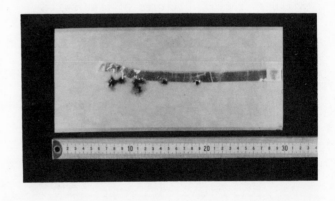

(a)

Fig. 5. Switch Electrodes After 1000
Switchings

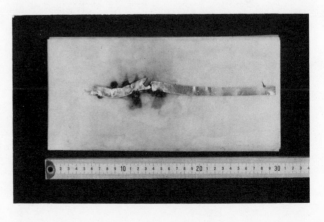

(b)

Fig. 7. Used Switch Packages

reduce switch package size and cost. A switch package is illustrated in Fig. 6. Layers A and B are composed of .127 mm Mylar sheets. The individual Mylar sheets are dipped into transformer oil during assembly to prevent tracking between the sheets during switching. When the oil is used, the arc channels occur at the trigger strip edges, directly through the package. The thickness ratio of layer A to layer B is 5/4. Dokopoulos[2] determined an optimum ratio of 6/1 for a similar switch. The total thickness should be the minimum necessary to hold off the switching voltage for repeatable switching. In the theta pinch system, however, the preheating transient produces a voltage 50 per cent greater than the switching voltage across the dielectric switch. The switch electrodes are pitted and scarred. The combination of rough electrodes and the large overvoltage require a larger total thickness and possibly explains the difference in the thickness ratio. Pictures of used switch packages are shown in Fig. 7.

The trigger generator is the small, fast erection Marx circuit shown in Fig. 8.

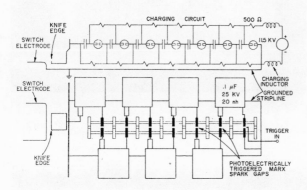

Fig. 8. Marx Trigger Generator

The Marx circuit produces a negative 80 kV, 50 J pulse with a rise time of less than 50 ns. The main feature is the photoelectrically triggered spark gap chamber. The successive Marx-stage spark gaps are placed in a common chamber. The first gap is externally triggered and the resulting radiation triggers the subsequent gaps with very little

delay. The average switching delay is .1 μs with an average jitter of ± .05 μs.

CONCLUSION

A simple, multiple-arc, dielectric switch that works well with pitted electrodes and a relatively slow trigger system is used in a theta pinch. The switch holds off 110 kV, and switches 70 kV at a maximum current of about 500 kA. The switch inductance is about 12 nH, most of which is due to the connecting stripline. The average delay is about .1 μs with an average jitter of ± .05 μs. The circuit resistance is 13 mΩ, most of which is due to the capacitor bank. This indicates a low switching resistance of 1-2 mΩ.

REFERENCES

1. J. D. Martin, A. MacAulay, Proc. 5th Symposium on Fusion Technology, Oxford, United Kingdom Atomic Energy Agency Authority, Abingdon, England, Culham Laboratory, 1968.

2. P. Dokopoulos, F. Lorbach, Proc. 6th Symposium on Fusion Technology, Germany, Center for Information and Documentation, Luxembourg, 1970.

3. H. Knoepfel, Pulsed High Magnetic Fields, North-Holland, 1970, p. 323.

4. F. E. Terman, Radio Engineer's Handbook, McGraw-Hill, 1943, p. 48.

*This work was supported by the AFOSR on Grant No. AFOSR-74-2639.

†Present address: Los Alamos Scientific Laboratory, P.O. Box 1663, Los Alamos, New Mexico.

HIGH CURRENT, FAST TURN-ON PULSE GENERATION USING THYRISTORS

John C. Driscoll
General Electric Company
West Genesee Street
Auburn, New York 13021 U.S.A.

ABSTRACT

It will be demonstrated that, contrary to popular belief, the turn-on time of thyristors can be orders of magnitude faster than that of transistors of equivalent voltage capability. This misconception stems from the conventional characterization of thyristors for wide pulse duty in inverter circuits where, of necessity, current densities must be limited to minimize power dissipation. In pulse modulator and GaAs injection laser applications, where pulse widths of at most a few microseconds are the norm, much higher current densities are permissible leading to dramatic improvements in switching speeds. Below, we tabulate the comparative specifications of a conventional inverter grade SCR against those of the device to be described in this paper.

	Typical SCR Specs	Specs for This Report
Current Density	$50\text{-}100 A/cm^2$	$100\text{-}10^4 A/cm^2$
Turn-on Time	$.2\text{-}2\ \mu s$	$5\text{-}40\ ns$
Method of Triggering	Gate	Gate and dv/dt
Pulse Widths	DC to $2\ \mu s$	$50\ ns$ to $4\ \mu s$
di/dt	$1\text{-}400 A/\mu s$	$>10^4 A/\mu s$ (circuit limited)
Inst. SCR (pulsed) Dissipations	~300 Watts	10^5 Watts

INTRODUCTION

To illustrate the considerable advantage the SCR enjoys over a similarly constructed transistor, Figures (1A) and (1B) depict current and voltage waveforms associated with these two devices in a simple capacitor discharge circuit. In both cases, a 0.005 μF capacitor charged to 500 volts is short circuited by the semiconductor switch. For purposes of comparison, the SCR and transistor structures are identical insofar as voltage capability and emitter design are concerned, the SCR being evolved from the transistor by addition of an extra n-type emitter region. It must be emphasized that this SCR was not optimized per se as a PNPN structure, but served as a vehicle for comparison, only. Its improved kperformance is attributable to the regeneration provided by the additional N base, that decreases switching speed by orders of magnitude and increases turn-on gain. In fact, further examination of

the basic circuit reveals that in the SCR case, its switching speed is strongly influenced by $\sqrt{LC}$, where L is the stray circuit inductance. Figure 2 portrays a plot of capacitor self resonant frequency as a function of capacitance, as determined experimentally for a series of high grade silver mica capacitors. Superimposed on this plot are three curves describing the performance of three different SCR designs discharging these same capacitors from various voltages close to the V_{BO}'s of the respective SCR's. It is interesting to note that an optimum value of capacitance exists. (Cop) for each case, where switching speed comes closest to that of the ideal shorted capacitor. As will become clear, this is a current density effect.

The theme of this paper is that a high SCR current density is essential to achieve an optimized turn-on performance, with current densities greater than 1000A per cm^2 usually required. To illustrate this point, the behaviour of several SCR limited circuits will be examined, with SCR current densities ranging from 50 to 800 amps/cm^2.

DISCUSSION

To better measure the SCR-limited rise times, the circuit of Figure (3) was constructed. The heart of this circuit is a Tektronix 50 ohm transmission line with inherent rise time of 0.1 ns. Current waveforms were monitored with a Tektronix model CT-1 current transformer.

Figure 4 illustrates the rise times recorded in this circuit with a General Electric ZJ353 experimental modulator device installed. This is a 600 volt design that operates at a peak current density of 50 amps/cm^2. Rise time is about 130 ns. To really dramatize the effect of increased emitter current density, Figure 5 shows the performance of yet another 600 volt device operated this time at 810 amps/cm^2. It is intriguing to note that the C106 used here, in fact is a commercially available 50-60 Hz thyristor intended for low cost consumer applications. That its performance is impressive is an understatement! Finally, to really drive home the lesson, a very small low voltage device with N-base width of only 0.8 mils and capable of blocking 100 volts only was also tried. Figure 6 indicates that rise times as short as 5 ns can be achieved. Current density under these conditions was also about 800 amps per sq. cm. The gain in speed over the C106 is, of course, due to the very narrow base width that significantly reduces transit times as compared to the 600 volt design. We may conclude from these observations that the switching time (rise time) of an SCR in SCR-limited circuits is determined predominantly by emitter current density and by base region width, this latter parameter directly influencing voltage capability and transit time in opposing directions. During the initial phases of conduction there is a large voltage drop across the device giving rise to a high electric field in the base regions. During this period, transit times in the lightly doped base regions are governed primarily by the field action. In the later stages of conduction, as voltage drop falls, transit times are proportionally increased as the electric field weakens; transit times now are influenced both by field drift and diffusion and both effects must be studied.

It is demonstrated analytically in the unabridged version of this paper that the improvement in rise time of an SCR over a comparable transistor may be represented by a turn-on function, F, which depends on the ratio of P base width to N base width. We have a rigorous means of computing how much faster the PNPN versus equivalent power transistor will be, given the ratio $\frac{W_p}{W_n}$ and W_n. Unfortunately, these solutions, attributable to Shockley and Davies, assume that minority carrier concentrations at distances remote from the junctions are vastly reduced by recombination effects. In pulse modulator work, where pulse widths of a microsecond are encountered, there is little time for recombination, when carrier life time is, for example, 100 μS. Current densities,

furthermore, are so high (10^4 amps/cm^2) that electron and hole densities are equivalent. Finally, the electric fields are extremely high. In short, a different solution is required to predict transit time in terms of current density and excess charge p. Fast transit velocity may be achieved by high total current and small p, but at the expense of low conductivity. Low conductivity implies that the equivalent resistance of the SCR is high, and in practice, this turns out to be essentially the unmodulated resistance of the wide N-base region. Experiments with a 700 amp SCR optimized for pulser duty (Figure 7) confirm this hypothesis. In the discharge circuit of Figure 8, the observed conditions are 300 amps peak at 580 volts across a 0.03µF capacitor. By determining L from self resonance of C and W by experimental observation, it is possible to solve for the average resistance of the SCR, which turns out to be 1.05 ohm. This is approximately equal to the known unmodulated N-base width resistance of the device, which is 1 ohm. Figure 9 shows that a fairly linear relationship exists between peak current and voltage in this circuit, confirming the maintenance of a constant resistance, or virtual lack of modulation. Since total current in the circuit will essentially determine current density in the SCR itself, intuitively a high voltage across the SCR will enhance switching speed by increasing current flow across this fixed resistance. If a voltage close to its V_{BO} could be maintained across the device for as long as possible during switching, switching speed would be maximized. To verify this conclusion, an SCR switched capacitor discharge circuit with a transformer load utilizing secondary voltage feedback to the SCR was tested. The object of the feedback winding was to hold up the discharge SCR voltage for a short interval after triggering. This feedback technique resulted in a reduction of 50% in switching time. The lesson here is that switching is definitely improved by maintaining high anode voltage.

One practical application of the pulser type SCR is to drive a GaAs type injection laser. P-N junction lasers require high threshold currents to initiate lasing action, but overheat rapidly at these current levels. Accordingly, current pulses must be very brief. A practical means of achieving these short conduction times is afforded by the capacitor discharge circuit using the special modulator device already described.

One other practical problem encountered in an SCR capacitor discharge circuit is the very high gate cathode voltage generated by the fast rising high amplitude anode current. This can cause undesirable negative gate currents at the instant of highest di/dt. A diode in series with the gate, or better yet, increased gate drive can be used to reduce this effect. High gate drive has the advantage of maintaining positive gate bias (0.6V) with respect to the emitter over a larger initial fraction of the P-base, resulting in larger initial turned on area. To improve gate drive in the basic capacitor discharge circuit, the arrangement of Figure 10 was devised as a means to easily generate high gate current pulses with fast rise times. This circuit produces 300 amps from a switching voltage of 480 volts, with pulse widths of 60 nS. Without the pilot SCR, voltage must be increased to 540 volts to attain the same current, resulting in a pulse width of 75 nS and a large reverse gate current.

Extending the results of this report to large area, highly interdigitated SCR's (employed to switch high Q (mica) capacitors with no load), the following current waveforms (sine wave) as a function of capacitance are possible:

Discharge Capacitance	Peak Current Frequency f	Current Per SCR * In Series String	Base Width (Half Period)
.01μf	10MHz	400A/SCR	50ns
.1μf	2MHz	857A/SCR	250ns
1μf	0.5MHz	2,000	1μs
10μf	0.1MHz	4,000	5μs
100μf	20KC	7,500	25μs
1000μf	3KC	15,000A	166μs

*Assumes a 600 volt potential on capacitor per SCR.

Use of saturable reactors to achieve full area turn-on may be necessary for the wider base width (>5 μs).

High peak current, narrow pulse applications, such as line modulators for radar or lasers may benefit from this work.

CONCLUSION

In conclusion, it has been demonstrated that a properly designed modulator thyristor can be orders of magnitude faster than an equivalent transistor when the device is operated at high current densities in the order of 10^3 amps/cm^2. In addition, far more circuit "gain" is available. Finally, adequate reliability also has been demonstrated. A more complete copy of this paper can be obtained by writing the author.

APPENDIX

Shockley (1) models the turn-on in terms of "hook collector gain M".

The first electrons emitted will initiate the injection of M more electrons after a time t_1 required for electron transit across the p base, plus t_2, the time required for hole transit across the n base. The cycle of operations just described now repeats itself with an amplified input to form a geometrically progressing series. A current and charge density plot of three such transits can be visualized by referring to Figure 7.

For n such transits at a time t:

$$p(t + n\tau_o) = M^n \, p(t)$$

where $\tau_o = t_1 + t_2$

therefore

$$\ln \, [p(t + n\tau_o)/p(t)] = n \ln M$$

The value of $n\tau_o$ for which

$$\frac{p(t + n\tau_o)}{p(t)} = e$$

is by definition, the time constant for the build-up or; $\tau = n\tau_o = \dfrac{\tau_o}{\ln M}$

A more detailed description describing the build-up of current by assuming an exponential time solution to the partial differential equation for hole continuity in the base gives (1):

$$\tau = \tau_d / [\ln 2\gamma M]^2 \qquad \text{for diffusion}$$

$$\tau = \tau_o / \ln \gamma M \qquad \text{for drift}$$

where τ = time constant for the build-up

τ_d = diffusion transit time

τ_o = drift transit time

γ = injection efficiency

M = current multiplication constant for the collector junction (Hook collector multiplication number)

An M of 11 is necessary for a ten fold improvement in rise time of the thyristor over the transistor for the diffusion solution. For the drift solution, an M of 20,000 is necessary. An M of 11 is more reasonable and implies a diffusion type solution for the continuity equation. R. Davies (2) has solved the two time dependent conservation equations for the build-up of excess charge in the p and n base of an SCR.

$$\frac{\partial m}{\partial t} = D_m \frac{\partial^2 m}{\partial x^2} - \frac{m}{\tau}$$

$$\text{where} \quad \frac{\partial m}{\partial x}\bigg|_{x=o} = -\frac{J_m(o)}{q\, D_m}$$

The time dependent solution is (by separation of variables):

$$m(x,\ ) = \sum^n e^{K_n t} \left(C_n e^{onx} + C_n' e^{-onx} \right)$$

$$\text{where} \quad \sigma_n = \sqrt{\frac{K_n + 1/m}{D_m}}$$

and $m = p$ and/or n

The following coupling conditions between the two bases are used as boundry conditions:

$$\frac{\partial n}{\partial y}\bigg|_{y=o} - \frac{\partial n}{\partial y}\bigg|_{y=W_p} = \frac{D_p}{D_n}\frac{\partial p}{\partial x}\bigg|_{x=W_p}$$

$$\frac{\partial p}{\partial x}\bigg|_{y=o} - \frac{\partial p}{\partial x}\bigg|_{y=W_n} = \frac{D_n}{D_p}\frac{\partial n}{\partial y}\bigg|_{y=W_p}$$

where $y=o$ and $x=o$ represent the position of the two emitter junctions. Assuming $n=p$, Davies obtains:

$$\sinh \frac{\sqrt{D_p}}{2\sqrt{D_n}}\sigma_p W_p \qquad \sinh \frac{\sigma_p W_n}{2} = \frac{1}{2}$$

Solving for $\sigma_p W_p$ will determine K.

i.e., $$K(W_p, W_n) = \frac{D_p f^2}{W_n^2} + \frac{1}{\tau}$$

For a ten fold increase in speed, over an equivalent voltage transistor (i.e., f=3.15) the thyristor ratio of

$$\frac{W_p}{W_n} = .27 \text{ gives an f of 3.0 which agrees with}$$

experiment (Figure I). We, thus, have a means to compute how much faster the pnpn vs. the equivalent pnp structure will switch; given the ratio W_p/W_n and W_n.

The Shockley & Davies solutions involve the assumption that minority carrier densities are vastly reduced (far from the junctions) by the process of recombination. At pulses shorter than .1 μs, (as compared to lifetime ~ 100 μs), there is little opportunity for recombination; which necessarily requires a much different solution. In pulser work where current densities approach 10 amps/cm (from both n & p injectors) in less than 20ns, electron and hole densities are equivalent. Also, under these conditions, electric fields are very high.

We, thus, wish to prescribe drift transit velocity in terms of current density and excess charge p. To do this, we will make the following assumptions:

1. Net charge density is zero, and that the donors and acceptors are fully ionized.
2. Recombination is negligible.
3. Diffusion currents can be neglected compared to currents produced by the field E.

From assumption 1:

$$n = p + n_e = p + N_a - N_d = p + ne$$

and, from 3:

$$I_p = q u_p\, pE$$
$$I_n = q u_n\, nE$$

The fraction of current carried by holes is then:

$$f(p) = \frac{I_p}{I_n + I_p} = \frac{\mu_p\, p/ne}{\mu_n\, (1 + (1 + 1/b)\, p/ne)}$$

where $b = \mu_n/\mu_p$

and, applying the condition of hole continuity, we obtain:

$$\frac{\partial p}{\partial t} = -\frac{1}{q} V \cdot Ip = -\frac{I}{q} \frac{\partial f(p)}{\partial X} = -\frac{I}{q} \frac{df}{dp} \frac{\partial p}{\partial X}$$

That the function $\frac{I}{q} \frac{df}{dp}$ represents a transit velocity (at a constant p) can be shown as follows:

$$p = p(X,t), \quad dp = \frac{\partial p}{\partial X} dX + \frac{\partial p}{\partial t} dt = 0$$

$$\text{(at } p = \text{const.)}$$

$$\text{or} \quad \frac{dX}{dt}\bigg|_p = -\frac{\partial p/\partial t}{\partial p/\partial X} = -V(I,p)$$

Solving for V (I, p) we obtain:

$$V(I,p) = \frac{I}{q} \frac{df(p)}{dp} = \frac{I}{q\mu_n ne} \frac{\mu_p}{1+(1+I/b)p/ne}^2$$

And, solving for the conductivity

$$\sigma(p) = \sigma(0) \quad 1+I/b) \, p/ne$$

We see that fast transit velocity V(I,p) can be achieved by high total current, and small p; but at the expense of a low conductivity. For example, the peak current for an RCL resonant discharge is:

$$I \max = V_o \sqrt{\frac{C}{L}} \, e^{-4\omega \tan^{-1}\left(\frac{\omega}{\alpha}\right)}$$

$$\text{where:} \quad \alpha = R/2L$$

$$\omega = \omega_o^2 - \alpha^2$$

$$\omega_o = \frac{1}{\sqrt{LC}}$$

Fitting observed pulse parameters (no load) namely 300 amps peak at 580 volts across a .03 capacitor, determining L by self resonance of C, and ω by experimental observation, we can solve for the average resistance of the SCR(R). Doing this we obtain: R ≅ 1.05 ohms (the switching SCR is the load). (Reference Figure 11).

This resistance is approximately equal to the unmodulated resistance of the wide N base:

$$R = \rho \frac{\ell}{\sigma A} \cong 1 \text{ ohm}$$

A fairly linear peak current/voltage relationship of the .03 μf discharge circuit is displayed in Figure 9. Thus, very little base modulation occurs during the ~ 30ns rise time interval (10 to 90%).

Diffusion/Drift Solution – We have discussed SCR turn-on speed, first by considering it a diffusion phenomenon, and then by describing transit velocity in terms of the electric field. A compatible resolution is suggested by the Haynes-Shockley experiments (3).

When localized, light pulses generate excess carriers in a semiconductor, the equation describing the subsequent transport of the pulse is:

$$\frac{\partial pn}{\partial t} = -\frac{P_{no} - P_n}{\tau_p} - \mu pE \frac{\partial P_n}{\partial x} + D_p \frac{\partial^2 P_n}{\partial x^2}$$

and the solution is:

$$P_n(x,t) = \frac{N}{\sqrt{4\pi D_p t}} e\left\{-\frac{(x - \mu pEt)^2}{4 D_p t}\right.$$

$$\left. + \frac{t}{\tau p}\right\} + pno$$

where N is the number of electrons generated per unit area.

The rise time of a current pulse will then depend both on the spatial distribution (in the semiconductor) of the advancing charge front at X=W (which is described primarily by diffusion) and the velocity of this front (which is described by electric field drift). However, the reasoning above implies a reduced carrier velocity which increases with total current density $(\mu E \rightarrow V(I,p))$.

REFERENCES

(1) W. Shockley and J. Gibbons, "Current Build-up in Semiconductor Devices", IRE Proceedings, December 1958.

(2) R. Davies, "Data and Calculations of the SCR Turn-on Mechanism," 8/63, internal G.E. communication.

(3) W. Shockley, "Electrons and Holes in Semiconductors," Van Nostrand, 1950.

(4) C. K. Chu, et al, "1,000 Volt and 800 Amp Peak Reverse Switching Rectifier," 1973, I.A.S. Conference Record, page 267.

(5) J. Quine, et al, internal G.E. publication.

438

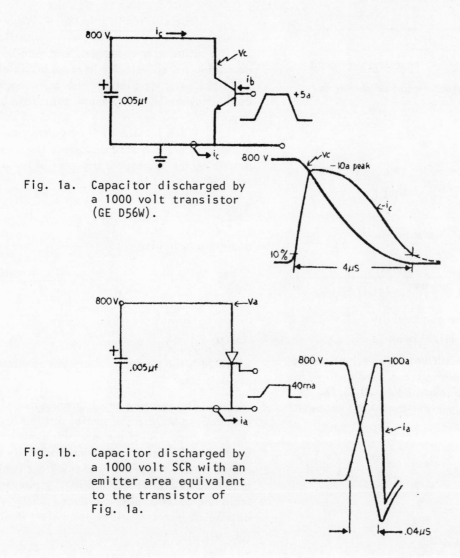

Fig. 1a. Capacitor discharged by a 1000 volt transistor (GE D56W).

Fig. 1b. Capacitor discharged by a 1000 volt SCR with an emitter area equivalent to the transistor of Fig. 1a.

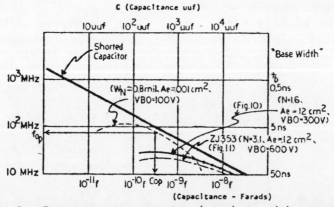

Fig. 2. Frequency vs. capacity (mica) for (1) shorted capacitor and (2) charged capacitor switched by three different SCR's at their VBO's.

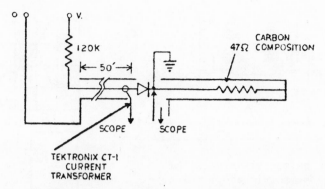

Fig. 3. Distributed parameter transmission line (delay line) discharged circuit. The line is a Tektronix #113 delay line of 60 ns length, 50 ohm with an inherent rise time of 0.1 ns.

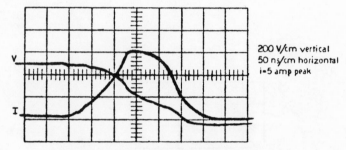

Fig. 4. Performance of ZJ353 in Tektronix #113 delay line circuit (J_p = 50 a/cm^2).

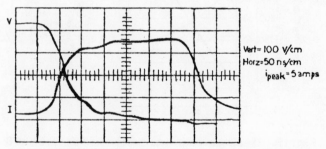

Fig. 5. Performance of C106 in Tektronix #113 delay line circuit (J_p = 810 a/cm^2).

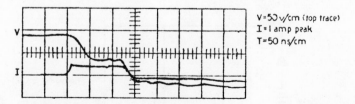

Fig. 6. The performance of the low voltage (100 volts) device (when tested in the circuit of figure 3) is shown here. An approximate 7 ns rise time is noted. However, the corrected rise time, (when considering the 4.4 ns rise time of the Tecktronics 585A oscilliscope with a type 82 plug in unit) is about 5 ns. At 1 amp, the peak SCR current density is about 800 a/cm^2.

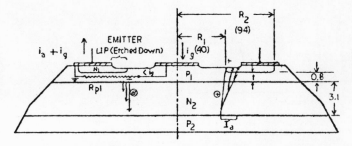

Fig. 7. The 700 A pulser structure (the device is symmetric about the center line). All dimensions are in mils.

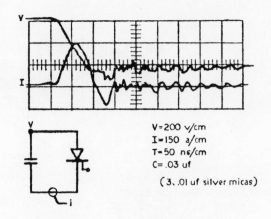

$V = 200$ v/cm
$I = 150$ a/cm
$T = 50$ ns/cm
$C = .03$ uf

(3, .01 uf silver micas)

Fig. 8. Negative current.

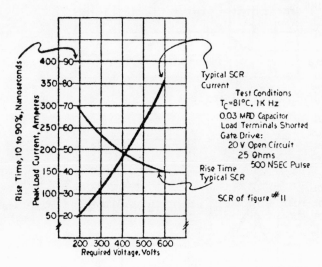

Test Conditions
$T_C = 81°C$, 1K Hz
0.03 MFD Capacitor
Load Terminals Shorted
Gate Drive:
    20 V Open Circuit
    25 Ohms
        500 NSEC Pulse

SCR of figure #11

Fig. 9. Peak current vs. voltage for 0.03 MFD.

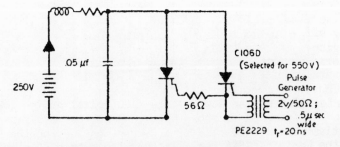

Fig. 10. SCR with pilot gate.

LASER TRIGGERED SWITCHING OF A PULSED CHARGED
OIL FILLED SPARK GAP [†]

A.H. Guenther, G.L. Zigler*, J.R. Bettis, and R.P. Copeland
Air Force Weapons Laboratory
Kirtland Air Force Base, New Mexico 87117

ABSTRACT

A focused, Q-spoiled laser, aligned along the interelectrode axis of a pulse charged switch assembly, was used to initiate the conduction of an overvolted transformer oil filled gap. Laser power varied between 5 and 200 MW and the voltage pulse exhibited a risetime of 500 nsec to a voltage of 700 kV. A parametric study of the factors affecting the delay between the laser pulse arrival at the gap and conduction of the gap was accomplished, in which the effect of the focal point location, laser power, switch polarity and voltage on the gap at laser arrival, was determined. Delay times as short as 12 nsec were recorded with jitter, a measure of reproducibility, in the low nanosecond region.

INTRODUCTION

Laser triggered switching (LTS), the technique of using a focused, giant pulse laser to initiate the conduction of a dielectric between charged electrodes, was primarily developed in order to meet the stringent requirements imposed on switches employed in the simulation of nuclear weapons effects by high voltage, high peak power electrical systems.[1] These simulators usually consist of an energy storage system, pulse forming and/or pulse shaping elements, one or more switching devices and some form of energy transducer which converts the stored energy into the desired output. The successful simulation of the effects produced by nuclear detonations can require voltages in the multi-megavolt region, peak powers of 10^{12} to 10^{14} watts with voltage rise times normally less than 10 nsec delivered to the transducer. Frequently, a major concern is the design of a switch capable of handling currents in the megampere range and of sufficiently low inductance to minimize the rise time, while still capable of holding off voltages in the multi-megavolt region. In many cases fast rise time generators operating at very low impedances ($<<1\Omega$) are dictated by other considerations. Thus, the reduction of switch inductance is extremely critical

since rise time is proportional to the inductance and inversely proportional to the impedance. This may require paralleling systems which then must be discharged synchronously in a time much shorter than the desired rise time.

The first paper on LTS published in 1965[2] exhibited delay to breakdown (the time from arrival of the laser beam in the gap to the onset of switch conduction) as low as 10 nsec. These early results were obtained by aligning the laser beam orthogonal to the axis of minimum separation of a sphere-sphere gap insulated by several different gaseous dielectrics at various pressures. Subsequent investigations[3,4] have shown that a coaxial geometry, where the laser beam is aligned along the axis of minimum separation of the gap yields much lower delays, as low as 1-2 nsec, with jitter (a measure of reproducibility) as low as $\pm$ 0.1 nsec.

Later, repetitive LTS in gases with jitter below 1 nsec was achieved[5] by employing a low energy (10's of millijoules) moderate power ($\sim$10 MW) Nd^{+3} YAG laser at a repetition rate of 50 pps. In addition, conduction in four individual gas gaps was initiated at 50 pps

---

*Submitted in partial fulfillment of the requirements for the Master of Science Degree, Air Force Institute of Technology, Wright-Patterson AFB, Ohio 45433.

[†]This paper was presented at the Torino Conference by Dr. Donald C. Wunsch, of the Air Force Weapons Laboratory, Kirtland AFB, NMex.

within 0.1 nsec of each other by beam splitting techniques[6] and the simultaneous LTS of two breakdown channels in a single gas gap reduced the voltage rise time of the output by almost a factor of two, when compared to single channel operation[7].

In addition, solid dielectrics[8] and high vacuum[9] have been used as dielectric insulation in LTS with delays to breakdown in the low nanosecond region in the former and in the low microsecond region in the latter. All of these results were aimed at achieving a very low switch inductance by either paralleling generators, paralleling switch channels or reducing the insulating dielectric length between switch electrodes.

Liquid dielectric filled gaps have been investigated for use in LTS by Marolda.[10] He used transformer oil in a sphere-sphere gap with a spacing of 0.365 cm and studied two triggering configurations for DC charged liquid LTS breakdown. Microsecond delays were observed when the laser was focused perpendicular to the interelectrode axis whereas delays as low as 30 nsec were observed when the laser was focused along the interelectrode axis. However, reproducible results were difficult to obtain under these DC conditions due to the erratic behavior in breakdown characteristics caused by measurable insulating material turbulence, entrapped gas bubbles or particulate inclusions. A detailed review of the present stage of development of LTS, coupled with an extensive and complete bibliography has been given by Guenther and Bettis.[7]

Pulse-charged, liquid dielectrics are of interest due to their inherently higher electrical insulating capability compared to compressed gases. Under fast pulse conditions they can hold-off as much as three to four times their DC breakdown value. This characteristic allows for a small gap spacing, hence, reducing the switch impedance and inductance. Furthermore, their breakdown characteristics for a given impulse waveform and geometry are much more reproducible than under DC conditions. Liquid dielectrics are also self-healing which allows for a quick restoration of their electrical insulation characteristics as compared to higher electrical strength solid dielectrics. Other characteristics of liquid dielectrics are the lack of the need for

a pressure vessel, the relatively lower cost of the dielectric and the capability of using one dielectric for insulating the energy storage system, transmission line and switch as is typical in many high power impulse generators. While there is no need for a pressure vessel as is normally used when employing gaseous dielectrics, recent work[11] indicates that pressurization of the liquid dielectric, particularly water, can lead to further improvements in voltage hold-off capabilities as well as the delay before self-breakdown.

The choice of using a pulse voltage at the electrodes, instead of DC, was due primarily to the problem associated with the instability in the DC breakdown characteristics. By pulse charging the electrodes, hence, overstressing the dielectric in the gap, the dependence of the self-breakdown potential on the previously noted effects is negligible. This is due to the inability of the impurities or nonuniformities in the dielectric to migrate substantially or develop hydrodynamically into breakdown regions on a time scale comparable to the impressed voltage pulse.

This research effort was an attempt to determine the suitability of using a giant pulse laser to initiate the conduction of an overvolted liquid dielectric. In particular, this effort involved a parametric study of the factors affecting the delay to breakdown, in which the effect of laser focal point location, laser power, switch polarity, and voltage on the gap at laser arrival was evaluated. The figure of merit was the jitter of the delay to breakdown as well as the reproducibility of the final breakdown voltage. Commercial transformer oil was chosen for this investigation because of its great utility and availability as a low cost insulating liquid.

EXPERIMENTAL APPARATUS

The apparatus used in this experiment consists of four major components: an electrically Q-switched ruby laser (0.69 μm), an optical conduit housing a lens system for focusing the laser radiation, a 2 megavolt pulsed charge 10 Ω coaxial line, and associated diagnostic equipment. The arrangements of these major components is shown in Fig. 1.

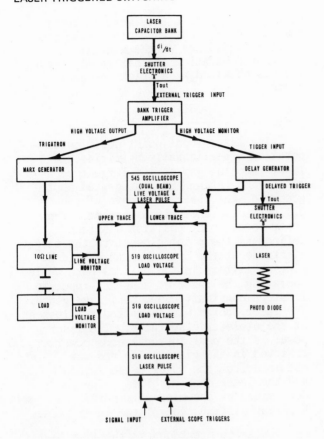

Fig. 1.   LTS System Trigger and Signal Flow Chart.

thickness for the ruby wavelength (0.69 µm). The influence of SBS will be addressed later in the paper.

The optical conduit contained a movable diverging lens to vary the focal position with the last optical element in direct contact with transformer oil dielectric in the switch chamber. This exit end was sealed by this last lens and a high pressure nozzle was provided to circulate the dielectric in the switch chamber over this final converging lens to insure that air bubbles or carbon caused by the laser breakdown of the dielectric would not degrade the performance of the focusing system during the passage of subsequent laser pulses.

The pulse generator used in this experiment was a Physics International model 214 2 MV Pulser capable of delivering a 2 megavolt pulse with a rise time of 700 nsec to a 24 inch diameter semielliptical electrode.

The Marx generator is connected to the target electrode by a 10 ohm coaxial line. A separate insulated chamber contains the electrodes forming the terminal switch electrode and load assembly. Connected directly to the 10 ohm line is the target electrode which can be remotely adjusted by a hydraulic system to vary the gap spacing. Both the target and load electrodes have semielliptical heads with a 24 inch diameter, and, a radius of about 22 inches, which allows for the electric field at the axis of the electrodes to be within 5 percent of that of a plane parallel gap. In the load electrode there is a 1/2 inch diameter aperture to allow passage of the laser beam into the interelectrode region.

The diagnostic equipment can be divided into two major categories: laser monitoring and electrical pulse monitoring. The laser pulse shape and time reference were obtained by an ITT S-20 vacuum planar photodiode exhibiting a 0.1 nsec rise time. The output was monitored on a Tektronix Model 519 scope and was used as an initial time reference. The laser pulse energy was monitored by a thin-foil calorimeter (foilometer)[12] which consists essentially of a piece of thin aluminum foil to which a thermocouple had been attached.

The voltage pulse was monitored at two different locations, one at the 10 ohm

The laser system used in this experiment was a Korad K-1500 laser. This system consists of ruby oscillator and amplifier rods Q-switched by means of a deuterated KD*P Pockels cell. Energy output of the laser system could be continuously varied from 1 to 8 joules with pulse widths (FWHM) from 40 nsec for the lower energy range to 30 nsec for the higher energy range. Peak powers varied from 20 to 220 MW.

The output of the laser system is allowed to pass through two optical beam splitters for diagnostic purposes (total energy and pulse shape) and an isolator to prevent the return of the Stimulated Brillouin Scattering (SBS) component back into the laser system. This system consists of a Brewster's angle tent polarizer horizontally polarized, followed by a quarter wavelength plate of appropriate

line and the other at the load. The voltage monitor at the 10 ohm line employed a capacitive voltage divider while the load voltage monitor consisted of a resistive voltage divider network. This last monitor was not calibrated as it was used strictly as a timing monitor for this experiment.

Through a series of delay and pulse generators (Fig. 1) the following sequence of events occurs: after charging the laser and Marx energy storage, the laser capacitor bank is discharged into the flashlamps; a di/dt sensing coil triggers the Marx trigger amplifier which in turn initiates the Marx voltage erection and, through proper delay circuits, initiates the laser shutter electronics. Because of the uncertainty in the delay between the high voltage monitor signal and the start of voltage erection (+ 50 nsec) and the uncertainty in the delay between the input to the Pockels cell and laser output (+ 15 nsec) the laser could not be placed on the voltage ramp more precisely than + 65 nsec. A discussion of how this overall delay and jitter has recently been reduced will be discussed later. From this basic arrangement one then selects the proper delays and cable lengths to operate the system and display the desired outputs on the recording oscilloscopes.

The raw data consisted of a set of four oscilloscope traces per shot. A trace from a Tektronix type 555 dual beam scope displayed both the voltage ramp and the laser output. Two type 519 scopes displayed the load voltage monitor on 20 nsec/cm and 100 nsec/cm sweep rates while a final Tektronix type 519 oscilloscope displayed the laser pulse. By calibrating the sweep initiation of the oscilloscopes versus the sequence of events with knowledge of the time of flight of the laser pulse to the target the absolute delay between arrival of the pulse at the gap and breakdown could be determined to within < 1 nsec.

PARAMETER DEFINITION

The main parameters monitored in this experiment were: gap spacing, laser focal point position, laser energy, laser power, delay to breakdown, delay to laser arrival, voltage and polarity at the target electrode at time of laser arrival, and breakdown voltage. Other secondary parameters, such as dielectric and optical system condition, were monitored on every

shot to verify that they were at their nominal operational condition. The gap spacing is defined as the minimum distance of separation of the two electrodes and was obtained by a thickness gauge block.

LASER FOCAL POINT POSITION

The laser focal point position was obtained experimentally by firing the laser into the gap at its largest separation and taking time integrated photographs of the laser breakdown of the dielectric in the switch chamber. The laser focal point position was then defined as the point closest to the target electrode at the breakdown threshold by placing filters in the optical path while operating the laser at constant input to insure a constant output beam divergence. The laser energy was corrected for losses at the dielectric surfaces. The laser power is the peak instantaneous power achieved in the giant pulse and was obtained from the laser pulse waveform and the laser energy. This value was determined by assuming a temporal Gaussian distribution of the laser pulse.

Delay to breakdown is the time between the arrival of the laser pulse at the target electrode and the conduction of the gap. The delay to breakdown was obtained by three separate experimental measurements and cross checked analytically. One of the experimental values was obtained by the dual beam trace depicting the line voltage and the laser pulse, both triggered by the Marx trigger amplifier. This trace is depicted in Fig. 2. The

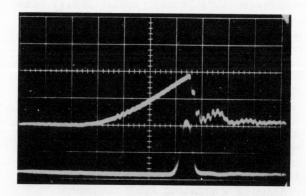

Fig. 2.   555 Dual Beam Oscilloscope Trace: Line Voltage Monitor and Laser Pulse.

other two values were taken from the
scopes monitoring the load voltage and
with the sweep initiated by the laser
pulse. The analytical cross-check was
made by using the delay to laser arrival
and adding the delay to breakdown. This
total time was then used to calculate the
voltage at which the gap went into conduction from the voltage pulse waveform.
The calculated breakdown voltage was
compared to the experimentally determined
breakdown voltage. All three values were
within 5.5 nsec of each other and the two
breakdown voltages compared within < 5%.

Delay to laser arrival is defined as
the time between the start of the voltage
pulse on the target electrode and the
arrival of the laser at the target electrode. The delay to laser arrival was
determined from the two traces of the
dual beam scope. This value was directly
read out from the trace as the time difference between the start of the voltage
pulse and the start of the laser pulse,
taking into account the laser time of
flight and the difference in cable
lengths. This value was cross-checked
analytically by using the voltage waveform and determining the voltage at the
target electrode upon laser arrival. The
analytical value and the experimental
value of the voltage on the target electrode compared favorably and were within
the 5% error associated with the line
voltage monitor. Voltage on the target
electrode at laser arrival and breakdown voltage are self-defining. The
first was determined from the dual trace
of Fig. 2 type and the latter was read
directly off the top trace of the aforementioned figure.

RESULTS

The experimental results of this
study are presented in Figs. 6 through
11. The dependent variable is the delay
to breakdown in nanoseconds, with the
independent variable being the voltage on
the target electrode at laser arrival
expressed in percent of self-breakdown.
The linear behavior depicted in these
plots are the result of least squares fit
to the data.

Each point on the plot represents an
average of 5 shots per polarity. The
self-breakdown voltage for the positive
pulse was found to be 700 + 35 kV while
the negative pulse exhibited a 600 + 35 kV
self-breakdown. Both values were deter-

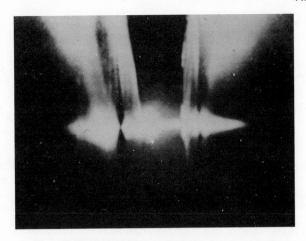

Fig. 3. Laser Pulse Interaction with the
Target Electrode(Target Electrode
is on the Left and the Laser Pulse
is travelling from Right to Left).

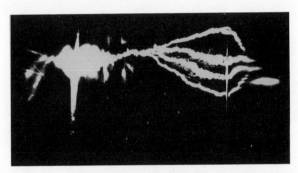

Fig. 4. Self-breakdown at 600 kV of a 20 mm
Gap (Left--Target Electrode;
Right--Load Electrode).

Fig. 5. LTS Breakdown at 580 kV of a 20 mm
Gap--Laser Travelling from Right
(Load Electrode) to Left (Target
Electrode).

mined employing a 20 mm gap. Figures 3, 4, and 5 are time integrated photographs of the laser interaction, self-breakdown, and LTS breakdown respectively. Note the linear arc when breakdown is initiated by a laser versus the branched breakdown observed for self-breakdown.

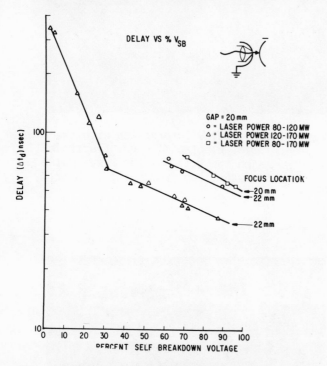

Fig. 6. Delay to Breakdown versus Voltage on Target Electrode at Laser Arrival--Negative Target Electrode.

Figure 6 illustrates the delay to breakdown versus voltage on a negatively charged target electrode in percent self-breakdown. Three important observations can be made from this plot. First, the plot of delay to breakdown for focus location of 22 mm and laser power from 120 to 170 MW results in two distinct slopes, with an inflection at a target voltage at laser arrival of 30% of self-breakdown voltage. This break can be attributed to the presence of the laser pulse at the onset of conduction of the gap. The laser pulse used in this experiment has an average width at its base of approximately 65 nsec which corresponds roughly to the marked increase in the delay to breakdown occurring at delays in excess of approximately 75 nsec. The importance of the laser pulse width to low jitter operation is discussed by Guenther and Bettis.[7] Their conclusions from DC gas dielectric LTS is that jitter is always minimum whenever the laser

pulse is still feeding energy into the gap region and that a marked change in the slope of delay and jitter vs SBV is evident when the delay is greater than the effective laser pulse duration. Second, the lower the laser power, the longer the delay to breakdown for the same focal position. This fact can be related to the initial number of charge carriers generated in the first few nanoseconds of the laser pulse. Finally, the optimum laser focus position was found to be at 2 mm inside the target electrode. Only the focal position of 20 mm is presented, but data points taken at a focal position of 27 mm from the load electrode (7 mm inside the target electrode) also demonstrated considerably higher delay to breakdown than at 22 mm.

Figure 7 illustrates the behavior of the breakdown voltage versus the voltage

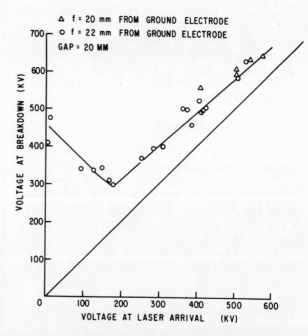

Figure 7. Breakdown Voltage versus Voltage on Target Electrode at Laser Arrival--Negative Target Electrode.

at laser arrival. The 45° line can be considered the zero delay to breakdown line. Note the bend on the RMS line joining the data points at a voltage at laser arrival of 180 kV (30% of self-breakdown), corresponding to the same phenomena associated with the presence of the laser pulse at the onset of conduction of the gap.

Figure 8 indicates the delay to breakdown versus voltage on a positively

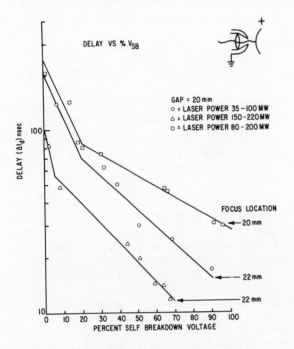

Fig. 8. Delay to Breakdown versus Voltage on Target Electrode at Laser Arrival--Positive Target Electrode.

charged target electrode at laser arrival in percent of self-breakdown. The same three conclusions as for the negative load electrode can be drawn from this plot. First, each of the lines for different focal positions and different laser powers present two distinct slopes associated with the presence of the laser pulse at the onset of conduction. Second, the delays to breakdown associated with the higher laser powers of the 22 mm focal are shorter than the delays associated with the lower laser powers for the same focal position. Finally, the optimum focal position was again found to be at 22 mm (2mm inside the target electrode).

Figure 9 indicates the final breakdown voltage versus voltage on the target

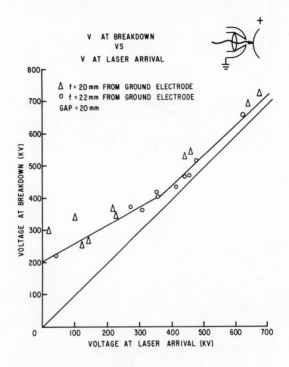

Fig. 9. Breakdown Voltage versus Voltage on Target Electrode at Laser Arrival--Positive Target Electrode.

electrode at laser arrival. Note that there seems to be a break in the slope of the lines joining the data points. This break is less distinguishable than the one present for a negatively charged electrode.

By comparing these four curves it is obvious that employing the target electrode at a positive potential leads to considerably less delay. The velocity of breakdown processes in liquids is always greater when initiation is at the positive surfaces. This is similar in behavior to solid and gaseous dielectrics LTS as well.[11,12]

Another interesting result was evident from this study. Figure 10 exhibits the delay time versus the difference between the time of arrival of the laser pulse at the gap and the start of the Marx Erection. This Marx Erection to Lase time (MEL) is a signed quantity

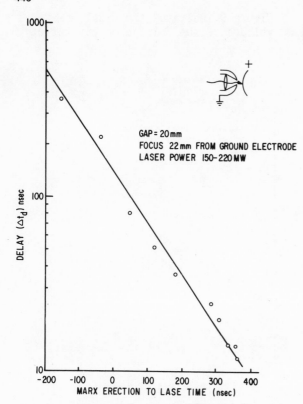

Fig. 10.   Delay versus Marx Erection
to Lase--Positive Target
Electrode.

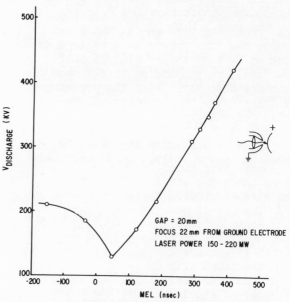

Fig. 11.   Breakdown Voltage versus Marx
Erection to Lase--Positive
Target Electrode.

with negative values when the laser
pulse arrives at the gap prior to Marx
Erection. It is seen that the first two
points on the curve were obtained by
laser pulses that reached the gap as much
as 150 nsec before any voltage appeared
across the gap. From this it is evident
that the laser beam is preparing the gap
by ionizing a path which retains its low
impedance for several tens of nanoseconds
since breakdown occurred at voltages well
below self-breakdown.

Finally we present in Fig. 11
what is perhaps the most significant
result from the standpoint of high-
power technology. We have replotted
the data in Fig. 10 as voltage at breakdown versus MEL. Since delay vs MEL
is logarithmic and the ramp voltage is
linear the curve in Fig. 11 is not
monotonic. However, for positive MEL
the curve is sensibly linear and it is
seen that by controlling the time of
entry of the laser pulse in the gap it is
possible to dictate the breakdown voltage
of the Marx generator. Just as the most

reliable way to vary output from a Q-
switched laser is to maintain fixed
operating conditions and insert variable
attenuators in the beam so it is that the
LTS technique supplies a "variable
attenuator" to the output voltage of the
Marx generator. Because it is possible
to command fire a laser within ± 1 nsec[13]
and because LTS exhibits jitter in the
low nanosecond range, it is possible to
quite accurately position a laser pulse
on a Marx ramp and deliver voltages within
± 2 kV out of 700 kV.

SPECIAL CONSIDERATIONS

Two problem areas were manifest
during this investigation--the mechanical
failure of the final lens and the
Stimulated Brillouin Scattering (SBS).
The final lens of the optical system is
placed three inches from the load electrode and has oil dielectric on one side
and air at atmospheric pressures on the
other. This glass lens normally failed
after five to six LTS events. When the
system was operated at gap spacings of
20 mm the electrical arcs, and presumably
the hydrodynamic shock, associated with
self-breakdown did not closely approach
the laser aperture. As a result, lens
failure occurred only when breakdown was

laser initiated, thereby directing the hydrodynamic shock into the aperture and onto the lens. This conjecture is borne out by the fact that when small gap spacings were employed, so that self-breakdown arcs terminated at or near the aperture, lens failure resulted. Recently, inexpensive acrylic lenses have been successfully used in the experimental arrangement to totally eliminate this problem.

Short life times of oscillator rods during the early stages of the experiment were attributed to backscattered SBS. The effect was revealed by the pulse shape detector noting that the time between several recorded pulses was the time of flight for the laser beam to cover the round trip distance between the photodiode, the dielectric, the rear reflector of the laser cavity and back to the photodiode. In addition, this back-reflected optical pulse is of the same polarization of the laser beam such that it is not reflected by the Brewster's angle faces of the amplifier rod. Since the SBS is within the fluorescent line width of the laser (albeit slightly shifted from its initial frequency due to the longitudinal sound velocity) the reflection can then cause stimulated emission of the inverted population still present at the amplifier rod. This amplified reflection is now further concentrated by the Galilean telescope between the amplifier and oscillator rods. The oscillator rod, still having an inverted population, further amplifies the reflection and is then subject to radiation power levels where damage to ruby crystal lattice can occur. This problem was sufficiently eliminated by placing a multiple plate tent polarizer and high quality $\lambda/4$ plate at the exit apertures of the laser system. Thus the backscattered SBS was effectively rejected by this combination of polarization sensitive optical elements.

SUMMARY

The versatility of LTS has been amply demonstrated by the variety of dielectrics and charging systems used. While the number of LTS events was too small to allow proper determination of jitter we can say that based on the excursion of data points from the straight lines of Figs 3 and 5 the jitter is no worse than + 2.2 nsec, and is most probably of the order of + 1.5 nsec. With careful attention to the problem of accurate command firing of the laser system the results of Fig. 6 indicate that very reliable voltage pulses can be triggered from a pulse charged liquid dielectric spark gap. Liquids such as water can be used to store enormous amounts of energy and on command discharge that energy through a laser triggered switch. Liquids also lend themselves well to multi-channel triggering as a method for lowering switch inductance. This study has demonstrated that laser initiated conduction of an overvolted transformer oil spark gap can effectively switch voltages up to 700 kV in times less than 15 nsec with jitter in the low nanosecond range. The use of liquid dielectrics is a compromise between the feature of nondestruction of the dielectric found in gases and the very small gap spacings possible with solid dielectrics while still retaining all the basic features of the LTS technique such as the absence of any hard wire electrical connection for command triggering.

REFERENCES

1. J.R. Bettis, "Laser-Triggered Megavolt Switching", M.S. Thesis, Air Force Inst. of Tech., Wright-Patterson AFB, Ohio, (1967).

2. W.K. Pendleton and A.H. Guenther, "Investigation of a Laser-Triggered Spark Gap", Review of Scientific Instruments, 36, 1546-50 (Nov 1965).

3. S. Barbini, "Coaxial Laser-Triggered Gap Study", Conference on Controlled Thermonuclear Reactions, Frascatti, Italy (1966).

4. A.H. Guenther and J.R. Bettis, "Laser-Triggered Megavolt Switching", IEEE Journal of Quantum Electronics, QE-3, 581-88 (Nov 1967).

5. A.H. Guenther, R.H. McKnight, and J.R. Bettis, "Nanosecond Jitter High Voltage Switching Using a Repetitively Pulsed Laser", NEREM Report, IEEE Publication F85, 36-37 (1967).

6.   A.H. Guenther, J.R. Bettis, R.E. Anderson, and R.V. Wick, "Low Jitter Multigap Laser-Triggered Switching at 50 pps" IEEE Journal of Quantum Electronics, QE-6, 492-95 (Aug 1970).

7.   A.H. Guenther and J.R. Bettis, "Laser-Triggered Switching", Laser Interactions and Related Plasma Phenomena, 131-71, Plenum Press (1971).

8.   D.M. Strickland, "A Laser-Triggered Switch Employing Solid Dielectrics", M.S. Thesis, Air Force Inst. of Tech., Wright-Patterson AFB, Ohio (1969).

9.   R.J. Clark, Laser-Triggered Switch Study, RADC Technical Report, TR-68-355, Rome Air Development Center, Griffiss AFB, New York (Dec 1968).

10.   A.J. Marolda, "Laser Triggered Switching in a Liquid Dielectric", IEEE Journal of Quantum Electronics, (Corresp) 503-5 (Aug 1968).

11.   A.P. Alkhimov, V.V. Vorob'ev, V.K. Klimkin, A.H. Ponomarenko, and R.I. Soloukhin, "The Development of Electric Discharge in Water", Soviet Physics Doklady 15 # 10, 959-61 (Apr 1971).

12.   C.N. Bruce and E. Collet, Laser Instrumentation, AFWL-TR-64-127, AD 364 551, Air Force Weapons Laboratory, Kirtland AFB, New Mexico (June 1965).

13.   Glenn A. Hardway, A.H. Guenther, and A.K. Graf, "Applications of High Power Pulsed Lasers", Annals of the New York Academy of Sciences, 168, 440-45 (Feb 1970).

A 2 MV, MULTICHANNEL, OIL-DIELECTRIC, TRIGGERED SPARK GAP*

K. R. Prestwich
Sandia Laboratories
Albuquerque, New Mexico 87115

ABSTRACT

A 2 MV, low jitter (< 2 nsec), multichannel (~ 10 channels), oil-dielectric spark gap has been developed to be used for switching the transmission lines of proposed new relativistic electron beam (REB) accelerators with substantially increased peak power (10^{13} W). The switch is designed for operation in a strip-line geometry with three 122 cm long electrodes, two smooth bars and one sharp-edged trigger blade. When a voltage pulse that rises to 1.5 MV in 20 nsec is applied to the trigger electrode, about 10 arcs will close simultaneously in 30 nsec with a standard deviation < 2 nsec. Data are presented on the variation of the switch closure time with applied voltages. The spark gap was installed on one of two 22 Ω, 20 nsec strip transmission lines connected as a Blumlein during test. The Blumlein output voltage pulse risetime decreases with increasing number of switch channels (13 nsec with 2 channels to 5 nsec for 14 channels). Data are presented on operation of the switch with the trigger electrode location varied from 20-40 percent of the gap spacing from the ground electrode. The effect of the interelectrode capacitance and isolation resistance on the operation of the switch is discussed. Several types of electrodes have been tested. Design data for a prototype REB accelerator using 12 of these switches in parallel are presented.

INTRODUCTION

Research efforts to decrease the pulse length and increase the peak power of relativistic electron beam accelerators (REB) for inertially confined fusion applications [1,2] have led to the development of a 2 MV, multichannel (10-12 channels in 120 cm) low jitter (< 2 nsec), oil-dielectric spark gap. The switch shown in Fig. 1 is commonly referred to as a rail switch and is designed for switching strip transmission line Blumleins as utilized in the proposed EBFA[3] and EBFA prototype designs. The development of low-jitter rail switches in gas dielectrics was initiated by Martin[4] at the United Kingdom Atomic Weapons Research Establishment and later extended to higher voltages (400 kV) by Champney[5] at the Physics International Company. Fifteen-megavolt field-distortion, multichannel, oil-dielectric switches having 5-6 channels with 90 cm diameter trigger electrode, and jitter of 5-10 nsec were developed by the Physics International Company for switching the Aurora pulse-forming transmission lines.[6,7]

EXPERIMENTAL CONSIDERATIONS

When the switch in Fig. 1 is attached to a transmission line, the trigger electrode is resistively biased for minimum electric field while the transmission lines are charging. Triggering is accomplished by applying a high-voltage, fast-risetime pulse to the blade. For low jitter operation the amplitude of the pulse should be large enough to overvolt both sides of the switch simultaneously. The fast-rise time pulse ($\dot{V} \sim 8 \times 10^{13}$ V/sec) highly overvolts the two gaps and large numbers of streamers are initiated at the trigger electrode. The upper photograph in Fig. 2 is an overexposed switch breakdown photograph of 90 percent of the wide gap and shows the very large number of channels that are initiated. The statistical processes involved in any dielectric breakdown result in a scatter in breakdown voltage. For fast rising pulses this voltage scatter is directly related to a time scatter in the closure of the switch. When the initial channels in a multi-channel switch close, the voltage across the gap(V) begins decreasing at a rate determined by the driving impedance, inductance and the rate of change of arc resistance of the individual channels. As V decreases, the closure time of the incomplete streamers increases (streamer velocity in oil is proportional to $V^{1.3}$), and only those channels that close while the voltage is within a few percent of

*This work was supported by the U.S. Atomic Energy Commission

its peak value will carry significant current. The highly field-enhanced, rapidly overvolted trigger electrode is a technique for minimizing the switch-closure-time scatter that is necessary for low jitter, multichannel operation

INITIAL EXPERIMENTAL SET UP

A sketch of the initial experimental set up is shown in Fig. 3 and a schematic is shown in Fig. 4. The 120 cm wide switch was installed on a 22 Ω, oil-insulated strip transmission line (Z_{01}) that was charged by a Marx generator to about 1.5 MV. The load (R_L) consisted of 8 copper-sulfate resistors. The 3 electrodes of the trigger Blumlein consisted of a pair of coaxial cylinders and a flat plate, and they were charged through an isolating inductor (L). The inductor was used to adjust the charge time of the trigger Blumlein to equal the Blumlein charge time. The trigger pulse must be opposite in polarity to the charge voltage and the reversal is accomplished by switching the trigger Blumlein as shown in Fig. 3; this switch is S_1 in Fig. 4. The trigger electrode is biased at approximately one-third of the charge voltage ($V/3$). The isolation switch (S_2) allows the trigger Blumlein to be fully charged while the trigger electrode is biased at $V/3$.

As indicated in Fig. 4, when the wide side of rail switch (d_1) closes, the trigger Blumlein and the strip transmission line (Z_{01}) are connected in series through the isolation resistor (R_I). Unless the value of ($R_I + Z_{T1} + Z_{T2}$) is substantially larger than Z_{01}, a precursor pulse will be propagated down the strip line (Z_{01}), decreasing the voltage across it while d_2 is closing. To limit the Z_{01} voltage decrease to < 10 percent of the charge voltage, R_I must be > 500Ω for this experiment. On the other hand, if d_1 and d_2 are adjusted such that they close nearly simultaneously, the need for the 500 Ω isolation resistor is relaxed.

The interelectrode capacitance (C_1 and C_2) of an oil dielectric rail switch is appreciable; C_2 is at least 70 pf for a 120 cm wide switch. The voltage risetime at the trigger electrode is not only dependent on the trigger Blumlein operation but also on the product $(R_I + Z_{T1} + Z_{T2})(C_1 + C_2)$, assuming

$Z_{01} \gg (R_I + Z_{T1} + Z_{T2})$. For a 10 nsec risetime, the value of $(R_I + Z_{T1} + Z_{T2})$ must be < 140 Ω. With this value of isolation resistance, the precursor pulse would be 0.3 V. In the most successful experiments, R_I was 100 Ω, and the geometry of the switch was adjusted such that d_2 closed within 4 nsec of d_1. With this arrangement the precursor pulse effect on the output risetime was negligible.

INITIAL EXPERIMENTAL RESULTS

With the experimental set up as shown in Fig. 3, the Blumleins were charged to 1.4 MV in 400 nsec. An average of 8 channels closed, giving a 10-90 risetime at the load of 13 nsec. A plot of the maximum slope output risetime (roughly a factor of two smaller than the 10-90 risetime), versus the number of channels carrying current is shown in Fig. 5. The number of channels was obtained from still photographs by counting only the bright channels. Some of the scatter in the data could be attributed to uncertainties in exactly which channels carry current. The average switch closure time was 50 nsec (65 nsec including the isolation gap) with a standard deviation of 7.6 nsec.

When the trigger blade was replaced by sixteen 0.64 cm diameter pointed brass rods, the average number of channels closing was 9. The operating voltage was lower because of increased field enhancement and imperfect biasing during charge. Therefore, the average 10-90 risetime increased to 18 nsec. The closure time decreased to 55 nsec including the isolation gap closure time with a standard deviation of 4.6 nsec. The standard deviation of the closure of isolation gap was < 2 nsec.

The closure time jitter with the set up shown in Fig. 3 was larger than the objective of 2 nsec. The trigger blade voltage excursion was smaller and slower (0.8 MV in 40 nsec) than desired. The capacitance to ground from the feed lines and isolation resistors is in parallel with the interelectrode capacitance (C_2). This large capacitance (~ 230 pf) and a relatively inductive trigger Blumlein initiating switch (S_1) limited the trigger pulse risetime. The low amplitude trigger voltage caused the relatively long closure of the wide side of the switch (40 nsec), and the trigger voltage was decreasing rapidly when closure occurred. Both of these conditions contributed to the jitter.

IMPROVED TRIGGER SYSTEM

Figure 6 is a sketch of an experiment that eliminated most of the problems associated with the trigger system shown in Fig. 3. The system has two Marx generators allowing independent adjustment of the trigger voltage amplitude. The trigger transmission line is a single, 40 Ω, 70 nsec coaxial transmission line and the isolation switch is a self-breaking, 3 MV, SF_6 insulated switch.

In this experiment, 4 voltage monitors were used. A capacitive divider was installed to monitor the trigger voltage and eliminate magnetic pick-up problems associated with the previously used resistive divider. A capacitive divider was installed in the transmission line (Z_{01}) 45 cm from the switch to monitor the switch risetime without the effects of load inductance. The RC time constant of this monitor was such that the 10 nsec pulse risetime could be measured, but it would droop to near zero voltage on the 700 nsec charging pulse from the Marx generator. Both resistive and capacitive dividers were used to measure the output pulse. In general, the two monitors agreed.

EXPERIMENTAL RESULTS

The voltage waveshapes for a typical shot are shown in Fig. 7. The trigger pulse rises 1.7 MV in 20 nsec. Breakdown of the wide side occurs when the pulse starts negative and the narrow side when it reverses again. The transmission line voltage monitor indicates a 10-90 rise of 6 nsec. Voltage reflections from the load appear 12 nsec into the pulse. The output voltage monitors indicate a 9 nsec, 10-90 risetime.

The lower six pictures in Fig. 2 are open shutter photographs of switch closures taken from the load side of the switch showing both gaps. Each pair of photographs show the complete 120 cm of the switch. In the lower two photographs the trigger electrode was replaced by a serrated blade with 20 points. Eighteen of the twenty appear in the photograph. Note that the streamers initiate from each of the points shown and that closure locations on

one side of the switch are completely independent of the locations on the other side of the switch.

The switch closure times decreased from 50-65 nsec to 30-35 nsec depending on the amplitude of the Blumlein charge voltage as shown in Fig. 8. The straight line is a least squares fit to the data and the RMS deviation from the line is 1.3 nsec. Therefore, the jitter of the switch closure for this set of data is 1.3 nsec.

Table 1 is a tabulation of data from experiments for several different switch configurations. All risetimes are the time differences between the 10 and 90 percent of peak voltages. Relative trigger voltage is the ratio of the trigger pulse amplitude to the Blumlein charge voltage. If closure times are expressed as the equation for the linear least squares fit to the data (i.e., $T = 57 - 15 V$), the σ is the RMS deviation from that line. In the cases where the closure times are expressed as the average time (i.e., $T - 35$), σ is the standard deviation. Figure 7 was derived from the same data as the first line in Table 1. The switch configuration for the data in the first two lines of Table 1 had a pipe for the high voltage electrode and a 6.4 cm gap between the main electrodes. Data for the essentially optimum bias are given. Data are also taken (but not shown) with this arrangement to determine the operating range of the switch. At 22 percent bias, the small gap (d_2) closed first and only one or two channels closed on the wide side (d_1). At 40 percent bias, the short gap closed 20 nsec after the wide side, and a 20 nsec precursor voltage was produced at the output.

The data in lines 3 and 4 of Table 1 were taken with a smaller field enhancement on the main electrodes and a 2.9 cm gap for 2 MV testing. The self-break voltage was 2.3 MV. The average number of channels decreased slightly but the 10-90 risetime improved at the higher voltage. On the average, the transmission line monitor indicated a 3 nsec faster risetime than the output monitors. The RMS deviation for the closure time was 2.1 nsec.

The fifth line is data taken with the trigger blade replaced by a serrated blade with 20 points. Substantially the same results were obtained as with the previous arrangement. The final three lines represent results of a test to determine the minimum trigger voltage amplitude while maintaining satisfactory low jitter operation. If the blade is located in the optimum position, the trigger voltage can be reduced to 44 percent of the switch voltage with very little change in switch performance. As the trigger voltage amplitude is lowered, it is necessary to locate the trigger blade nearer to the ground electrode for nearly simultaneous closing of the two gaps.

The risetime versus number of channels for the data in the first line of Table 1 is shown in Fig. 9. The curve is flatter than Fig. 5. The risetimes for 4 and 5 channels are faster than expected, probably because of effects of the high voltage trigger circuit on the resistive phase of the gap. Variations in average electric field across the gap will affect risetime and are the cause of some of the spread in the data.

Figure 10 is an artist's sketch of a 3 MV, 0.8 MA, 24 nsec electron beam accelerator presently under construction at Sandia Laboratories. It consists of two diodes with a common anode. Each diode is pulsed from 6 oil-insulated Blumleins in parallel. For 3 MV operation, intermediate storage capacitors with water dielectric are required for faster charge. Only one is shown on the sketch. Initial experiments will be at the 2 MV level and the Marx generator will charge the transmission lines in 700 nsec.

The 12 Blumlein switches will be rail switches described in this paper. The trigger pulse is generated with a separate Marx generator and a water-insulated transmission line. The pulse is transmitted to the 12 switches in oil-insulated transmission lines.

CONCLUSION

A 2 MV, 10-12 channel, oil-dielectric rail switch that produces an 8 nsec, 10-90% current risetime on a 22 Ω transmission line has been developed. The closure time has been shown to be voltage dependent. It is about 35 nsec

at 2 MV and the RMS jitter is < 2 nsec. An electron beam accelerator that utilizes 12 of these switches was briefly described. If < 2 nsec jitter can be maintained, the effect of this jitter on the output voltage risetime will be negligible.

ACKNOWLEDGEMENTS

The author wishes to acknowledge many helpful discussions with T.H.Martin and J.C. Martin. He also wishes to acknowledge J.C.Corley, R.S.Clark, A. W. Sharp and O. Marten for their assistance in design and construction of the experiment and in the collection of data.

REFERENCES

1. G. Yonas, J.W.Poukey, J.R. Freeman, K.R. Prestwich, A.J . Toepfer, M.M. Clauser, and E.H. Beckner, Sixth European Conference on Controlled Fusion and Plasma Physics, Moscow, p. 483, (1973).

2. J.Chang, M.J.Clauser, J.R. Freeman, D.L. Johnson, J.G.Kelly, G.W. Kuswa, T.H.Martin, P.A. Miller, L.P. Mix, J.W. Poukey, K.R. Prestwich, D.W. Swain, A.J. Toepfer, M.M. Widner T.P. Wright, G. Yonas, Fifth Conference on Plasma Physics and Controlled Nuclear Fusion Research, Tokyo (1974)

3. T.H. Martin and K.R. Prestwich, International Conference on Energy Storage Compression and Switching, Asti, Italy (1974).

4. J.C. Martin, "Multichannel Gaps," Internal Report: SSWA/JCM/703/27, AWRE, Aldermaston, England (1970).

5. P. Champney, "A Low-Inductance Triggered Multichannel Gas Switch," PIIR-23-70, Physics International Company, San Leandro, California (1970).

6. G. Bernstein and I. Smith, 1971 Particle Accelerator Conference on Accelerator Engineering and Technology, IEEE Trans. on Nucl. Sci. NS-18, No. 3, p. 294 (1971).

7. I. Smith, International Conference on Energy Storage, Compression and Switching, Asti, Italy (1974).

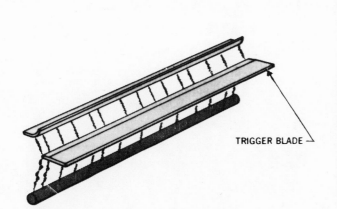

Fig. 1.   Rail switch electrode sketch.

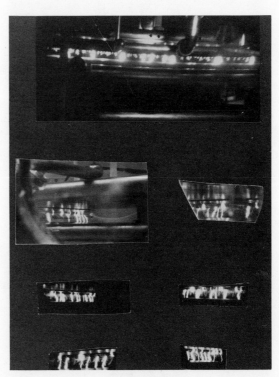

Fig. 2.   Multichannel switch photographs.

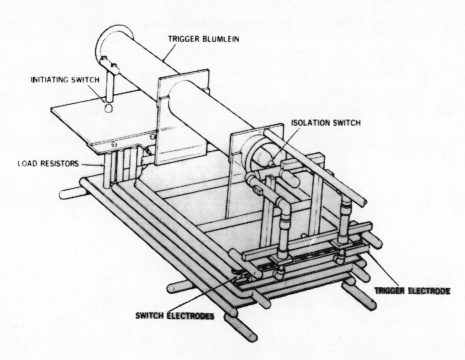

Fig. 3.   Initial multichannel switch test experiment.

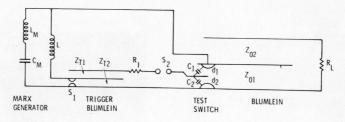

Fig. 4.   Schematic diagram of initial multichannel switch experiment.

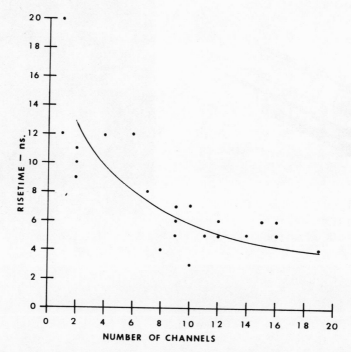

Fig. 5.   Output rise time vs. number of switch channels for initial experiments.

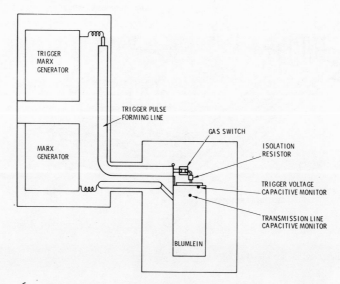

Fig. 6.   Test set up with improved trigger generator.

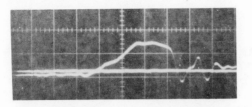

1.2 MV/CM   10 NS/CM

TRIGGER VOLTAGE

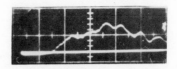

10 NS/CM

TRANSMISSION
LINE VOLTAGE

1.0 MV/CM   10 NS/CM

OUTPUT VOLTAGE
CAPACITIVE MON.

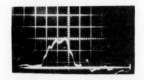

0.56 MV/CM   10 NS/CM

OUTPUT VOLTAGE
RESISTOR DIVIDER

Fig. 7.   Typical voltage waveshapes.

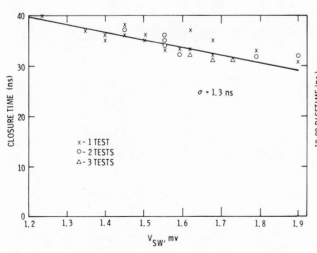

Fig. 8.   Triggered oil switch closure times
vs. voltage across main electrode.

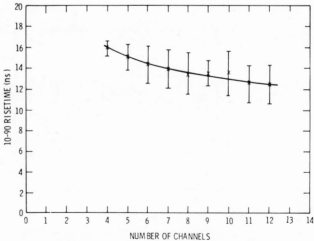

Fig. 9.   Output rise time vs. number of
switch channels with improved
trigger generator.

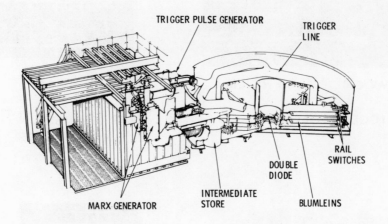

Fig. 10.   EBFA prototype.

Table 1.   Multichannel, triggered, oil switch
test data

ELECTRODES	BIAS	# TESTS	VOLTAGE RANGE MV	AVERAGE NUMBER CHANNELS	AVERAGE OUTPUT RISETIME ns	AVERAGE SWITCH RISETIME ns	RELATIVE TRIGGER VOLTAGE	CLOSURE TIME V−MV, T − ns σ − ns
STRAIGHT SHARP EDGE	0.31	38	1.3 − 1.9	10	13		0.94	T = 57 − 15V σ = 1.3
	0.28	14	1.6 − 1.8	12	14		1.06	T = 35 σ = 1.6
STRAIGHT SHARP EDGE	0.30	10	1.5 − 2.0	7	14	10	0.92	T = 63 − 16V σ = 1.6
	0.36	48	1.4 − 2.0	9	11	8	0.92	T = 67 − 15V σ = 2.1
SERRATED BLADE 20 POINTS	0.28	15	1.6 − 2.0	11	11	8	0.83	T = 33 σ = 1.6
	0.32	5	1.8 − 2.0	10	12	8	0.69	T = 32 σ = 1.4
	0.28	12	1.6 − 2.0	8	12	10	0.60	T = 39 σ = 1.6
	0.19	9	1.8 − 2.1	8	12	9	0.44	T = 37 σ = 1.3

A COMPACT, MULTIPLE CHANNEL 3 MV GAS SWITCH[*]

S. Mercer and I. Smith
Physics International Company
2700 Merced Street
San Leandro, California 94577
and
T. Martin
Sandia Laboratories
Albuquerque, New Mexico 87117

ABSTRACT

A triggered, pressurized sulfur hexafluoride spark gap has been developed in which multiple parallel spark channels are consistently generated in close proximity to each other. In a test in which the switch drove a 60 ohm, 20 nsec pulse line, an average of four to six channels were created within a diameter of about 3 inches. Each channel was about 3.5 inches long in 175 psig SF_6, and the total switch inductance was in the region of 90 nH. Triggering is effected by a field-enhanced electrode placed close to the ground electrode, a configuration known as the V/N switch. The trigger voltage at 3 MV pulse charge is 250 kV. The rms jitter of the switch breakdown is less than 1 nsec. It is expected that the high performance (low jitter obtained with a small ratio of trigger voltage to total voltage) will scale to much higher voltages. The switch used in the developmental tests was designed for oil immersion, and possessed a further novel feature, a pressure vessel that can withstand high electrical stress with good reliability. The pressure wall achieves this performance through the use of a segmented design, and withstood up to 3.75 MV over a total distance of about 8.5 inches.

INTRODUCTION

The work presented stemmed from a requirement for a compact, 3 MV gas switch with very low jitter and inductance, for possible use in the Sandia Laboratories (Albuquerque) Electron Beam Fusion Accelerator program. A prototype switch was developed and tested at Physics International Company.

The triggering scheme employed in the switch gives rise to the name "V/N switch," referring to the fact that the ratio of the voltage across the whole switch to that of the trigger electrode is a large number (N). However, this arrangement should be viewed as simply the logical result of extending the principle of the three electrode, field distortion gas switch to the multimegavolt range. Figure 1 illustrates the operation of a field distortion switch, typifying a gas switch in the range of tens to a few hundreds of kilovolts.

The trigger electrode is placed slightly off center (usually); it has a sharp edge, but there is initially no field enhancement there because the trigger electrode is maintained at a voltage corresponding to the equipotential on which it lies. The trigger pulse then changes the voltage of the trigger electrode, driving it towards, and usually behind, the potential of the nearest electrode. This distortion of the natural fields in the gap produces a very high field at the edge, launching a breakdown streamer. This usually closes first to the farthest electrode, then to the nearest.

To have a low jitter--on the order of 1 nsec was needed here--and stable operation, it is necessary to have a short breakdown time, of the order of 10 nsec. The breakdown time is essentially that of the streamer transit, and most of this occurs while the streamer is still close to the trigger edge; as the streamer

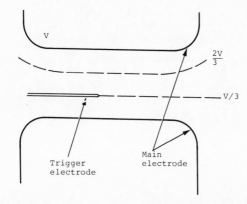

(a)   No trigger pulse

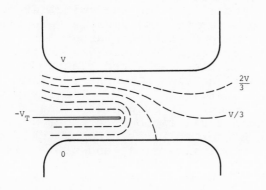

(b)   With trigger pulse

Fig. 1 Three electrode, field distortion switch. Trigger
electrode is at V/3.

lengthens, it accelerates rapidly.
Consequently, the switch performance is dominated by the field produced by the trigger pulse at the
trigger edge. This, in turn, is
largely determined by the trigger
pulse amplitude and the radius of
curvature of the trigger edge. As
the switch voltage is scaled upwards,
the overall switch spacing is scaled
up, but provided the trigger edge
radius is maintained constant the
trigger pulse need not be scaled in
proportion to the total voltage;
for constant performance the
trigger pulse need increase only
slowly.

    For a spark gap in the multi-
megavolt range, it would therefore
be acceptable to place the trigger
electrode near the center of the
gap and apply a trigger pulse that
drives it only a few hundred kilovolts off its natural equipotential. However, a higher field will
be obtained at the trigger edge if
the trigger electrode is located
on a natural equipotential only
a few hundred kilovolts from one
electrode; this is because the

presence of a nearby fixed equipotential helps to crowd the general field
changes produced by the trigger pulse
into a smaller region. Thus as field
distortion gas switches are used at
higher and higher voltages, the V/N
geometry becomes very appropriate; it
also eliminates the need to deal with
a trigger electrode several kilovolts
from ground.

    Achieving the desired short
breakdown times is also assisted by
the use of high average fields in
the switch, which also leads to
smaller dimensions and hence inductances. For this reason, sulphur
hexafluoride (SF_6) at a pressure of
175 psig was used here. In SF_6,
streamers propagate more rapidly
from positive than from negative
electrodes, and so the trigger electrode was placed on the ground side
of the switch with a negative high
voltage applied. A previous switch
of similar design used to trigger
the AURORA Blumlein[1] had achieved a
jitter of 2-3 nsec with a 1.8 mm
trigger-to-ground spacing; in order
to improve this to the sub-nanosecond range, it was decided to
increase the field at the trigger
edge by using a trigger-to-ground
spacing of about 5 mm, with a correspondingly larger trigger pulse.

SWITCH DESIGN

    Figure 2 shows the switch con-
figuration. The electrodes are
rounded cylinders of stainless steel,
4 inches in diameter with a 3 inch
separation. The trigger electrode is
a 2.75 inch diameter disk, curved to
follow the shape of an equipotential
near the ground electrode. Physical
support and connection to the trigger circuit are made through a hole
in the ground electrode. The connection is made via an isolation spark
gap, so that the trigger electrode
can assume the voltage of its natural
equipotential by capacitive division.
The design achieves the correct division by making the trigger-to-ground
capacitance eliminated by cutting the
hole in the ground electrode roughly
equal to that added in the connection to the isolation switch. Final
tuning is performed by externally
adjusting the spacing of the trigger
electrode to ground so that the
breakdown voltage of the entire
switch is maximized.

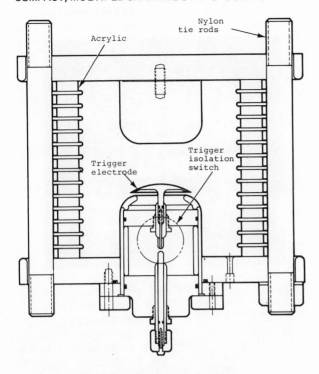

Fig. 2   Partial assembly drawing of triggered gas switch.

SWITCH ENVELOPE DESIGN

The prevention of tracking on the insulating solid/gas interface of the pressure envelope is often a concern in switch design. Tracking is an erratic process, which usually forces the designer to make the envelope much longer than is necessary on the average to withstand the voltage. Even then, tracking will occur from time to time, and require a repair to the insulating material. In the present switch, the envelope was designed to operate relatively close to breakdown. The envelope was made in 14 segments, separated by metallic annuli which enables occasional tracks to be tolerated; such tracks span only short small portions of the switch and they discharge only stray capacitances and thus do less damage. Total breakdown of the envelope on a single pulse requires several segments to track independently on the same shot, and the probability of this

can be very low even if individual tracks occur quite frequently. Using this principle, the switch was made only 20 cm long; single insulator switches in routine used at the same voltage are almost twice as long.

Breakdown tests were made on individual insulator segments to optimize their design. Various solid/gas interface shapes were tried, but none performed better than the right cylinder. The cylindrical interface tracked at fields ranging from 250 to 350 kV/cm.

In selecting the insulator material, only transparent materials with good resistance to shattering were considered. It was found that low energy tracks did not degrade the hold-off of acrylic insulators (some upward conditioning occurred), but polycarbonate degraded badly. Acrylic was therefore used in the prototype switch. The final conclusion of the single insulator tests was that at the field levels achieved, breakdown is more likely on the outside surface of an oil immersed switch and attention must be paid to the design of this region.

SWITCH TESTS

The gas switch was immersed in oil and charged in 0.5 μsec by a Marx generator. The trigger pulse was generated by shorting one end of a cable with a self-closing gas switch. The opposite end of the cable was connected to the trigger isolation switch of the main spark gap.

At a voltage of 3 MV, the trigger electrode potential is about 250 kV (N=12 here, in V/N terminology). The cable was pulse charged to about 250 kV. Ideally this should occur with a waveform similar to that charging the switch, so that little voltage appears on the isolation gap. The trigger pulse can then provide a huge over-voltage to the isolation gap, making its jitter negligible. This situation was not achieved in the present tests.

The switch was first tested without triggering to a maximum

voltage of 3.75 MV, an overtest of
25%. Occasional flashover of individual envelope segments was observed, but total breakdown only
occurred as a result of sparking
in the main electrode gap.

Triggered tests were then made
at voltages up to 3 MV. The time
between the trigger pulse application and switch breakdown was typically just over 20 nsec with a jitter of about 0.9 nsec rms including
measurement error. The isolation
gap breakdown time accounts for a-
bout half this delay, and may contribute most of the jitter. The
breakdown time is not strongly dependent on voltage (at a fixed
pressure), as is shown by Fig. 3.

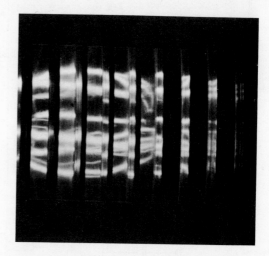

Fig. 4   Open shutter photograph of
multichannel discharge.

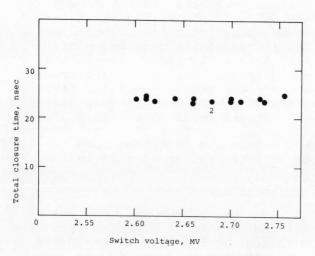

Fig. 3   Triggered closure times
versus switch voltages.

Finally, the switch was connected to a 20 nsec, 60 ohm open circuit pulse line. At voltages up to
3 MV it produced pulses rising in
about 3 nsec, implying an inductance
of about 85 nH.

This value is consistent with
the multiple channel breakdown of
the switch visible in open shutter
photographs. Five or six channels
were created on average, fairly
evenly distributed around the circumference of the trigger electrode.
A typical photograph is shown in
Fig. 4. The inductance with a single channel calculates to be approximately 150 nH.

SUMMARY

The goals of sub-nanosecond
jitter and compact design were
achieved. The inductance was well
below the target range because of
the multiple channel operation,
which should also help the switch
to carry large currents without excessive electrode wear. Although
the present tests were limited to
3 MV, the principle of the V/N
switch is such that simple scaling
up of the main gap spacing and envelope length, with no change in
the trigger region and trigger circuit, should result in much higher
voltage switches in which jitter
adequately low for almost all applications is obtained with trigger
voltages low enough to be handled
by readily available coaxial cables.

REFERENCES

*This work was supported by Sandia
Laboratories, Albuquerque, N.M.

1. B. Bernstein, I. Smith, Aurora,
An Electron Accelerator, IEEE Trans.
Nuc. Sci. NS-20, 3, June 1973, p. 294.

SOME RECENT ADVANCES IN THREE ELECTRODE FIELD ENHANCED TRIGGERED GAS SWITCHES

P. D'A. Champney
Physics International Company
2700 Merced Street
San Leandro, California 94577

ABSTRACT

The field distortion gap is a spark gap with a third (trigger) electrode having a sharp edge. The trigger electrode creates a high field at the edge, producing rapid breakdown. Gas spark gaps employing this principle have been widely used. Several recent advances are described which represent important progress in switch technology.

A dc-charged rail switch 1.8 meters in length with an inductance of 1.7 nH $\pm$ 2 percent has been developed. It approaches the performance levels hitherto available only with single shot solid dielectric switches. This has been used at a maximum voltage of 60 kV, to discharge a 55 kJ capacitor bank at currents up to 1.5 MA, with over 4 coulombs flowing in the switch. Typically about 15 channels are created over an operating range of over 3:1 in voltage.

The rail switch performance was achieved with a newly discovered gas mixture which gives much improved multiple channel breakdown characteristics and a wider triggering range. Two other switches are described that have utilized this mixture. One is an air-insulated 6 nH rail gap pulse charged up to 130 kV in 2 to 3 μsec. This switch carries 500 kA and 1 Cb. The other is a dc-charged single channel, 26 nH switch operable from 20 to 60 kV. This gap may be triggered down to 35 percent of self breakdown with a jitter of only a few nanoseconds.

INTRODUCTION

The development of energy sources capable of rapidly delivering their stored energy into a load upon command, has necessitated the design of a variety of spark gaps. Probably the most versatile type of spark gap is the field distortion gap. Field distortion gaps have three electrodes; a trigger electrode having a relatively sharp edge is positioned symmetrically or asymmetrically between two main electrodes. The trigger electrode is held at a proportion of the gap voltage corresponding to its physical spacing, so that the sharp edge does not appreciably disturb the equipotentials from that of a uniform field distribution. Normally the field distortion gap is employed in a cascade mode. In this mode the trigger electrode is often positioned midway between the main electrodes, which is convenient in a balanced charged system. A trigger pulse is applied to one half of the gap which breaks down, imposing the full charging voltage on the second half of the gap. This in turn breaks down, completing the closure of the gap. This geometry allows the gap to be triggered down to about 60 percent of the gap self-breakdown voltage with a typical rms breakdown jitter of a few nanoseconds. By off-setting the trigger electrode it is possible to arrange for both halves of the gap to break down simultaneously or very close to simultaneously. Under these conditions the gap may be triggered down to less than 40 percent of the gap self-breakdown voltage with an rms breakdown jitter on the order of a nanosecond.

In the past in order to achieve fast switching either the gap has been designed to have a small electrode spacing to minimize inductance and/or a number of gaps have been operated in parallel. The gaps have been triggered using the principle that each gap should be physically separated from the others such that the electrical transit time is comparable to the spread in breakdown times between gaps.

More recently the field distortion gap concept has been used in the form of continuous electrodes occupying the full width of the energy source feed. This allows a number of breakdown channels to be produced over the width of the switch. Normally the separation between the main electrodes is minimized to reduce the length of the individual spark channels. The voltage operating range of the gap, i.e., the range of voltages over which the gap may be triggered with an rms jitter of a nanosecond or so without prefire is about 1.9:1. When a number of such gaps are operated in parallel, the gaps require operating at a lower percentage of self-breakdown voltage to reduce the probability of prefire. Under these circumstances the voltage operating range may drop to 1.3:1. This range may be extended in the case of a gaseous dielectric medium by providing an envelope around the electrode structure so that the pressure of the gap may be varied. By this means the voltage operating range may be extended to over 3:1 while still operating the gap at approximately 60 percent of self-breakdown.

The passage of high currents (≥10's of kA) and/or large quantities of charge (≥100 mCb) requires consideration of electrode materials, dielectric media, and spark channel biproducts. The electrode material and chemical byproducts affect the surface condition of the switch electrodes which in turn influences the gap self-breakdown voltage, and ability to break down in multiple channels. Deposition of electrode material and/or chemical byproducts formed by the spark channels on the envelope wall can affect the voltage hold-off capability of the envelope.

One of the best switch material combinations has been found to be brass main electrodes, a brass or tungsten trigger electrode and a Lucite envelope with sulfur hexafluoride as the gaseous dielectric. Sulfur hexafluoride has an electrical breakdown strength 2.8 times that of air, and as a result allows small electrode separations to be used. The sulfur hexafluoride and its biproducts do not appear to electrically impair the switch voltage hold off or triggering characteristics, nor the multi-channeling capabilities.

When the spark channel forms, the voltage across the channel drops more or less exponentially. This fall in impedance, or resistive phase can, in some circuit designs, be the limiting factor on switch risetime. Sulfur hexafluoride exhibits an anomalously high and erratic resistive phase at pressures below 10 psig. To avoid this occurring switches operating at voltages below about 100 kV require very small switch gap spacings with high mechanical tolerances. These small spacings lead to electrical degradation of the gap when high currents and charges are passed due to electrode surface deformation by the spark channels and spark channel biproducts.

DESIGN AND EXPERIMENTATION

The following describes the development of three types of field distortion switch that use a gaseous dielectric that appears to exhibit all the required features of a dielectric for high current, high coulomb spark gaps.

The new features required in the development of an enclosed switch for capacitor bank modules operating in the region of 100 kV were: (1) the ability to hold off the charge voltage reliably, and (2) the use of small spacings and lower gas pressures commensurate with the lower voltage.

Three experimental programs to develop low inductance, low voltage rail switches have recently been conducted at Physics International Company.

The first was for Los Alamos Scientific Laboratories, New Mexico. Eight switches were to be back biased, -30 kV dc and then charged forward to about +125 kV in 2 μsec. Each switch had to fire with an rms jitter of less than 10 nsec when supplied with an 80 kV trigger pulse. The overall switch inductance including attachment to a parallel plate transmission line operating in a 50-cm-Hg air environment was not to exceed 10 nH. The switch was to be capable of passing 500 kA and 1 Cb reliably. The switch envelope consists of a Lucite block which is contoured to give adequate external flashover characteristics and has a 4.8-cm-diameter transversely bored hole into which the 2-cm-diameter main rail electrodes locate. The knife edge trigger electrode is positioned such that the gap spacing ratio is 1.5:1. A positive trigger pulse is supplied to the knife edge trigger so as to break the larger gap first. When triggered at a charge voltage of 140 kV with a switch pressure of 30 to 40 psig nitrogen, 15 to 20 breakdown channels are produced resulting in an overall switch inductance of 6 nH. The number of channels has recently been increased to 30 using the newly discovered gaseous dielectric.

The second experimental program was for the Air Force Weapons Laboratory, New Mexico. Two triggered gaseous dielectric switches were to be used in series, each occupying the full 1.8 meter width of a capacitor bank feed. The operating voltage range was to be 15 to 50 kV and the peak current 1.5 MA. The module design required the overall inductance of each switch to be ≤2 nH.

The switch design incorporates the following features: (1) the metal bolts securing the main brass rail electrodes to the transmission line plates assist in compressing a silicone rubber gasket beneath the epoxy switch envelope. Compressible silicone rubber gaskets provide the dc high voltage insulation throughout the module. Dependence upon corona-graded insulating sheets at conductor edges was eliminated resulting in a compact design

operable in air at ±60 kV charge independent of local atmospheric pressure or humidity and without the use of additional insulants such as sulfur hexafluoride or transformer oil. The metal bolts also carry the main discharge current and also provide a pressure seal by means of O-rings located around the bolts between the envelope and transmission line plates. (2) The knife edge trigger electrode is made of machinable tungsten, and is attached to a metal bar mounted parallel to the main electrodes. (3) In order that the inductance of the switch be minimized, the main electrodes are positioned as close as possible to each other and to the high voltage and load plates. This results in a high field at the epoxy-gas interface between the main electrodes. This interface is lengthened by grooving the surface of the epoxy. The grooving serves the additional function of breaking up the cylindrical shock wave that is launched as a result of the multi-channel breakdown of the switch. (4) An expansion volume is provided above the main electrodes to allow the hot gases produced by the breakdown channels to expand in a somewhat longer time scale, thereby considerably reducing the stresses that would otherwise build up in the epoxy pressure vessel.

Three 33 ohm trigger cables feed each rail switch via blocking capacitors and series post-breakdown cable terminating resistors. The cable feed points are equispaced along the length of the knife edge. The trigger cables impose a voltage excursion upon the knife edge approaching 140 kV in 15 nsec (10 to 90 percent).

Adjustment of the main electrode spacing is made possible by slots in the main electrodes through which the compression bolts pass. Initial testing of the switch was performed with a maximum spacing of 1.4 cm to (a) minimize gap tolerances and knife edge position tolerances over the 1.8 meter width and (b) enable the use of nitrogen or air as the gaseous dielectric. A large number of rail switch prefires or drop-outs occurred with both air and with nitrogen as the dielectric media. The standard deviation in breakdown voltage about the mean for a typical spark gap is on the order of 4 percent. A self-break or pre-fire occurring at less than 4 σ of the mean breakdown voltage is therefore regarded as a drop out. There appeared to be some correlation between frequency of drop outs and the magnitude of the peak current and total charge passed by the switch; the number of drop outs increased as the stored energy in the module increased.

Concurrent with these experiments an investigation was conducted for a gaseous dielectric compatible with low voltage, high current, high coulomb spark gap operation. Basically the problem was one of identifying an electrically weak gas (~30 kV/cm breakdown), that possessed a drop out probability as low as that of sulfur hexafluoride when used in a spark gap passing about 100 to 200 kA and about 0.5 coulombs per channel.

A test stand was fabricated comprising a 1.85 µF 60 kV capacitor discharging into a resistive load via a spark gap consisting of brass main electrodes, a machinable tungsten disk trigger electrode, and a Lucite envelope. The circuit enabled 155 kA, 0.53 Cb to be passed at 50 kV dc, and 190 kA, 0.63 Cb at 60 kV dc. The trigger electrode was resistively held at a potential appropriate to the physical spacing. A power supply charged the capacitor and spark gap exponentially to a maximum of 60 kV. The pressure in the spark gap was adjusted to ensure that the gap normally fired in the region of 50 to 60 kV. A self-breakdown rate of two shots per minute was typical. The breakdown voltage was recorded on a chart recorder, and the ringing discharge monitored periodically by an oscilloscope. Various gases and gas mixtures were tested in the spark gap. The data was plotted as an integral probability distribution, in which the probability that the spark gap would fire below a given voltage was plotted against that voltage. With probability as a logarithmic ordinate against a linear voltage abscissa, the results usually took the form

of a line with an approximately constant slope--indicating that, over a considerable range, the probability of breakdown decreased exponentially with decreasing voltage. As the maximum breakdown voltage of the gap was approached, the slope decreased to zero, while at lower voltages the slope appeared to increase. From this plot, a parameter could be defined which, like the standard deviation, gave a measure of the switch performance. This parameter was the change in voltage (Δ) which reduced the probability of breakdown from 10^{-1} to 10^{-2}, as a fraction of the median value.

Various percentages of sulfur hexafluoride and argon, ranging from 5 to 20 percent sulfur hexafluoride were tested. Approximately 200 shots were fired with each gas. All gave a pre-fire probability significantly less than nitrogen and comparable to pure sulfur hexafluoride. The 8 percent sulfur hexafluoride and 92 percent argon was chosen as having an equivalent dielectric strength to air and nitrogen. Further testing showed that failure to vent the gas after each shot significantly increased the pre-fire probability. (It should be noted that the volume of gas within the envelope was only about 300 cc.)

Normalizing the switch performances with the various gaseous dielectrics to that of sulfur hexafluoride was based upon the value Δ, the voltage which changes the firing probability from 10^{-1} to 10^{-2} as a percentage of the median. The 8 percent sulfur hexafluoride mixture is 0.89, the air 0.87, and the nitrogen 0.54. These results must be viewed with care. In the case of the pure sulfur hexafluoride, and the sulfur hexafluoride and argon mixture, no degradation in breakdown voltage, or in the condition of the switch occurred during the test sequences. However, with the air and nitrogen a steady voltage degradation appeared to take place after about 50 shots, and in some instances terminated the test sequence. With air as the dielectric a brown discoloration was observed on the inside of the Lucite envelope; a severe blackening occurred on the inside of the envelope with nitrogen, ultimately leading to small arcs developing.

Testing was resumed with the module rail switches, using a variety of gases and gas mixtures: pure sulfur hexafluoride, 13 percent sulfur hexafluoride and 87 percent air, 10 percent sulfur hexafluoride and 90 percent nitrogen, sulfur hexafluoride/argon mixtures with 20, 10, 8, and 5 percent sulfur hexafluoride. Pure sulfur hexafluoride gave very poor results--sub-atmospheric pressures being required to ensure switch triggering. (One or two breakdown channels were typical.) The sulfur hexafluoride/air, and sulfur hexafluoride/nitrogen mixtures were again required, resulting in about 3 to 4 breakdown channels.

The sulfur hexafluoride/argon mixtures gave very encouraging results. Two general observations may be made of the data collected: (1) It was possible to operate the rail switch much closer to self-breakdown without pre-fire occurring. (2) The lower the percentage of sulfur hexafluoride in the mixture, the lower it was possible to operate the rail switch in terms of percentage of self-breakdown for a constant number of breakdown channels. The limitation in this respect was the allowable working pressure within the rail switch envelope (40 psig maximum). For this reason a gas mixture of 8 percent sulfur hexafluoride and 92 percent argon was chosen as the preferred rail switch environment.

A further improvement was made to the switch performance in terms of increasing the number of breakdown channels, and hence lowering the switch inductance and minimizing electrode erosion. The position of the knife edge trigger electrode was changed such that the gap ratio was increased from 1.5:1 to 2:1. A switch inductance of 1.7 nH was obtained which proved to be extremely consistent with an rms shot-to-shot inductance fluctuation of 2 percent while passing 1.5 MA and

4 Cb. Typically the current was carried by 15 channels. It was not necessary to vent the gas after each shot, and in fact sequences of 10 to 20 shots were fired in between gas venting. The total volume of gas was 17 liters corresponding to about 1.1 liters per breakdown channel, i.e., 4 times the volume in the self-breakdown switch tests.

It is extremely significant to note that as the pressure in the rail switch was increased for a given charge and trigger voltage, the inductance of the rail switch decreased. For example, at a charge voltage of 35 kV dc the overall inductance of the circuit fell from 36.6 nH at 5 psig 8 percent sulfur hexafluoride, 92 percent argon to 35.3 nH at 15 psig. This reduced inductance was maintained up to 40 psig (the maximum envelope working pressure). This corresponds to a switch inductance of 1.7 nH. This is significant in that a rail switch can be made to breakdown in multiple channels giving a consistently low inductance while passing 1.5 MA and 4 Cb and while being triggered at less than 35 percent of the self-breakdown voltage.

A further type of field distortion switch has been developed by Physics International Company, designed to couple the energy from many low inductance capacitors into a load transmission line. One switch is used for each capacitor. The switches and capacitors are designed to be independently unplugged from the system for ease of maintenance. The bank operates over a voltage range of 20 to 60 kV dc, and over a switch pressure range of 0 to 40 psig 8 percent sulfur hexafluoride and 92 percent argon. The inductance of each switch is 26 nH, and the switches may be triggered down to less than 36 percent of self-breakdown with an rms breakdown jitter of a new nanoseconds. Each switch passes about 160 kA and 0.7 Cb under typical operating conditions. The inductance of the total module is 28 nH.

REVERSIBLE ENERGY TRANSFER BETWEEN INDUCTANCES[*]

S. L. Wipf[**]
University of California
Los Alamos Scientific Laboratory
Los Alamos, New Mexico 87544

ABSTRACT

Repeatedly-pulsed magnetic fields of the order of 10 GJ with rise times of 10 ms, as they would be used in a theta-pinch fusion reactor, need energy storage and transfer devices of high efficiency. Inductive storage would be favourable but needs a suitable transfer system. The transfer device must be capable of storing about half the total energy during transfer. Two possibilities are suggested, both using kinetic energy as interim storage. One is a homopolar machine, in lieu of a capacitor, and operates with liquid metal as the only movable medium. The other is switchless and works on the principle of a single phase alternator; during transfer the rotor performs half a rotation between two unstable equilibrium positions. The feasibility of the liquid metal transfer system will depend on the results of research into the losses; the other system has worked on a small scale in a fully superconducting device.

INTRODUCTION

This study investigates conceptual solutions to the following problem: A full sized conceptual theta-pinch reactor [1] will need pulses of 65 GJ with a rise time of between 10 and 100 ms to supply the compression field of approx. 10 T, diameter 1.5 m and length 1000 m. At the end of the burning cycle this energy should be stored again ready for the next cycle, 3 to 5 s later; the efficiency of transfer and storage should exceed 95%.

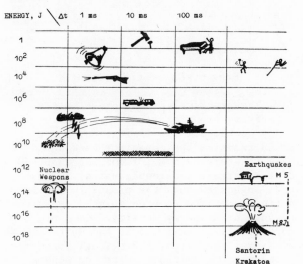

Fig. 1. Order of magnitude of "energetic" events

A feeling for the size of the problem is best gained by comparing various "energetic" events in our experience. This is attempted with Fig. 1. There is, from top to bottom, a progression from the pleasant to the disagreeable. Comparing our problem to a serious car crash is equivalent to comparing the same car crash to the hammering of a household nail.

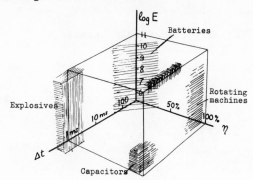

Fig. 2. Fast energy supplies; alternatives

Conventional alternatives cannot satisfy all three major parameters of the posed problem simultaneously, as demonstrated in Fig. 2. Neither explosives nor batteries are efficient enough; capacitor batteries are limited in size to <20 MJ; conventional rotating machinery barely reaches rise times below one second, although it is nearest to satisfying the conditions. Inductive storage, though not yet conventional, may be suitable for this kind of task.[2]

[*] Work performed under the auspices of the U.S.A.E.C.

[**] On leave of absence from Max-Planck-Institut für Plasmaphysik, 8046 Garching, W. Germany. (present address)

HEURISTIC PROBLEM

To illustrate the problems to be dealt with, the physics of energy transfer between inductances will be investigated. The simplest case is an irreversible transfer as shown in Fig. 3. Throughout this paper superconducting circuits are assumed or else time constants $\ll L/R$. This means that the flux linkage of a closed circuit remains unchanged during any transient process. In the simple case we must conserve the flux but we cannot also conserve the energy in the circuit. At least half the initial energy is lost during transfer. It is converted to a lower form by dissipation during the switching process.

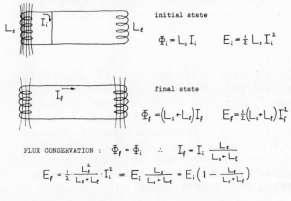

Fig. 3. Irreversible transfer.

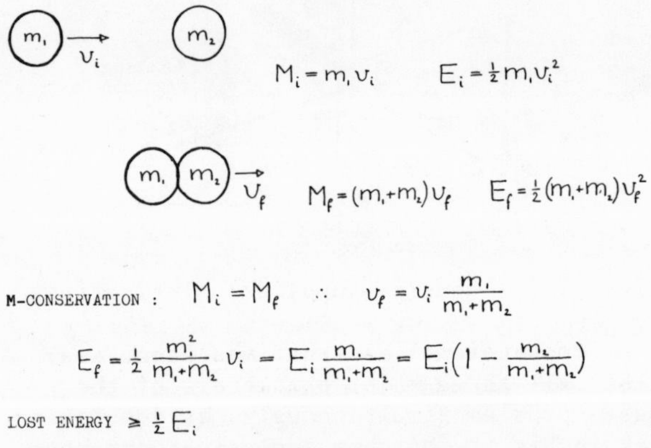

Fig. 4. Inelastic collision.

Better known is the mechanical equivalent of this process, the inelastic collision, Fig. 4. Mathematically the process is exactly the same, only momentum is conserved instead of flux, and the energy is lost in plastic deformation. The remedy is also well known: the otherwise lost energy is stored elastically and the transfer of the energy can be completed, as in Fig. 5.

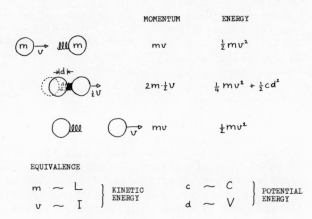

Fig. 5. Elastic collision.

The electrical equivalent of the spring is, in this case, the capacitor, and the inductive transfer is similarly improved by using a capacitor as a transfer element; see Fig. 6. The crowbar switch serves to freeze the otherwise oscillating process at the right moment.

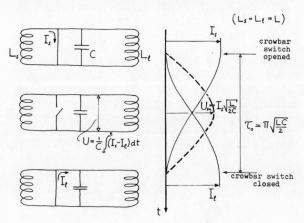

Fig. 6. Capacitive transfer.

The practical disadvantage of the capacitive transfer is given by the previously mentioned size limitation. For a 65 GJ transfer the capacitor bank would have to store 37.5 GJ, requiring about 2000 of the largest manageable units! It has been suggested that the principle of magnetic insulation be used to make capacitors of very high energy density.[3] However, a more practical solution is the multiple use of a small capacitor for a single transfer by means of a suitable switching scheme.[4-6]

It is desirable to find alternative simple transfer elements capable of replacing the capacitor. In order to do so some basic principles, summarized in Fig. 7, should be kept in mind. The importance of the conservation laws, especially the flux conservation, has already been demonstrated.

The principle of least action says that the difference between the two complementary forms of energy integrated over the duration of the dynamic process should be minimal. In practical terms this means that the transfer element should be able to accommodate about half the initial energy in a complementary form. This is not a hard rule: energy transfer is possible in a quasistationary way with arbitrarily small complementary energies (see Fig. 17). In mechanics T and V are known as kinetic and potential energy, in electrodynamics as magnetic and electrostatic energy. If, as in the present case, T is magnetic energy, then V can also be any other energy such as kinetic or potential mechanical energy, or electrochemical energy, or adiabatically compressed gas, etc. In order to ensure reversibility the energy conservation law should be applied to only the high grade energies, excluding heat; or more generally, there should be no entropy production. This rules out diffusion processes and with those, batteries, resistors, explosives etc.[7]

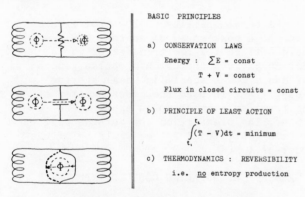

BASIC PRINCIPLES

a) CONSERVATION LAWS

Energy : $\sum E$ = const

$T + V$ = const

Flux in closed circuits = const

b) PRINCIPLE OF LEAST ACTION

$$\int_{t_1}^{t_2}(T - V)dt = \text{minimum}$$

c) THERMODYNAMICS : REVERSIBILITY

i.e. no entropy production

Fig. 7. Summary of transfer problem. Flux movement and basic principles.

The transfer can be seen as a flux movement from the circuit containing the storage inductor to the one with the load. In this sense a resistor can be seen as a leak, a capacitor as a gate for flux. An other way of changing the flux linkage is by means of suitably deformable circuits. Two such methods, different from each other but equally promising, will be described. The one functions on the basis of the homopolar machine, the other on the basis of a single phase alternator.

LIQUID METAL HOMOPOLAR TRANSFER ELEMENT

The MHD principle is explained in Fig. 8. The speed v of the MHD medium and consequently the voltage V are proportional to the charge q flowing through the device. It can therefore be employed just like a

capacitor with an effective capacity proportional to the density δ of the medium and to B^{-2}. For comparison a homopolar disk machine is sketched.[8]

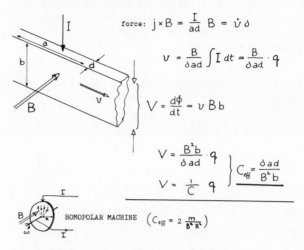

force: $j \times B = \dfrac{I}{ad} B = \dot{v}\,\delta$

$v = \dfrac{B}{\delta ad}\int I\,dt = \dfrac{B}{\delta ad}\cdot q$

$V = \dfrac{d\phi}{dt} = v\,B\,b$

$V = \dfrac{B^2 b}{\delta ad}\,q$

$V = \dfrac{1}{C}\,q$ $\quad\left\}\quad C_{eff} = \dfrac{\delta ad}{B^2 b}\right.$

HOMOPOLAR MACHINE $\left(C_{eff} = 2\,\dfrac{m}{B^2 R^2}\right)$

Fig. 8. MHD - principle.

Taking advantage of the fact that the transfer element is only operating for a short time, it is possible to consider solutions which would not qualify as energy stores over a longer period. Liquid metal is proposed as an MHD medium. A first selection as to suitability is made on the basis of resistive loss, see Fig. 9.

$$\Delta E_{resistive} = \frac{b}{ad}\,\rho\int_0^{\tau_0} I^2\,dt = \frac{\delta\,\rho}{B^2 C_{eff}}\,I_0^2\,\frac{\tau_0}{2} = \frac{\delta\,\rho}{B^2 C}\,I_0^2\,\frac{\pi LC}{2\sqrt{LC}}$$

$$= E_{stored}\cdot\frac{\delta\,\rho}{B^2}\cdot\frac{\pi^2}{2\tau_0}$$

$\tau_0 = \pi\sqrt{\dfrac{LC}{2}}$

$I = I_0\cos(\pi\dfrac{t}{\tau_0})$

$\int_0^{\tilde{\tau}} I^2\,dt = I_0^2\int_0^{\tilde{\tau}}\cos^2\!\left(\dfrac{\pi t}{\tau_0}\right)dt = I_0^2\,\dfrac{\tau_0}{2}$

$$\frac{\Delta E_{resistive}}{E_{stored}} = \frac{\pi^2}{2}\,\frac{\delta\,\rho}{B^2\,\tau_0}$$

Fig. 9. Resistive loss in MHD medium.

The quality ratio, resistive loss to energy stored, is proportional to the resistivity ρ and the density δ of the medium. This ratio can be kept below 1%; but then Hg and also Ga would be ruled out, according to Table 1 which lists approximate data.

More important, however, will be the viscous losses in the boundary layer (Hartmann layer, or Ekmann-Hartmann layer if rotational motion is important) which may also have enhanced current density. Unfortunately, the nonsteady state of high Hartmann number flow has not been investigated yet.

With Hartmann numbers of the order of 10^3 one could expect to stay in the laminar flow regime up to Reynolds numbers of 10^6, which would allow speeds up to 10^3 m/s. The viscous loss might be kept to a few percent of the total transferred energy, provided some research towards this goal proves successful.

Table 1. Properties of liquid metals

metal	ρ $\nu \Omega m$	d kg/m^3	$\rho \cdot d$ V^{2}s^3/m^4	viscosity kg/m$\cdot$s
Na	150	976	1.46×10^{-4}	5×10^{-4}
Li	450	534	2.4×10^{-4}	9×10^{-5}
K	200	830	1.6×10^{-4}	3×10^{-4}
NaK	350-500	800-900	$2.8\text{-}4.5 \times 10^{-4}$	4×10^{-4}
Ga	300	5900	1.8×10^{-3}	8×10^{-5}
Hg	1000	13600	1.36×10^{-2}	1.5×10^{-3}

Figure 10 gives an artist's impression of a conceptual supply for an equally conceptual reactor, suitably landscaped for popular acceptance. The reactor is substantially confined to a tube surrounding the compression coil, here represented by a single turn coil.

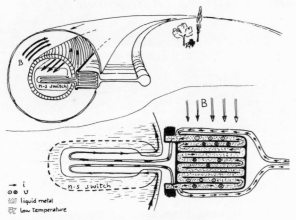

Fig. 10 Conceptual energy supply for theta-pinch reactor.

Several ideas are incorporated: The energy store has force free windings, at 45° to the torus axis, creating storage fields in- and outside. The outside field is shielded by a toroidal current in series with the storage current. For a 65 GJ store the shield would have a diameter of 3 - 4 m. The outside field is also used to drive the transfer element. This consists of many horizontal liquid metal layers going uninterrupted around the whole reactor.

The layers are a few mm thick and confined by some suitable insulating material; velocity and current flow during transfer are illustrated in the inset detail. The connection to the storage coil is by individual current leads optimized for low heat leak. The liquid metal connects the windings of the storage coil in parallel which otherwise are connected in series by the normal-superconducting (n-s) switch. As the field runs down the transfer becomes slower, ending with the energy locked in the compression coil and the transfer element, with the liquid metal at rest again, acting as a crowbar. The transfer can of course be stopped by making the switch superconducting again. If series connection is required (periodically for recharging the store) insulating pistons are inserted to remove the liquid metal from the connections; these can be operated by changing the pressure of the liquid. High pressure in the liquid is almost certainly needed during operation to maintain good contact both to the leads and in regions of high shear flow.

A total quantity of 200 t of liquid metal at a peak speed of approx. 600 m/s would suffice for the transfer of 65 GJ (For comparison the Garching pulse generator has a 200 t steel fly wheel storing 4 GJ). If Li is used the cross section of the transfer element would approximate 0.5 m^2. The pressure on the outside wall to contain centrifugal forces would be of the order of 1 MPa.

TRANSFER BY ROTATING COUPLED INDUCTANCE

The principle of this fully switchless transfer method is best explained by first discussing the kinetic - inductive transfer sketched in Fig. 11.

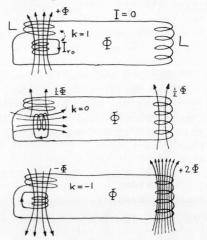

Fig. 11. Principle of kinetic-inductive energy transfer.

The rotor carries a shorted inductance with linked flux Φ. It is coupled to a stationary inductance which is connected to an equal load inductance. Initially the load current is zero but the flux linked with the load circuit is also Φ because the coupling is assumed to be ideal, $k=1$. If the coupling is made zero by rotating through $\pi/2$ the load current increases to maintain the total flux linkage in the circuit. The transfer is completed by further rotation through $\pi/2$. Now the load current becomes four times as large to maintain the original flux linkage. Naturally the current in the rotated inductance is also increased to compensate for the load current which flows in the opposite direction. The necessary energy is provided by the torque effecting the rotation.

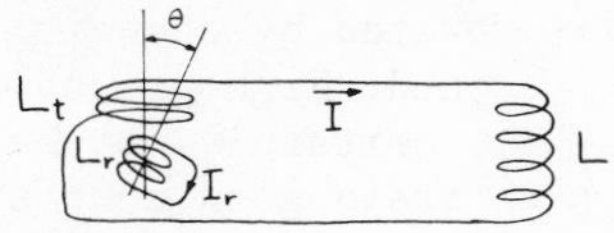

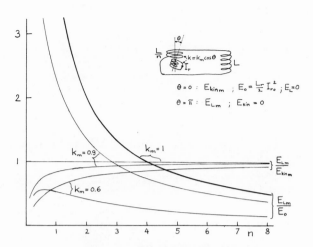

Fig. 13. Maximal transferred energy versus n; kinetic-inductive transfer.

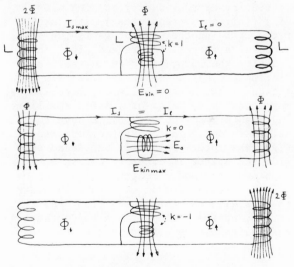

Flux conservation:

Load circuit: $I \cdot L \left(1 + \frac{1}{n}\right) + k \frac{L}{\sqrt{mn}} \cdot I_r = \Phi \, k_m \sqrt{\frac{m}{n}}$

Rotor circuit: $I_r \frac{L}{m} + k \frac{L}{\sqrt{mn}} \cdot I = \Phi$

$$I = I_{ro} \sqrt{\frac{n}{m}} \, \frac{k_m - k}{n+1-k^2} \quad ; \quad I_r = I_{ro} \frac{n+1-k k_m}{n+1-k^2}$$

$$E_{tot} = E_o \left[1 + \frac{(k_m-k)^2}{n+1-k^2}\right] ; \quad E_L = E_o \, n \frac{(k_m-k)^2}{(n+1-k^2)^2} ; \quad E_{kin} = E_{tot\,max} - E_{tot}$$

$$\left(I_{ro} = \Phi \frac{m}{L} \quad ; \quad E_o = \frac{1}{2} \Phi^2 \frac{m}{L} \right)$$

Fig. 12. Analysis of kinetic-inductive energy transfer.

An analysis of the general case is presented in Fig. 12 together with a few results of interest. The size of the rotor is determined by the energy it must be capable of storing; especially its magnetic energy E_o at the beginning of the transfer indicates the requirements for the super-conducting windings. In Fig. 13 the maximum transferred energy E_{Lm} in comparison with E_o and also in comparison with the initial kinetic energy E_{kinm} (assuming $E_{kin}=0$ at the end of the transfer) is given. Their dependence versus the ratio between load inductance and stationary transfer inductance, n, shows that an optimum value for n is probably to be found for $2 < n < 5$ depending somewhat on the degree of coupling. The reader is reminded, that two nested coils of mean radius $r_1 < r_2$ have $k_m = (r_1/r_2)^{3/2}$. In a practical device: $0.6 < k_m < 0.9$.

The transfer element just described could be used as a kinetic store and the load circuit fitted with a switch to be closed at $\theta=0$ in order to obtain a very fast transfer. This principle has in fact been used in a very similar manner to produce pulses of order 10 ms duration up to MJ size.[9-10] It should be mentioned that also the homopolar machine is suitable for a kinetic-inductive transfer and an elegant version of similar size has been built.[11]

Fig. 14. Principle of switchless transfer by variably coupled inductance.

The complete inductive transfer system is shown in principle in Fig. 14. An analysis follows the same method as before with results shown in Fig. 15. The ratio E_{Lm}/E_o is practically unchanged from the kinetic - inductive case (dotted line), whereas E_{Lm}/E_{kinm} is independent of k_m and approaches 2 towards high values of n. In the homopolar transfer element this ratio is exactly 2, provided storage and load

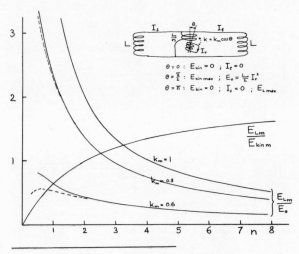

$$I_r = I_{ro}\frac{n+2}{n+2-2k^2} \; ; \; I_\ell = I_{ro}\sqrt{m}(p_m-p) \; ; \; I_s-I_\ell = 2I_{ro}p\sqrt{m}$$

$$E_\ell = E_o n(p_m-p)^2 \; ; \qquad E_{kin} = E_o 2(n+2)(p_m^2 - p^2)$$

$$\left(I_{ro} = \Phi\frac{m}{L} \; ; \; E_o = \frac{1}{2}\Phi^2\frac{m}{L} \; ; \; p = \frac{k}{n+2-2k} \; ; \; p_m = \frac{k_m}{n+2-2k_m^2} \right)$$

Fig. 15. Maximal transferred energy
versus n and some analytical results of switchless transfer.

inductance are equal. It is an advantage
of the inductive system that adaptation
between different storage and load inductance is easily achieved by a stator coil
consisting of two ideally coupled windings (air core transformer) with turn
ratio L_s/L. Another advantage is the absence of switches. Before and after the
transfer the rotor is in an unstable equilibrium and has to be secured in this position either mechanically or, more likely,
magnetically by means of a small auxiliary
inductance. Many more problems have to be
solved in designing a practical device.
E.g.: The rotor could well be placed outside the stator in order to obtain both
high coupling and strength; the stator may
have a certain freedom to rotate in order
to cope with the torque reaction, and so
on. The optimal value of n also depends on
the admissible ac losses in the superconducting rotor.

Taking into consideration reasonable
limiting values of pinning strength in
superconducting windings ($jxB \approx 5\times10^8$ N/m^3)
and mechanical strength (hoop stress $\approx$
500 MPa) transfer times of less than milliseconds should be achievable. A single
transfer element may well approximate a
weight of 100 tons and be capable of handling transfers up to several GJ. Compared
to the homopolar device the total inductive
system is heavier by at least an order of
magnitude. On the other hand the fact that

the homopolar device as sketched in Fig. 10
cannot easily be broken down into several
independent units makes it inconsistent
with a modular design.

As a final exercise a comparison is
made between the presented transfer system
by rotating coupled inductances and a similar system as proposed by P. F. Smith
for power supplies of superconducting
synchrotrons with pulse rise times of seconds.[12-13] The use of a mechanical equivalent is most suitable for this purpose.
This time springs are taken to represent
inductances; the deformation is equivalent to the current (and the force equivalent to the flux; there is, however, no
easy analog to flux conservation). The

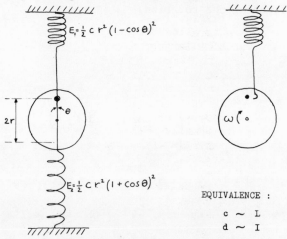

Fig. 16. Mechanical equivalent using
springs.

transfer element is a rotatable body to
which the springs are fastened at a distance r from the rotation axis. The presented
transfer system as well as the kinetic -
inductive transfer are readily recognized
in Fig. 16. Figure 17 shows an equivalent

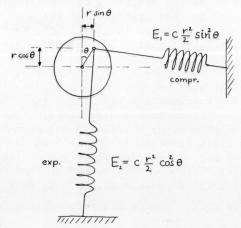

Fig. 17. Mechanical equivalent of
P. F. Smith transfer system.

of the Smith system. The energy is transferred between two springs, at right angle to each other attached to the transfer element. The springs are relaxed when their length is on a level with the axis; rotation causes alternate compression and expansion while the torque on the transfer element remains always zero and the total energy stored by the two springs stays constant. Each of the two schemes represents an extreme case: The fastest method of transfer is shown in Fig. 16, the slowest, which is also the quasi stationary case, in Fig. 17.

In conclusion it should be mentioned that a small (of the order of 1 Joule) fully superconducting model was constructed and successfully tested, demonstrating the principle in Fig. 14. Present state-of-the-art would allow the design and construction of a larger model.

REFERENCES

1 R. A. Krakowski, F. L. Ribe, T. A. Coultas and A.J.Hatch,"An Engineering Design Study of a Reference Theta-Pinch Reactor" Los Alamos Scientific Laboratory Report LA-5336, March 1974, 137 pp.

2 S.L.Wipf,"Supraleitende Energiespeicher" Max-Planck-Institut für Plasmaphysik, Garching, Report IPP 2/211, Feb. 1973. NASA Tech. Transl. F-15109, Report N73-31676, Sept. 1973, 43 pp.

3 F. Winterberg,"The Possibility of Producing a Dense Thermonuclear Plasma by an Intense Field Emission Discharge" Phys. Rev. 174, 212-220 (1968).

4 E. Simon and G. Bronner, "An Inductive Energy Storage System using Ignitron Switching", IEEE Trans.NS-14,33-40 (1967)

5 H.A.Peterson, N.Mohan, W.C.Young, R.W. Boom,"Superconductive Inductor-Converter Units for Pulsed Power Loads", These Proceedings

6 E.P.Dick, C.H.Dustmann, A.Ulbricht, "Inductive Energy Transfer using a Flying Capacitor", ibid.

7 For a more comprehensive discussion, see the review paper by O. Zucker and W. Bostick, ibid.

8 K.I.Thomassen,"An Inductive Energy Storage System with Capacitive or Homopolar Transfer", Proc. 5th Symp. on Engin. Problems of Fusion Res.,Princeton(1973), IEEE 73 CH 0843-3-NPS, pp. 444-446.

9 P. Kapitza,"Further Developments of the Method of Obtaining Strong Magnetic Fields", Proc.Roy.Soc.(London) 115A, 658-683, (1927).

10 G.Gauchon, P.H.Rebut, A.Torossian,"First Experiments on Synchronous Pulse Generators", Report EUR-CEA-FC-485; 5th Symp. on Fusion Techn. Oxford, July 1968, 15pp.

11 C.Rioux,"Experiment of Pulsating Unipolar Generator without Ferromagnetic Materials", 4th Symp. Engin. Problems of Fusion Res., Frascati, 1966, 14 pp.

12 P.F. Smith and J.D.Lewin,"Superconducting Energy Transfer Systems", Particle Accel. 1, 155-172 (1970).

13 K.I.Thomassen, "Reversible Magnetic-Energy Transfer and Storage Systems", Technology of Controlled Thermonuclear Fusion etc., Austin, Tex., 1972, USAEC CONF-721111, pp. 208-225 (1974).

CIRCUIT BREAKING BY EXPLODING WIRES IN MAGNETIC ENERGY STORAGE SYSTEMS

J. Salge, U. Braunsberger, U. Schwarz

Hochspannungsinstitut Technische Universität Braunschweig
ASSOCIATION EURATOM-KFA-PROF. KIND
3300 Braunschweig, Postfach 3329, Germany

ABSTRACT

Usually the application of exploding wires or foils for current breaking in magnetic energy storage systems is only efficient if the time of current flow through the wire or foil can be limited to around 100 μs. A remarkable improvement in voltage generation by slowly exploding wires can be achieved by a special containment of the wire. Thus voltages of a few kV per cm wire length can be produced even after some ms of current flow through the wire. The properties of such wires allow a parallel operation with commercially available a. c. breakers with mechanically moved contacts. After a description of such a breaking unit results of tests with "tamped wires" are discussed. At currents of 500 A and a current flow of 1 ms voltages of 140 kV could be generated with copper wires, 95 cm long. Recovery behaviour tests with mainly resistive loads show, that the exploded wire gap could withstand voltages for 50 ms with peak values of 1.3 kV/cm wire length. The voltage steepness generated by the exploded wires depends on the time of current flow through the wire, the wire length, and its containment.

INTRODUCTION

For power amplification in magnetic energy storage systems circuit breakers are needed, which are able to generate rapidly high voltages at high current rates.

It has been shown in the past, that the use of exploding wires or foils for this breaking operation is a very simple but efficient method, if the charging process of the storage coil is comparatively short[1-4]. There are no severe difficulties to generate voltage pulses of more than 1 MV if the storage inductance is charged by a capacitor bank[5].

Figure 1 shows for example a voltage pulse of 1.7 MV peak at a current of 5 kA, generated by the explosion of a copper wire with a length of 120 cm. In this case the charging voltage of the capacitor C was 720 kV; thus a voltage amplification factor of 2.4 has been achieved.

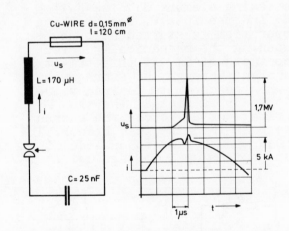

FIG.1: MV-PULSE GENERATED BY AN EXPLODED WIRE

One disadvantage of systems like this is the use of capacitor banks for charging the storage coil. In addition the power amplification factor is relatively small. For a more economic application of magnetically stored energy the use of cheap d. c. charging sources is desirable. To keep the power of these sources as small as possible, the storage

477

coil should allow charging times
of more than 100 ms and the resistance of the breaking elements
should be small compared to the
resistance of the storage coil
during the charging period.

It has been demonstrated,
that in principle these requirements can be met by shunting the
exploding wire during the charging
period with a commercially available a. c. breaker[6],[7]. For a
successful and efficient operation
a proper matching of exploding
wire and a. c. breaker is necessary
and high voltages per cm wire
length are desirable. The results
of investigations given in this
paper show, that wires tamped
with a special containment are
qualified for this purpose.

TEST CIRCUIT

Figure 2 shows the basic
diagram of an inductive energy
storage system with the corresponding current and voltage shapes.
The breaking unit consists of the
breaking element S_1, the switch
S_2 and the fuse F. The breaking
operation is started at t_2 by
opening the mechanically moved
contacts of S_1. At this time the
current through the storage coil
L has reached its desired value.
At t_3 S_2 is closed. The arc
voltage of S_1 forces the current
to pass into the fuse branch. The
fuse wire begins to explode at t_5
and builds up a voltage u_2. At t_6

the load is switched into the
circuit by S_3. Now the current is
transferred into the load by the
voltage u_2. According to the
properties of the used a. c.
breakers the time of current flow
through the wire $t_4 - t_7$ should be
in the ms-range in order to give
S_1 enough time for recovery so
that S_1 can be stressed by u_2
without danger. A more detailed
description of this circuit with
the important requirements to be
met by the single elements of the
breaking unit is given in[8].

PROPERTIES OF TAMPED EXPLODING WIRES

PEAK VOLTAGES

If a wire is heated by a
nearly constant current for a short
time its explosion mainly depends
on the current density. The peak
voltage across such a wire is
proportional to its length and is
depending on its explosion
velocity, its material and the
wire surrounding.

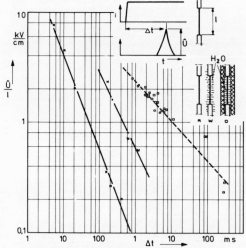

FIG.3: PEAK VOLTAGE PER LENGTH Û/l OF COPPER WIRES
VERSUS TIME TO EXPLOSION Δt
PARAMETER: SURROUNDING MEDIUM

In Fig. 3 this is shown for
copper wires. The shorter the time
to explosion the higher the peak
voltage per cm wire length.
Furthermore Fig. 3 indicates that
in the ms-range the highest peak
voltages per length so far are
achieved by tamped wires with a
special containment. It consists
of a polyethylene tube filled with
water and corresponds to that
described in our previous paper[8].

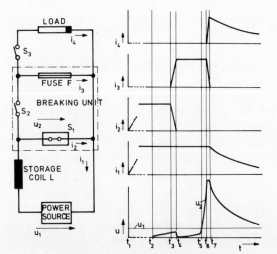

FIG.2: BASIC DIAGRAM OF AN INDUCTIVE ENERGY STORAGE
SYSTEM WITH BREAKING UNIT AND CORRESPONDING
CURRENT-AND VOLTAGE SHAPES

Only the inner diameter of the
tube was changed to 12 mm. This
results in higher peak voltages
than given in[8]. The optimization
of the tube is not yet finished.

Recent investigations have
shown, that the high peak voltages
of such wires are mainly due to
the high pressure generated in the
containment by the expanding
metallic plasma. This leads to an
intensive cooling process of the
exploded wire gap so that the
resistance of the gap is increasing rapidly. For a sucessful
operation it is important, that
the current flow through the
exploded gap is finished by transferring the current into the load
before the tube will be destroyed
by the increasing pressure.

Up to now in our test circuit
we were able to generate peak-
voltages of more than 140 kV with
tamped exploding wires at times
to explosion of 1 ms. In these
cases 5 tubes with wires 19 cm
long each were connected in
series. The storage coil of 1 H
was charged to 500 A with a
voltage of 2 kV d. c. Because the
current did not change remarkably
during the wire explosion the
power amplification factor was
around 70. The produced voltages
were only limited by the breakdown
voltages of the used elements of
the test circuit.

RECOVERY BEHAVIOUR

Another important property of
such wires is their recovery
behaviour. Figure 4 shows the
voltage stress on the whole
breaking unit by different loads.
It can be seen that the highest
stress is given by a resistive
load and by a combination of a
resistance and an inductance. In
order to get information about
the recovery behaviour of tamped
wires we have carried out a
number of tests.

Figure 5 shows a typical
result. In this case the load was
a current dependent resistor to
make the voltage stress across the
breaking unit for a long time as
high as possible. At t_4 a current

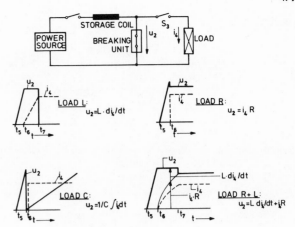

FIG.4: VOLTAGE u_2 ACROSS THE BREAKING UNIT AND LOAD-
CURRENT i_4 FOR DIFFERENT LOADS

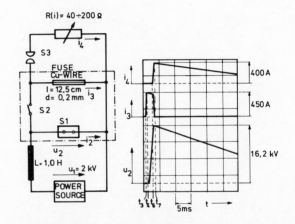

FIG.5: STRESS ON THE BREAKING UNIT BY THE VOLTAGE
ACROSS A RESISTIVE LOAD

of 450 A is transferred immediately
into the fuse by the arc voltage
of S_1. At t_6 the voltage produced
by the explosion of the tamped
wire ignites the spark gap S_3 and
the current is commutated very
rapidly into the load $R(i)$. The
recovery voltage with a peak value
of 16.2 kV decays to zero in about
50 ms. The maximum voltage stress
across the exploded gap was 1.3 kV
per cm. In summary, these tests
have demonstrated the ability of
tamped exploded wires to withstand
such voltage stresses for remarkably long times. If the current
transfer to the load has been
successful we never obtained a
breakdown of the exploded gap.

VOLTAGE STEEPNESS

In order to keep the losses
during the wire explosion at a
minimum the steepness of the

voltage should be as high as possible. Under constant current conditions the voltage steepness across an exploding wire is directly proportional to its length. But to ease the current commutation into the wire from the breaker S_1 by its arc voltage this length should be small. Therefore one has to look for other methods of increasing the voltage steepness.

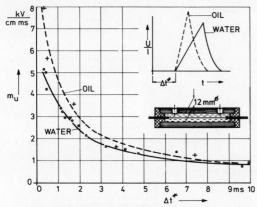

FIG.6: VOLTAGE STEEPNESS PER cm WIRE LENGTH m_u VERSUS Δt^* FOR COPPER WIRE IN MINERAL OIL AND WATER

Figure 6 shows the results of tests with different embedding media. For mineral oil and water the voltage steepness m_u per cm length is plotted versus the time Δt^*, i. e. the time of current flow through the wire to the start of voltage rise. These test results indicate first the shorter the time to explosion, the higher the voltage steepness, and secondly an increase of the steepness when oil is used. The reason for the latter may be the higher viscosity of the used mineral oil compared to that of water. Therefore the leakage of oil through the holes of the tube is smaller and the pressure is built up more rapidly.

CONCLUSIONS

The results of the investigations indicate:

1. Tamped exploding wires in connection with commercially available a. c. circuit breakers are a simple and cheap method for circuit breaking in magnetic energy storage systems.

2. It has been demonstrated, that voltages of more than 100 kV can be generated by this method.

3. After a successful current transfer into the load tamped exploding wires are able to withstand voltage stresses of more than 1 kV per cm length.

REFERENCES

1. H.C. Early, F.J. Martin, Method of Producing a Fast Current Rise from Energy Storage Capacitors, Rev. Sci. Instrum., Vol. 36, 1000 (1965)

2. Ch. Maisonnier, J.G. Linhart, C. Gourlan, Rapid Transfer of Magnetic Energy by Means of Exploding Foils, Rev. Sci. Instrum., Vol. 37, 1380 (1966)

3. J.N. DiMarco, L.C. Burkhardt, Characteristics of u Magnetic Energy Storage System Using Exploding Foils, Journ. Appl. Phys., Vol. 41, 3894 (1970)

4. D. Kind, J. Salge, L. Schiweck, G. Newi, Explodierende Drähte zur Erzeugung von Megavolt-Impulsen in Hochspannungsprüf-kreisen, ETZ-A Bd. 92 (1971)H.1

5. G. Schenk, Metallfolien zur schnellen Kommutierung hoher Ströme aus induktiven Energie-speichern, Dissertation TU Braunschweig 1972

6. R.C. Walker, H.C. Early, Half-Megampere Magnetic-Energy-Storage Pulse Generator, Rev. Sci. Instrum., Vol. 29, 1020 (1958)

7. J. Salge, N. Pauls, Schaltein-richtung für die Einschaltung von Nebenwegen in Stromkreise mit großer Induktivität, ETZ-A, Bd. 91 (1970) H. 10

8. U. Braunsberger, J. Salge, U. Schwarz, Circuit Breaker for Power Amplification in Poloidal Field Circuits, Proc. VIII Sym. Fus. Techn. Noordwijkerhout 1974

THE COMMUTATION OF THE ENERGY PRODUCED BY A HELICAL
EXPLOSIVE GENERATOR USING EXPLODING FOILS

B. Antoni, Y. Landuré, C. Nazet
Commissariat à l'Energie Atomique, Centre d'Etudes de Limeil
B.P. n° 27, 94190-Villeneuve-Saint-Georges, FRANCE

ABSTRACT

The generator has an outer diameter of 200 mm, is 800 mm long and is filled with 5 kg of a high explosive. So as the energy rise time in the load be of the order of a few microseconds which is requested in various applications, a switching is included, consisting mainly of the opening of the circuit through the vaporization of a thin foil (100 μm). In the 60 nH load the current intensity rises to 2 MA in 6.5 μsec. Five identical shots have been made. All were very satisfactory. The same combination generator and switch have been successfully used with a plasma Focus device protected from the exploding generator by a concrete building. The present experiments at the level of 100 kJ in 60 nH with risetimes of a few microseconds show the reliability and reproductibility of the devices developed.

INTRODUCTION

To power plasma physics experiments where energy pulses of a few megajoules are to be delivered in a few microseconds, magnetic field compression generators can present a more economic solution than capacitor bank sources.

These generators convert the chemical explosive energy into magnetic energy. The magnetic energy can be delivered in times of the order of one hundred microseconds. To shorten this energy pulse, it must be stored in the final generator inductance and subsequently switched by a system of fast circuit breakers.

We have studied the energy supply to a "plasma focus" type discharge where the implosion lasts some 5 μsec, using a helical generator switched by exploding wires [1] or foils. Actually we use a foil which explodes only with the aid of the generator current.

PRINCIPLE OF THE COMMUTATION

The electrical circuit shown in Fig. 1 represents the helical generator by a variable impedance ($L_H(t)$ and $R_H(t)$) ; the copper foil by its variable resistance $R_F(t)$ and the load impedance L_u into which the magnetic energy is to be transferred. A dielectric breakdown closing switch is used.

During the field compression phase the evolution of the current is as shown in Fig. 2. This current passes through the copper foil as the closing switch S_1 is open. The foil heats up, its resistance increases but remains low compared with the generator resistance. Towards the end of the compression phase the foil vaporises ; the resistance increases suddenly, becoming large compared with that of the generator. The voltage peak $R_F I_F$ (which attains some 20 kV) (Fig. 3) causes the dielectric breakdown switch to close and the current transfers to the load at a rate given by

$$\frac{dI_u}{dt} = R_F \frac{I_F}{L_u}$$

Evidently the transfer rate is increased if L_u/R_f is decreased.

EXPERIMENTAL UNIT

The system is shown in Fig. 4. It shows the dimensions of the helical generators and the copper foils used which permit the maximum current to be transferred to an inductive charge of 60 nanohenries.

Figure 4 shows the 2 copper foils mounted in parallel, surrended by fiber glass. The fiber glass tissue seems adequate to absorb the gaseous metallic vapors and the physical shock of the explosion. Several thicknesses are used (with a total thickness of a few millimeters) and properly fitted to the copper foil.

THEORETICAL DETERMINATION OF THE
FOIL PARAMETERS

We have used a series of calculations by simulation to determine the best parameters for the system. These have been based on the values of the injection current for which the generator function is well known.

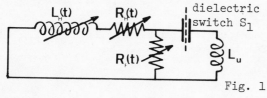

Fig. 1

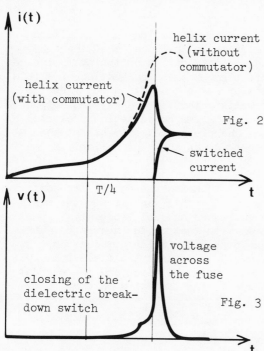

Fig. 2

Fig. 3

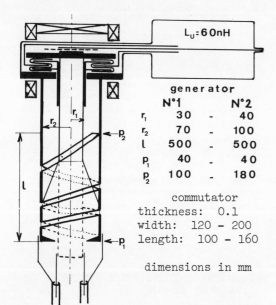

Fig. 4

be calculated. Initially the cross section S must be chosen in order to determine approximately the instant of explosion.

From the cross section which corresponds to the maximum current transferred, the vaporization energy, and the energy required for the commutation, a new length can be calculated. With this length a better determination of the optimum S is possible. The process can thus be reiterated until only small variations are found for successive reitterations and the values thus obtained are considered as optimum.

The following approximations are made for the calculation of the fuse resistance : the resistance depends little on the rate of Joule heating, only relatively thin foils and slowly increasing current are considered. We suppose that the solid and liquid phases are homogenous and neglect the acceleration (kinetic energy) due to their thermal expansion. For the vaporization phase, the acceleration of the gaseous mass due to the sudden change in volume is taken into account by using the hypothesis of the vaporization wave of F.D. Bennett [2].

EXPERIMENTAL RESULTS

For the Helix 2 (with a load $Lu = 60 \times 10^{-9}$ H) we have found the following parameters theoretically

$$I_o = 380 \text{ kA}$$

2 copper foils — thickness 0,1 mm, width 200 mm, length 160 mm

Figure 5 shows the theoretical and experimental results.

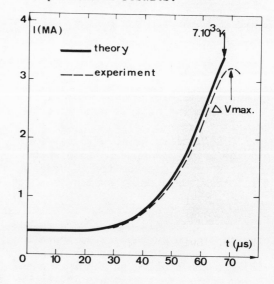

Fig. 5

The optimum cross section S and length L which is to permit the current maximum to be switched to a given charge must

The graphs correspond to the current pulse delivered by the helix during the magnetic compression. The two curves are relatively similar.

The instant at which the voltage peak appears across the fuse corresponds to within a few microseconds the calculated instant when all the foil is vaporised. Figure 6 shows the oscillographs taken of the current pulse and its dI/dt (derivative).

Figure 7 shows the voltage peak across the commutators. The small break in the curve at 10 kV is due to the closing of the dielectric breakdown switch.

Figure 8 shows the current transferred to the inductive charge of 60×10^{-9} H. 2 megaamps are reached in 6 microseconds.

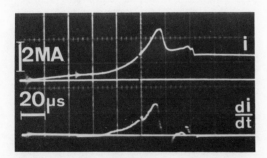

Fig. 6. Helix current and its derivative

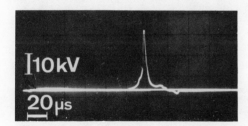

Fig. 7. Foils voltage

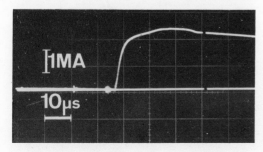

Fig. 8. Switched current

EXPERIMENTALY REPRODUCIBILITY

Before coupling the generator commutator assembly to prove the reproducibility of the results. We have performed 5 identical experiments. The 5 transferred currents obtained, are shown in **Fig.** 9. It is noted that in spite of erratic discontinuities in the progression of the sliding contact 3, the transformed current increases are all very similar.

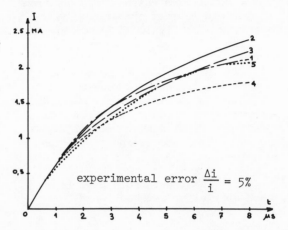

experimental error $\frac{\Delta i}{i} = 5\%$

Fig. 9. Switched currents reproducibility

THE COUPLING OF AN EXPLOSIVE GENERATOR TO THE PLASMA FOCUS EXPERIMENT

We have used the system, placed behind concrete protective walls to drive a focus type discharge.

This arrangement reduces the cost of operation as only the explosive generator and commutator are destroyed. Besides it is possible to use a conventional plasma head with all diagnostics judged necessary. However the connections are long and inductive.

They are shown on Fig. 10. The helix and its commutator are connected to a condensor bank by 16 coaxial cables. A flat triaxial cable passes through the wall to connect the dielectric closing switch to a series of 100 coaxial cables which in turn connect to the focus head.

The total inductance of the connections is 30 nanohenries which is to be added to the initial inductance of 30 nanohenries of the Focus discharge.

This explains why our initial experiments on the commutator were carried on a 60 nanohenries fixed charge.

EXPERIMENTAL RESULTS

We achieved a very satisfactory coupling between generator and plasma. It can be seen from the evolution of the current time derivative obtained has sufficiently good characteristics to obtain a pinch under conditions which give a neutron yield comparable to that obtained with a condensor bank. The Figure 11 shows the current pulse delivered by the helix and the current transferred to the plasma. In Figure 12, it is possible to compare the evolution of the transferred current in Focus and signals obtained by photo-multipliers, located at a few meters from the focus head. The first signals correspond to the X-ray emission, the second to the neutron emission.

Four identical shots have been done, each delivering between 2 and 5×10^{10} neutrons.

CONCLUSION

These experiments, which are our very first in this domain, show the efficiency and reliability of the system. The use at a later date of a planar "bellows" type generator which avoids the erratic discontinuities in the progression of the moving contact, and release more efficiently the chemical energy of the explosive should permit even better performances to be obtained.

REFERENCES

[1] J. Bernard, J. Boussinesq, J. Morin, C. Nazet, C. Patou, J. Vedel, An explosive generator-powered plasma Focus, Phys. Let. 35A, 4, 14.6.1971, p. 288-289

[2] B. Antoni, Etude d'interrupteurs à explosion de feuilles en cuivre pour des courants de l'ordre du méga-ampère, Le Journal de Physique, Tome 9, janvier 1974, p. 331-340

[3] B. Antoni, C. Nazet, L. Pobé, Theoretical and experimental study of explosive driven magnetic field compression generators (these proceedings)

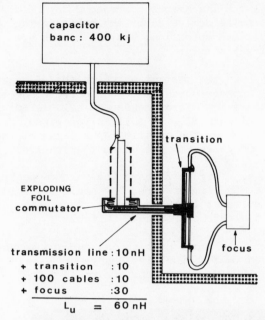

transmission line : 10 nH
+ transition : 10
+ 100 cables : 10
+ focus : 30

L_u = 60 nH

Fig. 10

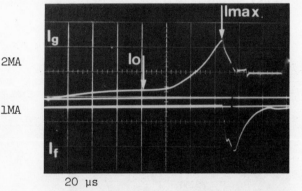

20 µs

Fig. 11.   Helix and switched current

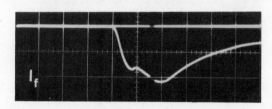

5 µs

Switched current

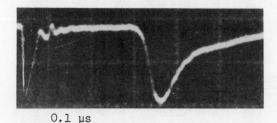

0.1 µs

Fig. 12.   X and neutrons

INDUCTIVE ENERGY TRANSFER USING A FLYING CAPACITOR

E.P. Dick, C.-H. Dustmann

Gesellschaft für Kernforschung Karlsruhe

Institut für Experimentelle Kernphysik III

7500   Karlsruhe/Germany

Postfach 3640

ABSTRACT

A new inductive energy transfer system for fusion reactor technology using superconductive coils is proposed. Transfer is achieved stepwise by a small thyristor-switched capacitor between the inductances. The system is highly efficient and allows return of the transferred energy. Transfer times are in the one-second range. A superconductive 10 kJ model was designed and built. In tests, 80% of the stored energy was transfered into a load within 0.5 seconds.

INTRODUCTION

Inductive energy storage with superconducting coils will be attractive with respect to the energy density for those applications where large systems have to be pulsed. For example it has been shown that a future thermonuclear fusion reactor can be realisable only with superconducting magnets.[1] For these GJ range systems recycling of the energy is necessary. One of the main challenges is to design a transfer system for the required power which is efficient and economic. Several methods have been previously proposed. Parallel resistor transfer[2] is limited to 25% transfer efficiency. Parallel capacitor transfer[3] allows almost 100% efficiency with the trade-off that the capacitor bank must store half the total energy. P.F. Smith[4] proposed a system of constant total energy with mechanically rotated superconductive coils. In the Princeton transfer device[5] an ignitron-switched capacitor system transports incremental quantities of the energy to the load. R.W. Boom et al.[6] from the University of Wisconsin proposed a system that converts the dc storage current to three phase ac which is reconverted with conventional methods to the load.

A new method, the flying capacitor transfer system, is proposed for one to ten second transfer times needed for fusion power reactors.

PRINCIPLE OF OPERATION

The storage inductor is discharged by a small thyristor-switched capacitor. Figure 1a shows the basic circuit in which two thyristors shunt storage and load inductors. In contrast to other proposals, the ungrounded capacitor (hence called flying) is in a series connection. Initially switch S, mechanical or superconducting, is closed for storage of energy with i_0 circulating in L_S. C and L_L are uncharged. When transfer is desired, S is opened and thyristor T_2 fired. Current i_s rapidly charges C to the maximum voltage V_m whereupon T_1 is fired shorting L_S. This voltage V_m now appears across L_L (note polarity) with T_2 opened by current reversal. The capacitor discharges into L_L and the voltage is allowed to reach a small negative value $V\Delta$ through the resonating action of the parallel L_L-C as shown in Fig. 1b. Thus thyristor T_2 is biased in the forward direction by $V\Delta$ and is again fired. Simultaneously T_1 is opened by the

reverse voltage of $V\Delta$. The discharge
of inductor L_S continues into C until
the V_m limit is again reached. Thus
the process is repeated with decreasing i_S and increasing i_L, Fig. 1c,
until L_S is completely discharged and
the energy in L_L is maintained by the
crowbar action of T_2. Note that T_1 is
always fired with voltage V_m on L_S
and T_2 with $V\Delta$ on L_L. Thus the firing
circuits are controlled by simple voltage comparators. Reversing the firing
levels for T_1 and T_2 allows return of
the energy to L_S as seen by the symmetry of the circuit.

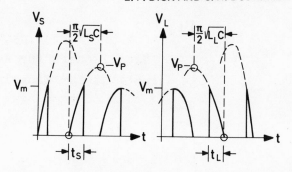

Fig. 2.   Variable step time

TRANSFER SPEED

The transfer period is composed
of a series of many steps. In Fig. 2
the step length t_S varies over the
transfer time and becomes progressively longer with decreasing storage current. The rise of V_S is part of a sine
wave with variable peak amplitude V_p
and a constant quarter cycle determined

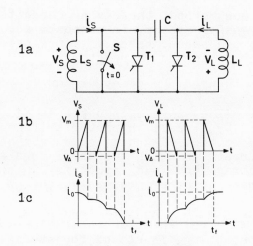

Fig. 1.   Flying capacitor transfer

by the square root of L_S and C.

$$\frac{V_m}{V_p} = \sin\left(\frac{t_S}{\frac{\pi}{2}\sqrt{L_S C}} \cdot \frac{\pi}{2}\right) \qquad (1)$$

The peak voltage V_p would result if
the energy $E(K)$ available in L_S were
completely transfered to C, where K
is the step number.

$$\frac{1}{2} C V_p^2 = E(K) \qquad (2)$$

Assuming no losses, $E(K)$ varies linearly with K since a constant energy
is transfered in each step with V_m
and C fixed.

$$E(K) = \frac{1}{2} L_S i_o^2 \left(\frac{N-K+1}{N}\right) \qquad (3)$$

with $1 \le K \le N$ and N the total number of
steps. The fully-charged capacitor
stores a factor N less energy than
the initial inductor energy.

$$\frac{1}{2} C V_m^2 = \frac{1}{2N} L_S i_o^2 \qquad (4)$$

Equations (1) to (4) are solved for
t_S with V_p, C, and $E(K)$ eliminated.

$$t_S = \frac{L_S i_o}{V_m} \cdot \frac{1}{\sqrt{N}} \sin^{-1} \frac{1}{\sqrt{N-K+1}} \qquad (5)$$

A similar equation can be written
for t_L. The total transfer time is
given by the summation of t_S and t_L
with K varying from 1 to N, which
is approximated by eq.(6)

$$\sum_{k=1}^{N} \sin^{-1} \frac{1}{\sqrt{N-K+1}} \simeq 2\sqrt{N} - \frac{1}{2} \qquad (6)$$

Assuming equal inductances the final
transfer time is

$$t_f = \frac{L i_o}{V_m}\left(4 - \frac{1}{\sqrt{N}}\right) \qquad (7)$$

This analysis neglects the small time
lengthening effect of $V\Delta$. The effect
of N on t_f is graphed in Fig. 3. Small
capacitors result in large N with
negligible effect on transfer time.
The capacitor size is, however, constrained by commutation limits discussed in the following section.

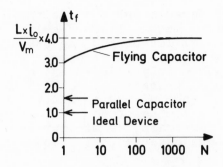

Fig. 3.   Theoretical transfer time

COMMUTATION LIMITS

Two constraints must be met to ensure that thyristor T_1 opens when T_2 is fired by V_Δ, see Fig. 4. Sufficient energy must be available in C to force a current greater than i_0 though the stray capacitor inductance L_c which achieves current reversal of T_1.

$$C \geq L_c \frac{i_0^2}{V_\Delta^2} \qquad (8)$$

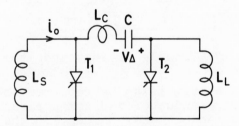

Fig. 4.   Commutation limits

Secondly C must maintain a charge of the polarity shown while T_2 takes on a current increasing to i_0 during the thyristor switching time t_c.

$$C \geq \frac{i_0 t_c}{2 V_\Delta} \qquad (9)$$

L_c and t_c are determined by the quality of the components. V_Δ must remain small relative to V_m for high transfer speed and efficiency. Hence the capacitor must exceed a threshold size.

10 kJ MODEL

The main components of a 10 kJ experiment are shown in Fig. 5.

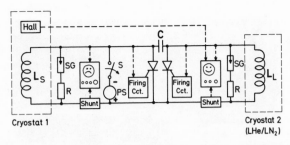

Fig. 5.   Block diagram of the experiment

L_S - superconducting storage coil[7] with L = 17.7 mH, critical current of 1440 A, 22 strands, 1000 filament NbTi Cu-matrix wire with 2 mm twist pitch of filaments

L_L - a) normalconducting load coil L = 19.7 mH in LN_2 bath, R = 47 mΩ
     b) superconducting load coil with L = 17.7, critical current 704 A

PS - 1200 A, 10 V current regulated power supply

S - mechanical circuit-breaker

C - 1-10 parallel units 384 µF, 400 V metallized paper capacitors

SCR- 4 parallel units of 290 A, 800 V thyristors

SG - 350 V limiters with dumping resistors R

Voltage and current waveforms were monitored by camera-equipped oscilloscopes. Sample pictures for transfer into a normalconducting and a superconducting load are given in Fig. 6 and Fig. 7 respectively.

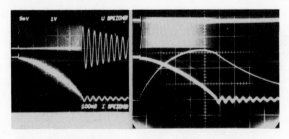

V_S: 100 V/div. V_L: 100 V/div.
i_S: 333 A/div. i_S: 400 A/div.
                  i_L: 200 A/div.

Fig. 6.   Transfer into normal conducting load

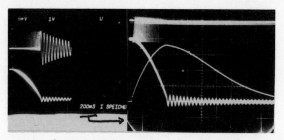

V_S: 100 V/div. V_L: 100 V/div.
i_S: 333 A/div. i_S: 200 A/div.
                   i_L: 200 A/div.

Fig. 7. Transfer into superconducting
         load

The results appear in table 1 with
an example extrapolated to fusion
reactor size.

Table 1. Results and extrapolation

		Fig. 6	Fig. 7	Extrapolated
Initial stored energy	(J)	$10.3 \cdot 10^3$	$5.7 \cdot 10^3$	10^{10}
Storage current	(A)	1080	800	10^5
Transfer time	t_f(s)	0.51	0.45	10
Maximum voltage	V_m(V)	180	130	10^5
Transfer steps	N	202	157	1000
Transfer efficiency	(%)	30.0	77.1	785
Resistive losses	(%)	52.4	1.2	0.1
Thyristor losses	(%)	15.4	19.0	10
Capacitor losses	(%)	1.6	1.7	1
Superconductive losses	(%)	0.2	0.4	?
Untransferred	(%)	0.4	0.6	0.1

All losses have been calculated from
component characteristics and experimental data. Superconductive losses
include the effects of hysteresis and
eddy currents.[8]

Measurements with various voltages V_m and currents i_o are shown in
Fig. 8 and indicate that eq. (7) is
valid.

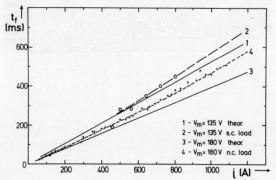

Fig. 8. Flying capacitor transfer
         time

The difference between the theoretical
and experimental lines result from
effects of losses and the commutation
voltage V_Δ. As expected the transfer
time was mainly independent of the
number of transfer steps N which were
in the range of 100 to 200. The commutation limits of eq. (8) and (9)
which constrain the minimum capacitor
size were observed.

FURTHER DEVELOPMENT

Future development will concentrate on increasing the system size
possibly to that mentioned in Tab. 1
and in modifying the circuit to improve performance. For example the
ripple may be reduced by doubling the
flying capacitor as shown in fig. 9.

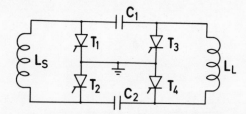

Fig. 9. Double flying capacitor

When energy is to be returned to the
storage inductor two symmetrical systems can be used. In fig. 10 the relatively small power supply is left
continuously in the circuit for top-
ping up losses. No mechanical switch

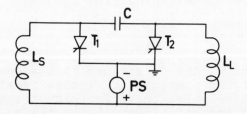

Fig. 10. Symmetrical circuit for
          short storage times

is required because the capacitor is
charged to the power supply voltage
during storage periods and provides
the commutative voltage V_Δ necessary
to open the appropriate thyristor.
When transfer is infrequent superconducting switches[9] may be used as

shown in fig. 11.

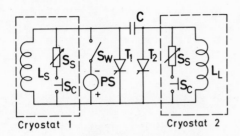

Fig. 11. Symmetrical circuit for long storage times

CONCLUSIONS

The feasibility of transfering inductive energy using a flying capacitor system has been demonstrated. With superconducting coils energy in the kJ range was transferred with nearly 80% efficiency in 0.5 seconds. No limitations were encountered which would prevent extrapolation to larger sizes.

ACKNOWLEDGEMENTS

The authors wish to thank H. Andexinger, H. Kiesel, H. Kornmann, G. Noether, E. Süß and A. Ulbricht for their assistance in completing the experiment.

REFERENCES

1. F. Arendt et al.:'Energetic and Economic Constraints on the Poloidal Windings in Conceptual Tokamak Fusion Reactors', Proc. of 8th. Symp. on Fusion Technology, Noordwijkerhout, Netherlands, 1974

2. R.R. Hake: 'Single-Shot Pulsed Magnetic Fields from Inductive Energy Stores', Los Alamos Report LA-4617-MS, 1971

3. S.L. Wipf: 'Supraleitende Energiespeicher', Inst. f. Plasmaphysik, München/Germany, IPP-Report No. 2/211, 1973

4. P.F. Smith: 'Multi-Stage Energy Transfer Schemes', Proc. of the 8th. Int. Conf. on High Energy Accelerators, CERN, 1971,p.213

5. E.D. Simon, G. Bronner: 'A One MJ Inductive Energy Storage System using Ignitron Switching', Princeton PPL Report, MATT-741, 1970

6. R.W. Boom, H.A. Peterson, W.C. Young: 'Pulsed Power Supply for the U.W. Tokamak Reactor', Wisconsin Superconducting Energy Storage Project, Vol. I, July 1974, Appendix VI-H

7. H. Kornmann: 'Ein supraleitender 12 kJ Energiespeicher mit kurzer Entladezeit', Diplomarbeit Universität Karlsruhe, Germany, 1974

8. H. Brechna: 'Superconducting Magnet Systems', Springer-Verlag Berlin, 1973, p. 246, 251

9. K. Grawatsch et al.: 'Investigations for the Development of Superconducting Power Switches', Proc. of Appl. Superconductivity Conf. Argonne, USA, Sept. 1974

INDUCTIVE ENERGY STORAGE USING HIGH VOLTAGE
VACUUM CIRCUIT BREAKERS

R.B. McCann   H.H. Woodson   T. Mukutmoni

Department of Electrical Engineering
University of Texas at Austin
Austin, Texas 78712, USA

Fusion Research Center
University of Texas at Austin
Austin, Texas 78712, USA

ABSTRACT

Controlled thermonuclear fusion experiments currently being planned require large amounts of pulsed energy. Inductive energy storage systems (IES) appear to be attractive for at least two applications in the fusion research program: high beta devices and those employing turbulent heating. The well known roadblock to successful implementation of IES is the development of a reliable and cost-effective off-switch capable of handling high currents and withstanding high recovery voltages. The University of Texas at Austin has a program to explore the application of conventional vacuum circuit breakers designed for use in AC systems, in conjunction with appropriate counter pulse circuits, as off-switches in inductive energy storage systems. The present paper describes the IES employing vacuum circuit breakers as off-switches. Since the deionization property of these circuit breakers is of great importance to the design and the cost of the counter-pulse circuit, a synthetic test installation to test these breakers has been conceived, designed and is being installed in the Fusion Research Center, University of Texas at Austin. Some design aspects of the facility will be discussed here. Finally, the results of the study on a mathematical model developed and optimized to determine the least cost system which meets both the requirements of an off-switch for IES Systems and the ratings of circuit breakers used in power systems has been discussed. This analysis indicates that the most important factor with respect to the system cost is the derating of the circuit breakers to obtain satisfactory lifetimes.

INTRODUCTION

High voltage AC circuit breakers are attractive candidates for the current interrupters in Inductive Energy Storage (IES) systems with energy transfer times of 0.1-10 ms. They are reasonably priced, reliable, highly developed and they are commercially available in large quantities. A fusion test reactor (FTR) based on a staged theta pinch is being planned at the Los Alamos Scientific Laboratories. The compression phase of this device will require a power supply capable of delivering 300 MJ in 1.0 ms. The design of the IES system for this application using vacuum circuit breakers operated within their a.c. commercial ratings is the subject of this paper.

This system, shown schematically in Fig. 1, will consist of a storage inductor, a load inductor which is the compression coil of the theta-pinch, a transfer capacitor, a power supply and the circuit breakers. In addition, auxiliary circuit elements are required to insure proper operation of the circuit breaker. The transfer capacitor is required in order to avoid the dissipation of large amounts of energy in the circuit breaker. Even though this capacitor must be capable of storing one half of the total energy, the overall cost of the system is less than for a completely capacitive system.

Vacuum circuit breakers are AC devices and require a current zero before interruption can occur; however, there is no natural current zero in an IES application. Therefore, the current must be artificially driven to zero by counterpulsing a current through the circuit breaker in opposition to the initial current. The design of this counterpulse portion of the circuit is critical to the proper functioning of the circuit breaker as an off-switch. The current must be driven to zero with a time rate of change of no more than the a.c. short circuit rating of the breaker. Although it can not be maintained throughout the counterpulse, this slow di/dt must be maintained for a period greater than the deionization or "memory" time of the circuit breaker. This allows the plasma formed during the high current portion of the arc to dissipate and can be accomplished by placing a saturating inductor in series with the switch.

The deionization property of the vacuum circuit breaker at these current levels is of great importance to the design and consequently to the cost of the counterpulse circuit. The data available from existing publications on vacuum circuit breakers at this point are not sufficient to design a reliable and cost-effective counterpulse circuit. Therefore, to investigate thoroughly the deionization property of a vacuum interrupter, a synthetic test installation capable of applying a wide range of dynamic stress to the breakers has been conceived, designed and is being built at the University of Texas at Austin. The results of tests on vacuum breakers in the synthetic test circuit will reveal the limitations of off-switching elements and help to optimize the design of the counterpulse circuit.

A mathematical model has been developed and optimized to determine the least cost system which meets both the requirements of the LASL experiment and the ac short circuit ratings of the breakers. This study has indicated that inductive energy storage systems using high voltage vacuum circuit breakers are feasible not only from an engineering point of view, but also from an economic standpoint. Although the use of these circuit breakers introduces some special problems, these problems can be easily solved by the proper design of the overall system.

The fact that vacuum breakers appear economically attractive as switches for IES operating within their 60 Hz ratings is conservative. The ratings for 60 Hz operation are based on adverse field conditions; whereas, IES switching will be under carefully controlled laboratory conditions. Hence, it is expected that in IES switching service, vacuum breakers should perform somewhat better than their 60 Hz ratings would indicate. This hypothesis is to be tested and evaluated quantitatively using the synthetic test circuit.

DESCRIPTION OF IES CIRCUIT

The basic IES circuit proposed for use with the fusion test reactor (FTR) being planned at the Los Alamos Scientific Laboratories is shown in Fig. 1.

This application requires that 300 MJ be initially stored in L_1 and then transferred in about 1.0 ms to the load inductor L_2. The load is the multiturn copper compression coil of a staged theta pinch. The storage inductor may or may not be superconducting. The capacitor C_C is required to transfer the energy from L_1 to L_2 without unacceptable losses in the switch S. The operation of the circuit is straightforward. The switch is initially closed and a current i_0 is established in the series combination of the switch and storage inductor L_1. The power supply, PS, which drives i_0 through the coil is a low impedance device such as a homopolar generator.[2,3] The storage inductor L_1 may be superconducting. If L_1 is superconducting, the current i_0 can be easily induced in L_1 over a long period of

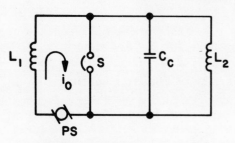

L1 ENERGY STORAGE INDUCTOR
L2 LOAD INDUCTOR
PS CURRENT SOURCE
C_C ENERGY TRANSFER CAPACITOR
i_0 INITIAL CURRENT IN L1
S   OPENING SWITCH

Fig. 1: Basic Inductive Energy
Storage Circuit

time. However, if L_1 is made of a normal conductor, i_0 will have to be induced in a time long compared to the energy delivery time t_E, but short compared to the L/R time constant of the coil if substantial energy loss and heating are to be avoided.

In this simplified circuit, it can be shown that when the switch S opens, the energy initially stored in L_1 is transferred to the load coil L_2. The energy transfer is maximized when $L_1 = L_2$. Under this condition the current i_1 and i_2 in L_1 and L2 respectively are given by:

$$i_1 = \frac{i_0}{2} (1 + \cos \omega t)$$

and

$$i_2 = \frac{i_0}{2} (1 - \cos \omega t)$$

where

$$\omega = \frac{1}{\sqrt{.5 \, L_1 C}} \quad .$$

The voltage across the capacitor C or the switch S is

$$V_C = i_0 \sqrt{\frac{L_1}{2C}} \ \text{Sin} \ \omega t \quad .$$

It will be apparent from ω that the energy transfer time $t_E = \pi \sqrt{0.5 \, L_1 C}$. The capacitor voltage is maximum at $t = 1/2 \, t_E$, and at this time exactly one-half of the energy stored in L_1 is resident in the capacitor. The maximum dv/dt across the capacitor and hence the switch is

$$\text{dv/dt}_{max} = \frac{i_0}{C} \quad .$$

A straightforward dc switch as shown in Fig. 1 capable of interrupting thousands of amperes and subsequently capable of withstanding hundreds of kilovolts is not, as we know, available at the present state of technology.

DIFFERENT OPTIONS
FOR AN OFF-SWITCH

Studies made at the University of Texas at Austin[4] on the performance and cost-effectiveness of different kinds of off-switching elements that may be used in IES circuits reveal the cost dependence of different switching elements alone with energy transfer time as shown in Fig. 2.

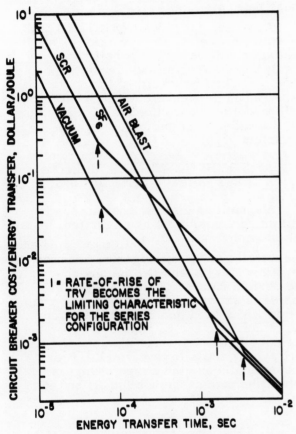

Fig. 2: Curves Showing the Cost of Different Breaker Types

In arriving at this figure, the cost of various switch candidates has been taken from their respective catalogs. This cost figure as tabulated in Table 1 is only approximately what the actual switch would cost because the catalog price includes many auxiliaries that would not be needed in the IES system.

However, it serves as an upper bound thus making calculation conservative. Furthermore, in this study, only those switching elements have been taken into consideration which have established cost and performance characteristics. Developmental

TABLE 1

Circuit Interruption Candidates

	COST ($)	V_{MAX} (kV)	I_{MAX} (kA)	RRRV (kV/μs)
SF$_6$ (230SF20000,WEST.)	22,000	500	60	1.0
AIR BLAST (ATB7, GE)	23,000	500	60	0.5
VACUUM (R-4, WEST.)	2,200	100	25	5
SCR (60IPS, GE)	100	1.7	12	0.1

devices like the Hughes Plasma Valves[5] have not been considered.

In Fig. 2, the cost of the switch alone in dollars/joule as a function of energy transfer time for application in the 300 MJ theta-pinch circuit has been shown. The cost is determined by the number of switches needed which is determined by the current (parallel switches) and voltage (series switches). The break in the curves occurs when the rate of rise of transient recovery voltage determines the number of switches in series instead of the recovery voltage itself. It can be noted from the figure that the vacuum breaker is more economical at energy transfer times less than about 1.2ms.

IES CIRCUIT USING
VACUUM INTERRUPTERS

When vacuum interrupters are used as off-switches, the basic IES circuit in Fig. 1 is modified to the circuit shown in Fig. 3.

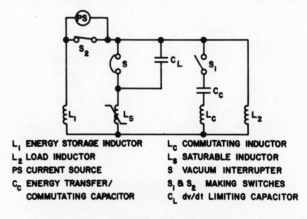

L$_1$ ENERGY STORAGE INDUCTOR	L$_C$ COMMUTATING INDUCTOR
L$_2$ LOAD INDUCTOR	L$_8$ SATURABLE INDUCTOR
PS CURRENT SOURCE	S VACUUM INTERRUPTER
C$_C$ ENERGY TRANSFER/ COMMUTATING CAPACITOR	S$_1$ & S$_2$ MAKING SWITCHES
	C$_L$ dv/dt LIMITING CAPACITOR

Fig. 3: Modified Inductive Energy Storage Circuit

The following characteristics of this circuit can be noted:

1. The same energy transfer capacitor of the basic circuit (C_C in Fig. 1) has been also used as the commutating capacitor.

2. A fixed value commutating inductance L_C together with making switch S_1 has been introduced in series with the energy transfer capacitor.

3. A dv/dt limiting capacitor C_L has been placed across the off-switch (Vacuum Interrupter).

4. A saturable inductor L_S has been put in series with the vacuum interrupter such that

$$L_S = L_{SO} \text{ for } |i| \leq i_S$$

and

$$L_S = L_{SS} \text{ for } |i| > i_S$$

where i is the current through the inductor and i_s is the current which saturates the inductor L_S. The ratio of L_{SO} to L_{SS} is typically 10^3. The modification of the basic circuit to incorporate the commutating circuit has been effected keeping the following major constraints in mind:

1. The energy transfer time T_E must not change substantially due to introduction of additional circuit components.

2. The rate of decay of current during the off-commutation phase must be maintained sufficiently low for the duration of the "memory time" of the interrupter to comply with the commercial (60 Hz) ac duty of the breaker.

3. The rate of rise of voltage across the vacuum interrupter (both inverse and forward) should be maintained within the prescribed limits of commercial breakers.

4. The bias voltage to the energy transfer capacitor C_C for commutation should be as low as possible.

The circuit of Fig. 3 has been analyzed extensively to determine the relations between L_S and i_S, and di/dt during commutation, the required bias voltage, the length of time from the end of saturation of L_S until the current zero and the maximum dv/dt across the vacuum interrupter. These analyses have confirmed that an IES system with an energy transfer time of 1 ms as required by the FTR can be reliably and economically realized using vacuum interrupters as off-switches in the circuit shown in Fig. 3.

DEIONIZATION PROPERTIES OF VACUUM INTERRUPTER

The deionization properties of circuit breakers play a significant role in the design of IES Systems. This affects the so-called "memory-time" and "the deionization margin" of the circuit breaker. Unfortunately, these parameters are the most uncertain of all of the circuit breaker ratings because they are not specified by the manufacturer. Although the value which has been assumed for this study appears to be reasonable[6] investigations using different types of commercial vacuum circuit breakers addressed specifically to these properties will be required to definitely establish the values of these parameters.

A synthetic test facility is being built for this purpose at the University. This installation will have the capability to vary the dynamic stress on the breaker over a wide range and is described in short below. Experiments on breakers in this installation will reveal relevant information relating to the deionization properties of vacuum interrupters.

SYNTHETIC TEST CIRCUIT TO TEST VACUUM INTERRUPTERS

A simplified diagram showing the important sections of the synthetic test installation is shown in Fig. 4.

The working of the circuit is as follows: The capacitor C_L, C_C and C_H are pre-charged to the desired values. The polarity of the charge is shown in Fig. 4. At $t = t_1$, the ignitron IG_1 is fired and current builds up in the circuit C_L, L_S, B_B, and B_T. When the current reaches its peak value and the polarity in C_L tends to reverse, IG_2 fires automatically and the stored energy in L_S drives the current through B_B, B_T and IG_2.

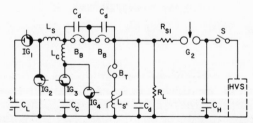

C_L LOW VOLTAGE/HIGH CURRENT CAPACITOR	B_T BREAKER UNDER TEST
C_C COMMUTATING CAPACITOR	C_d DAMPING CAPACITOR
C_H HIGH VOLTAGE/LOW CURRENT CAPACITOR	$L_{s'}$ SATURABLE INDUCTOR
HVS HIGH VOLTAGE SOURCE	R_{SI} SERIES RESISTANCE
$IG_1...IG_4$ SWITCHING IGNITRONS	R_L LOAD RESISTANCE
L_S L.V. SOURCE INDUCTOR	G_2 TRIGGERED GAP
L_C COMMUTATING INDUCTOR	S CHARGING SWITCH
B_B BLOCKING BREAKER	

Fig. 4: Simplified Diagram of the Synthetic Test Set-Up at the University of Texas at Austin

At some point after this, at time $t = t_2$, B_B and B_T open and continue conducting current in the arcing mode. At $t = t_3$, IG3 is fired and the counterpulse from the capacitor C_C drives the current through the breakers B_B and B_T to zero as shown in Fig. 5. The residual voltage in C_C after current zero keeps the breaker B_T back-biased for a while when at $t = t_4$ the triggered gap G_2 is fired (see Fig. 5) which puts high voltage across B_T.

The rate of rise of this voltage can be controlled by RS_1 and C_d. Provision has been made in the circuit to insert a saturable reactor L_S in series with B_T. The circuit of Fig. 5 is sufficiently flexible that by manipulating the initial conditions and parameter values, the duty on the vacuum interrupter B_T can be varied to a great extent. The quantities that can be varied are listed below:
1. Forward current i_o
2. Rate of decay of current $[di/dt]_o$ near current zero.
3. Peak of inverse voltage, V_1
4. Deionization margin, τ_o
5. Rate of rise of forward voltage $[dv/dt]_o$ at voltage zero.
6. The peak magnitude of the forward voltage, V_f.

OPTIMIZATION OF THE INTERRUPTER IN THE IES CIRCUIT

To optimize only the interrupter portion of the IES circuit (Fig. 3), the cost of the following major components have been taken into account:
1. The circuit breakers
2. The power supply to bias C_C (not shown in Fig. 3)
3. dv/dt limiting Capcitor C_L
4. The saturating inductor L_S (the cost of L_C is small and therefore, has not been taken into account)
The sum of these costs is minimized for a specified stored energy W_L and transfer time t_F.
The model considers five variables:
1. Peak voltage across C_C
2. Current interrupted by each circuit breaker
3. The capacitance of C_L
4. The unsaturated inductance L_{SO}
5. Saturation current i_S
The cost of each individual circuit breaker has been obtained by consulting the literature of circuit breaker manufacturers. A commercially available circuit breaker with $I_{nominal} = 2kA$, $I_{interrupt}(Sym) = 25kA$, $V_{nominal} = 14.4kV$ and $V_{withstand} = 50kV$ has been chosen for this costing exercise. In addition to the initial cost, the cost of replacement of the vacuum bottles over the lifetime of the IES

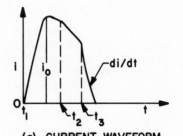

(a) CURRENT WAVEFORM

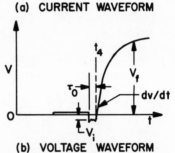

(b) VOLTAGE WAVEFORM

i_0	PEAK FORWARD CURRENT
di/dt	RATE OF DECAY CURRENT
dv/dt	RATE OF RISE OF FORWARD VOLTAGE
V_f	PEAK FORWARD VOLTAGE
V_i	PEAK INVERSE VOLTAGE
τ_0	DEIONIZATION MARGIN
$t_1 \ldots t_4$	SWITCHING INSTANTS

Fig. 5: Current and Voltage Waveform in Breaker under Test (B_T)

system was considered. A typical vacuum breaker for this application has a mechanical lifetime at zero current of 10,000 operations. At its nominal current, the lifetime is at least 5,000 operations, but at its maximum interrupting current the life is only 4 operations.

In this study, it has been estimated that the IES System shall be operated 2,000 to 3,000 times over a period of about two years[7] and will require the full 300 MJ (W_L) capacity of the system for each operation.

The total cost Q_{CB} of vacuum interrupters for the IES System can now be calculated as

$$Q_{CB} = \$CB N_M + \$CBR N_M \left[\frac{3000}{N_{CB}} - 1\right] \text{dollars}$$

where $\$_{CB}$ = initial cost of the circuit breaker
$\$_{CBR}$ = replacement cost of one circuit breaker
N_M = number of modules required to transfer the energy W_L
N_{CB} = minimum number of unit operations before replacement of the vacuum bottle (estimated from manufacturer's data).

The cost of other items mentioned above has been computed in similar fashion. The optimization problem was solved numerically using the sequential unconstrained minimization technique (SUMT) of Fiacco and McCormick[8]. The details of this optimization exercise are available elsewhere[4].

The results of the study can be summarized as follows:

1. For an IES System of 300 MJ and $t_E = 1ms$ the cost of the interrupting system is primarily dictated by the replacement cost of interrupter bottles.

2. For the pessimistic case, the cost of the interrupting system may be about 3.05¢/joule.

3. Taking this cost figure for interrupting system, the total cost of IES system as defined under item 1 (300 MJ & $t_E = 1ms$) is probably not very much less than that for an all capacitive system. This does not take into account the money-value of the advantage of the compactness of IES System over the capacitive energy storage system.

4. By properly derating the vacuum interrupters the replacement cost can be reduced to an extent that the overall cost of the interrupting system will only amount to 1.28¢/joule.

5. Under the above mentioned circumstances the overall IES System will be substantially cheaper than an all capacitive system.

CONCLUSIONS

Inductive energy storage (IES) systems using appropriate switching techniques can be effectively used to supply large amounts of pulsed power required for fusion experiments. The studies performed up to date indicate that an IES system with commercial circuit breakers used now a days for ac power systems is technically feasible. However, as indicated in the preceding section, experimental investigation of the breakers is necessary in order to determine their deionization properties. This will help to optimize the design of auxiliary equipment in the switching circuit. A synthetic test facility is being built for this purpose at the University of Texas at Austin.

Costing analyses of different types of energy storage systems (300 MJ, $t_E = 1ms$) indicate that an IES System using properly derated (60 Hz rating taken as the base) vacuum circuit breakers is substantially cheaper than an all capacitive system.

REFERENCES

1. F. L. Ribe, "Parameter Study of a Long Separated Staged Theta-Pinch with Superconducting Inductive-Energy Storage," LASL Report No. LA-4828 MS, December 1971.

2. H. G. Rylander, H. H. Woodson, E. B. Becker and R. E. Rowberg in "Technology of Controlled Thermonuclear Fusion Experiments and the Engineering Aspects of Fusion Reactors," ed. by E. L. Draper, Jr. (USAEC: Conf. - 721111, 1974) p. 260.

3. W. F. Weldon, M. D. Driga, H. H. Woodson, H. G. Rylander, "The Design, Fabrication and Testing of a Five Megajoule Homopolar Motor-Generator," Companion paper in this conference.

4. R. B. McCann, "An Assessment of the Applicability of High Voltage ac Circuit Breakers to Inductive Energy Storage." Dissertation presented to the Faculty of the Graduate School of the University of Texas at Austin for the degree of Ph.D., 1974.

5. T. Mukutmoni, "Synthetic Test Installation for Converter Valves for hvdc Transmission." IEEE Transactions on PAS, Vol. PAS-93, No. 4. July/August (1974).

6. G. A. Farrall, Proc. of the IEEE 61, 1113 (1973).

7. K. I. Thomassen, Los Alamos Scientific Laboratories (personal communication).

8. A. V. Fiacco and G. P. McCormick, "Nonlinear Programming: Sequential Unconstrained Minimization Techniques," (John Wiley and Sons, Inc., New York, 1968).

FAST CIRCUIT BREAKERS FOR 200 kA CURRENTS

C.A. Bleys, D. Lebely, C. Rioux, F. Rioux-Damidau
Laboratoire de Physique des Plasmas
Groupe Electrotechnique et Fusion Contrôlée
Université Paris-Sud, 91 405 Orsay - France

ABSTRACT

A circuit breaker operated by mechanical rupture, capable of opening very high currents in times of the order of ten microseconds has been realized. A copper conductor in the form of a T, fragilized at the joint of the arms and the leg, is ruptured by oil under a pressure of 3 kbars. This pressure is obtained by the magnetic forces of a 600 kA current pulse flowing in a coaxial conductor system. Current flowing in an inductive storage coil is transferred onto a fuse in 9 microseconds. The subsequent explosion of the fuse can either open the circuit and absorb all the stored energy or transfer it to any desired load. The switch is sufficiently reisolated to withstand the voltage peak thus created. A 200 kA current has thus been opened in 20 microseconds and a 9 kV overvoltage sustained. The system, because of its inherent symetry, can be lengthened to obtain any desired current carrying capacity and requires a relatively small command energy.

INTRODUCTION

Homopolar generators which convert the inertial energy of a flywheel to electrical energy show great promise as a cheap means of powering plasma physics and fusion experiments. The delivery time however is relatively long and, to be useful, must be compressed. This can be achieved by first storing the energy in a magnetic storage coil and at the current maximum transferring the energy to a load by means of a high speed circuit-breaker. This unit must be capable of carrying the full coil current during the whole of the build-up time and subsequently open in an extremely short time. It must then reinsulate to withstand the voltage peak when the energy is transferred from the inductive store to the load and absorbe all the energy which must be dissipated during the transfer. We here describe a circuit breaker that can be adapted to the existing homopolar generator system in our laboratory.

THE HOMOPOLAR GENERATOR SOURCE

An iron-free, self-excited, homopolar generator coupled to a 1 µH storage coil, which can deliver 1 MJ of energy has been in operation in our laboratory since 1967. The current pulse in the storage coil reaches a maximum of 1.6 MA and has a half-value width of 0.2 seconds. At present we are engaged in the construction of a much larger 100 MJ unit. We however here only consider a circuit breaker which is destined for the smaller of the two.

A complete description of the two machines can be found in existing articles.[1,2]

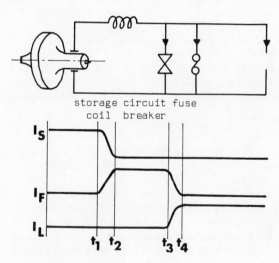

storage circuit fuse
coil breaker

Fig. 1
Basic circuit & current waveforms

THE BASIC MODE OF OPERATION

To open the storage coil current a mechanical circuit breaker is used in parallel with a very fast fuse.Fig.1. Initially the full storage coil current flows in the mechanical switch (S). When the current maximum is reached the switch S opens, an arc is established between the open contacts, and the current is transferred to the fuse in time (t_2-t_1).

During the latence time (t_3-t_2) the arc has been extinguished (t_2) and insulation can be built up between the contacts. Between the contacts. Between time t_3 and t_4 the fuse explodes and all the current is transferred to the load. The exploding fuse causes a voltage peak to appear across the contacts of the switch which must be sufficiently insulated if it is not to break-down. If the load is inductive, energy must be absorbed during the transfer and this is dissipated in the fuse, not the mechanical switch.

The switch must have a sufficient time to transfer the current and completely insulate. The obtained transfer time to the fuse (t_3-t_2) is governed by the arc voltage of the circuit breaker, the initial fuse resistance and the total inductance of the fuse circuit. Insulation time is determined by the speed at which insulation can be introduced between the contacts. The latence time of the fuse must therefore be longer than the total transfer and insulation time of the switch if this is not to break down during the transfer to the load. With the present state of the art, fast fuses [3] can be constructed which have latence times 5 to 10 times the explosion time. The obtainable transfer time to the load is therefore largely dependent on the achievable performance of the mechanical switch.

ANTERIOR DESIGNS

The first really successful circuit breaker built by us used the explosion of a gaseous mixture of $2H_2 + O_2$ to cut a circular ring, 10 mm internal and 16 mm external diameter out of a 0.1 mm thick copper plate and forced insulating grease into the opening[4]. Currents of upto 16 kA have been opened in 1 µsec and a reinsulation of 30 kV achieved. The system however proved inefficient and would have used prohibitively great amounts of energy if designed to carry the current of the homopolar generator system. Subsequent switches using exploding wires as the command energy source showed promise but were difficult to manipulate.

THE MECHANICAL SWITCH

In the present switch, the command is obtained by magnetic pressure, created when a capacitor bank is discharged into a coaxial conductor system Fig.2 . The central conducting element is a bar in the form of a T (4). The current is made to flow symetrically through the "leg" and divides equally between the two 'arms'.

Electrical contact is made to the two sides of the "leg" by series of teeth (3), pressing into the conductor, and to the "arms" of the T by ridges cut into the back plate which press into the conductor when the unit is assembled. Oil is used

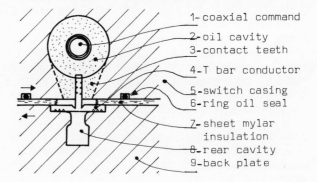

1- coaxial command
2- oil cavity
3- contact teeth
4- T bar conductor
5- switch casing
6- ring oil seal
7- sheet mylar insulation
8- rear cavity
9- back plate

Fig.2 Crossection of circuit breaker

as the insulating medium and fills the sealed cavity (2). It can flow freely between the teeth (3) that make contact with the "leg". The fuse is connected on one side of the switch, between the switch casing (5) and the back plate (9). By using a wide foil fuse and reducing the distance between the conducting surfaces, the total fuse circuit inductance can be very much reduced.

When the command current is discharged into the coaxial system, the pressure due to the magnetic field created between the two conductors causes the outer tube to explose. The oil in cavity (2) is thereby pressurised and the T bursts open (as shown in Fig.3). To ensure that the rupture is properly localised, lateral grooves are cut into the T arms near the center. The conducting circuit is thus opened up and an arc established between the ruptured ends. The resultant arc potential transfers the current to the

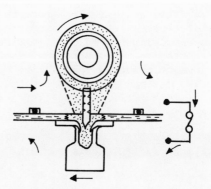

Fig.3 the opened circuit breaker

fuse. When all the current has been transferred, the arc extinguishes and new,
clean oil flows around the central contact,
replacing the burnt arc products and insulating the switch.

THE CURRENT CARRYING CAPACITY

The current carrying capacity of the
switch is limited by the crossection of
the copper conductor. This can be increased by either increasing the thickness of
the T or the length. Practically, however,
the thickness is limited to about 1 mm.
If this dimension is increased, the burs-
ting pressure required becomes too large
and may cause structural damage within
the switch. Further-more, the additional
pressure will cause the leg of the T to
slip out from between the "teeth". The
total length of the bar is not limited.
The command pressure is independent of
the length and only the actual energy
lost within the command tube is determined by the total length.

FUSES

Only simple flat exploding foil fuses,
placed between two layers of adhesive tape
have been used for this experiment. It
is possible that markedly better results
could be obtained if a better insulated
fuse system were to be used.

EXPERIMENTAL RESULTS

Two prototype circuit breakers have
been constructed. Ultimately a switch will
be constructed, adapted to the homopolar
generator but here the preliminary experiments have been carried out at a reduced
scale and a capacitor bank used as energy source. The experimental circuit diagram is shown in Fig. 4. Magnetically 5
operated closing switches and crowbars
have been used to control the currents
in the circuit. The stored energy is not
transferred to a load but is instead
completely dissipated in the exploding
fuse. This allows the reinsulation of the
switch to be tested more thouroughly as
the voltage peak thus produced is much higher than if a load was present.

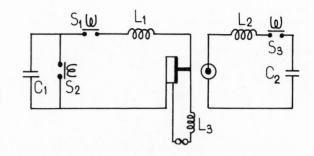

Fig. 4   Experimental   Circuit
C_1 - source 1000µF at 5kV max
S_1 & S_3 - magnetically closed switches
S_2 - crowbar switch
L_1 - storage coil 440 nH
L_2 - 60 nH
L_3 - 10 nH stray inductance
C_2 - command 2000µF at 5kV max

THE 25mm CIRCUIT BREAKER

Initially a switch using a 25 mm
long T element was tested. Two different
diameter commands were tried. Copper was
used in all cases and tubes with 8 and
12 external diameters with 1mm wall thickness were taken. For the 8 mm tube, a
command current of 550 kA seemed optimum
but for the 12mm tube a much higher value
of 760 kA was required. The results obtained for the two cases were otherwise
very similar. Subsequent experiments have
been carried out with 8mm tubes because
of the lower current requirement. It is
not possible to use even smaller command
tubes as the inner conductor fails due
to the magnetic pressure exerted on it.

Several experiments have been done
under identical conditions. For these, a
current of 30kA has been transferred to
the fuse in 4 µsec. After a latence time
of 20 µsec the fuse explodes and opens
the storage coil circuit. Voltage peaks
of upto 9 kV have been sustained, however
the actual insulation level has not been
tested. A more sophisticated fuse system
would be required.

THE 160mm CIRCUIT BREAKER

Subsequently a new circuit breaker
with a 160mm T has been constructed to
see if the results for the 25mm unit
could be extrapolated. An 8mm command
tube has been retained and a current
of 550 kA used as this has proved optimum for the smaller unit. For this switch
several identical experiments have also
been carried out.

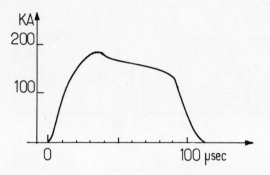

REFERENCES

1   R. HAHN, J. LUCIDARME, C. RIOUX,
    R. GUILLET.
    Revue Phys. Appl. $\underline{8}$,53,1973.
2   J. CHABOSEAU, R. GUILLET, J. LUCIDARME,
    C. RIOUX.
    Proc. Third International Cont. Magnet.
    Technology. Hamburg 1970 - Pages 1305-
    1312.
3   J.N. DI MARCO, L.C. BURKHARDT.
    J. Appl. Phys. $\underline{41}$, 3894,1973.
4   A. DELMAS, C. RIOUX, F. RIOUX-DAMIDAU.
    Revue Phys.Appl. (to be published).
5   C.A. BLEYS, D. LEBELY,
    F. RIOUX-DAMIDAU.
    Rev. Scient. Instrum. $\underline{46}$, 180 (1975).
    Jan   75).

Fig. 5

Current in L_1 for a 160 mm T expt.

Figure 5 shows a typical trace of the current in the storage coil. Just before
the circuit was opened, 180 kA were flowing in the circuit breaker. The current
was transferred to the fuse in 9 µsec
and after a latence time of 35 µsec the
circuit opened in 20 µsec. For these
experiments, the fuse was very badly adapted to the switch and further experiments
are being undertaken.

DESIGN OF CIRCUIT BREAKER ADAPTED TO THE

HOMOPOLAR GENERATOR

     The circuit breakers tested upto
the present time have not been tried for
the total current   carrying capacity.
It is essential that the unit which is
to be mounted on the generator be capable of carrying the full storage current
during the whole of the build-up time.
For our generator we have calculated that,
allowing for no heat loss by diffusion and
a maximum temperature increase of 600°
above ambiant, a copper conductor must
have a minimum crossection of 10 cm^2
if it is to support the buildup of a 1 MA
current pulse. This corresponds to a circuit breaker T element that is 1 meter
long, has a 1 mm arm thickness, 2 mm leg
thickness and 0.5 mm deep rupture grooves.
The resultant current density in the T is
approximately that chosen for the prototype models during the reduced scale
tests. Further-more, if local heating
occurs near the rupture grooves, the
opening of the T will be facilitated and
it is possible that even thicker T elements
would eventually be possible, thus reducing the total length required.

STORAGE CAPACITORS, START AND CROWBAR SWITCHES AND INDUCTORS FOR HIGH BETA PLASMA EXPERIMENTS

L. FELLIN, A. MASCHIO, P.L. MONDINO, G. ROSTAGNI, A. STELLA

Centro di Studio sui Gas Ionizzati del Consiglio Nazionale delle Ricerche e dell'Università di Padova (Ass. EURATOM–CNR)

Via Gradenigo 6a -- PADOVA (Italy)

ABSTRACT

An introductory description of the "Eta--beta" device, built in Padua for the production of toroidal axisymmetric plasmas, is given. The design of the main components is briefly discussed and the results of a number of design and operating tests performed are reported. Some problems of "reliability" and "life" are also considered.

INTRODUCTION

A device for the production of pulsed toroidal discharges has been built in Padua and it started regular operation a few months ago. The available energy is only 120÷150 kJ , but the system is designed to be very flexible by the choice of small modular units and interchangeable connections.The purpose is to study axisymmetric pinch configurations with various combinations of toroidal and poloidal fields.

We are presently interested to short rise time (~1μs) discharges, but the long term plan is to compare similar discharges with different time scales. Therefore a convenient trasformer with changeable turn ratio, already described in principle [1] , will be added to the system.

In this paper, after a short description of the device , we wish to present some comments on the main design choices and recall some of the results of preliminary tests. More detailed information have already been reported in [2, 3] , while [8] gives the first scientific results obtained.

SYSTEM DESCRIPTION AND PARAMETERS

The quartz discharge vessel, placed in a vertical plane, has a major radius R = 40 cm and a minor inside radius r= =5 cm. It is surrounded by 24 alluminium alloy shields , 4mm thick, individually encapsulated with 2mm of epoxy for electrical insulation (Fig.1). Their purpose is to reduce the transverse field components. Outside the shields is placed the toroidal field (B_z) inductor, divided in six sectors, with twelve feeding lines connected at the back side. A slow

"bias" bank (to produce an initial field B_{zo}), when used, is

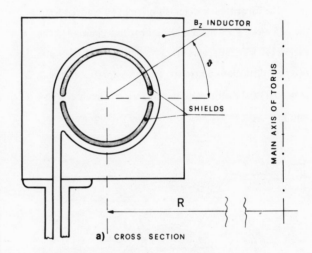

a) CROSS SECTION

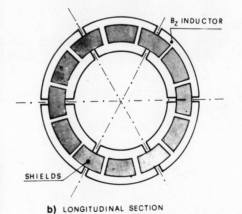

b) LONGITUDINAL SECTION

Fig. 1

connected to the same lines through 12 separate transformers with 100:1 turn ratio. Polithene sheet insulation is used .

The main reason for this structure was diagnostics access. The longitudinal plasma current (I_z) is induced by 36 insulated cables, laying on the two cylindrical surfaces (inner and outer) of the B_z coil: leakage inductance and transverse field at the gaps between sectors are thus kept to reasonably low values. Each cable has six feeding points, in order to increase the loop voltage and match the I_z and B_z rise times (Fig.2). All the feeding lines are connected together by a coaxial cable system that forms a closed equipotential ring of about 3m diameter: it acts as a protection for the I_z capacitor bank in case of misfiring and enables a good earthing of all the circuits. A toroidal preionization voltage is sometimes applied by a separate coil loosely coupled to the main I_z circuit.

The capacitor bank consists of up to 60 modules arranged in 5 racks. Each module comprises a low inductance capacitor (40 kV, $3\mu F$, mean life 10^5 shots at 85% reversal), two field distortion compressed air spark gaps (for start and crow–bar) and their polarization and trigger circuits. The start gap has been designed to withstand without maintenance more then 10^4 shots at full voltage, while keeping a low premature discharge rate. The crow bar is similar to the start gap, but the capacity between trigger and main electrodes has been increased in order to reach a ± 100% working range. The inductance of a complete unit is 40 nH. Parallel plate lines are used in the B_z circuit, bundles of 50 Ω coaxial

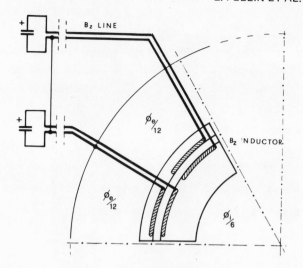

Fig. 2 b

cables in the I_z circuit. Fig. 3 and 4 show a picture of the capacitor bank and load assembly respectively.

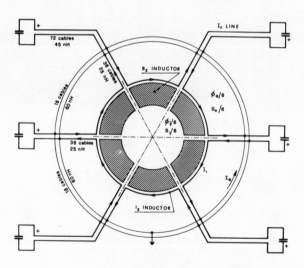

Fig. 2 a

Fig. 3

TECHNOLOGICAL DEVELOPMENTS

It may be wortwhile for the purpose of this conference to mention, among a number of technical problems solved,

Fig. 4 a

Fig. 4 b

the following:

1)   **Low inductance capacitors**: current technology employs paper foil windings arranged as low impedance transmission lines; the inductance is inherently low, but the foil
resistance cannot be neglected. An attempt to use extended
foil elements with welded connections proved satisfactory ,
provided the current density in the connections was kept
low enough (to avoid excessive electrodynamical forces during the discharge and the resulting saw-effect): a   convenient design had to be developed in order to obtain in the sa
me time low inductance and high energy density (the resi -
stance naturally results very small). After some initial failure due to manufacturing errors, life tests gave figures comparable to those of transmission line capacitors.

2)   **Spark gaps**

Extensive development tests have been performed on
three electrode, field distortion, pressurized air spark gaps.
The results are generally in agreement with those of   James
et al. [4, 5] , in particular the breakdown delay and   jitter ,
the dependance of the premature discharge rate from   the
volume of the gap chamber and from the overpressure ,   or
the dependance of the erosion rate from Q^* (the total charge flow per shot). Some influence of the discharge period on
the gap life has also been observed.

In addition, an appreciable influence of the   flushing
air throughput on the premature discharge rate has been measured (e.g. at an overpressure $p_w/p_b = 1.3$ and $Q^* = 3$ cou-
lomb/shot, the premature rate decreased from 0.4% to 0.1%
as the air flow increased from 10 to 50 liters). The final choice of the various parameters brought in our case to 10 liters/
coulomb: about 10 times as much as in high voltage switchgear. Although a better design of the gap chamber could reduce the necessary quantity, the cost of the dry air   plant
will be high.

The considered gap has a low interelectrode capacity
(Fig.5 a) and works in the cascade mode, with a   working
range limited, but sufficient for usual applications as a start
gap. In order to extend the working range to ±100% ,   as
necessary for crow bar, a high capacity version (Fig.5 b) has
been developed and the results of a number of tests   have
already been reported [3] .

The alternative choice of a low capacity gap with a ve-

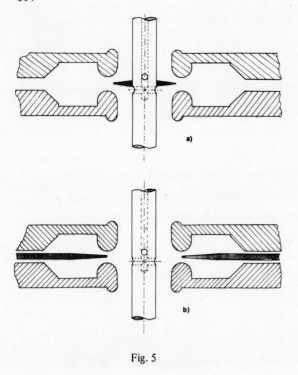

Fig. 5

ry high voltage, very steep trigger pulse [6] results more expensive and tends to produce excessive interference on the plasma diagnostics.

3)   Insulations

The design life of all the insulation has been chosen of the order of 10^5 ringing shots at full bank voltage, 85% reversal. Accelerated life tests have been carried out under "equivalent" A.C. voltages (being careful to avoid thermal stresses) on a few specimens of each item.

Polithene sheets have been used whenever possible (plane or cylindrical geometries). Due to their toroidal   shape , the shielding sectors had to be incapsulated. To this purpose, different materials and techniques have been considered and tested. Silicon rubber needed a too large thickness; the use of dust bath technique gave non uniform results for   a thickness larger than $1 \div 1.5$ mm , both with nylon and   epoxy. Finally a moulding technique with pure epoxy $2 \pm \pm 0.1$ mm thick was chosen. The mould was also in   epoxy with an higher Martens point. The shield elements were centered in the mould by a number of calibrated epoxy spacers; they did not give any local reduction in the withstand voltage if accurately cleaned.

A similar technique has been used also to produce the

shaped caps designed for the longitudinal insulation   between the B_z coil sectors.

COMMENTS ON SOME DESIGN CHOICES

1)   **Relative position of I_z and B_z inductors**: it can easily be shown [7] that the magnetic energies of both components, B_z and B_θ , in the toroidal space delimited by   the first conducting wall (radius r_o) around the discharge vessel, are proportional to the total energy of the plasma to be confined. Moreover, in a stabilized pinch:

$$\frac{W_z}{W_\Theta} = \frac{1}{\Theta_0^2} \frac{K_z}{K_\theta}$$

where K_z and K_θ are shape factors of order unity (usually $K_z > K_\theta$), and $\Theta_0 = B_\theta(r_0)/B_z(O)$ is a parameter controlling the plasma behaviour, usually comprised between 1 and 2 . In a reversed field pinch, the energy necessary to reverse B_z in the outside region must be added.   Therefore W_z and W_Θ tend to be of the same order.

On the other end, the total B_z flux within the   coil and leakage inductance up to the crow-bar point (a   nearly flux conserving circuit) is a critical parameter for plasma stability: it can be shown that it is essential to minimize the inductance external to the plasma region.

It may be concluded that the choice to keep the B_z inductor nearer to the plasma appears as the more   convenient.

2)   **Transverse fields shielding**

In order to reduce the transverse field components inside the main B_z coil (having one longitudinal feeding gap and six transverse gap), a screening shell divided in 24 elements has been used (Fig. 1).

This solution is clearly convenient from the insulation point of view: if m is the number of gaps along a major circumference and V_z the total induced voltage, the   voltage across one gap is V_z/m . A reduction of m obviously increases the voltage per gap, but also the voltage between the shields and the corresponding coil sectors (shown in   Fig. 6 for the cases m = 12 and m = 6 , while Fig. 7 gives a   diagram of the shield and sector voltages referred to a mean value). In particular with m = 12 the central shield of   each

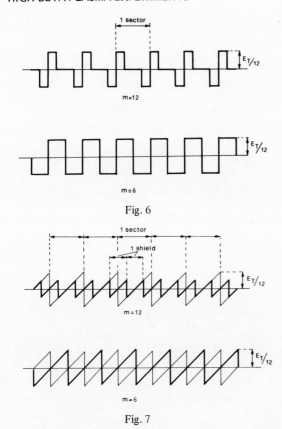

Fig. 6

Fig. 7

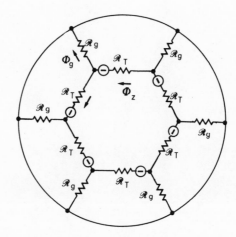

Fig. 8

sector remains "equipotential" with the sector, thus simplifying diagnostic access. The number of gaps has also a marked effect on the electrostatic field, before the space charge
regime is reached, in the discharge vessel. More difficult is to
compare the effect on plasma equilibrium and stability of
the residual transverse magnetic field as a function of the
number of gaps: each perturbation tends to be smaller, but
their number increases with m .

3) Magnetic coupling between sectors

If $\mathscr{R}_T$ is the magnetic reluctance of each coil sector,
in the toroidal direction, and $\mathscr{R}_g$ the reluctance of a transverse gap, the equivalent circuit of Fig.8 can be drawn for
the complete toroidal coil. With a gap a few millimeter thick
and our torus dimensions, $\mathscr{R}_T \sim \mathscr{R}_g$. It is easy to see that
the coupling coefficient between sectors is small: $M/L \sim 1/3 \div$
$\div 1/2$ (in absence of transverse gaps, the longitudinal feeding
gap must be considered and a similar result is obtained).The
refore, if the effect of the shields is neglected, a good balance of all the circuits is necessary to reach negligeable values
for Φ_g : it has been obtained by an adjustment of the feeding lines inductances. Still more important is a very low jitter between the spark gaps of the different circuits, but well

reproducible results have been obtained. An advantage of the
small coupling degree is that the capacitors are automatically
protected against overvoltages, in case of any misfiring.

Similar considerations apply also to the I_z circuit if
the external ring connection of the different sections is present.

4) Use of spark gaps for the crow-bar

They have been preferred to metal to metal switches,
because of the low jitter necessary with our short risetimes
and the absence of maintenance work between the shots.The
drawback is that the arc voltage V_a gives a linear decay of
the current (neglecting the resistive decay) to zero in a time
t_d proportional to the current itself:

$$t_d = \frac{LI}{V_a} = \frac{2}{\pi} \, t_r \frac{V_o}{V_a}$$

(where $t_r = (\pi/2)\sqrt{LC}$ is the current rise time and V_o the
charging voltage of the capacitor bank). In our case $V_a \sim 200V$
and t_d results $\sim 100\mu s$ with maximum flux: some difficulty arises in the study of low energy discharges, or when
very small degrees of field reversal are desired in the B_z circuit.

REFERENCES

[1] MONDINO P.L., ROSTAGNI G.; Alta Frequenza, XLII , 6
(1973) and UPee Report 71/08 (1971)

[2] MONDINO P.L., ROSTAGNI G.; 7th SOFT, Grenoble (1972)

[3] FELLIN L., et al.; 8th SOFT, Noordwijkerhout (1974)

[4] BURDEN R.A., JAMES T.E.; 7th SOFT, Grenoble (1972)

[5] GRUBER J.E., JAMES T.E.; Proc. IEE, No 10, 1530 (1968)

[6] KLEMENT G., WEDLER H.; 7th SOFT, Grenoble (1972)

[7] BODIN H.A.B., et al.; IAEA Workshop on Fusion Reactors,
Culham (1974)

[8] BUFFA A., et al.; IAEA CN-33/E9-3, 5th Conference PPCNFR,
Tokyo (1974).

STUDY OF THE HIGH VOLTAGE CIRCUIT-BREAKERS BEHAVIOUR AROUND CURRENT ZERO

E. Carnevali   V. Villa

Centro Elettrotecnico Sperimentale Italiano
Via Rubattino,54 - Milano - Italy

ABSTRACT

The paper gives a short description of the interrupting phenomena in high voltage circuit-breakers and the various measuring techniques are described which are at present used to record the voltage and the current around zero.

An analysis of a method to interpret such measurements is given which permits to derive the arcing characteristics and to calculate the arc-behaviour around zero.

INTRODUCTION

The design of high-voltage circuit-breakers generally is arrived at by theory and experiment. The designer's hypothesis has to be experimentally verified in laboratory circuits that simulate network conditions.

The most difficult and important switching operations are certainly those related to the breaking of heavy short-circuit currents. Maximum short-circuit current values are now about 50-80 kA r.m.s. and will be about 100-120 kA in the near future. Maximum system voltage values are at present 765 kV r.m.s. and will be around 1000-1200 kV in the near future.

For this reason, a lot of laboratory experiments on circuit-breakers, are made with the object of studying the interrupting phenomena.

INTERACTION PHENOMENA

Specific measures to study interruption phenomena are the recording of voltage and current around zero. Infact, in a short interval around current zero, there is an interaction between the breaker and the circuit. Here the actual interaction conditions determine the clearance or failure results. The instantaneous voltage and current values give the results of this interaction.

Generally, the interaction intervals are very short compared with the whole breaking operation. Arcing times can be from 1 to 30 ms; interaction intervals are generally from 5 to 50 μs, depending above all on the type of extinction medium.

Figure 1 shows a simplified laborato-ry circuit for tests on circuit-breakers, and typical test values. It also shows interaction intervals, which are important in the breaking.

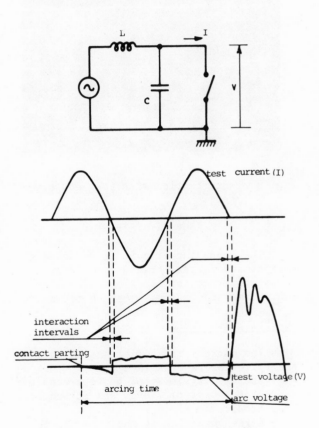

Fig. 1. Simplified test circuit.

Due to shortness of interaction intervals, a high speed of recording, from 1 to 10 μs/cm is required. Consequently the current and voltage measurements have to be very sensitive compared to test cur-

rents and test voltages. For example: the
sensitivity of current measurements may be
10 A/cm with a 50 kA r.m.s. test current;
the sensitivity of voltage measurements
may be 1 kV/cm with 100 kV r.m.s. test
voltage.

In this way, it is possible to measure the post-arc current which passes
through the breaker soon after current zero, when the transient recovery voltage
occurs between the breaker contacts. This
current is due to residual ionization between the breaker contacts, and generally
has an amplitude of some amperes and a duration of few microseconds.

Figure 2 shows a typical interruption
with post-arc current.

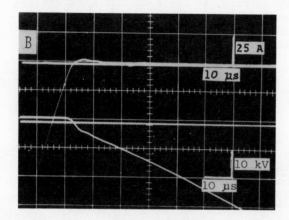

Fig. 2. Interruption with post-arc current.

MEASURING SYSTEMS

The measuring circuit is composed of
current and voltage transducers, a transmission device, signal limitations, and
dual-beam oscilloscopes for the recording.

CURRENT TRANSDUCERS

The current transducer has to have
the following requirements :
a) - the ability to stand up to a consider-
     able amount of energy and strong elec-
     trodynamic forces;
b) - a response time as short as possible,
     in any case, around a hundred of ns;
c) - insensitivity to the magnetic and e-
     lectrostatic fields of the power cir-
     cuit.

At present the current transducer
most used is a coaxial shunt. We have
also been using a coaxial shunt with satisfactory results for several years[1]. It

is shown in fig. 3 and its performances
are :
a) - 50 kA r.m.s. for 0.2 s;
b) - 125 kA peak value;
c) - 100 ns response time;
d) - 2 mΩ resistance.

Fig. 3. 50 kA r.m.s. coaxial shunt.

VOLTAGE TRANSDUCERS

Requirements for the voltage transducer can be summed up as follows:
a) - it has to bear the whole test volt-
     age;
b) - it has to have a large frequency
     bandwidth;
c) - it must modify the test circuit as
     little as possible.
So the voltage transducer has to measure
the arc voltage before zero current, which
is a low frequency signal, and the initial
transient recovery voltage after zero current, which may have a high frequency, up
to approximately 1 MHz.

Pratical needs make it advisable to
use capacitive dividers and mixed, capaci-
tive-resistive, dividers.

MEASURING CIRCUITS

Figure 4 shows the voltage and zero current measuring system that we normally

use. Usually, this method is used on single high voltage breaker units.

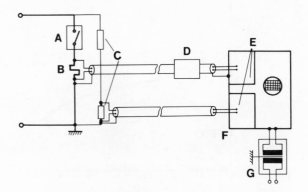

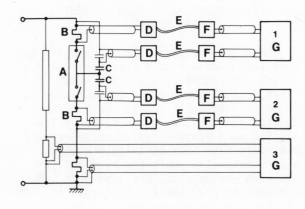

Fig. 4. Usual measuring system for high
         voltage circuit-breaker units.
         A – test circuit-breaker
         B – shunt
         C – voltage divider
         D – limiting voltage device
         E – differential preamplifiers
         F – dual-beam oscilloscope
         G – shielded transformer

Fig. 5. New light –link telemetry system
         for voltage and zero current
         measurements.
         A – circuit-breaker
         B – insulated current shunt
         C – grading capacitance
         D – transmitter
         E – optic fibre
         F – receiver
         G – dual-beam oscilloscope

It is worth mentioning briefly, that
high voltage circuit-breakers are normally
composed of several breaker units in series. The number of breaker units increases when the voltage of the system is
higher. With this method, the transmission of the current and voltage signals is
made through shielded cables.

Figure 5 shows a new measuring method
recently developed, which enables voltage
and zero current measurements to be made
from live points to the grounded oscilloscopes on several single breaker units at
the same time. The diagram shows an example of this method on a circuit-breaker
composed of two units.

Light-link telemetry systems, with a
bandwidth from d.c. to 2 MHz, are employed
to transmit the current and voltage signals from live points to ground. With
this method the transmission of the current and voltage signals is made through
optical fibres. Each current shunt is in
series of its breaker units and insulated
from the ground. Voltage signals are obtained by capacitive dividers, using the
grading capacitance of each breaker unit
as high voltage impedance. Generally,
high voltage circuit-breakers have grading
capacitances which distribute the voltage
in each unit.

These measurements make it possible
to observe the interaction phenomena on
each breaker unit.

An example is shown in fig. 6 on circuit-breaker composed of two units.

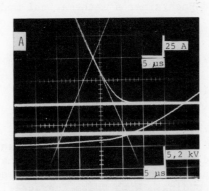

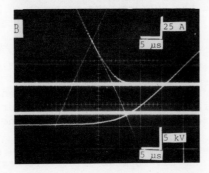

Fig. 6. Current and voltage measurements
         on two circuit-breaker units.
         A – unit one; B – unit two.

INTERPRETATION OF ZERO CURRENT MEASUREMENTS

These measurements make it possible to judge the behaviour of a circuit-breaker through the voltage and current wave-shapes, and to see whether the stresses on the breaker bring it to its thermal limit or, alternatively, to its dielectric lim-it. For many years research engineers and physicists have been studying the arc, using models of circuit-breakers, and in-teraction phenomena. They have made some mathematical arc models, applicable to circuit-breakers around zero current. Theories by Mayr, Cassie, Browne, Rieder and Hochrainer are the ones most commonly applied.

Recently we have begun the same stud-ies with reference to high voltage circuit breakers, using Hochrainer's mathematical arc model [2], and Brakelmann-Hochrainer's computer method [3]. This method is both theoretical and experimental, so that the actual arc in the circuit-breaker around zero current can be simulated more effec-tively. Generally, arc characteristics are different for a number of circuit-breakers, and depend on the breaker design and on the extinction medium.

The Hochrainer's mathematical model is:

$$g'(t) = \frac{1}{\tau(i)} \left[G(i) - g(t) \right]$$

$g(t) = i(t) / v(t)$ dynamic arc con-ductance
$i(t)$ arc current
$v(t)$ arc voltage
$g'(t)$ derivate of $g(t)$
$G(i)$ steady state conductance
$\tau(i)$ steady state time constant.

$G(i)$ and $\tau(i)$ are the steady state charac-teristics computed using voltage and cur-rent recordings. With these characteris-tics and the current ramp response of the circuit, it is possible to simulate tests by computer.

Figure 7 shows an example of how the method is used. The continuous lines are the actual voltage and current around zero as measured during a test. The dotted lines are the computed voltage and current around zero using Brakelmann-Hochrainer's method.

These studies will be continued, and so will the development of circuit meas-urement, in order to improve test current and test voltage performances.

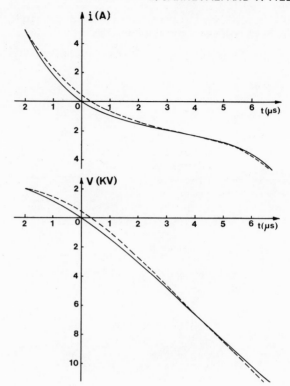

Fig. 7. Comparison between test and cal-culation.
Continuous lines: measured values
Dotted lines: computed values

CONCLUSIONS

The measurements around zero are im-portant to investigate the circuit-breaker behaviour in the interaction intervals.

It is possible to simulate tests by computer.

REFERENCES

1 - P. Pezzi V. Villa
A very low response-time 50 kA shunt
for current zero measurement.
Current Zero Club Meeting 1967.

2 - A. Hochrainer A. Grütz
Study of arcs in breakers with the
help of a cybernetic model.
CIGRE 13.10, Part A, 1972 Paris.

3 - H. Brakelmann
Berechnung des Verhaltens von Leist-
ungsschaltern mit Hilfe der Lichtbogen
bedingungen von A. Hochrainer unter
Benutzung der unbeeinflussten Ein-
schwingspannung.
ETZ-A Bd.93 (1972) H.7 S.419-420.

A FAST HIGH CURRENT MAKING SWITCH FOR SYNTHETIC TESTING OF CIRCUIT BREAKERS

Romano Ballada
CESI
Via Rubattino, 54 - Milano - Italy

ABSTRACT

The report briefly analyzes the making operations occurring during short circuit direct and synthetic tests, with particular consideration to the main characteristics required to the making-switch for synthetic make circuits. Two different types of ionization making-switches have been experimented on at CESI laboratories with satisfactory results and a multi unit 150 kA r.m.s. making-switch recently built and tested is described, particular attention being paid to the performances and the interesting possibilities offered by this device in the testing procedure.

INTRODUCTION

The circuit-breakers making test is usually made by means of direct circuits, which employ a full short circuit power having to provide (fig. 1) to the circuit breaker the full test voltage (at the nominal network frequency), before the making operation and the full short-circuit current after the electrical closing of the circuit breaker. New tests methods have been developed with the aim to extend the testing facilities to more and more high power circuit-breakers, by means of two different electrical sources which give the above described stresses in a subsequent way. These methods are usually called synthetic test methods.

SYNTHETIC MAKE TESTS

The synthetic make test (fig. 2) is usually made applying to the test circuit breaker (P) a high-voltage source (1) which provides the voltage and the pre-arcing energy in the early moments of the making operation, and, subsequently a low-voltage high-power source 2 (applied to the test circuit-breaker with a short delay), which provides the full short-circuit current (Ip).

This test synthetic method gives a test power gain, in comparison with a direct test. In fact the short circuit power required to the source 2 is developed at the low voltage Vg, lower than the test voltage Vp.

The fast making-switch (CH) has the important task of giving the full current a start (Ip), supplied by the source 2, after the electrical closing of the test circuit-breaker P, which has determined

a pre-arc, and an initial transient making-current, supplied by the source 1.

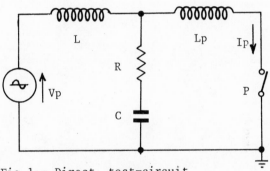

Fig.1 - Direct test-circuit
Vp= Test voltage
Ip= Short Circuit Current
P = Test Circuit Breaker

The making time of the switch (CH) should be very short (order of some hundred microseconds) in order to avoid current discontinuity during the commutation of the two circuits.

Another important characteristic of the making-switch is its ability to withstand the high voltage Vp minus Vg (see fig. 2) and to close the low-voltage source 2, even near the voltage zero (in asymmetrical tests condition) when the test circuit breaker is electrically closed.

IONIZATION MAKING SWITCH

From the fundamental characteristics of the synthetic making test it appears that the performances required to the "making-switch" are very severe and that it is extremely difficult to achieve such performances by using a mechanical "making-switch" even of advanced technology.

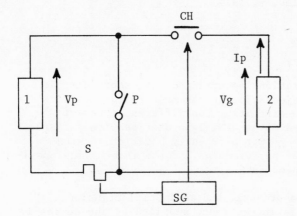

Fig.2- Synthetic make test: simplified
        scheme
        1-High voltage source
        2-Power source
        P-Test circuit-breaker
        CH-Fast making switch
        SG-Automatic control device of the
           switch CH
        S-Shunt
        Vp-Test voltage supplied by Source 1
        Ip-Test current supplied by Source 2
        Vg-Power frequency low voltage sup-
           plied by Source 2

Making techniques based on spark-gap ionization have been developed in several laboratories and two methods appear particularly suitable:
a) - the detonation ionization method[1], where the ionization is obtained firing an explosive cap
b) - the plasma gun ionization method[2], described later on.

At the CESI laboratories, one alternative solution to the detonation method has been tested in an attempt to solve the problem of explosive-cap exchange, which involves a delay in the test procedure and to solve the problem of handling the explosives.

A proper hydrogen and oxygen detonation mixture was fired by means of a spark produced on one electrode surface (200 J) in a two electrodes insulated chamber.

The detonation front propagates rapidly to the opposite electrode, thus involving the whole gas volume.

The propagation of the detonation front is associated with propagation of the ioni-

zation throughout the chamber's lenght, and a real short circuit throughout the whole gap is thereby obtained.

Fig. 3 shows the device used for the test.

Fig. 3 - Gas-blend experimental device
          used for electrical closure of a
          circuit

Some making tests were carried out with an applied minimum voltage of 25kV d.c. on a distance between the two electrodes of 500 mm: the making delay was of the order of 250μs, corrisponding to an ionization speed of 2000 m/sec.

Though the results achieved were satisfactory, this method was not developed any further, mainly for safety consideration.

The plasma gun ionization method (see fig. 4) is often used on triggered spark-gap. A powerful jet of ionized air (plasma) is generated into a chamber (plasma chamber) by means of a precharged capacitor and is injected in the main gap through a little hole.

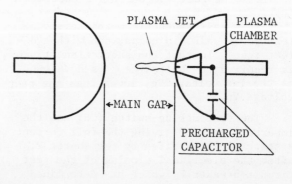

Fig.4 - Plasma gun ionization principle

PLASMA-GUN MAKING SWITCH

This method was applied to the making switch[3] designed and costructed at CESI laboratories for specific making synthetic tests operations taking into account the following main factors:
- simplicity of concept
- few maintenance problems
- low cost
- modulars elements
- safety in service

A single chamber prototype shown in figure 5, was built in 1972 and preliminary tests were conducted in order to check that the features were as planned.

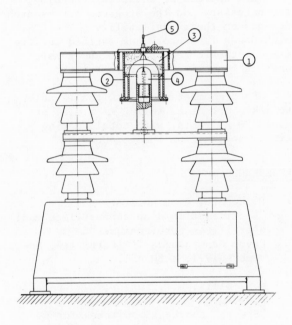

Fig.5 - Prototype of 150kA plasma-gun making switch
   1 - structure
   2 - making chamber
   3 - 1st electrode
   4 - 2nd electrode
   5 - plasma-gun

The strong structure able to stand up to the high current electrodynamic stresses supports a cylinder-shaped chamber, closed on the top side, containing one electrode, the second electrode is inserted in the open bottom side and the insulation is provided by an air ring of appropriate wide.

The arc produced between the two electrodes is contained in the chamber (this reduces the noise produced), while the arc gases are expelled through the bottom aperture. The chamber withstand voltage is 95kV (impulse voltage) and 52kV r.m.s., while the minimum control voltage is about 2kV (istantaneous value).

The making delay times depend on the istantaneous applied voltage value and are between 100 and 400μs, the accuracy is about ± 20μs.

Short circuit making tests were carried out on this making switch prototype up to the present power limits of the laboratory: 150kA r.m.s. for a duration of 200ms and peak value of 280 kA were achieved.

In all tests, with current between 10kA r.m.s. and 150kA r.m.s., the arc voltage was lower than 100 V, with a minimum value of 30 V at about maximum current.

After all tests, the contacts and chamber's conditions showed that the wear was not a significant factor.

The ratings of this single chamber prototype are summarized in the table 1, and cover all the requirements for making medium-voltage, direct test circuit and synthetic test current circuits; but for performing a synthetic test according to the scheme in fig. 2, an higher withstand voltage is necessary.

For this purpose a two chamber prototype (fig. 6) was built, with a triggered control device on each chamber.

On this prototype all the most significant tests were repeated, and it was verified that the making time was within the previous limits even with very low (5-6kV) istantaneous applied voltage.

Table 1 - Rating of the single chamber making switch

Nominal current:	150 kA r.m.s.
Short circuit duration:	200 ms
Making time:	100 ÷ 400 μs
Minimum firing voltage:	2 kV
Arcing voltage:	30 ÷ 100 V
Accuracy:	± 20μs
Withstand voltage:	52 kV r.m.s. 95 kV (impuls)

The possibility of increasing the working voltage of the making switch by means of a series of triggered chambers

was than established on condition that the
voltage distribution in the chambers is
controlled and the plasma injection in all
the chambers occurs simultaneously.

Fig.6 - Two chambers making-switch in the
CESI synthetic current circuit

The two chambers making switch was
put in service with satisfactory results
since the beginning of 1974 as making
switch in the synthetic current circuit.

The use in direct tests to control
the initiation of short circuit current
was also tested.   The possibility
to achieve high precision performances in
the study of the making transient phenomena appeared very usefull.

APPLICATION OF THE CESI MAKING SWITCH

The more common applications of the
above described making switch may be identified in the following matters:
- synthetic making circuits as heavy current fast making switch
- current circuit in synthetic tests and
  direct tests circuit as device which
  provides precise control in the current
  initiation
- high voltage circuit synthetic tests as

large voltage range spark gap
- as high performances short circuit shunt
  device.

But the introduction into the circuit
technique of this device able to withstand
the maximum current faults and to operate
istantaneous commutations in power circuits
may open the way to the use of new circuits
in the future.

CONCLUSIONS

The increasing of the requested testing power determines the developement of
synthetic make circuits.

New devices of advanced technology
are necessary and the progress of research
depends on their availability. The CESI
high power laboratory has carried out the
make synthetic test problem and a multi
unit making-switch has been built.

Each chamber of the switch is able
to make current up to 150 kA r.m.s. at
about 50 kV r.m.s. with total making
times ranging from 100 to 400μs.

REFERENCES

1 - P. Heroin, C. Benoist
    "L'utilisation d'un appareillage muni
    d'artifices pyrotecniques"
    Revue Générale de l'Electricité, No-
    vembre 1971, t.80 n.11

2 - N.S. Ellis, W.T. Lugton, C.W. Powell,
    H.M. Rayn
    "Special spark-gap switches for use
    in synthetic test circuit"
    IEEE Transaction paper, February 18,
    1972

3 - Italian Patent n. 31007-A/73

UNTRIGGERED MULTICHANNEL OIL SWITCHING*

D. L. Johnson

Sandia Laboratories

Albuquerque, New Mexico 87115

ABSTRACT

Experiments on a fast risetime, self-breakdown, multichannel oil switch have given an average switch risetime (10-90%) of 5.7 nsec and an average of 12 arc channels per pulse. The apparatus consisted of a 140 kJ Marx generator which charged an intermediate energy store, an SF_6 switch and a series inductor. The inductor was varied to change the rate of voltage rise on the Blumlein and knife edge switch. Data are presented on the simultaneity of breakdown of two switches, the dependence of switch risetime and number of arc channels on dV/dt, and on comparisons with empirical oil breakdown formulae.

INTRODUCTION

To provide fast risetime, short pulse accelerators, low-inductance switches (either triggered or self-closing) must be employed. Self-closing oil switches have long been used on high voltage accelerators;[1,2] however, they have generally been single channel, high-inductance switches. J. C. Martin[3] has shown that fast-charged switches can have multichannel breakdowns. It was felt that switches charged at a very fast rate could be made to have enough breakdown channels (hence, low inductance) and sufficiently low jitter to compete with triggered oil switches. Experiments were performed to investigate the feasibility of using untriggered switches on an EBFA prototype.[4]

EXPERIMENTAL SET UP

The apparatus used for the experiments included the following:

1. A 140 kJ, 2.5 MV Marx generator;

2. An intermediate storage capacitor;

3. An SF_6 gas switch;

4. A parallel plate Blumlein transmission line with a self-closing knife-edge oil switch.

Figure 1 is the electrical schematic of the experiment. The subscripts on the elements denote the associated lines shown in Fig. 2.

---

*This work was supported by the U.S. Atomic Energy Commission.

The Marx generator charged the 0.5 nF intermediate storage capacitors C_{1-2} and C_{2-3} to a negative potential. At nearly peak voltage (400 to 500 nsec), the SF_6 switch closed, transferring charge to C_{3-4} and C_{4-5}, which form the 44 Ω, 24 nsec Blumlein transmission line. The self-closing oil switch was located across C_{4-5}. Each line was 120 cm wide by 240 cm long and 10 cm thick. High density polyethelene supports maintained a 10 cm separation between the lines. Figure 3 is a drawing of the positive electrode of the oil switch. The edge had a 0.05 to 0.10 mm flat on the tip. The electrode extended 5 cm on each side beyond the negative electrode.

EXPERIMENTAL RESULTS

Figure 4a is an open shutter photograph of the switch breakdown. It shows 5 "bright" channels and 4 "dim" channels. The classification into "bright" and "dim" categories is somewhat arbitrary, but is an indication of the current carried in the channel. Figure 4b is an oscillograph of the signal from the capacitive voltage divider located between lines 4 and 5, and corresponds to the breakdown photograph, Fig. 4a. The horizontal sweep is 5 nsec/div and the vertical deflection is 1.8 MV/div. A breakdown voltage of 2.6 MV and a 10 - 90 percent switch risetime of 9 nsec were recorded.

A tabulation of the results of tests on this switch is shown in Table 1. The charge voltage was nearly a ramp, so dV/dt was obtained by dividing the breakdown voltage by the charge time.

Table 1. Results of first switch tests

SWITCH SPACING (cm)	10-90% RISETIME nsec	V_{BD} (MV)	CHARGE TIME nsec	dV/dt kV/nsec	Number of Bright Channels
6.4	9.6	2.6	145	18	5
6.4	9	2.8	120	23	6
5.0	9.7	2.7	95	28	6

Table 2. Results of tests on repositioned switch

SWITCH SPACING (cm)	10-90% RISETIME nsec	V_{BD} (MV)	CHARGE TIME nsec	dV/dt kV/nsec	Number of Bright Channels
5.0	5.8	2.18	73	30	11
5.7	5.7	2.34	78	30	12

Table 3. Results of timing measurements of the two-switch tests

SWITCH SPACING (cm)	V_{BD} (MV)	T_{Charge} (nsec)	dV/dt kV/nsec	ΔT (nsec)	STANDARD DEVIATION (nsec)
5.7	2.3	80	29	.3	1.4
5.7	2.1	130	16	1.7	2.1

Only about 10 shots each were taken but were enough to point out that the results were affected by the inductive switch configuration. By reconfiguring the switch, the inductance was reduced by 30 nH and, hence, reduced the 10 - 90 percent risetime by ~ 3 nsec.

Figure 5 shows the new position of the switch. The rest of the experiment remained the same. Figures 6a and 6b show the improvement in number of "bright" channels and risetime with the repositioned switch. The 10 - 90 percent risetime is 5 to 6 nsec. The overshoot of the base line is due to the RC decay of the monitor voltage during the charge of the Blumlein. Table 2 tabulates the results of approximately 50 shots in the new position.

To test the switch jitter, a second switch was placed at the opposite end of the Blumlein and the load moved to the center as shown in Fig. 7. A second capacitive monitor was also installed near the second switch. The two monitors were separated by 9 nsec which determined the upper limit on the simultaneity timing resolution. Table 3 gives the results of the jitter measurements with ~ 20 shots for each charge rate.

The average time difference in breakdown is ΔT. The standard deviation in breakdown of a single switch can be obtained by dividing by $\sqrt{2}$. This yields a standard deviation for both cases of slightly greater than one percent (either in time or voltage due to the nearly linear charge rate).

The following empirical formulas[3] can be used to estimate the switch risetimes:

$$\tau_R = \frac{5}{E^{4/3}(nZ)^{1/3}} \quad \text{(nsec)} \qquad (1)$$

$$\tau_L = \frac{L}{nZ} \quad \text{(nsec)} \qquad (2)$$

where

E = mean electric field (MV/cm)

Z = source impedance (ohms)

L = inductance of one arc channel (nH)

n = number of arc channels

and, τ (total e-fold risetime) equals $\tau_R + \tau_L$.

To convert from an e-fold risetime to a 10 - 90 percent risetime for a purely exponential waveform, multiply by 2.2.

Table 4 gives the calculated risetimes for given numbers of arc channels. These risetimes are in good agreement with those measured.

Table 4. Calculated switch risetimes from Eqs. (1) and (2).

# Channels	τ_R	τ_L	2.2 τ
6	3.3	0.8	9.0
9	2.9	0.6	7.7
12	2.6	0.4	6.6

Another empirical switching formula,[3]

$$2\sigma \frac{V}{dV/dt} \leq 0.1\tau + 0.8t_{trans} \qquad (3)$$

where

σ = standard deviation in switch breakdown voltage

V = breakdown voltage

$\frac{dV}{dt}$ = slope at breakdown

τ = total e-fold risetime

t_{trans} = transient time between arcs.

can be used to determine the number of channels for a given situation or, in this case, to find a σ to fit the observed results. Using the calculated risetimes from Table 4, the standard deviations required for a given number of channels are shown in Table 5.

Table 5. Calculated σ from Eq. (3).

# Channels	σ (percent)
6	0.97
9	0.69
12	0.54

The value of σ for 12 channels is roughly a factor of 2 lower than measured. However, this is still a reasonable agreement for multichannel breakdown.

CONCLUSIONS

It has been demonstrated that switch risetimes ($\sim$ 6 nsec) and jitter (1.0 - 1.5 nsec) sufficiently low for short pulse, high power applications can be produced with two electrode switches on rapidly charged transmission lines. Six such switches will be tested on an EBFA prototype.

REFERENCES

1. T. H. Martin, IEEE Trans. on Nucl. Sci., NS-16, No. 3, p. 59 (1969).

2. D. L. Johnson, Record of the 11th Symposium on Electron, Ion, and Laser Beam Technology, p. 445 (1971).

3. J. C. Martin, "Multichannel Gaps," AWRE Report, SSWA/JCM/703/27 (1970).

4. K. R. Prestwich, "A 2 MV Multichannel, Oil-Dielectric, Triggered Spark Gap," International Conference on Energy Storage, Compression and Switching, Torino and Asti, Italy (1974).

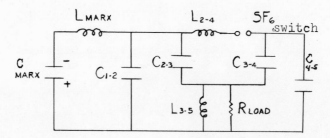

Fig. 1.   Electrical schematic of the
          switching experiment.

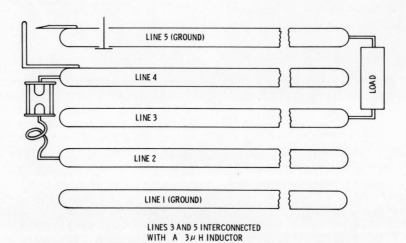

Fig. 2.   Physical arrangement of the flat plate
          transmission lines showing the knife edge
          switch position.

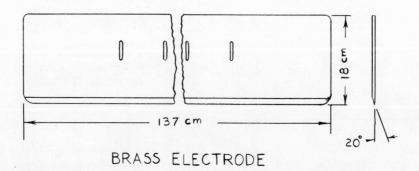

BRASS ELECTRODE

Fig. 3.   Positive electrode of knife edge switch.

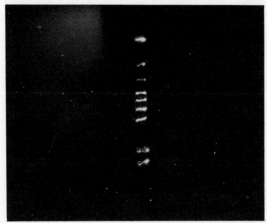

Fig. 4a. Breakdown channels across switch.

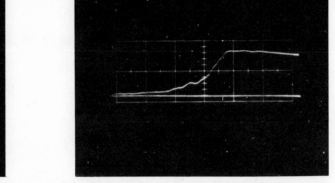

Fig. 4b. Corresponding oscillograph showing
         the switch rise time.

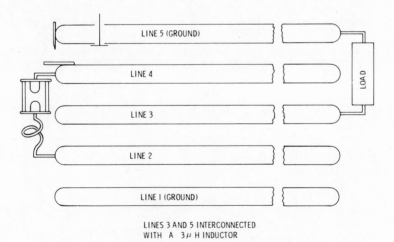

Fig. 5. Transmission lines showing repositional
        knife edge switch.

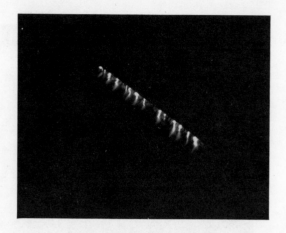

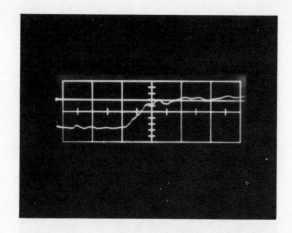

Fig. 6a. Breakdown channels across repositioned switch.

Fig. 6b. Corresponding oscillograph showing
the switch rise time.

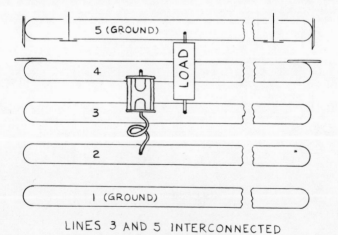

LINES 3 AND 5 INTERCONNECTED
WITH A 3 μH INDUCTOR

Fig. 7. Transmission lines with two switches installed.

Aaland, K.
Lawrence Livermore Laboratory
Electronics Engineering - Univ. of
    California
P. O. Box 808
Livermore, California 94550 (U.S.A.)

Andreani, R.
Laboratorio Gas Ionizzati
Ass. Euratom - CNEN
Casella Postale 65
00044 - Frascati, Roma (Italy)

Aslin, Harlan K.
Physics International Corporation
2700 Merced Street
San Leandro, California 94577 (U.S.A.)

Ballada, R.
Laboratori C.E.S.I.
Via Rubattino, 54
20100 - Milano (Italy)

Bariaud, André
Commissariat à l'Energie Atomique
Centre d'Etudes Nucleaires
Boite Postale 6
92260 - Fontenay-Aux-Roses (France)

Barnes, P. M.
Culham Laboratory
UKAEA Research Group
Abingdon, Oxon OX14 3DB (England)

Bleys, Cyriacus
Laboratoire de Physique des Plasmas
Université de Paris XI - Batiment 214
91405 - Orsay (France)

Bostick, W. H.
Stevens Institute of Technology
Physics Department
Castle Point Station
Hoboken, New Jersey 07030 (U.S.A.)

Caini, Vasco
C.A.M.E.N.
San Piero a Grado
56100 - Pisa (Italy)

Catenacci, Giorgio
Laboratori C.E.S.I.
Via Rubattino, 54
20100 - Milano (Italy)

Cavalleri, Giancarlo
Laboratori C.I.S.E.
P. O. Box 3986
Via Redecesio, 12
20100 - Milano (Italy)

Cortese, Cesare
I.E.N.G.F. - Lab. Plasmi
Corso Massimo d'Azeglio, 42
10125 - Torino (Italy)

De Bernochi, Cesare
I.E.N.G.F. - Lab. Plasmi
Corso Massimo d'Azeglio, 42
10125 - Torino (Italy)

Delmas, Alain
Laboratorie de Physique des Plasmas
Université de Paris XI - Batiment 214
91405 - Orsay (France)

Dick, Eugene Peter
Kern Forschungzentrum Karlsruhe
IEKP, Postfach 3640
7500 - Karlsruhe (West Germany)

Dustmann, Covol-Henrich
Kern Forschungzentrum Karlsruhe
IEKP, Postfach 3640
7500 - Karlsruhe (West Germany)

Fellin, Lorenzo
Istituto di Elettrotecnica
Via Gradenigo, 6/A
35100 - Padova (Italy)

Ferro Milone, Andrea
I.E.N.G.F. - Director
Corso Massimo d'Azeglio, 42
10125 - Torino (Italy)

Fitch, Richard A.
Maxwell Laboratories, Inc.
9244 Balboa Avenue
San Diego, California 92123 (U.S.A.)

Freytag, E. Karl
University of California
Lawrence Livermore Laboratory
P. O. Box 808 - L-153
Livermore, California 94550 (U.S.A.)

Gourlan, C.
Laboratori Gas Ionizzati
Ass. Euratom - CNEN
Casella Postale 65
00044 - Frascati, Roma (Italy)

Gribble, Robert F.
Los Alamos Scientific Laboratory
P. O. Box 1663
Los Alamos, New Mexico 87544 (U.S.A.)

Guillot, Maurice
Laboratoire de Magnetisme - CNRS
Cedex 166, Grenoble-Gare (France)

Harrison, John
Maxwell Laboratories, Inc.
9244 Balboa Avenue
San Diego, California 92123 (U.S.A.)

Hicks, J. B.
J.E.T. Design Team
Culham Laboratory
Abingdon, Oxon OX14 3DB (England)

Hofmann, I.
Max-Planck-Institut für Plasmaphysik
8046 - Garching (West Germany)

Hunt, T. W.
British Insulated Callenders Cables, Ltd.
Helsby, Warrington WA6 0DT (England)

Hutzler, John R.
Aerovox Industries, Inc.
740 Belleville Avenue
P. O. Box 970
New Bedford, Massachusetts 02741 (U.S.A.)

Kolb, A. C.
Maxwell Laboratories, Inc.
9244 Balboa Avenue
San Diego, California 92123 (U.S.A.)

Lanzavecchia
Tecnomasio Italiana Brown-Boveri
Piazza Lodi, 3
20100 - Milano (Italy)

Laquer, H. L.
Los Alamos Scientific Laboratory
University of California
P. O. Box 1663
Los Alamos, New Mexico 87544 (U.S.A.)

Leloup, Christian
Commissariat à l'Energie Atomique
Centre d'Etudes Nucleaires
Boite Postale 6
92260- Fontenay-Aux-Roses (France)

Linhart, G.
Biras Creek
P. O. Box 54
Virgin Gorda, British Virgin Islands
(U.S.A.)

Manintveld, P.
FOM - Institut voor Plasmafysica,
    Rijnhuizen
P. O. Box 7
Nieuwegein (Holland)

Martin, J. C.
UKAEA - A.W.R.E.
Aldermaston, Berk. (England)

Mezzetti, Franco
Istituto di Fisica dell'Università
Via Paradiso, 12
44100 - Ferrara (Italy)

Mohan, Narendra
University of Wisconsin
539 E R B
Madison, Wisconsin 53706 (U.S.A.)

Muller, Gerhard
Max-Planck-Institut für Plasmaphysik
8046 - Garching (West Germany)

Mukutmoni, T.
University of Texas
E N S 541
Austin, Texas 78722 (U.S.A.)

Murray, John
Forrestal Corporation
Princeton University
Building 1/4 - P. O. Box 451
Princeton, New Jersey 08540 (U.S.A.)

Nardi, Vittorio
I.E.N.G.F. - Lab. Plasmi
Corso Massimo d'Azeglio, 42
10125 - Torino (Italy)

Nazet, Christian
C. Energie Atomique
Boite Postale 27
94190 - Villeneuve-Saint-Georges (France)

Page, Derrick
Westinghouse Electric Corp.
Research Laboratories
Beulah Road, Churchill Borough
Pittsburgh, Pennsylvania 15235 (U.S.A.)

Payton, Daniel N.
Air Force Weapons Lab.
4516 Andrew NE
Albuquerque, New Mexico 87109 (U.S.A.)

Passari, Luigi
Istituto di Fisica dell'Università
Via Paradiso, 12
44100 - Ferrara (Italy)

Pecorella, F.
Laboratori Gas Ionizzati
Ass. Euratom - CNEN
Casella Postale 65
00044 - Frascati, Roma (Italy)

Picci, Guido
Passoni & Villa
Via Olofredi, 43
20100 - Milano (Italy)

Pillsticker, M.
Max-Planck-Institut für Plasmaphysik
Garching 8046 (West Germany)

Raboisson
Commissariat Energie Atomique
Boite Postale 14
21120 - Is-Sur-Tille (France)

Rager, J. P.
Laboratori Gas Ionizzati
Ass. Eruatom - CNEN
Casella Postale 65
00044 - Frascati, Roma (Italy)

Raudorf, Walter R.
Sir George Williams University
570 St. Leon Avenue
Dorval, P.Q. H9P 1Z9 (Canada)

Robson, Anthony E.
Naval Research Laboratory
Code 7708
Washington, D. C. 20375 (U.S.A.)

Rocci, Ivano
ENEL
Corso Galileo Ferraris, 2
14100 - Asti (Italy)

Rostagni, Giorgio
Istituto di Elettrotecnica
Via Gradenigo, 6/A
35100 - Padova (Italy)

Rioux, Christian
Laboratoire de Physique des Plasmas
Université Paris XI - Batiment 214
91405 - Orsay (France)

Rioux-Damidau, Françoise
Laboratoire de Physique des Plasmas
Université Paris XI - Batiment 214
91405 - Orsay (France)

Rumi, Giancarlo
I.E.N.G.F.
Corso Massimo d'Azeglio, 42
10125 - Torino (Italy)

Salge, Jurgen
Hochspannungsinstitut
Technisch Universitat Braunschweig
Postfach 3329
D-33 - Braunschweig (West Germany)

Schenk, G.
Commissariat à l'Energie Atomique
Centre d'Etudes Nucleaires de Grenoble
Boite Postale 85
38041 - Grenoble-Cedex (France)

Seacat, Russel H.
Department of Electrical Engineering
Texas Technology University
Lubboch, Texas 79409 (U.S.A.)

Selin, Karl
Royal Institute of Technology
Stockholm (Sweden)

Shearer, James W.
University of California
Lawrence Livermore Laboratory
P. O. Box 808
Livermore, California 94550 (U.S.A.)

Schluter, Walter
Siemens AG Erlangen, ABT. E484
Werner von Siemenstrasse 50
8520 - Erlangen (Germany)

Smith, Robert
University of California
Lawrence Livermore Laboratory
P. O. Box 808
Livermore, California 94550 (U.S.A.)

Smith, Ian
Physics International Company
2700 Merced Street
San Leandro, California 94577 (U.S.A.)

Taquet, Bernard
Centre d'Etudes Nucleaires de Grenoble
Boite Postale 85 - Centre de Tri
38041 - Grenoble-Cedex (France)

Tessitore, Elio
I.E.N.G.F. - Laboratorio Plasmi
Corso Massimo d'Azeglio, 42
10125 - Torino (Italy)

Thomassen, Keith
Los Alamos Scientific Laboratory
P. O. Box 1663
Los Alamos, New Mexico 87544 (U.S.A.)

Toniolo, Sergio Bruno
I.E.N.G.F.
Corso Massimo d'Azeglio, 42
10125 - Torino (Italy)

Turchi, Peter J.
Naval Research Laboratory
Plasma Physics Division - Code 7760
Washington, D. C. 20375 (U.S.A.)

Vecchiotti, Enrico
I.E.N.G.F. - Laboratorio Plasmi
Corso Massimo d'Azeglio, 42
10125 - Torino (Italy)

Villa, Giorgio
Passoni & Villa
Via Olofredi, 43
20100 - Milano (Italy)

Villa, V.
Laboratori C.E.S.I.
Via Rubattino, 54
20100 - Milano (Italy)

Wavre, Jean-Jacques
Emile Haefely Co.
353 Lehenmattstrasse
4028 - Basel (Switzerland)

Wedler, Helmut
Max-Planck-Institut für Plasmaphysik
8046 - Garching (West Germany)

Van Ingen, A. M.
FOM - Institut voor Plasmafisica,
    Rijnhuizen
P. O. Box 7
Nieuwegein (Holland)

Van Wees, A. C. A.
Jet Design Group, Culham Laboratory
Abingdon, Oxon (England)

Weigand, Wolfgang
Siemens AG Erlangen, Abt. E484
Werner von Siemenstrasse 50
8520 - Erlangen (Germany)

Weldon, William F.
University of Texas
E N L 219
Austin, Texas 78722 (U.S.A.)

Wells, Daniel R.
University of Miami
Dept. of Physics
P. O. Box 248284
Coral Gables, Florida 33146 (U.S.A.)

Winterberg, F.
P. O. Box 267
Black Springs, Nevada (U.S.A.)

Wipf, S. L.
Max-Planck-Institut für Plasmaphysik
8046 - Garching (West Germany)

Wunsch, Donald C.
AFWL, Kirtland Air Force Base
907 Stagecoach Road SE
Albuquerque, New Mexico 87123 (U.S.A.)

Zucker, Oved S. F.
University of California
Lawrence Livermore Laboratory
Electronics Engineering Department
P. O. Box 808
Livermore, California 94550 (U.S.A.)

Date Due

			UML 735